HEALTH RISKS FROM HAZARDOUS SUBSTANCES AT WORK

Assessment, Evaluation and Control

Other Pergamon titles of related interest

Books

CRONLEY-DILLON et al:
Hazards of Light

DODGSON et al:
Inhaled Particles VII

VINCENT:
Ventilation '88

Journals

Accident Analysis and Prevention

Annals of Occupational Hygiene

Annals of the ICRP

Food and Chemical Toxicology

Journal of Aerosol Science

Toxicology in Vitro

HEALTH RISKS FROM HAZARDOUS SUBSTANCES AT WORK

Assessment, Evaluation and Control

by

STAN ROACH

Fordham Ely, Cambridgeshire

PERGAMON PRESS

OXFORD · NEW YORK · SEOUL · TOKYO

U.K. Pergamon Press Ltd, Headington Hill Hall,
Oxford OX3 0BW, England

U.S.A. Pergamon Press Inc., 660 White Plains Road,
Tarrytown, New York 10591-5153, U.S.A.

KOREA Pergamon Press Korea, KPO Box 315, Seoul 110-603,
Korea

JAPAN Pergamon Press Japan, Tsunashima Building Annex,
3-20-12 Yushima, Bunkyo-ku, Tokyo 113, Japan

First edition 1992

British Library Cataloguing in Publication Data

A catalogue record for this book is available from the British Library.

Library of Congress Cataloging in Publication Data

Roach, S. A. (Stanley Alec)
Hazardous substances at work/by S. A. Roach.
p. cm.
1. Hazardous substances—Safety measures. I. Title.
T55.3.H3R63 1992
615.9′02—dc20 91-48254

ISBN 0-08-040837 0

Printed in Great Britain by BPCC Wheatons Ltd, Exeter

Contents

Preface

Hazardous substances are found in large quantity in the chemical industry, mining, steel mills, agriculture, cotton and wool mills, furniture production, automobile manufacture and many other branches of industry. Furthermore, in small factories, trade, commerce and offices not infrequently one finds there are isolated individuals exposed to hazardous substances. This book deals with the assessment of health risks that arise at work from the hazardous substances being handled there and how they can be prevented. It has been written primarily for the use and information of those engaged in industry, particularly occupational hygienists, physicians, safety officers, nurses, chemists and engineers. Hopefully it will also be instructive to those contemplating research and education in the subject. Likewise it is evident that there is no sharp distinction between exposure to hazardous substances at work as against exposure at home, school or in leisure pursuits. Exposure to hazardous substances spreads beyond the factory fence.

Formal assessment of health risks from exposure to hazardous substances at work is increasingly being required under law. In the UK the Control of Substances Hazardous to Health Regulations 1988 and Approved Code of Practice require an assessment to be made of work places for health risks created by work with all substances which may be hazardous to the health of employees. The assessment is further required to include the steps needed to be taken to control exposure so as to avoid these risks.

West German Regulations of Hazardous Substances, made under the German Chemicals Law, require assessment of the risks connected with the handling and use of all hazardous substances and the determination of the necessary precautions to avoid them. An example from Canada is the Ontario Health and Safety Act 1978 under which Regulations regarding 'designated' substances require that an assessment be made of the likelihood of worker exposure and that a control programme be developed and be put into effect. In each case an integral part of the assessment would be an assessment of the extent of employee exposure. In the USA the Occupational Safety and Health Act of 1970 requires every employer to provide employment free from recognised hazards. Prudence dictates that this condition be confirmed by a systematic appraisal of exposure to hazardous substances at work.

The exposure assessment identifies populations exposed to the agent, describes their composition and size, and presents the types, magnitudes, frequencies and durations of exposure to hazardous substances. The results of the exposure assessment are combined with appropriate exposure limits to make a quantitative assessment of the health risks. Risk assessment as an activity has been gathering definition over the past two decades.

The subject matter embraces numerous technical terms which are in common use in occupational hygiene, industrial medicine and engineering control so a rather full glossary of terms has been included at the beginning.

The main part of the book is in four parts.

The biological effects of exposure to hazardous substances are discussed in Part I. This opens in Chapter 1 with an analysis of the various ways hazardous substances may enter the body. An outline of the nature of the principal occupational diseases from hazardous substances which succeed in entering the body is given in Chapter 2. The reaction to the substance in the body compartments may be rapid, slow or delayed. Prior to clinical disease are changes which represent the threshold of adverse effects on the body. Furthermore the reaction of the most sensitive individuals may be many times more severe than the average. These considerations are discussed in the context of effect thresholds in Chapter 3. An important conclusion is that the time scale of exposure causing occupational diseases can range from a few seconds to a lifetime. The biological half time of a hazardous substance governs the way short-term or long-term exposures at high concentration may influence the body burden. There is a growing interest in this aspect of toxico-kinetics. Therefore in Chapter 4 compartmental models of human toxico-kinetics are introduced, which in recent years have become enshrined in mathematics and tested against observation in animals and man.

All substances are harmful if taken in excessive amounts. However, even the most hazardous substances can be tolerated in sufficiently small doses. It is necessary to evaluate at the earliest possible moment whether or not the controls over hazardous substances at work are adequate and effectively used. But without satisfactory exposure limits, doing a risk assessment can mean little more than running through a check list of things to think about. Occupational exposure limits are considered in Part II. Fortunately the hazards posed by exposure at work have already been quantified for about 1,000 important substances. For these it is possible to assess the potential risks resulting from exposure by direct reference to published exposure limits and from these it is possible to foresee the risks from many others. Chapter 5 contains information about lists of published limits and their limitations.

In order to proceed with the same certainty when handling all the other substances it is necessary to devise exposure limits which are comparable with those which have public recognition. Published exposure limits cover an important but very small fraction of the hazardous substances and exposure limits have to be devised in-house for the remainder. Over the years the task of devising new exposure limits has become wrapped in mystery through lack of information about how they are derived from epidemiological and toxicological data, or by invoking 'analogy'. The way it is done has been variously described as intuition, 'guesstimating' or pure conjecture. Perhaps all these techniques could play a part in the process but it is possible to be more scientific about the matter. In this book an attempt is made to set forth some of the basic scientific principles and practical considerations which come into play when setting exposure limits. This is presented first in Chapter 6, which deals with the epidemiological approach and human experience generally. Chapter 7 deals with how to extrapolate from animal toxicology to practical exposure limits for employees. It is shown how published data on animal toxicology are sufficient to enable exposure limits to be specified for at least 25,000 different substances.

Most published occupational exposure limits are for single substances, but hazardous materials and products are mixtures of several, and sometimes many different substances. Furthermore, employees move about in their jobs and this alone ensures that most are exposed to a mixture of hazardous substances. It is imperative to formulate ways of approaching the issue of exposure limits for mixtures. A detailed analysis is given in Chapter 8.

Exposure limits are mostly designed for employees working a nominal 8-hour day, 5-day week and exposed on the average for this same period each week. However employees in some jobs are exposed for only a brief period of just a few minutes or an hour during the work shift. Others are exposed cyclicly and another common problem is how to deal with extremely high excursions either side of the mean. Moreover, 12-hour shift systems combined with long week-end breaks are becoming increasingly common. The adaptation of exposure limits to employees who experience unusual exposure schedules is in Chapter 9.

Regulations, Codes of Practice, Trade Association Guidelines and similar documents encapsulate good ideas. These ideas have to be converted into practical schemes by each individual employer. A practical scheme for assessment of health risks from exposure to hazardous substances at work is covered in Part III. This part is based on a valedictory report by the author to health professionals in Imperial Chemical Industries PLC. The widespread interest created by that report encouraged the author to expand on the subject and give a broader treatment.

There are many who see the need to conduct health risk assessments but are not sure exactly how far to go into it. For these Part III has two interconnected aims. First it offers a broad view of how an assessment of the nature and degree of health risk from hazardous substances can be made. Whereas in complex situations it may be advisable to seek professional advice on occupational hygiene, for many situations simple precepts can be applied. The precepts put forward for making an assessment of hazardous substances are flexible rules-of-thumb and are in no way exhaustive. There are no hard and fast rules. Amendments to the procedures put forward could readily be accommodated to suit individual circumstances.

Companies handling hazardous substances have assessments made of possible health risks to their employees in the normal course of things and try to ensure that essential control measures are in place to reduce exposure to a minimum. The principles of assessment of health risk to employees exposed are described in Chapter 10.

Central to the issue of assessment of health risks is the evaluation of atmospheric exposure. Much has been written about the measuring of air contamination, indeed for many industrial hygienists this is a full time job. The main types of instruments for measuring atmospheric exposure and their calibration are introduced in Chapter 11. In practice the measuring of atmospheric exposure seems to be either done so intensely that one is drowning in data, as with vinyl chloride monomer, asbestos and lead, or not done at all. It is easy in a book on this subject to lose the principles in a welter of detail. Low level but regular monitoring is advocated here, based on the occasional intensive reconnaissance, which is developed at length in Chapter 12.

In Chapter 13, the continuous surveillance of health risk from exposure to hazardous substances at work is described. This does not necessarily involve regular medical examinations, although these do play an important part, where needed. The surveillance is of those factors which may lead to and precede significant health risks; the engineering and personnel control, the monitoring of the air contamination and biological monitoring.

The second main aim of Part III is to offer practical guidance about how to organise health risk assessments in a systematic fashion. A systems approach is recommended. Allied to this is communication of data and reports within a company. Central to the task of organising the information flow is the preparation and dissemination of hazard data sheets. These matters are the subject of Chapter 14.

Assessment of health risks at work is not conducted for its own sake. The purpose is to identify the how, why, where and when of better control over exposure. Methods for getting control over health risks are described in Part IV. Industrial diseases are liable to be concentrated in particular occupations where exposure is difficult to control. The outcome of an assessment may be entirely favourable, but the assessment should be so designed as to pinpoint those features which need better control over employee exposure. Based on practices found satisfactory in establishments which have considerable experience, this part is mainly intended to be informative to those with limited occupational hygiene experience. The control is of two different kinds: engineering control and personnel control. Control of the physical environment, engineering control, is dealt with broadly in Chapter 15. Within this framework the basic principles of ventilation systems for control of atmospheric exposure are explained in Chapter 16. A whole chapter is devoted to that most important control over health risks from hazardous substances, local exhaust ventilation, and this is the subject of Chapter 17. The detailed investigation of ventilation systems is considered further in Chapter 18.

An integral part of the control of exposure is personnel control, in all its facets; education, training, job placement, respiratory protective equipment and clothing. This is in the final chapter, 19.

The overall aim of this book is not so much to contribute to the detection and treatment of ill health caused by hazardous substances, nor to aid in the winning of compensation for impairment of health. Rather, the aim is to introduce those findings, principles and practices upon which to base effective measures for preventing industrial diseases from exposure to hazardous substances. The bibliography in the references at the end of each chapter would make a fine start to an occupational hygiene library on hazardous substances.

Many of the technical conclusions about the control of hazardous substances at work are naturally embodied in a comprehensive and concise form by mathematical results and formulae. The mathematics is, however, for the most part elementary and need not deter the reader. Worked problems are provided throughout the book. The mathematical background has been added in Appendix 1.

Conversion factors are in Appendix 2.

In the interests of keeping the length of the book within reasonable bounds some subjects are not covered. Two particular topics only given passing mention are radioactive substances and viable organisms, both well covered elsewhere.

The author is grateful to all the many colleagues in the chemical industry in Britain and overseas who freely criticised the early drafts, made many useful suggestions and picked out the worst of the errors.

Glossary of Terms

There are technical terms employed in occupational hygiene theory and practice which need explanation. Some are special to occupational hygiene, some are borrowed from related disciplines and some are common words which have taken on a specific meaning in this field. Furthermore some of the terms employed are given different meanings by various authorities. In this book the following terms, definitions and explanations apply.

absolute pressure. Pressure measured with respect to zero pressure. See **gauge pressure**.

absorption. 1. The taking-up of gases, liquids and solids by the tissues and vessels of the body, especially the skin and lungs. 2. In chemical engineering, a unit operation in which a soluble component of a gas mixture is dissolved in a liquid. Also includes gas collection by chemical reaction with the liquid. See **adsorption**. 3. In physical chemistry, the removal of visible light, infra-red or ultra-violet by passage through a medium. Each layer of equal thickness absorbs an equal fraction of the light which traverses it. See **Lambert's law**.

ACGIH. American Conference of Governmental Industrial Hygienists.

action level. A level of exposure to air contamination, which is below the exposure limit but necessitates the application of specified precautionary measures, usually defined to be half or one third the permissible exposure limit. These measures include measuring the level of air contamination and may also include biological monitoring and medical examinations.

activated carbon. Carbon obtained from vegetable matter by carbonisation in the absence of air, preferably in a vacuum. Activated carbon has the property of absorbing large quantities of gases and vapours. Used in respirator canisters and in sampling gaseous air contamination.

accuracy. Difference between the mean of a large set of measurements of a parameter and the true value of that parameter. In the measurement of air contamination accuracy generally varies with concentration. See **precision**.

acute effect. Effect which rapidly follows commencement of exposure, possibly within seconds, commonly within minutes or hours and in any event within two weeks. In the absence of further exposure an acute effect is usually of short duration. Examples are irritation, corrosion and narcosis. See **chronic effect**.

additive. An inclusive name for any of a wide range of chemical substances which are added in low percentage to stabilise certain end products, such as antioxidants in rubber.

adsorption. Gas adsorption is a unit operation in which the gas is taken up at the surfaces of a solid. See **absorption**.

aerodynamic diameter. Refers to the size of a particle in terms which indicate how and where it will deposit from suspension in air. It is the diameter of the unit density sphere which would fall at the same terminal velocity in air as the particle in question. This velocity is directly related to the aerodynamic properties of the particle suspended in air and takes into account its size, shape and density.

aerosol. All the various disperse systems in air, such as dust, fog, clouds, mist, fumes and smoke.

aetiology. The medical study and knowledge of the causation of disease.

AIHA. American Industrial Hygiene Association.

air. The mixture of gases and vapours which surrounds the earth. The four major components of dry air are, by volume; nitrogen 78.08%, oxygen 20.95%, argon 0.93% and carbon dioxide 0.03%.

air contamination. The presence of substances in the air, which can either be solid, liquid or gaseous in nature, put there by acts of man or otherwise in concentrations sufficient to interfere with the health, safety, wealth or comfort of man.

airborne dust. Solid particles suspended in air. Commonly formed by grinding, crushing, drilling or milling the parent materials. See **total dust, inhalable dust, respirable dust, fume**.

airways. 1. Those portions of the respiratory tract which are not lined by alveolar epithelium. 2. In a mine, the underground roadways or tunnels through which air is conveyed to or from the workings.

albumins. The name given to a group of simple proteins which are soluble in water, the solutions being coagulated by heat; not to be confused with albumen, the particular albumin found in egg white. See **albuminuria**.

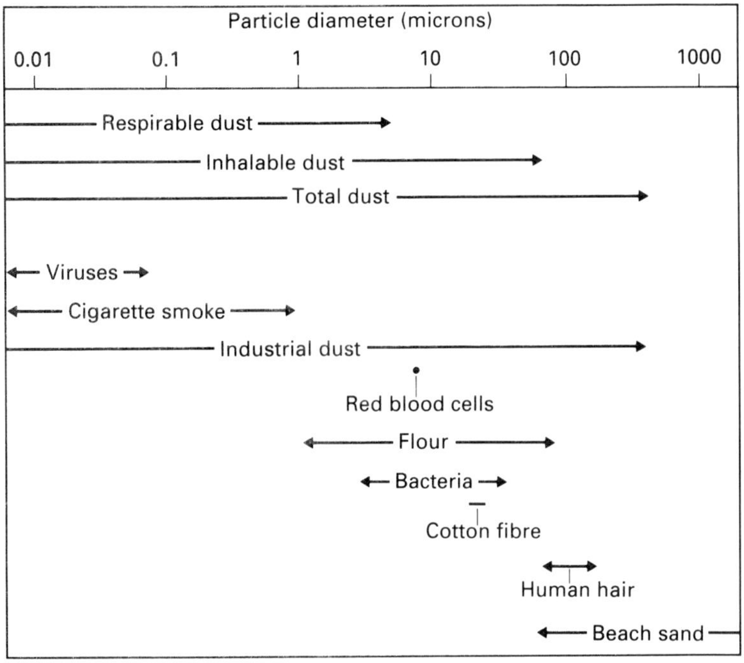

FIG. G.1. Airborne dust—a comparison with other fine particles.

albuminuria. The presence of albumin in the urine.

allergen. Substance which, on being introduced into the body, is capable of making it especially sensitive, so that it will respond with enhanced reactivity when later challenged by that same substance. See **allergy**.

allergy. A state in which the cells of the body are hypersensitive to substances introduced into it which have little or no effect on most individuals. The reaction of the body to a substance to which it has become sensitive is characterised by oedema and inflammation. See **allergen, asthma**.

alveoli. Air sacs of the lung found clustered at the end of each bronchiole. Each alveolus is 100 to 200 microns across. Blood takes up oxygen through the walls of the alveoli and gives up carbon dioxide. See **bronchiole**.

anaemia. Diminution of the amount of haemoglobin in the blood, due to a reduction in the quantity of the red blood cells or their haemoglobin content.

anemometer. Instrument for measuring air velocity. Hot wire anemometers have a small electrically heated element whose resistance is registered by the meter. The anemometer probe is held in the air stream which reduces the temperature and hence resistance of the element. Another common type in

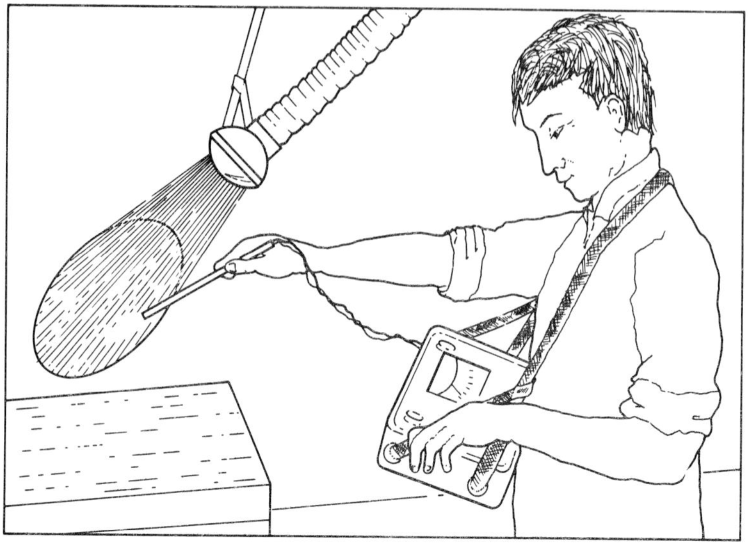

FIG. G.2. Anemometers are instruments for measuring air velocity, based on a variety of different physical principles. A hot wire anemometer is illustrated in the sketch, being used to measure air velocity into an exhaust hood. There is a small element at the tip of the probe, which is heated electrically. The draught of air cools the element whose change of resistance is registered on the dial, calibrated in terms of air velocity.

industry is a swinging vane anemometer (Velometer) which has a pivoted vane inside the meter. See **rotameter**.

angstrom. Unit of measure equal to 10^{-7} mm or 0.1 millimicrons. Named from the Swedish physicist, A J Angstrom (1814–74). See **micron**.

anodising. An electrolytic process whereby a hard, non-corroding oxide film is deposited on aluminium and light alloys generally. See **pickling**.

antagonism. Action of two or more substances whose total biological effect is less than the sum of their separate effects. See **synergism**.

aplastic anaemia. A condition in which the bone marrow fails to produce an adequate number of red blood corpuscles. One of the effects which may occur on chronic exposure to high concentrations of benzene vapour.

asbestos. A generic term for a wide variety of hydrated mineral silicates which break into long, relatively soft, silky fibres. The fibres can be spun, woven into fabric or moulded into the required shape. Asbestos is found in two mineral species; serpentine and amphibole. The most common variety of serpentine asbestos is chrysotile or white asbestos. The most important variety of amphibole asbestos is crocidolite or blue asbestos which has a distinctive

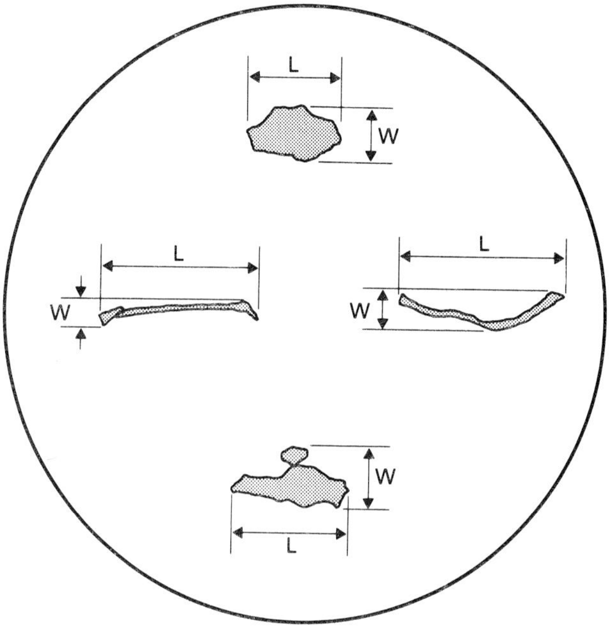

FIG. G.3. Aspect ratio of dust particles ($L:W$). L = longest diameter. W = width of the surrounding rectangle of length L.

blue-grey colour. Another amphibole asbestos of particular commercial importance is amosite. The name embodies the initials of the company exploiting this material in the Transvaal, viz. the 'Asbestos Mines of South Africa', AMOSite. Amosite asbestos is off-white, slightly brownish in colour.

aspect ratio. 1. In the design and operation of local exhaust systems, the ratio of the length to the width of a slot or rectangular hood opening. 2. In the design of ventilation air distribution outlets, the ratio of the width of the core of a grille, face, or register to its length. 3. In dust particle microscopy, the ratio of the length of a particle to its width.

asphyxiant. Substance which causes asphyxiation, that is to say suffocation caused by deficiency of oxygen in the blood. A simple asphyxiant such as helium, neon and argon does not itself affect the body but causes asphyxia through replacing the oxygen in the air. A chemical asphyxiant is a substance such as carbon monoxide which has a great affinity for haemoglobin and thereby reduces the blood's capacity to transport oxygen.

asthma. A disorder in which there are attacks of difficult breathing due to spasmodic constriction of the bronchial muscles. May be in response to inhalation of irritants, allergens or other stimuli. See **allergy, allergen**.

atmospheric pressure. The pressure exerted by the atmosphere at the surface of the earth. The average atmospheric pressure is 1.013×10^6 dynes per square centimetre, or 14.7 pounds per square inch, which is equal to the pressure exerted by a column of mercury 760 mm high. See **absolute pressure, gauge pressure**.

autocorrelation. The correlation between sequential values of a time series.

bag house. Many different trade meanings. Term commonly employed for the housing containing bag filters for the recovery of dust and fumes from process air or exhaust ventilation. The dust-laden gases are passed through long, woven fabric bags, which filter out the dust, allowing the gases to pass on.

benign. Not malignant. A benign tumour is one which does not metastasise. See **metastasise**.

biological half time. A means of classifying the rate of elimination of substances from the body according to the time it takes for the body burden to fall to half its initial value. Time it would take for half the body burden to be excreted, exhaled or otherwise eliminated from the body by natural processes. Half times of hazardous substances range from a few seconds to many years. The term is also used in relation to particular organs or parts of the body.

biological monitoring. Systematic collection of human specimens such as blood and urine and including exhaled air. The analysis of the specimens for hazardous substances, metabolites and biotransformation products is normally of immediate application; analysis and evaluation will generally be performed within a period of weeks after collection.

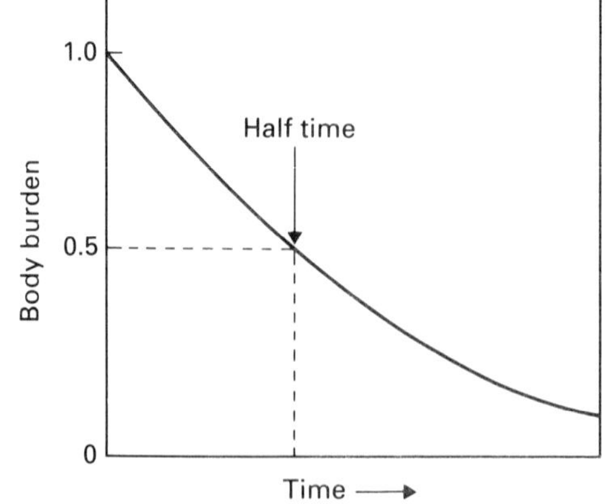

FIG. G.4. Biological half time. The time it takes for the body burden of a hazardous substance or metabolite to fall to half its initial value.

blackhead itch. Also called 'cable rash' by workmen but medically known as chloracne. A form of acne resulting from exposure to the chlorinated naphthalenes or chlorodiphenyl.

blast gate. A sliding damper placed in a duct and used to reduce the effective cross-sectional area of the duct.

blood count. A count of the number of corpuscles per cubic millimetre of blood. Separate counts may be made for red and white corpuscles.

body burden. Quantity of substance in the body, organ or tissue. It is equal to the dose to the body, organ or body, respectively, less the amount which has been exhaled, excreted or otherwise removed. See **dose**.

BOHS. British Occupational Hygiene Society.

boiling point. Temperature at which the vapour pressure of a liquid equals the atmospheric pressure and boils. Boiling point is usually given at a pressure of 760 mm mercury.

breakthrough time. A test of ease of penetration of a material for making protective clothing, respirator canisters or arrestment plant. It is the time it takes for a specified liquid, vapour or gas to first appear on the other side of the material after it makes contact on one side under defined standard conditions.

bronchioles. The most slender tubes, being the terminal subdivisions of the respiratory tree, immediately preceding the alveoli. See **alveoli; bronchus**.

bronchus. One of the two branches into which the trachea divides and which lead to the lungs.

Brownian motion. A continuous agitation of particles in suspension in a gas or liquid. It is caused by impacts with molecules of the surrounding medium. The movement is very slight with larger particles because of their inertia and the high frequency of impacts on opposite sides. The motion is sensible for particles suspended in air which are smaller than 0.1 micron diameter.

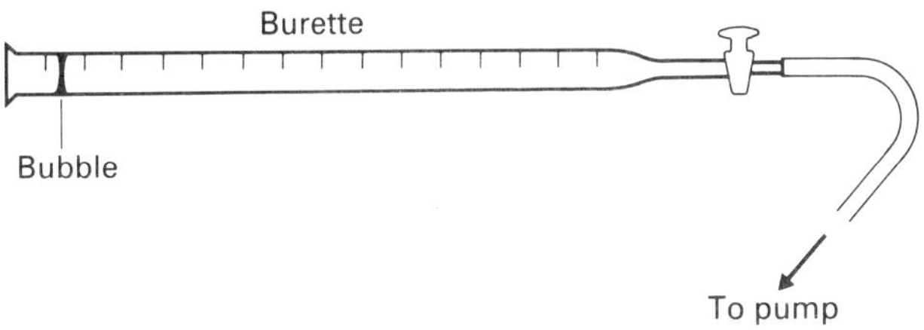

FIG. G.5. Bubble flow meter.

bubble flow meter. A simple air flow meter. The device consists essentially of a graduated tube such as a laboratory burette held horizontally. The inside is first wetted with detergent solution and a bubble is formed at the open end. The pump or other device to be checked is connected to the other end. With the aid of a stop watch the time taken for the bubble to move between two marks is measured. The volume between the marks divided by the time taken is the air volume flow rate drawn by the pump.

byssinosis. Lung disease caused by inhaling vegetable dusts, especially cotton and flax. Clinically the disease is characterised by shortness of breath on Mondays or on the first day back at work after a break. The disease gets progressively worse with continued exposure and the shortness of breath extends to other days in the week.

capture velocity. A term employed in exhaust ventilation. The minimum air velocity induced by an exhaust at a certain point relative to the hood entrance which would capture air contamination liberated at that point. The magnitude of the capture velocity is dependent primarily upon the momentum of the contaminated air, its direction relative to the induced streamlines and the turbulence in the vicinity.

carborundum. Not a natural product. This is a registered trade mark designating a proprietary range of products, among them being silicon carbide, the basis for abrasives which are substituted for the hazardous, natural sandstone.

carcinogen. A substance capable of causing cancer.

carcinogenesis. The production and development of cancer.

carcinoma (cancer). A disorderly growth of new cells in the body which invade and can destroy adjacent tissue and are carried to other parts of the body where they form new growths. See **neoplasm**.

carding engine. A machine used extensively in the process of preparing textile fibres, such as cotton, flax, and wool for conversion into sliver, which is then progressively drawn and twisted into yarn. In the cotton and flax industry the carding engine combs the fibres and cleans them of soil and unwanted vegetable matter. The dust given off is a major cause of byssinosis.

cardiovascular. Pertaining to the heart and the blood vessels.

case control study. An epidemiological term for a study in which the investigator studies the exposure of individuals with the disease in question in comparison with others closely matched for age, sex, place of residence, income group and so on but without the disease. See **cohort study, cross-sectional study, longitudinal study**.

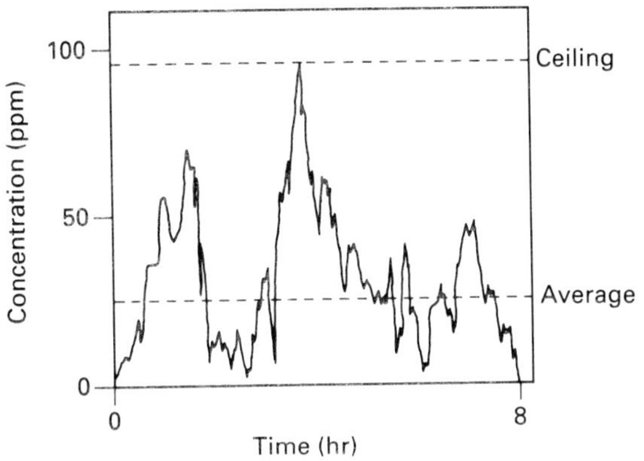

FIG. G.6. Ceiling concentration. The highest value indicated on the trace from a continuous recorder of the concentration of hazardous substance present in work place air.

catalyst. 1. In chemical engineering the term refers to a substance which changes the rate of a chemical reaction but itself undergoes no permanent chemical change. Commonly used catalysts greatly increase reaction rates. 2. In respirator use, a substance which converts a hazardous gas or vapour into one which is less hazardous.

caustic. Something which strongly irritates, burns, corrodes or destroys living tissue.

cell. A structural unit of which tissues are made. There are many types; nerve cells, muscle cells, blood cells, connective tissue cells, fat cells and others. Each has a special form to serve a particular function.

ceiling concentration. Maximum concentration of atmospheric contamination at any time during a work shift.

CEN. Committee for European Normalisation.

chemical asphyxiant. A substance which halts the respiratory process by chemical means. An example is carbon monoxide which displaces oxygen in the blood by its much greater affinity for haemoglobin. Another example is hydrogen cyanide which inactivates cytochrome oxidase, a catalyst essential in cellular respiration.

chemiluminescence. The production of light without heat in certain chemical reactions.

chronic effect. Long lasting effect which persists indefinitely, even in the absence of further exposure. Examples are carcinogenicity, teratogenicity, mutagenicity, pneumoconiosis, chronic bronchitis, liver atrophy and, of course, death from any cause. See **acute effect**.

cilia. Small whiplike appendages of cells lining the trachea and extending down to the lower ends of the bronchioles. They beat rhythmically and create a current of fluid which carries upward any foreign bodies that chance to touch the linings of the respiratory passages. The nasal passages likewise are bathed in mucus and lined with cilia. All the mucus moves towards the exits of the nose and mouth and is never stagnant. The mucus is swallowed or spat out.

coefficient of entry. The ratio of the actual volume flow rate of air into an exhaust hood to the ideal flow rate calculated by assuming that the velocity pressure in the exhaust duct is equal and opposite to the hood static pressure.

$$\text{Coefficient of entry} = \frac{\text{actual flow rate}}{\text{ideal volume flow rate}}.$$

$$\text{Ideal volume flow rate} = 4.04\ (\text{hood static pressure})^{1/2},$$

where volume flow rate is measured in cubic metres per second of air and hood static pressure is measured in millimetres of water gauge. See **hood static pressure**.

coefficient of variation. Standard deviation of a population expressed as a percentage of the mean of the population. See **standard deviation**.

cohort study. An epidemiological term for a study in which a group of healthy employees who are exposed is followed up for a defined period of time with regard to disease status and exposure. They are compared with regard to disease status against a similar group who are not exposed. See **case control study, cross-sectional study, longitudinal study**.

colic. A severe, cramping, gripping pain in or referred to the abdomen.

conjunctivitis. Inflammation of the conjunctiva, that is, the mucous membrane covering the inner surface of the eyelids and front of the eyeball.

contamination. A substance (gas, vapour, dust, fume, or mist) whose presence in air is harmful, hazardous, or deleterious.

COSHH. Control of Substances Hazardous to Health Regulations 1988, UK.

cristobalite. A crystalline form of free silica, extremely hard and chemically inert, used extensively in precision casting by the hot wax process and certain speciality ceramics. Quartz in refractory bricks and amorphous silica in diatomaceous earth are altered to cristobalite when exposed to temperatures in excess of 1470C. Cristobalite dust is more fibrogenic than quartz dust, which in turn is much more fibrogenic than amorphous silica. See **quartz, diatomaceous earth, tridymite**.

critical organ. Sometimes called 'target' organ or 'effector' organ. That organ or organ system of the body receiving the hazardous substance which results in the first or most serious effect on the body. The effect is the 'critical effect'. The mean concentration of hazardous substance in the critical organ at this point is the 'critical organ concentration'.

cross-sectional study. An epidemiological term for a study in which a group of employees is selected at a single point in time without regard to disease or exposure status. Each employee is then evaluated with regard to the presence and severity of disease and as to the level of his or her exposure to hazardous substances. See **case control study, cohort study, longitudinal study**.

cross-sensitivity. A term employed to express quantitatively the interference by one substance in the determination of another. The ratio of the measured value of an interfering substance to the measured value of the substance to be determined. See **specificity**.

cumulative effect. The result of repeated exposure to a substance. The cumulative effect arises from an accumulation somewhere in the body of the substance itself or a metabolite of the substance or of the biological effects from it.

cyclone separator. 1. In air cleaning practice, a conical dry-air primary dust collector of high efficiency for coarse dust but low efficiency for fine dust. 2. In dust sampling instruments a cyclone is used to separate coarse dust prior to collection of the finer 'respirable' fraction. See **bag filter, respirable dust**.

demography. The study of population statistics and the estimation of their variation with time.

dermatitis. Inflammation of the surface of the skin or epidermis. See **dermatosis**.

dermatosis. Affection or inflammation of the deeper layers of the skin. A broader term than dermatitis: it includes any cutaneous abnormality, thus encompassing acne, pigmentary changes and tumours. See **dermatitis**.

diatomaceous earth. A soft, gritty amorphous silica composed of minute siliceous skeletons of small aquatic plants called diatoms. Used in filtration and decolorisation of liquids, in insulation, also as a filler in dynamite, wax, textiles, plastics, paint and rubber. See **cristobalite**.

diffusivity. A quantity which determines the gradient of concentration of a substance at a point in space due to liberation of the substance from some other point. The units of mass migration rate through space, mass per second per

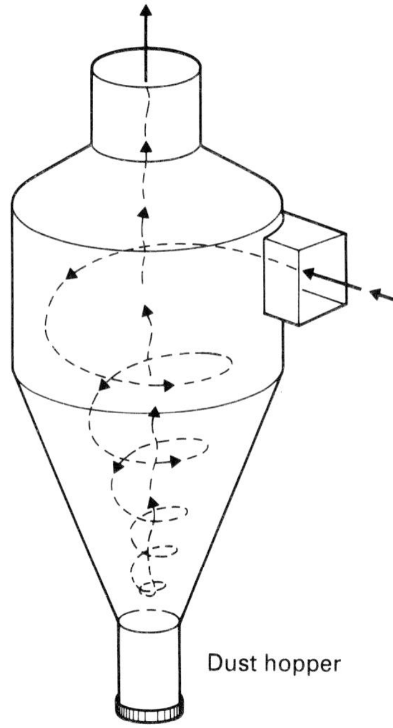

FIG. G.7. Cyclone separator.

square metre, are $MT^{-1}L^{-2}$ and of mass concentration gradient, change of concentration per metre, are ML^{-4}. The units of diffusivity, which is the ratio of the two, reduce to L^2T^{-1}, or square metres per second, and the same is true if migration of contamination is expressed in terms of volume units. This expression of the magnitude of migration due to diffusion may be termed the 'effective' diffusivity when embracing both molecular and turbulent diffusion.

$$\text{Diffusivity} = \frac{\text{rate of migration of contamination}}{\text{concentration gradient}}.$$

dose. Quantity of substance administered to an individual or experimental animal. It may refer to a single means of administration or several, but especially ingestion, inhalation and dermal absorption. Dose rate is the dose delivered per unit of time. See **body burden**.

dose response. The relationship between the dose and the biological effects from it. In the individual the response is invariably a graduated one although the rate of increase of response with increasing dose is rarely proportionate and is sometimes quite sharp. Individuals vary widely in their susceptibility to dose with the result that an overall group response increases much less sharply than does the response in an individual.

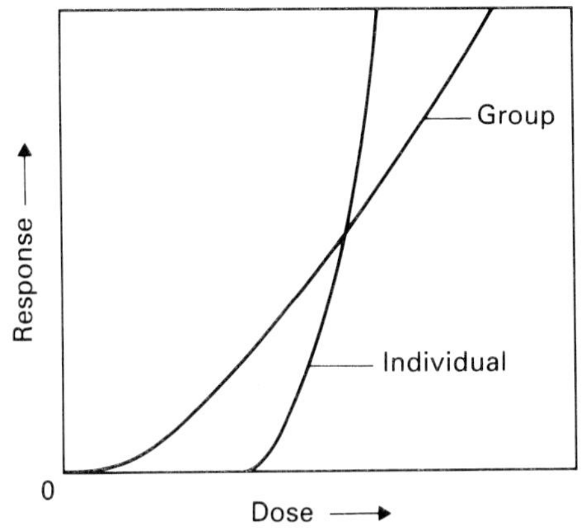

FIG. G.8. Dose response. Response to increasing dose of hazardous substance in an individual is gradual although sometimes quite sharp. Response in a group is less sharp because of natural variation of tolerance between one individual and another.

dust. Solid particles commonly formed by reducing larger rock-like materials by processes such as grinding, crushing, hammering, drilling, and blasting. Examples are mineral dusts; quartz, coal and asbestos. Also includes organic dusts like those from sugar, wood, grain, cotton and tea. See **mist, fume, smoke**.

dyspnoea. Shortness of breath. Difficult or laboured respiration.

eczema. Itching inflammatory condition of the epidermis occurring as a reaction to irritants.

engineering control. Refers to all the physical controls over the environment which aim to reduce employee exposure to hazardous substances. See **personnel control**.

epidemiology. The study of epidemics. Branch of medical science concerned with the study of patterns of disease amongst large groups of people. It includes, in the present context, determination of the relationships between exposure of a group of employees to hazardous substances and the resultant effects on their health.

equivalent ventilation. A measure of the ventilation between two points. When atmospheric contamination is released at point A and the concentration of that contamination is measured at point B the ratio of the rate of release to the concentration is the equivalent ventilation of B with respect to A.

$$\text{Equivalent ventilation in m}^3\,\text{s}^{-1} = \frac{\text{rate of release in mg s}^{-1}}{\text{concentration in mg m}^{-3}},$$
$$= \frac{\text{rate of release in ml s}^{-1}}{\text{concentration in ppm}}.$$

exhaust hood. A structure to enclose or partially enclose an industrial operation or process, or to guide air flow in an advantageous manner so as to capture air contamination. The exhaust hood is connected to a duct which removes the contaminated air.

exposure. Quantity of substance at the interface between the individual employee or experimental animal and the substance in the environment, in whatever form. The boundary between the individual and the 'atmosphere' is to some extent arbitrary because inhaled air changes composition as it penetrates further into the respiratory system. In this book the interface between the lungs and the atmosphere is arbitrarily defined as being at the entrance to the nose and mouth.

exposure limit. The average concentration of atmospheric concentration and duration of exposure at which specific effects to health occur. The precise definition of an exposure limit differs according to the authoritative source of the limit and the substance in question.

extrapolation from animals to humans. 1. Application of results from studies of the effects of specified doses or exposures on experimental animals to the probable relationship between dose or exposure and their effects on humans. 2. A specific calculation based on data from experimental observations on animal populations which aims to estimate the dose vs response and hence exposure vs response relationships in human populations.

face velocity. Average air velocity over the plane marking the entrance to an exhaust hood. The velocity is measured in the direction perpendicular to the plane.

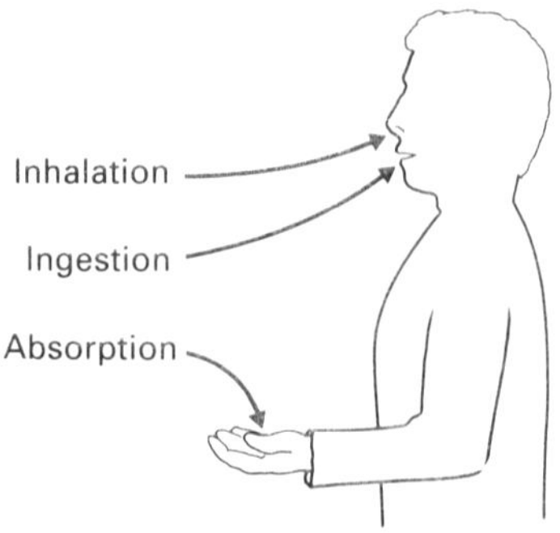

FIG. G.9. Exposure. There are three main means of exposure to hazardous substances at work; inhalation, skin absorption and ingestion.

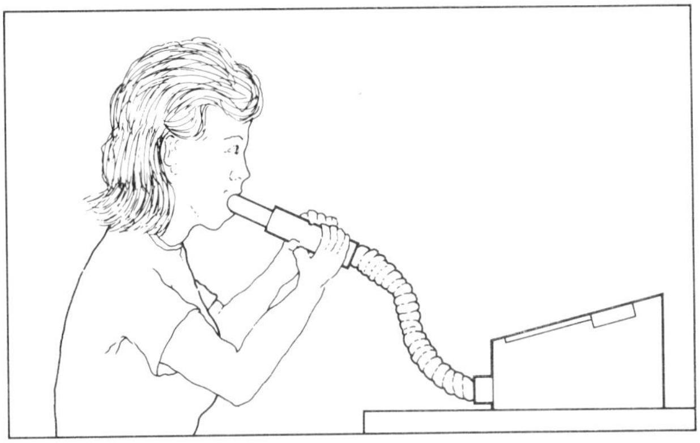

FIG. G.10. Forced expiratory volume in one second (FEV_1) is the maximum volume of air that can be expelled in one second after a full inspiration with maximum effort. It is commonly measured with a bellows spirometer.

fan static pressure. Fan total pressure diminished by the fan velocity pressure.

$$\text{Fan SP} = \text{Fan TP} - VP_{\text{outlet}}$$

See **fan total pressure, fan velocity pressure.**

fan total pressure. The increase in total pressure through or across the fan.

$$\text{Fan TP} = TP_{\text{outlet}} - TP_{\text{inlet}}$$

See **fan static pressure, fan velocity pressure**.

fan velocity pressure. The velocity pressure corresponding to the mean air velocity at the outlet of the fan. See **fan static pressure, fan total pressure**.

FEV_1. Forced Expiratory Volume in one second. On taking a deep breath, a maximum inspiration, the air is breathed out as fast as possible through a device which measures the volume expired in one second. This is believed to be a good test of respiratory function.

fibre. A particle having a length/width ratio (aspect ratio) greater than 3:1.

fog. Aerosol formed by the condensation of water vapour upon suitable nuclei.

fume. Aerosol of solid particles resulting either from condensation of the vapour given off from the heating of a solid body such as lead or zinc, or from their combustion such as by burning magnesium or phosphorus, or from sublimation of a solid such as iodine crystals. Generally the individual particles of the fume have diameters less than 1 micron. Fumes flocculate; clumps or aggregates of particles often adhere together to form a single, much larger airborne entity. In common parlance it is usual to speak of acid fumes, for example, as denoting a mixture of gas and mist. Throughout this book, however, to avoid confusion the word 'fume' is used to designate only solid particles. See **smoke, dust**.

gas. Word invented by Belgian chemist, van Helmont (d. 1644). A state of matter in which a substance completely fills the region in which it is contained; has very low density and viscosity; can expand and contract greatly in response to changes in temperature and pressure; and easily diffuses into other gases. A gas can be changed to the liquid or solid state only by the combined effect of increased pressure and decreased temperature. Gases of interest to the occupational hygienist include carbon monoxide, chlorine, hydrogen cyanide, ammonia and arsine. See **vapour**.

gauge pressure. Pressure measured with respect to atmospheric pressure. See **absolute pressure**.

geometric mean. Where a variate has values $C_1, C_2, C_3, \ldots C_n$

$$\text{geometric mean} = [C_1 \times C_2 \times C_3 \ldots \times C_n]^{1/n}$$

hazard. 1. That dangerous condition, potential or inherent, which can bring about an interruption or interference with the expected orderly progress of an activity. 2. Said of the inherent or potential harm which could arise on being exposed to a substance in unfavourable circumstances. See **risk**.

hazardous substance. Substance which is toxic, corrosive, flammable or reactive. All substances are hazardous to health if taken in sufficient quantity and the line which separates a hazardous one from a non-hazardous one is arbitrary. Substances presenting little or no hazard include food-stuffs, milk, water, non-toxic metals, formed polymers, freon, sulphur hexafluoride, lubricating oils, greases and paper. As a working definition, in this book a hazardous substance is defined as a substance whose acute toxicity meets one or more of the following conditions

(a) for solids, one whose acute oral LD50 in rats or mice is less than 15 g per kg body weight;
(b) for liquids, one whose oral LD50 in rats or mice is less than 5 g per kg body weight;
(c) for solids or liquids, one whose dermal LD50 in rabbits is less than 2 g per kg body weight;
(d) for solids, liquids or gases, one whose inhalation LC50 in rats or mice is less than 50 g m^{-3}.

Henry's law. The amount of a gas absorbed by a given volume of a liquid at a given temperature is, at equilibrium, directly proportional to the (partial) pressure of the gas. In gaseous mixtures, by Dalton's law, the partial pressure of a gas in a mixture is equal to the pressure which it would exert if it occupied

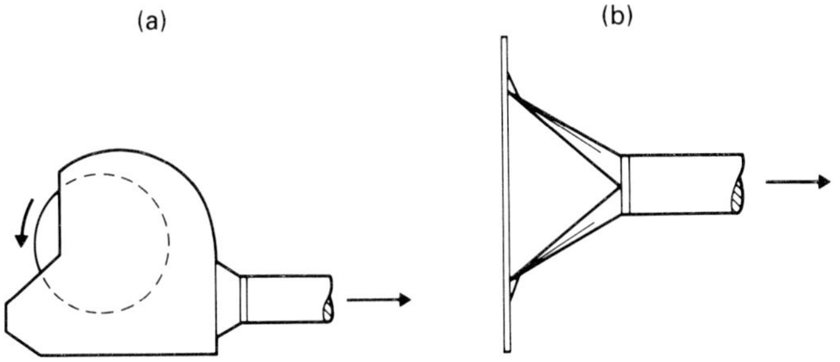

FIG. G.11. A local exhaust hood partially encloses a source of air contamination or is located as close as practicable to it. The flow rate of air extracted from the hood should be sufficient to capture virtually all the air contamination. (a) Grinding wheel hood. (b) Flanged hood.

the same volume alone at the same temperature. Thus Henry's law applies for the individual components of a gaseous mixture as well as for a single pure gas, the concentration of each component absorbed being, according to the law, proportional to its own partial pressure and not the total gas pressure. Consequently, provided the law holds, the equilibrium partial pressure over the liquid is unaffected by the presence of an inert gas.

homeostasis. A bodily process which permits complete return to a steady state of a cell or organ system after a change induced by imposed environmental conditions. Changes associated with homeostasis do not lead to morbid states. This may be contrasted with adaptation in which the alteration does not necessarily result in a return to the original state.

hood. 1. Entrance to a local exhaust ventilation system, often, but not necessarily hood-shaped or shaped like a fish-tail or cow-horn. 2. A device that completely covers the head, neck and portions of the shoulders as part of respiratory protective equipment.

hood static pressure. Static pressure measured downstream of an exhaust hood, relative to room air. Hood static pressure is negative and somewhat greater in magnitude than the velocity pressure of the air in the duct, the difference being the hood entry loss.

$$\begin{aligned}\text{Hood static pressure} &= \frac{\text{velocity pressure}}{(\text{coefficient of entry})^2},\\ &= \text{velocity pressure} + \text{hood entry loss.}\end{aligned}$$

See **static pressure, velocity pressure, coefficient of entry**.

hygiene. From the greek 'hugieine', '(art) of health'. Science and art of maintaining the health of the individual or community, especially a society of people with common occupations.

IARC. International Agency for Research on Cancer. Lyon, France.

ILO. International Labour Office. Geneva.

in vitro. Said of experiments conducted outside the body. See **in vivo**.

in vivo. Said of observations of processes in the body. See **in vitro**.

inclined manometer. Manometer with one leg inclined, usually with a slope of 1 in 10 for use in studying industrial ventilation systems. The commercial inclined manometer has a built-in level, a means of adjusting the zero and is usually filled with dyed paraffin which has the advantage of reduced evaporation and ease of wetting the tube.

inert dust. Dust which when inhaled and deposited in the lungs will not produce significant collagenous fibrosis. Nuisance dust.

inhalable dust. Fraction of airborne dust which is normally inhaled through the nose and mouth. Also known as 'inspirable' dust. The inhalable fraction is markedly less than total dust since 50% or more of the coarser airborne particles, larger than 30 microns aerodynamic diameter, fail to be inhaled. See **total dust**.

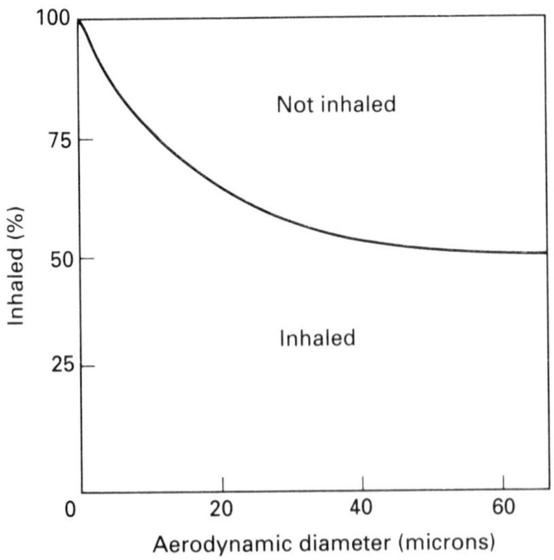

FIG. G.12. Inhalable dust. During normal activities at work, coarse airborne particles, larger than about 10 microns in aerodynamic diameter, are progressively less likely to be inhaled. The inlet of a dust sampling instrument may be designed to simulate the size cut-off characteristics of the human nose and mouth.

insidious. Refers to the biological activity of hazardous substances. Means acting in imperceptible steps; having a gradual cumulative effect.

intermediate. A general term for any chemical compound which is manufactured from a substance obtained from natural raw materials, and which serves as a starting material for the synthesis of some other product, especially in the manufacture of organic dyes and pigments. In many cases, it may be isolated and used to form a variety of desired products. In other cases, the intermediate may be unstable or used up at once. See **primary**.

involuntary risks. Risks which impinge on an individual without his or her consent.

IOH. Institute of Occupational Hygienists.

irritant. Substance which causes a local inflammatory reaction on contact.

ISO. International Standards Organisation. Geneva.

isokinetic sampling. Maintaining the air velocity in a sampling nozzle equal in magnitude and direction to the air velocity in the surrounding atmosphere. Used when sampling aerosols in high velocity air streams and when the aerosol contains a significant quantity of particles larger than 5 microns. Commonly used when sampling aerosols in a ventilation duct.

jaundice. Icterus. Yellow coloration of the skin, the whites of the eyes and other tissues of the body, by excess of bile pigment present in the blood and the lymph.

lag time. Period of time after exposure and before any response occurs in the body. See **latent period**.

Lambert's law of absorption. Where I_0 represents the original intensity of electromagnetic radiation and I is the intensity after passing through a thickness x of a medium,

$$I = I_0 \exp(-kx),$$

where k is a constant. The law applies to contamination in air. Where C is the concentration of gas or vapour in air and α is the molar absorption coefficient of the contamination medium of interest,

$$k = \alpha C$$

laminar flow. Streamline flow in viscous fluid. A type of fluid flow in which there is continuous steady motion of the particles, the motion at a fixed point always remaining constant. See **turbulent flow**.

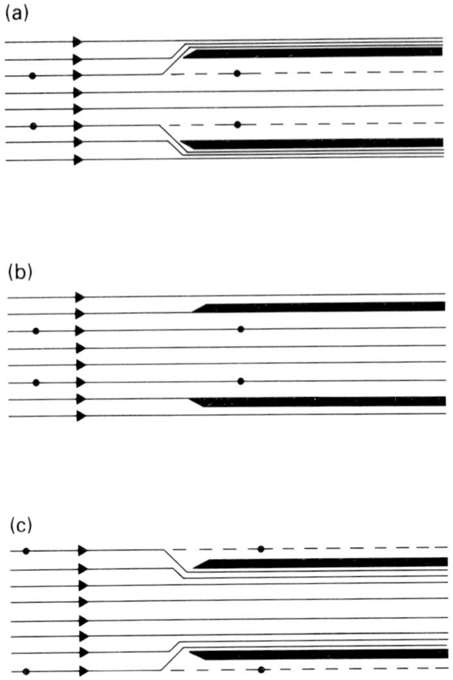

FIG. G.13. Isokinetic sampling. Aerosol particles larger than about 5 microns in diameter have a significant inertia at air velocities of the same velocity as employed in common sampling nozzles. By aligning the nozzle along the air streamlines, and by adjusting the sampling volume flow rate or nozzle diameter the velocities may be matched to within 10% and prevent over- or under-sampling from this cause. (a) Sampling velocity too low. Coarse particles over-sampled. (b) Isokinetic sampling. (c) Sampling velocity too high. Coarse particles under-sampled.

latent period. Period of time between the very beginning of exposure and the first manifestation of a response. It is crudely estimated as the time from first exposure to the detection of a response. See **lag time**.

LC50. Median lethal concentration. The atmospheric concentration, exposure to which, for a specified length of time, would cause the death of 50% of the entire population of specified animals.

LD50. Median lethal dose. The dose which would cause the death of 50% of the entire population of specified animals. Other lethal dose percentiles such as LD1, LD10, LD30, LD99 and so on are quoted occasionally

leukaemia. A fatal condition in which there is excessive multiplication of the tissues producing white cells, with consequent increase in the numbers of white cells in the blood and great enlargement of the spleen.

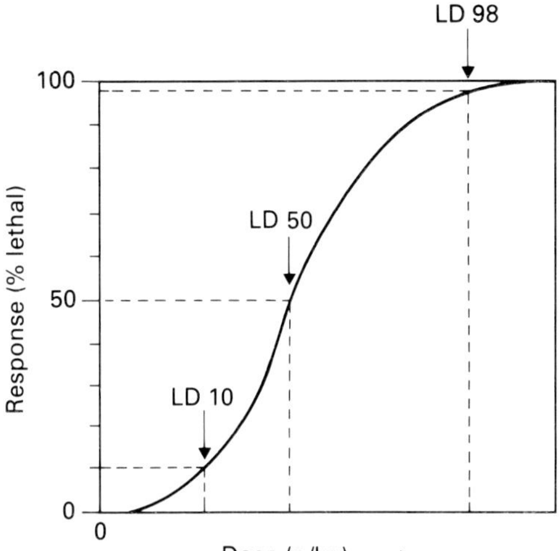

FIG. G.14. Median lethal dose (LD50). The dose of a substance which is expected to kill 50% of the animals to which it is administered. The dose is usually expressed in terms of weight of substance per kilogram of the body weight of the animal.

life expectancy. The number of years remaining for a certain attained age. For example, the life expectancy at age 62 is 18 years.

lifetime risk. The risk which results from lifetime exposure.

longitudinal study. An epidemiological term for a study in which the investigator begins by selecting groups of employees expected to be at risk, determines whether disease is present or absent and looks back in time to ascertain the level of exposure to which they have been exposed. Alternatively the investigator may follow up the employees in the future, measuring their exposure periodically and their disease status. See **case control study, cohort study, cross-sectional study**.

lung allergy. An acquired sensitivity of the lung to exposure to certain agents. Subsequently the body's reaction (wheeziness, shortness of breath, cough etc) usually occurs within minutes or hours of the start of atmospheric exposure.

make-up air. Clean, tempered outdoor air supplied to a work space in a controlled manner to replace air removed by exhaust ventilation or some industrial process.

manometer. An instrument used to measure the pressure of a gas. The usual form of manometer consists of a U-tube containing a liquid (water, oil, or mercury), one limb being connected to a point in the ventilation system or enclosure whose pressure is to be measured, while the other limb is either open to the atmosphere, or to another point in the ventilation system or is closed.

The open pattern reads the difference between the required pressure and atmospheric pressure as the difference in level of the liquid in the two limbs. See **static pressure**.

maximum use concentration. A measure of the performance of respiratory protection equipment, expressed in terms of a multiple of the exposure limit of specific substances.

metal fume fever. A fever associated with other systemic signs and symptoms and is caused by the inhalation of fumes from molten metals. The most common of these fumes is composed of zinc oxide.

metastasis. Transfer, by lymphatic channels or blood vessels, of diseased tissue (especially cells of malignant tumours) from one part of the body to another. See **benign**.

material. That of which a thing consists. It is the whole thing including any impurities. See **substance**.

melting point. Temperature at which a solid is in equilibrium with its liquid phase.

metabolism. The sum total of the chemical and physical changes constantly taking place in all living matter.

metal fume fever. Also known as 'brass founders' ague'. A peculiar malady from inhaling high concentrations of freshly formed zinc oxide fume as in brass foundries. A few hours after exposure the victim experiences chills followed by fever said to be flu-like or not unlike fever of malaria. Workers develop an immunity but this is readily lost in the absence of regular exposure.

mg m^{-3}. Milligrams per cubic metre. Common expression of concentration of aerosols suspended in air. Refers to the concentration expressed as the number of milligrams of aerosol per cubic metre of air. For purposes of standardisation the concentration refers to air at 25 degrees Celsius and 760 mm mercury pressure. See **ppm**.

1 micron = 10^4 Angstrom units

= 10^{-3} millimetre

= 10^{-4} centimetre

= 10^{-6} metre

FIG. G.15. A micron is a unit of length, used to express the dimensions of dust particles.

micron. Unit of length used for expressing the diameter of fine particles. A micron is equal to a millionth of a metre. It is also equal to a thousandth of a millimetre. A millimicron is equal to one millionth of a millimetre. See **angstrom**.

mill fever. A mild transient malady, not unlike metal fume fever; common in people first starting work in dusty cotton or flax mills. An attack confers temporary immunity for a day or so. See **Monday tightness**.

minute volume. Volume of air inhaled (and exhaled) in one minute.

mist. Finely divided liquid droplets suspended in air. May be formed by the atomisation of liquids by bubbling, boiling, foaming, spraying, splashing or otherwise agitating a liquid and also by condensation from the vapour phase. Droplet sizes vary widely depending on the prevailing conditions.

mole. That quantity of a substance whose weight in grams, pounds or any other convenient unit is numerically equal to its molecular weight. If the substance is a gas the gram-mole occupies 22.4 litres at NTP. See **mole fraction**.

mole fraction. The mole fraction of a component in a homogeneous mixture or solution is the number of moles of that component divided by the sum of the number of moles of all components. For example, if N_1, N_2, N_3 are the number of moles of the components of a mixture, then

$$\text{mole fraction of component number 1} = \frac{N_1}{N_1 + N_2 + N_3}.$$

See **mole**.

Monday tightness. Chest tightness experienced by employees exposed to high concentrations of dust in the cotton and flax industries on Mondays or on the first day back at work after an absence. It is an early sign of byssinosis and if high dust exposure continues the tightness extends to Tuesdays and other days of the week. Eventually the employee may be permanently disabled. See **mill fever**.

monitoring. Continuous or discontinuous measurement of a variable in the course of time. The variable may be the concentration of air contamination, biological indices, or control measures.

monomer. A compound of relatively low molecular weight which, under certain conditions, either alone or with another monomer, forms various types and lengths of molecular chains called polymers or copolymers of high molecular weight. An example is the monomer styrene which polymerises readily to form polystyrene. Another example is vinyl chloride and polyvinyl chloride (PVC). See **polymer**.

mutagenic. Prone to induce adverse heritable variation.

narcosis. A state of drowsiness, sleepiness, stupor and possibly unconsciousness.

necrosis. Death of groups of cells while still part of the living body.

NIOSH. National Institute for Occupational Safety and Health.

neoplasm. A tumour. A non-inflammatory mass formed by the growth of new cells in the body and having no physiological function. It may be benign or malignant. A 'malignant' tumour invades adjacent tissues and spreads to other parts of the body. See **carcinoma**.

nuisance dust. Generally innocuous dust, not recognised as the direct cause of a serious pathological condition.

oedema. Pathological accumulation of fluid in the tissue spaces.

OEL. Occupational exposure limit.

olfactory. Pertaining to the sense of smell.

osteosclerosis. Abnormal thickening of the bone.

partition coefficient. The ratio of the equilibrium concentrations of a substance dissolved in two immiscible liquids. Provided no chemical interaction occurs, it is independent of the actual values of the concentrations and equal to the ratio of the solubilities of the substance in each solvent. Also known as 'distribution' coefficient.

peak exposure. A transient high exposure extending over a period much shorter than one work shift.

pellet. In various industries the powder form of a material may be made into pellets or briquettes for convenience. The pellet is a distinctly small briquette.

personnel control. Refers to protective clothing to be worn, respiratory protection, bathing arrangements, education, training and personnel management generally insofar as it leads to a reduction in employee exposure to hazardous substances. See **engineering control**.

phagocyte. Cell which exhibits amoeboid phenomena, ability to engulf foreign bodies and to migrate from place to place.

pharmacology. Scientific study of drugs and their action.

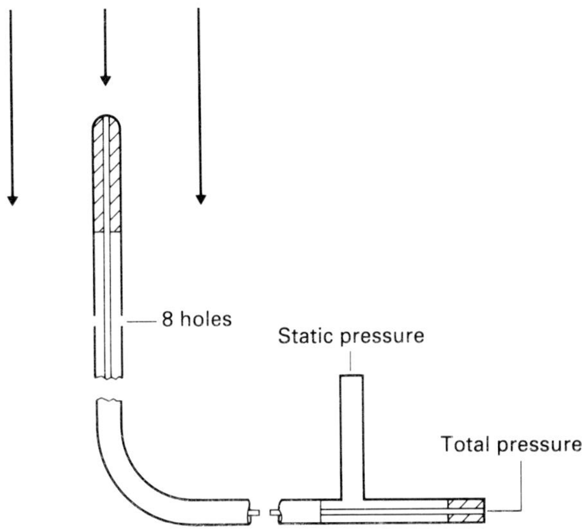

FIG. G.16. The Pitot-static tube is used for measuring the air velocity in ventilation ducts. It is actually made of two tubes, one inside the other. The front end of the device is inserted in a hole in the duct and pointed upstream. The inner tube senses the total pressure of the air, being the sum of velocity pressure and static pressure. Holes drilled through the wall of the outer tube sense the static pressure. The inner tube is sealed from the outer tube. The velocity pressure is the difference between total pressure and static pressure, determined by connecting the inner tube and outer tube to either side of a manometer.

pickling. The process of removing a coating of scale, oxide, tarnish and other impurities from metal objects so as to obtain a chemically clean surface prior to plating or other surface treatment. Pickling is effected by immersing the objects in an acid bath. Sulphuric acid is commonly used, although hydrochloric, phosphoric, nitric, chromic or hydrofluoric acid are sometimes used according to the metal being pickled. See **anodising**.

Pitot-static tube. Device for measuring the velocity of air, especially in ventilation ducts. It consists of two concentric tubes; the inner one, with its open end facing the air stream, serving to measure the total or impact pressure of an air stream and the outer tube, with small holes drilled at right angles to the air stream measuring the static pressure only. A manometer having the two tubes connected to either limb registers the algebraic sum of the total pressure and static pressure, that is, the velocity pressure. See **total pressure, static pressure, velocity pressure**.

plasma. The watery fluid part of the blood containing salts, protein and other organic compounds. The cells of the blood are suspended in plasma.

plethysmograph. An apparatus for measuring variations in the size of bodily parts.

pleura. The delicate, serous membrane lining the pulmonary cavity in mammals (and birds).

pneumoconiosis. Term applied to various dust diseases of the lung caused by the inhalation of dust particles in such occupations as coal mining and asbestos textile manufacture. An excess of fibrous tissue forms around the dust particles in the lungs.

population. Total number of people in a country, or the total number of employees in an industry, company, works, job, etc.

polymer. A high molecular weight material formed by the combination of many molecules to form a more complex molecule having the same empirical formula as the simpler ones. There may be hundreds or even thousands of the original molecules linked end to end and often crosslinked. Rubber and cellulose are naturally occurring polymers. Most resins are chemically produced, man made polymers. See **monomer**.

potency. Power, strength, efficiency or potentiality of a substance for causing effects on the body.

ppm. Parts per million. Common expression of concentration of gaseous contamination of air. Refers to the concentration expressed as the number of parts of gas, by volume, per million parts of air. For purposes of standardisation the concentration refers to air at 25 degrees Celsius and 760 mm mercury pressure. See **mg m**$^{-3}$.

precision. Sometimes termed reproducibility. The repeatability or scatter of the measurements within a set of ostensibly identical ones. See **accuracy**.

prevalence. Common occurrence; widespread existence. Thus one refers to the prevalence of coal workers' pneumoconiosis in South Wales. See **prevalence rate**.

prevalence rate. Frequency of occurrence. Thus one refers to the prevalence rate of asbestosis among male employees at a certain factory as being, say, 10%. See **prevalence**.

primary. A substance which is obtained directly from natural raw material by extraction and purification; for example, benzene, phenol, anthracene are coal-tar primaries. See **intermediate**.

primary airborne dust. Solid particles rendered airborne during the crushing, grinding or attrition of hard, rock-like materials. Such dust particles generally have irregular shapes. See **secondary airborne dust**.

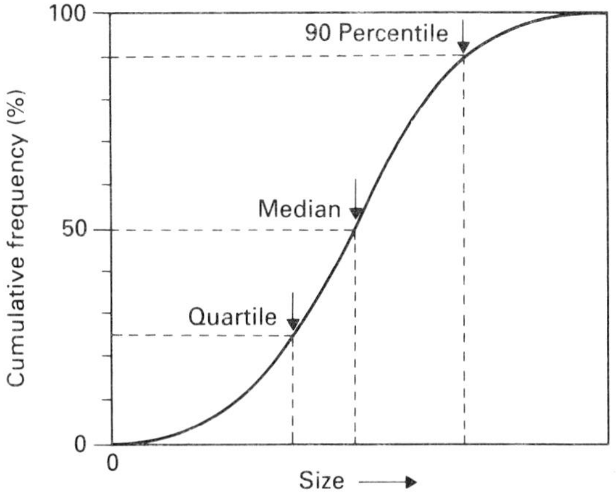

FIG. G.17. Quantile.

primary irritant. A substance whose irritant action is in excess of any systemic toxic action. Examples are hydrochloric acid gas and sulphuric acid mist. See **secondary irritant**.

pruritus. Medical term for intense itching of the skin.

quantile. The quantiles are the parameters of position in a cumulative probability distribution. They mark the values of the variable corresponding to a specified cumulative fraction below that value. The quantiles most commonly used are named median (50%), quartile (25%), decile (10%) and percentile (1%).

quartz. The common, naturally occurring form of crystalline silica (SiO_2). Widely distributed in rocks of all kinds, particularly sandstone, igneous rocks and common sand. Usually colourless and transparent but often coloured by minute quantities of impurities. See **cristobalite, tridymite**.

random variation. Exposure from a continuous process which features a fluctuating concentration with no regular pattern discernible except for a dip during the mid-shift break. See **systematic variation**.

Raoult's law. For dilute solutions, the relative lowering of the vapour pressure of a liquid by a dissolved substance is approximately equal to the mole fraction of the latter, independently of the temperature and of the nature of both the solvent and the solute. While Raoult's and Henry's laws are closely followed by many substances and approximately followed by others with accuracy sufficient for most applications, in numerous cases the deviations are large. They should therefore be regarded as ideal solution laws to which real solutions only approximate. See **Henry's law**.

RD_{50}. The atmospheric concentration of a hazardous substance necessary to reduce the respiratory rate of rodents (mice) by 50%. This parameter is highly correlated with the exposure limit of respiratory irritants.

reference period. Averaging period to which an atmospheric exposure limit refers. An exposure limit has little meaning without its associated reference period. In national exposure limits the reference period is commonly 8 hours but may be anything up to a year or down to a few seconds.

relative humidity. The ratio of the amount of water vapour in the air to the amount which would saturate it at the same temperature; or the ratio of the pressure of water vapour present to the saturated vapour pressure at the same temperature. Usually expressed as a percentage, viz.

$$\text{R.H.} = \frac{\text{partial pressure of water vapour}}{\text{saturated vapour pressure of water}} \times 100\%$$

reproductive toxicity. Adverse effect on the nature, quality or rate of reproduction.

respirable dust. Airborne dust of sizes which are carried into the alveolar region of the lungs. It consists mostly of particles less than 5 microns in aerodynamic diameter. Larger ones are deposited in the upper respiratory tract. The very finest particles, smaller than 0.5 to 1.0 microns diameter, are mostly exhaled. See **inhalable dust, total dust**.

Reynolds number. A dimensionless parameter of fluid flow, the value of which indicates whether the flow is laminar ($Re < 2{,}000$) or turbulent ($Re > 3{,}000$).

risk. A statistical concept. Probability, likelihood or chance that harm will arise in particular circumstances. May be measured directly, although not necessarily accurately by the relative frequency of cases in exposed and unexposed groups. See **hazard**.

rotameter. A variable area fluid flow meter for measuring volume flow rate, much used in occupational hygiene instruments and air sampling trains. The rotameter consists of a vertical, slightly tapered tube in which a metal or plastic float is free to move vertically. The fluid flows upwards, causing the float to rise. The position taken up by the float in the tube is governed by the equilibrium between the weight of the float and the velocity pressure of the fluid flowing through the tube. The tube has a a flow rate scale etched directly on it. See **anemometer**.

secondary airborne dust. Airborne dust produced by dispersion into the air of fine powder from a bulk source or from previously settled primary airborne dust. Airborne particles of secondary airborne dust, on close examination, are

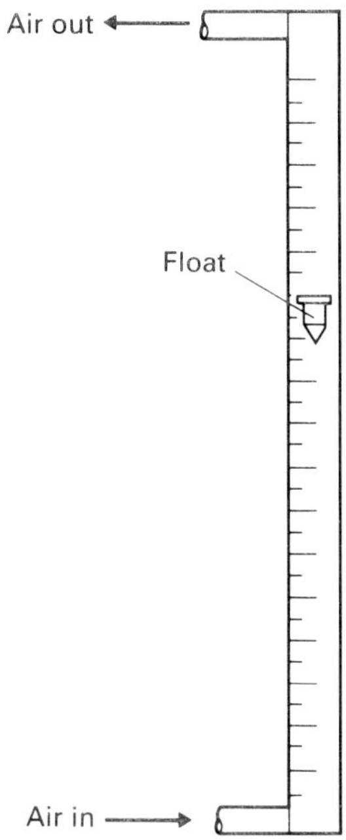

FIG. G.18. A rotameter consists of a float inside a tapered vertical tube increasing in cross section from bottom to top. The range of flow between maximum and minimum reading on the tube is usually 10:1 and different tube sizes cover the range of gas flow from 1 ml min^{-1} to 25 m^3 min^{-1}.

often found to consist of clumps or aggregates of smaller particles adhering together. See **primary airborne dust**.

secondary irritant. Substance whose systemic effects exceeds its irritant action. An example is hydrogen sulphide which is an eye irritant in low concentration but a systemic poison which affects the nervous system and may cause respiratory paralysis within seconds in the highest concentration. See **primary irritant**.

sensitiser. Substance which does not necessarily cause immediate irritation or other damage to the skin but causes the skin to become very sensitive to further exposure to that substance. A sensitiser causes specific changes in the skin so that contact with that substance at a later time will cause dermatitis. Often an employee will work with a skin sensitiser for a long period of time and then suddenly develop a rash.

sensitivity. The least alteration in a factor which will cause a change in its measurement. The sensitivity of instruments for measuring atmospheric concentration generally varies with the magnitude of the concentration. Sensitivity near zero concentration is the least detectable concentration.

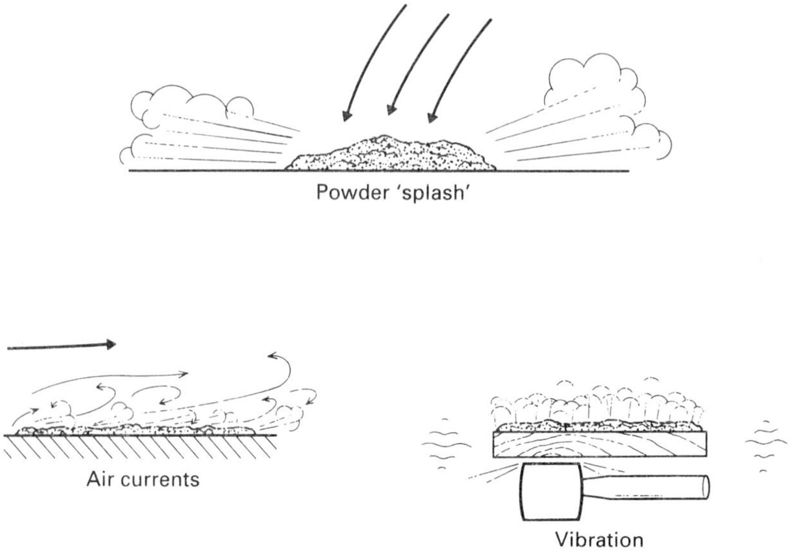

FIG. G.19. Secondary airborne dust is produced by the dispersion of fine powders of bulk materials or from previously settled dust. Dispersion is caused by a variety of mechanisms, including powder 'splash', air currents and vibration.

sign. A medical term referring to any objective evidence of disease or bodily disorder. See **symptom**.

significant health risk. A risk to health is significant when it is known or foreseeable and a reasonable employer is more likely than not to take steps to reduce it, for whatever weighting of reasons. By 'reasonable' employer is not meant the most reasonable nor, for that matter, the least reasonable, but rather an employer of average reasonableness. The proverbial bus passenger on the top deck of a No. 19 bus would be expected to be neither more reasonable nor less reasonable than this employer.

silicosis. Pneumoconiosis due to the inhalation of particles of silica by gold miners, masons, pottery workers and others who work in the presence of free silica.

smoke. A suspension of a solid in a gas, generally organic in origin and visible, such as the smoke from burning tobacco, wood, oil or coal. The particle size is generally below 0.5 micron diameter. See **fume**.

sol. A colloidal solution. Suspension of solid particles of diameter lying between 0.1 and 0.001 micron.

specific gravity. Relative weight of a given volume of a substance as compared to an equal volume of a standard substance at a standard temperature. Water at a temperature of 4C is the standard usually referred to for solids and liquids. Dry air at the same temperature and pressure as the gas in question is usually taken as the standard reference substance for gases.

specificity. Intrinsic power of a method to respond to only one substance or effect. Applies to analytical methods, to methods of measuring concentration and to methods of measuring biological effects. The specificity of a method in a particular environment or individual also depends on the likelihood of interfering substances or biological effects being present in that environment or individual. See **cross-sensitivity**.

standard deviation. Square root of the average of the squares of the deviations of a number of observations from their mean value. A common measure of the scatter of a set of observations about their mean. See **co-efficient of variation, variance**.

static pressure. Fluid pressure exerted in a direction perpendicular to the direction of fluid flow through the measuring point. It is normally measured in a ventilation duct by drilling a hole through the side, 1–2 mm in diameter, at right angles to the air stream. Inside the duct, for a correct measurement the entrance to the hole should be flush with the surface, smooth and with any burr removed. In order to measure the static pressure the open end of a flexible tube or pad is held against the hole on the outside, the other end of the tube being

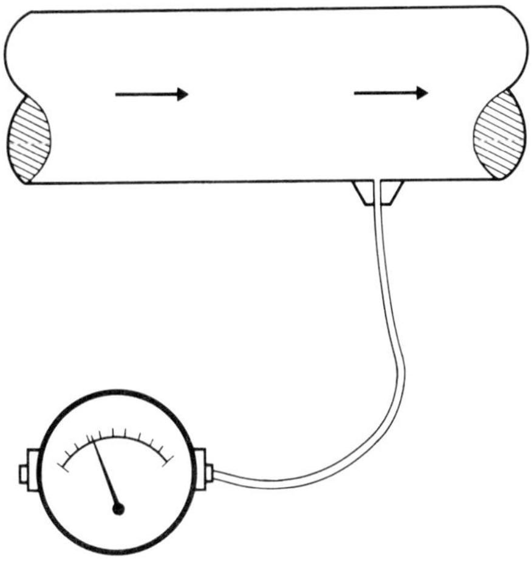

FIG. G.20. Static pressure in ventilation systems is conveniently measured with an aneroid type pressure gauge connected by tubing to a small hole (2 mm dia.) in the duct.

connected to a manometer. To make a permanent fitting a somewhat larger hole with a capped metal tube screwed in place on the duct is more convenient. In ventilation studies static pressure is normally measured relative to room air. Static pressure may be positive or negative. See **total pressure, velocity pressure**.

substance. The essential, most important element(s) or component(s) of anything. Thus, for example, commercial trichloroethylene is the material which contains a number of impurities besides the essential component, the substance trichloroethylene. See **material**.

symptom. Subjective evidence of disease or disorder as experienced by the patient, such as pain, weakness or dizziness. Includes any abnormal sensation or emotional expression or thought accompanying disease or disorder. See **sign**.

synergism. Cooperative action of substances whose total biological effect is greater than the sum of their separate effects. See **antagonism**.

systematic variation. Exposure with a distinct pattern, usually following a cycle of operations. This type includes cyclic exposures of various periodicity. Brief inhalation of high concentration of air contamination with no other exposure during the day is just an extreme example of this general type, as is brief inhalation of a high concentration during prolonged exposure to low concentrations, etc. See **random variation**.

systemic effect. Effects spread through several or all body systems and not localised in one spot or area.

target organ. Organ in the body which, on administration of a dose of hazardous substance, first experiences an adverse effect. Sometimes called the 'critical' organ.

teratogen. Substance which causes defects in development of an embryo or foetus during pregnancy.

terminal velocity. The constant velocity acquired by a particle falling in air or water when the drag resistance is equal to the gravitational pull. The terminal velocity in air of spherical particles between 0.5 and 20 microns diameter is given by

$$V = 0.0026(\text{s.g.})d^2$$

where V is terminal velocity in cm s^{-1}, s.g. is specific gravity of particle and d is particle diameter in microns.

therapeutic substance. Substance administered with the intention of treatment, remedy, cure or prevention of disease.

throw. In air distribution, the distance an air stream travels from an outlet to a position at which air motion along the axis reduces to a velocity similar to the velocity of random air currents. Also called the 'blow' of a stream.

tidal volume. Difference in volume of the lungs between the end of a normal inspiration and the end of a normal expiration. The volume of air inhaled and exhaled during a normal respiration.

total dust. Dust of all types and sizes suspended in air. Sizes vary from a few angstrom in diameter up to particles over 200 microns diameter. See **inhalable dust**.

total pressure. Driving force of fluid flow. Sometimes known as dynamic pressure. In a ventilation duct it is the air pressure, relative to room air, exerted in the downstream direction. Generally measured with a Pitot tube held with its open end facing into the air stream and the other end connected to a manometer in the work-room. It is the algebraic sum of the static pressure and velocity pressure. See **static pressure, velocity pressure**.

toxico-dynamics. Study of the time course of toxicological response to a substance present in the body. It includes metabolism, cellular reaction and functional response. See **toxico-kinetics**.

toxico-kinetics. How the body handles a foreign substance. Study of the time course of toxic substance levels in different fluids, tissues and excreta of the body resulting from exposure to the substance in the gaseous, liquid or solid state. It includes study of the mathematical relationships required to develop models to interpret such data. See **toxico-dynamics**.

trachea. The windpipe leading from the glottis to the lungs, where it divides into the two main bronchi, the left bronchus and the right bronchus. See **bronchiole**.

transport velocity. Sometimes called conveying velocity or minimum design duct velocity. The minimum velocity in a duct which is necessary to prevent settlement of particulate material on the floor of the duct. Ranges from 5 to 20 $m\ s^{-1}$ depending on the physical nature of the material, especially its particle size and density. For example, the transport velocity of sawdust is 6 $m\ s^{-1}$ whereas that of rock dust is 20 $m\ s^{-1}$.

tremor. Involuntary shaking, trembling, quivering or shivering due to disease of the nervous system.

tridymite. Vitreous, colourless form of free silica formed when quartz is heated to 870C. See **quartz**.

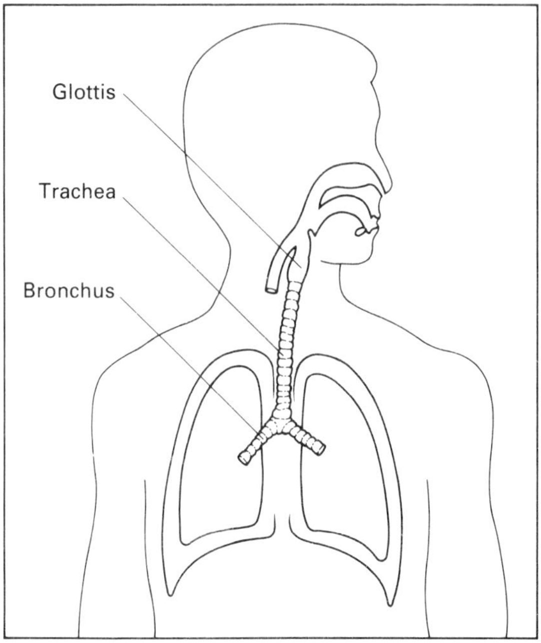

FIG. G.21. Trachea. The windpipe, leading from the glottis to the lungs, where it divides into two main bronchi.

turbulent flow. Also called eddy flow or sinuous flow. A type of fluid flow in which there is an unsteady motion of the fluid particles, the motion at a fixed point varying in no definite manner. Turbulence increases with fluid velocity, fluid shear and surface roughness. In the immediate vicinity of a solid object, eddies generated by fluid flow over the object are properly visualised as highly ordered vortices. Further from the object the motion is increasingly disordered.

Tyndall effect. The light scattered by very fine particles in suspension. Blue light is scattered to a much greater degree than red light. This accounts for the blue colour of the sky, which is produced by sunlight scattered by gas molecules and very fine particles. Conversely, red light is transmitted preferentially through a suspension. This accounts for the red colour of the sky at sunset which is essentially transmitted light. The Tyndall effect is employed in the Tyndallometer and similar devices by which the intensity of light scattered from a parallel beam is viewed by eye or a photocell and the concentration is thereby estimated.

vapour. The gaseous state or form of a substance which is normally in the liquid or solid state at room temperature and pressure. The liquid or solid may be reduced to the vapour, that is, the gaseous, elastic condition by the action of heat. A vapour can be liquefied by a suitable increase in pressure. See **gas**.

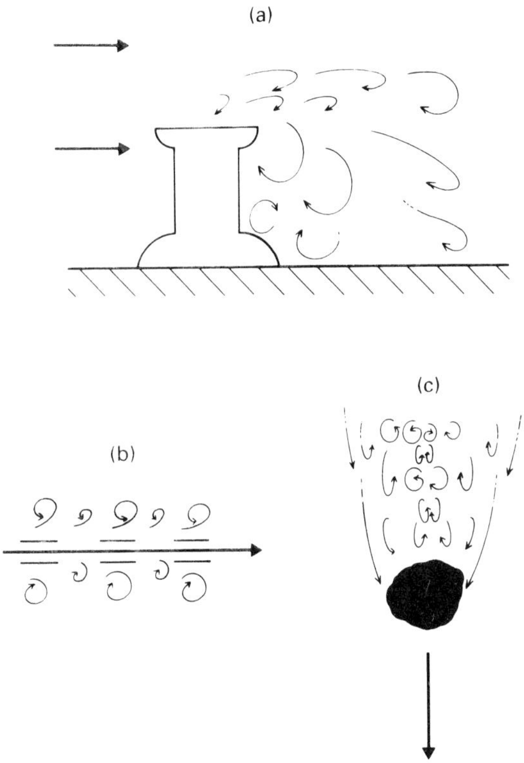

FIG. G.22. Turbulent motion in the wake of air flow over objects. A. Fixed object. B. Moving belt. C. Falling particle.

vapour pressure. Molecules escaping from the surface of a liquid in a closed container are confined to the space above the liquid. At equilibrium the rate of return of molecules to the liquid equals the rate of escape and the pressure exerted by the vapour in the container at this point is called the vapour pressure of the liquid. It is usually expressed in mm of mercury.

vapour specific gravity, air = 1. Relative weight of a given volume of vapour as compared to the weight of an equal volume of air.

variance. Average of the squares of the deviations of a number of observations of a quantity from their mean value. See **standard deviation**.

velocity pressure. Component of pressure due to the velocity of a fluid. Velocity pressure is exerted in the downstream direction. It is the difference between total pressure and static pressure and is readily found by sensing these pressures either side of a manometer. Velocity of air at standard temperature and pressure is given by,

$$V = 4.04\ (\text{velocity pressure})^{1/2},$$

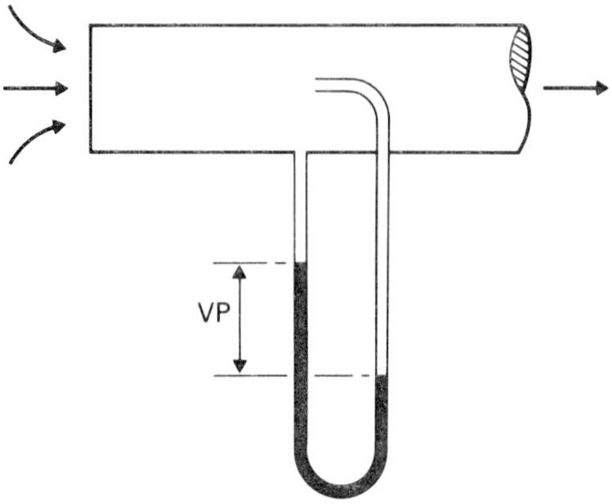

FIG. G.23. Velocity pressure (VP) in an air duct is the difference between the total pressure (TP) exerted in the downstream direction and the static pressure (SP) at the same point. VP = TP − SP.

where V is in m s^{-1} and velocity pressure is in mm of water gauge (mm WG). See **total pressure**.

WHO. World Health Organisation. Geneva.

vital capacity. On taking a deep breath, a maximum inspiration, the air is blown out completely, through a device for measuring the volume expired, a maximum expiration. The measured volume thereby expired is the vital capacity.

work place. Well defined, relatively small area of a building or other place of work which is occupied by an employee in the course of his or her work.

PART I

BIOLOGICAL EFFECTS

OF

EXPOSURE TO HAZARDOUS SUBSTANCES

Introduction

Mild adverse effects on health caused by excessive exposure to hazardous substances at work, such as dermatitis, respiratory irritation, headaches or narcosis are readily apparent but many of the most serious effects do not at first have obvious signs. Occupational diseases from exposure to hazardous substances are often insidious in nature. Unlike accidents, there is no blood to be seen; there are no broken bones. The consequences, however, may be just as severe, are frequently long lasting and there may be little prospect of a return to normal health. Early detection of developing disease is essential but often made more difficult because of its insidious nature.

Exposure of an individual to a substance at work includes every type of contact between the substance and the physical boundary of the individual's body. The most important means of exposure are inhalation of air contamination, ingestion and skin contact with the substance in question. The most important means of elimination from the body are by excretion in the urine or faeces and, for gases and vapours, exhalation. The principal pathways taken by hazardous substances through the body are illustrated schematically in Figure I.1.

At work, inhalation is the primary means of entry to the body. The substance may affect the point of first contact; upper respiratory tract, lungs and skin. Inhaled materials such as irritant gases or acid mists produce a rapid response from the respiratory tract. Other gases, vapours and aerosols may pass into the lungs, become absorbed or deposited there, become distributed about the whole organism in body fluids and exert a systemic toxic action at an organ remote from the respiratory system. Depending on the properties of the substance various amounts are accumulated in different organs, bones, tissues and fat. These substances or their metabolites are eventually eliminated from the body in the urine, faeces or exhaled air. Insoluble gases and the finest airborne particles which are inhaled penetrate to the alveoli and if they are not absorbed or deposited there and remain in suspension they are eliminated directly by being exhaled.

The principal methods of access of hazardous substances to the body are threefold:

(1) Inhalation of the substance suspended in air. The net input of substances by inhalation, after taking account of the amount exhaled, is:

$$V_r(C_i - C_e)t \text{ mg min}^{-1}$$

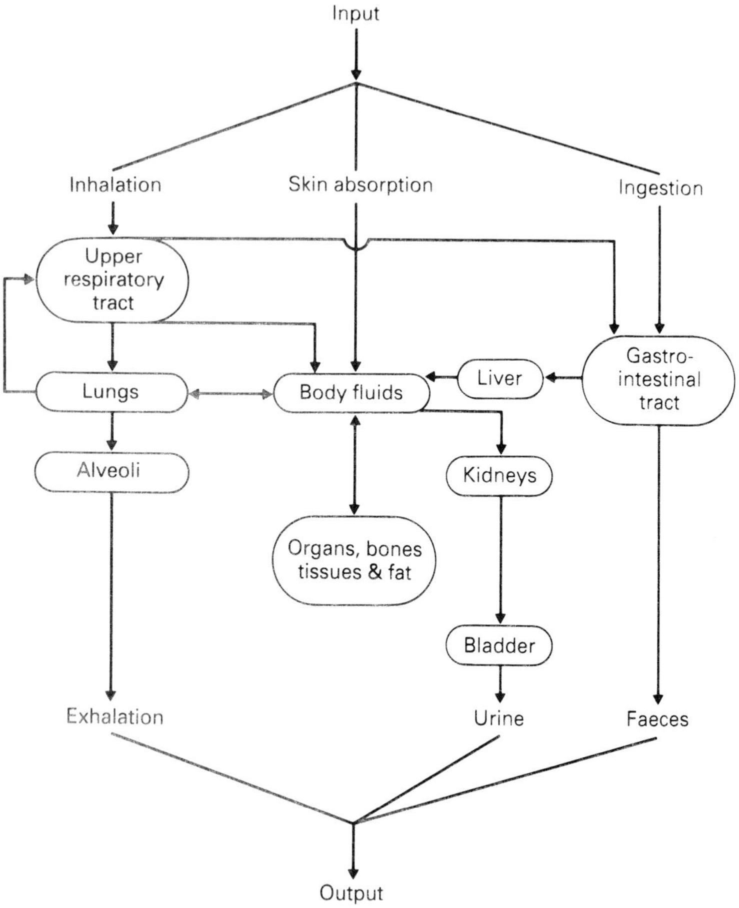

FIG. I.1. Main pathways through the body followed by hazardous substances.

where V_r = respiratory minute volume ($m^3\ min^{-1}$)
C_i = concentration in inhaled air ($mg\ m^{-3}$)
C_e = concentration in exhaled air ($mg\ m^{-3}$)
t = duration of exposure (min)

(2) Ingestion.

(3) Contact with skin and eyes. Exposure by skin contact is proportional to the surface area exposed and concentration of hazardous substance in contact.

Poisoning from inhaling hazardous dusts is much more likely than from swallowing them. Dusts which deposit in the alveoli can pass directly into the blood stream, thence to the heart, and immediately be pumped all over the body. Distribution to the body tissues is therefore brought about rapidly and effectively. On the other hand, dust taken in with the food goes to the stomach and the major part passes out in the faeces. Some is picked up by the portal blood circulation and moves on to the liver. That portion which causes poisoning must first pass through the liver, which is an effective filter and detoxifier; only then can it enter the general circulation.

Solids which deposit but are not dissolved in the lungs are mostly removed along the ciliated epithelium and are expectorated or swallowed. Some of substances swallowed are subsequently absorbed from the gastro-intestinal tract. In the case of a substance with a significant vapour pressure some, at least, is returned to the lungs from venous blood and is exhaled. The fraction of material which is not quickly removed from the lungs will accumulate and will damage the lungs after lengthy exposure.

Identifying the materials which are hazardous to health, the places they are found and how they may gain ingress to the body are considered in Chapter 1. A simplified outline is given in Chapter 2 of the main types of occupational disease from excessive exposure to hazardous substances. Only the bare outlines are included here to enable the reader to understand the elementary principles. Nevertheless the anatomy and function of the organs or organ system involved is described briefly to explain broadly how substances accumulate and leave or become otherwise eliminated.

The reaction by the body to the hazardous substances in different parts of the body varies from one substance to another and also from one person to another. There is at first a gradual increase in body burden without any accompanying cellular or functional change, then, before there is any detectable functional change there will be some slight biochemical changes. The changes, whether biochemical, cellular, functional or clinical increase imperceptibly. At some point these changes will constitute what is believed to be an adverse effect and this state, however defined, becomes the first recognised step leading to the threshold of disease in the individual. The exposure needed to reach this defined adverse effect threshold will vary from one individual to another around the norm. People's tolerance of exposure to a chemical differs widely. The consequences which flow from these essentially toxico-dynamic factors are discussed in Chapter 3.

Within the body foreign substances accumulate and are removed from different organs at various rates so that there is a lag between exposure and output. This lag may be a relatively brief one, lasting a few minutes or hours, with other substances it may be weeks or months and some may stay in the body for a life time. The time course of these events may be studied wherever it is feasible to monitor the substance or metabolite in excretion, exhaled air, blood or other body fluids. Compartmental models may be constructed to explain the washout curves. The toxico-kinetics of the matter are considered in Chapter 4.

CHAPTER 1

How Hazardous Substances Enter the Body

A necessary feature of a thorough assessment of possible health risks would be at an early stage to compile a list of the principal materials in the vicinity of employees. They may consist of gases, vapours, liquids or solids. This does not mean that all these materials will necessarily pose a health risk to those working on a process. The steps taken to abate emissions to atmosphere and otherwise reduce employee exposure may be so effective that exposure is negligible. In any case employees who work far away from production areas are not likely to be exposed to the materials being used there. At the other extreme minor materials are by no means necessarily harmless. A peculiar type of work or manner of handling or storage of a particular material could also result in hazardous exposure to a minor material or ingredient. These possibilities should be considered in a methodical manner. If it is judged that at some particular stage in a process such adverse conditions are liable to occur, then the materials being handled there, albeit in small quantities, should be added to the list.

A material may be intrinsically hazardous or become so by the nature of its use. Health risk at work is created principally by inhalation of air which has become contaminated with hazardous substances. Aside from those materials liable to give significant atmospheric exposure, materials which are absorbed through the skin or irritant on contact with the skin or eyes should also be included for consideration of their potential hazard by this means.

The three principal routes of entry to be discussed are inhalation, ingestion and skin absorption.

What is a substance hazardous to health?

Substances hazardous to health are all those which have a potential for causing harm to employees' health. Whether or not they do so in practice is dependent on the form they are in, on the way that they are used and on the precautions that are taken to limit exposure. Basic industrial chemicals are produced primarily for use in subsequent chemical or process applications. They are not generally sold as such directly to the ultimate consumer, but rather to other manufacturers or formulators. They range from the relatively simple basic commodities like chlorine, caustic soda and sulphuric acid to extremely complex dyes and bulk medicinals. Other industrial chemicals are used as solvents, as intermediates in further chemical manufacture and as the basis for making synthetic rubber, plastics and man-made fibres. Consequently the hazardous materials on the inventory of a company may be few in number in service industries or a small manufacturing company where hazardous substances, whether they be 'chemicals' or otherwise, are incidental to the main products, but there could be hundreds or even thousands of hazardous substances in a large multi-product chemical company.

There is a large number of different materials in common use, all of which are, to some extent, mixtures and an immense number of impurities in them, albeit in small or trace quantities. When compiling a list of hazardous substances on the inventory it is necessary to recognise at the outset that it is not feasible to account for every minor substance and impurity down to the last molecule present. Many factors could influence the decision on a particular material; tonnage produced/used/held; its composition, toxicity of the component substances, volatility if liquid, friability if solid and so on. Whichever criterion, formula or combination were employed the choice of cut-off point would necessarily be to some extent an arbitrary one. Above all it is important to employ factors which are known or can be readily determined. A list of hazardous materials includes ones which are conveniently classified as:

(a) present in substantial quantity, or

(b) present in small quantity but are especially hazardous, or

(c) present in small quantity, are moderately hazardous materials but are subject to an especially hazardous use.

On the other hand, materials lacking any propensity to become airborne or brought in contact with the skin may be excluded.

Materials in substantial quantity

For practical purposes, the simple tonnage of material produced/used/held at a works or process is a factor which is known or can be discovered for most materials and despite its weaknesses, repeatedly attracts a consensus of opinion as being the most ready means of defining the initial list of materials for inclusion in the assessment (Health and Safety Executive, 1989; Ionising Radiation Regulations, 1986). Toxicological properties are taken into account in formulating exposure limits and these can form a sound basis for subsequent prioritisation.

The locality of each process is identified on a layout plan and a list is compiled of hazardous materials handled there in significant quantities. This includes materials brought into the work place, stored there and produced there. By way of guidance, the author suggests that these might usefully include at least materials in excess of 50 kg per annum through-put or 1 kg inventory which contain more than 1% of substances hazardous to health.

The materials included in the list of those handled in a work place could be either raw materials, basic chemicals, ancillary chemicals, intermediates, finished products, by-products, impurities, residues or wastes. All are possible candidates for inclusion. A process flow chart would be helpful. This would identify the various stages of a process or an activity in which the material is involved, where it is, what happens to it at each stage and what gases, fumes, mists or dusts are given off. The flow chart would start at the point where materials enter the unit and continue to the point where they leave. The physical form or forms of the material (liquid, paste, powder, pellets, etc) should be given and the chemical names of the substances in it which are hazardous to health. Molecular structure diagrams, where appropriate, should be added. Chemical Safety Data Sheets from suppliers or in-house should also be appended as should labels on packaging and containers. Chemical Safety Data Sheets for chemicals in common use are published by the Royal Society of Chemistry (Bretherick, 1981).

Suitable criteria have to be adopted to decide which substances are hazardous to health. To the layman it seems that almost all substances are labelled toxic at some time or another. Water from the tap is sometimes toxic—alcohol is toxic—coffee is toxic—milk can be toxic—some foods are toxic—too much oxygen is toxic—everything we eat, drink and inhale is potentially toxic—the whole world is a toxic substance. The only difference between substances is that some are more toxic than others.

Yet some materials are so innocuous they can in reality pose little threat to health even through being inhaled. A definition based on substance potency or toxicity is widely used. The degree of toxicity possessed by a solid or liquid is most commonly measured by its median lethal dose or LD50 for experimental animals, usually rodents. The lower the LD50 the greater the toxicity. The manner of administration of the dose must also be defined; whether it be oral, intravenous, inhalation, skin contact or otherwise. The acute oral LD50 is the oral dose administered in one or more individual portions within a period of 24 hours which causes death in 50% of the animals within 14 days. The required dose to cause death increases more or less in proportion to the weight of the animals, so the results are expressed as dose per kilogram of animal weight. As a guide, liquids whose acute, oral LD50 in rats or mice is less than 15 g per kg body weight and solids whose acute, oral LD50 in rats or mice is less than 5 g per kg body weight are 'hazardous' for the present purposes. Substances whose acute oral LD50 is greater than this are not usually classified as hazardous by national and international authorities (Organisation for Economic Cooperation and Development, 1984; Oliver, 1986).

Gases are not administered orally but the toxicity of any substance, whether a gas, vapour or aerosol, may be expressed in terms of its median lethal atmospheric concentration, or LC50. The acute inhalation LC50 is the concentration inhaled for a brief period, which is normally 8 hours or less, usually 4, and which causes death in 50% of the experimental animals within 14 days. There is no correction for animal weight since LC50 does not increase with animal weight as does LD50. Larger animals inhale more air. It is recommended that gases and vapours are classified as hazardous whose LC50 in rats or mice is less than 50 g m^{-3}.

Finally, it is recommended that substances whose dermal LD50 in rabbits is less than 2 g per kg body weight are also classified as hazardous in this book. This is the value in the EEC classification of dermal biological hazards (Garlanda, 1983). It is also the value which has been selected by the American Conference of Governmental Industrial Hygienists as a suggested cut-off value for percutaneous 'absorbable' substances (1988)

Especially hazardous materials

Materials which are in small quantities but are known to be especially toxic should be added to the main lists. Materials handled in high quality, high value batch processes are candidates for consideration under this heading. A small quantity, as defined earlier, is less than 50 kg annual throughput and less than 1 kg inventory. The degree of toxicity possessed by a gas or vapour is used to define those which are especially toxic. It is most commonly measured by the median lethal concentration in inhaled air, or LC50 for short. As a working definition for the purposes of this book especially toxic liquids are ones whose toxicity in rats or mice is represented by an acute oral LD50 less than 10 mg per kg body weight, gases or vapours whose acute inhalation LC50 is less than 40 mg m^{-3} and solids whose acute oral LD50 is less than 2.5 mg per kg body weight or inhalation LC50 is less than 10 mg m^{-3}, or dermal LD50 is less than 1.5 mg per kg body weight. These definitions are summarised in Table 1.1.

Known carcinogens and teratogens should also be included under this heading.

Propensity to become airborne

Some materials which, by their nature, lack any significant propensity to become airborne in the work place could be considered for exclusion. Examples which make the point would be; a stack of lead ingots, a pan of engine oil at room temperature, a tin of grease, alcohol in a stoppered Winchester and

TABLE 1.1. *Definition of hazardous materials by their acute toxicity indices*

Physical state	Toxicity index	Materials: Hazardous	Materials: Especially hazardous
		Acute toxicity	
Solid	Oral LD50, rodents	< 5 g kg^{-1}	< 2.5 mg kg^{-1}
	Inh. LC50, rodents	—	< 10 mg m^{-3}
Liquid	Oral LD50, rodents	< 15 g kg^{-1}	< 10 mg kg^{-1}
	Dml. LD50, rabbits	< 2 g kg^{-1}	< 1.5 mg kg^{-1}
Gas or vapour	Inh. LC50, rodents	< 50 g cm^{-3}	< 40 mg m^{-3}

Note. LD50 expressed in terms of dose per kg body weight.

paint in a lidded tin. Specific guidance could be formulated related to measurable physical properties of the material as, for example, a sieve analysis, which is often available for quality control of powders. Or there may even be available a standard index of its dustiness (Andreasen, Hofman-Bang and Rasmussen, 1939; Wells and Alexander, 1978; American Society for Testing Materials, 1984: BOHS Technical Guide No. 4, 1985; Davies, Hammond, Higman and Wells, 1988; Cowherd, Grelinger, Englehart, Kent and Wong, 1989). A component vapour pressure above a liquid might be tabulated in the technical literature or calculable. Process evaporation rates or emission rates may be known, or it may be calculable from material consumption (Hoy, 1969; ACGIH, 1986).

Exclusion on grounds of lacking significant airborne tendency might thus be made on grounds of the material meeting one of the following criteria:

(a) The material as used is a clean, dry solid or is in granular form of which less than 1% would pass a square mesh sieve of 0.2 mm width opening.

(b) The material is a solid wetted at all times by the addition of more than 10% water or otherwise such that dust is not given off during work.

(c) At the process working temperature and bearing in mind the composition of the liquid being handled, the combined saturated vapour concentration of every chemical is less than the occupational exposure limit of the vapour mixture, and the process does not give off a mist.

A component whose concentration in air is less than one-tenth its exposure limit may be safely ignored. Thus, by application of Raoult's law, components may be ignored whose mole fraction in the liquid,

$$N_0 < \frac{\text{OEL} \times 7.6 \times 10^{-5}}{P}$$

where N_0 is the mole fraction of the component in the liquid, OEL is the atmospheric exposure limit of the pure component expressed in ppm, P is the vapour pressure of component at room temperature in mm mercury.

(d) Under working conditions the rate at which hazardous gas or vapour is given off is less than 0.2 mg min^{-1}.

(e) The material is held in unbroken, hermetically sealed containers.

Hazardous use

Consideration should be given to activities, processes or operations in which the materials are present only in small quantities (<1 kg) to assess whether the type of work activity is one which, despite the size of the source, could give rise to significant emissions, whether in the form of dust, mist, fume, fog, vapour or gas. This may be termed a hazardous 'use'. Peculiar uses which may give rise to excessive exposure by ingestion or skin contamination should also be identified. Any use should also be included which, for whatever other reason, is suspected of being associated with adverse health effects in the operatives.

There are well known examples of operations which characteristically give rise to a high concentration of contamination in a relatively small volume of air, including hand grinding, sieving, drilling, sawing, melting, boiling, welding, spraying and sweeping-up amongst many others. Materials incorporated in the fabric of the buildings, plant and machinery or used in their maintenance should be considered for employees undertaking dismantling, repair and maintenance operations. Materials used for insulation such as those containing asbestos and those in wall, ceiling and floor coverings, including paint, may also give rise to significant air contamination during regular maintenance and repair operations. Where it is judged that the work or the circumstances of use of the material might conceivably give rise to employee exposure to a substance in excess of $0.1 \times$ exposure limit such a material should be added to the initial list.

A chemical laboratory with its attendant exhaust hoods is a particular example where chemicals are characteristically in small amounts but the manner in which they are handled or the reactions produced between them are such that exposure to the substances or their reaction products may be excessive unless properly controlled. It is part of normal laboratory management to undertake many chemical procedures in fume cupboards or under ventilated hoods and to group together whole classes of like chemicals such as volatile liquids, cylinder gases, 'poisons', acids, alkalis and so on. Work activities with such a class of like chemicals could properly be considered together for assessment purposes and for devising appropriate operating procedures and other preventative control regimes.

Skin/eye irritants

Splashes and spills which result in contact of a hazardous substance with the skin and eyes are an important consideration because of the relative frequency with which they occur. The skin is rarely affected by gases and vapours but is vulnerable to irritation by liquids and solids. Acids, alkalis and organic solvents are obvious examples but there are many others. The effects of accidental eye contamination can range from mild irritation to complete loss of vision. Aerosols and, in addition, some gases and vapours in air may produce acute or chronic eye irritation (Grant, 1962).

A supplementary list should be compiled of any materials handled which are particularly irritant to the skin or eyes. Such materials are listed by the UK Health and Safety Commission (1985, onwards). With some substances significant absorption through the skin can result from a splash of liquid or by exposure to a high vapour concentration. Some of these substances are in published lists of exposure limits (American Conference of Governmental Industrial Hygienists, 1988; Commission for the Investigation of Health Hazards of Chemical Compounds in the Work Area, 1988; Health and Safety Executive, 1991). When present they should also be included on the list for special attention to the precautionary measures taken to prevent skin and eye contact.

Keeping the list up-to-date

Maintaining the list of hazardous materials intact and keeping it up-to-date and correct can be an onerous task in companies where many different materials are handled. Computerised records may be recommended. It is advisable to allocate responsibility for maintaining the list to a named department or individual. Systematic checking and updating of the complete list is probably sufficient once a year, except for major product changes. Such changes can be accommodated as they occur by making amendments to the previous annual up-date.

Portals of entry to the body

The rate at which hazardous substances are taken up by the body is proportional to the intensity of exposure. At this point it may be inquired in precisely what sense is the term 'exposure' employed here. There are many technical terms employed in industrial hygiene and related fields and an extended Glossary of these is given at the beginning of this book. Some are common words which have taken on a special meaning in this field. One of these is the term 'exposure' in relation to hazardous substances at work. Unfortunately this has been given various meanings by different authorities. In this book 'exposure' is the concentration and time at the interfaces between the body and the hazardous substance in the environment at work. The substance may be in solid, liquid or gaseous form. The interfaces include the lungs, skin, eyes and mouth. Air gradually changes its composition as it penetrates further into the respiratory system so that with respect to the lungs the boundary between them and the atmosphere is somewhat arbitrary. For the present purposes; the interface between the lungs and the atmosphere refers to the entrance to the nose and mouth. When a respirator is donned the interface is inside the respirator, not outside.

The original materials which enter a works may be gaseous, liquid or solid (Walton, 1976). They may come into direct contact with the employees. The liquids may be splashed or sprayed to make a mist or may evaporate and become inhaled in that fashion. The solids may be broken down to dust and become dispersed in the air. When a hazardous liquid or solid is broken up into finely divided droplets or particles the available surface area is greatly increased, which increases the potential hazard to health.

Gases and vapours are equally gaseous. A vapour is the gaseous state or form of a substance which is normally in the liquid or solid state at room temperature and pressure. A gas can be changed to the liquid or solid state only by the combined effect of increased pressure and decreased temperature. The liquid or solid may be reduced to the vapour, which is its gaseous condition, by the action of heat.

Dry air is a mixture of gases; principally nitrogen, oxygen, argon and carbon dioxide. The water vapour in air is variable; the relative humidity being the partial pressure of the water vapour expressed as a percentage of the vapour pressure at the same temperature. In the same way, the partial pressure of the vapour of any other liquid lies between 0% and 100% of the vapour pressure of the liquid. Highly volatile liquids have a high vapour pressure whereas those which do not evaporate readily have a low vapour pressure. As temperature is reduced vapour pressure falls. The temperature at which water vapour in a given sample of air becomes saturated is known as the dew point. Below this, nucleation and condensation take place: hence the appearance in the air of liquid droplets visible as mist or fog.

Gibbs (1924) gave the general name 'aerosol' to all the various disperse systems in air, such as mist, dust, fume, smoke and fog. A mist is a dispersion of finely divided liquid droplets suspended in air. It may be formed by the atomisation of liquids by bubbling, boiling, foaming, spraying, splashing or otherwise agitating the liquid and also by condensation from the vapour phase. Condensation from the vapour phase occurs when the temperature falls below the dew point. Droplet sizes in a mist vary widely depending on the prevailing conditions.

Airborne dust is a dispersion of solid particles. That resulting from the fracture of larger masses of material such as in drilling, crushing or grinding operations is referred to as 'primary' dust. Dust in bulk and settled dust which is re-dispersed by brushing, by compressed air, by wind, vibration and so on is referred to as 'secondary' airborne dust.

Dust in bulk, a very fine powder, has physical properties in many ways very like a liquid. It splashes like a liquid. It can be poured through a pipe. Ripples are formed when a stone is dropped into a tank of dust. When air is blown through bulk dust the air becomes 'saturated' with dust. When the air speed through the dust reaches a certain critical value the whole body of the dust 'boils' over. This phenomenon is employed to good effect in the pneumatic conveying of powders.

When dust in suspension in air is collected and examined it is found to consist of individual particles of a range of sizes from 0.2 mm diameter down to the smallest that can be resolved under the electron microscope. Dust suspensions in air have considerable stability and may persist for long periods. The very smallest particles behave in air the same as large molecules of gas or vapour. They are almost indistinguishable but for the fact that when particles strike a surface they usually adhere.

Industrial aerosols generally have long been known to be mixtures of particles of vastly different size from a few angstrom in diameter up to 100–250 microns (Drinker and Hatch, 1954). Mist droplets 200 microns or less in diameter are spherical in suspension in air. The particles of airborne dust, however, are generally of irregular shape. Some dusts are distinctly fibrous, containing many long and thin particles, as for example, dust from asbestos, carbon fibre, glass fibre and rock wool. When airborne dust deposited on a glass slide is viewed under the microscope many individual dust particles are seen to consist of aggregates or clumps of still smaller particles.

A fume is formed by evaporation from a molten solid and condensation of the vapour phase. It is an aerosol of solid particles resulting either from condensation of the vapour given off from the heating of a solid body such as lead or zinc, or from their combustion such as by burning magnesium or phosphorus, or from sublimation of a solid such as iodine crystals. Generally the individual particles of the fume have diameters less than 1 micron. Fumes have a strong tendency to flocculate; clumps or aggregates of particles often adhere together to form a single, much larger airborne entity. In common parlance it is usual to speak of acid fumes, for example, as denoting a mixture of gas and mist. Throughout this book, however, to avoid confusion the word 'fume' is used to designate only solid particles.

Smoke is a suspension of a solid in a gas, generally organic in origin and visible, such as the smoke from burning tobacco, wood, oil or coal. The particle size is generally below 0.5 micron diameter. Fog is an aerosol formed by the condensation of water vapour upon suitable nuclei.

The relative significance of three main types of input to the body in any exposure situation must be taken into account when assessing the level of risk to health; inhalation, skin contact and ingestion. In general, the respiratory system offers the most rapid route of entry and the dermal the least rapid, although much depends on the distribution of exposure between the three routes. Sometimes the contribution from one or other input is negligible. Sometimes it is very uncertain. In such cases the level of risk may be ascertained by a different strategy; measuring the output from the body. Exhaled air, urine and faeces give indications of the output of hazardous substances and their metabolites and the output reflects the body burden of such substances whatever the method of their input. Sampling of blood may also be employed for this purpose.

The main routes of input and output are indicated in Figure 1.1. Inhaled gases and vapours are partly absorbed in different regions of the respiratory tract and the remainder exhaled. Fine airborne particles are more readily inhaled than coarse particles. Some dusts, once inhaled are readily soluble in fluids bathing lung surfaces but many others are only poorly soluble and are partially removed by a combination of phagocytic action in the alveolar regions and ciliary action in the respiratory airways. These latter particles become

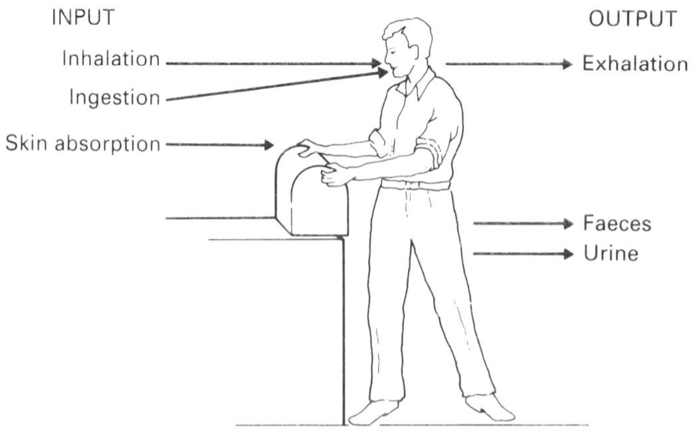

FIG. 1.1. Principal routes of input and output of hazardous substances at work.

swallowed or spat out. The amount of dust removed and at what speed is intimately connected with the regional deposition in the lungs, which is governed in turn by the size, density and shape of the airborne particles. Output of a substance from the body, by whatever means, is an indirect and more or less belated index of exposure, whether it be a gas, vapour, liquid or solid.

Improved methods for assessment of gases and vapours are being developed continually. In recent times exposure to vapours has been measured primarily by their concentration in the atmosphere in the breathing zone, and these measurements are now increasingly supplemented by analysis of samples of exhaled air. Skin absorption from the associated liquid phase is often suspected. In adverse circumstances intake through the skin can be high but it is difficult to measure directly. Substances absorbed into the body from the liquid and vapour phase are subsequently returned to alveolar air and exhaled in measurable quantities so that analysis of exhaled air may be the best guide to what the total exposure has been.

The surface area of the skin is 1.6–1.8 m^2 as compared with the surface area of the lungs, which is 70–90 m^2, and absorption of gases and vapours through unit area of skin is much slower than through the lungs (International Commission on Radiological Protection, 1974). Consequently skin absorption from the vapour phase is usually insignificant compared with its absorption in the lungs. Even so, in certain circumstances absorption of vapour through the skin can be relatively important, such as, for example:

(a) when respiratory protection is being worn, thereby protecting against lung absorption and there is a particularly high concentration of vapour around the rest of the body, or

(b) when clothing has become saturated with liquid so that beneath the clothing there is a particularly high concentration of vapour over the skin.

A few solids have a significant vapour pressure at room temperature and present a health hazard in the vapour state, although most exert their health risk through aerosol form. The potential hazards of metals in elemental form are particularly interesting as it is again vapour pressure which directly or indirectly plays the key role in determining the hazard they present through being available for uptake by the body. Examples are elemental mercury, cadmium and lead (Table 1.2).

Mercury

This is an element whose vapour poses a health hazard at work since the metal is liquid at room temperature. When elemental mercury is handled at work harmful atmospheric concentrations of the vapour of mercury may be easily produced and inhaled. Its saturated vapour concentration at room temperature is many times the level of its atmospheric exposure limit.

TABLE 1.2. *Saturated vapour concentration of some metals*

Temperature (Celsius)	Saturated vapour concentration (mg m^{-3}) Mercury	Cadmium	Lead
0	1.94		
25	19		
50	111		
100	2870	<0.01	
150		0.10	
200		2.01	
300		290	
400		7100	<0.01
500			0.13
600			5.5

Note. Melting points: mercury −39C, cadmium 321C, lead 327C.

Cadmium

Most metals are solid at room temperature and without significant vapour pressure. Cadmium is one of these metals. However, when heated to a temperature just below its melting point cadmium develops a significant vapour pressure whilst still in the solid state. Its high vapour pressure around its melting point is the basic reason for the liberation of very considerable quantities of cadmium oxide fume when it is melted. Residual airborne cadmium fume then presents a potential health hazard. The fume is inhaled and deposited in the lungs.

Lead

Chemically, closely allied to the element cadmium is another metal, lead. Their melting points are about the same, but lead does not have a significant vapour pressure until its temperature is well above its melting point. Molten lead can be poured without special ventilation provisions. A health risk from direct exposure to elemental lead in air is rare, although only because airborne metal dust is so uncommon. A risk can occur from the grinding, sanding or drilling of metallic lead and its alloys. Health risks at work from exposure to lead derive principally from lead compounds and comes mostly from airborne dust of compounds soluble in the lungs and from the vapour of organic lead compounds.

Historically, exposure to dusts of sizes causing different kinds of pneumoconiosis has been characterised by atmospheric exposure alone. In pneumoconiosis input through the skin and by ingestion does not contribute anything of significance to lung burden. In contrast, whilst exposure to metals and their compounds has long been measured by atmospheric exposure, in this case it is often supplemented by analyses of urine, blood and sometimes faeces. The contribution to body input by ingestion of metallic compounds may well be significant and particularly difficult to quantify in any other way.

A substance which is of the type which affects only the lungs or only the skin generally does this through atmospheric exposure or skin contact respectively. However a substance which gives rise to systemic effects may do so from atmospheric exposure, skin contact, ingestion or a combination of all three. Different compounds in a family may pose a health risk from absorption by one, two or all three routes. For example, lead and some of its compounds, listed in order of decreasing toxicity are given below. Tetraethyl lead is the only compound of lead in the list which is liquid at room temperature and the only one which is absorbed through the skin in significant quantities:

Tetraethyl lead—used in leaded gasoline, $Pb(C_2H_5)_4$
Lead carbonate—white lead, $2PbCO_3.Pb(OH)_2$
Lead mono-oxide—litharge, PbO
Metallic lead—Pb
Lead chromate—chrome yellow, $PbCrO_4$
Lead oxide—red lead, Pb_3O_4
Lead dioxide—used in the chemical industry, PbO_2
Lead sulphide—galena, PbS

Save for tetraethyl lead the compounds in the above list are all solids, lead carbonate being the most toxic and lead sulphide the least. Lead sulphide is almost insoluble in body fluids. In lead mines the ore dust (lead sulphide) apparently produces no risk of poisoning (Belden and Garber, 1949).

Exposure by inhalation can be expressed as the sequence of concentration values in successive inhalations of air. Exposure by skin contact and ingestion is very much less easy to quantify, although in certain cases analysis of urine, blood or exhaled air may be a useful guide to the amount absorbed. The lack of a means to measure a type of exposure is no excuse to ignore it. In certain countries occupational dermatitis has been found to affect one third or more of workers who come in direct contact with mineral oils, cement and other substances. Personal sampling of atmospheric exposure of many substances is done daily but occupational hygienists have yet to devise really effective means to measure skin contact routinely.

Getting through the respiratory tract

Examination of the Reynolds number for the nature of the flow during respiration shows that air flow conditions are turbulent in the nose and upper respiratory tract when the employee is under light to moderate exercise. This encourages enhanced absorption of gases and vapours and deposition of aerosols. Under heavy exercise fully developed turbulence extends to the bronchi and for a few branches further into the lungs (Weibel, 1963; Pedley, Schroter and Sudlow, 1977). Branching of the airways induces standing eddies and turbulence may extend deep into the lungs, although as the terminal bronchioles are approached undisturbed streamline flow conditions prevail even under heavy exercise. Absorption of gaseous substances by molecular diffusion and deposition of aerosols under gravity are the predominant mechanisms in this region.

On transient exposure the depth of penetration of gases and vapours into the respiratory tract is determined by their aqueous solubility. The convoluted, moist passages in the nose and at the back of the mouth absorb the most soluble gases such as ammonia from inspired air and little of these may penetrate any further (Brain, 1970). Penetration of gases into regions deeper in the respiratory tract, including the alveoli, is dependent on their having limited solubility in aqueous media. For gases with the same solubility those with lower molecular diffusivity, that is, higher molecular weight, also tend to penetrate furthest. Personal factors play a part. High minute volume, in particular, also favours deep penetration by gases and vapours (Davies, 1985).

Inhaled aerosols deposit at various points in the respiratory tract from the nose down to the alveoli and the exact place the particles deposit influences its subsequent fate. Coarse particles are deposited in the nose and upper respiratory tract. Particles which penetrate to the alveoli are principally those smaller than 5 microns in diameter. Differences in site, nature and magnitude of hazard posed by aerosols are associated with differences in particle size. As with gases and vapours when the employee is engaged in heavy work his or her minute volume increases and the regional deposition of particles shifts markedly. Deposition by impaction of particles increases so that more are deposited in the upper respiratory tract.

Absorption of inhaled gases and vapours

Uptake of foreign gases and vapours in the respiratory system occurs firstly by solution. The higher the concentration in air the higher the rate of solution. Different gases and vapours have widely different solubilities. A gas or vapour with a high solubility is one defined arbitrarily as one with a solubility of more than 10 ml of gas in 1 ml water at 20C.

Solubility	Water : air partition coefficient
High	Over 10
Intermediate	0.1–10
Low	Under 0.1

It is the solubility of hazardous gases and vapours in the fluids lining the respiratory tract which is the principle property governing their subsequent absorption in the rest of the body (Schroter and Lever, 1980). Furthermore, in those instances that the gas or vapour is non-reactive, that is, does not combine chemically with a liquid, it generally obeys Henry's law in ordinary ranges of pressure and temperature. According to Henry's law, at equilibrium the amount of gas absorbed by a given quantity of liquid is directly proportional to the partial pressure in air above the liquid.

Solubility is also important in the absorption of reactive gases and vapours. When a reactive gas of high solubility, such as sulphur dioxide, is inhaled it is mostly absorbed in the upper respiratory tract and produces its effects there (Davies, 1985). This remains true even after long exposure, since the reaction and consequent destruction of the gas prevent the attainment of equilibrium. The rate of uptake of such gases is approximately proportional to the respiratory minute volume and is only slightly influenced by changes in blood circulation.

After prolonged exposure and rising tissue concentration less high solubility gas is absorbed in the upper respiratory tract and more penetrates further. Some gas and vapour of slight or low solubility such as methanol, that is a solubility of less than 1 ml gas or vapour in 10 ml water at 20C, penetrates to the alveoli even on transient exposure. Once there the maximum amount of absorption of non-reactive gas from alveolar air is determined by its partial pressure, its solubility in blood and pulmonary blood flow (Bird, Stewart and Lightfoot, 1960). A virtually complete equilibrium of gas pressures between the alveolar air and arterial blood is attained almost instantaneously. Complete saturation of the whole of the blood is rapid, depends chiefly on the rate of blood circulation and is little influenced by the respiratory minute volume.

Reactive gases of low solubility cause different actions according to the nature of the reaction. Nitrogen dioxide, for example, penetrates to the alveoli because of its relatively low solubility but hydrolyses on contact with alveolar tissue. Inhaled in lethal concentrations, it causes oedema of the lungs which gets progressively more extensive several hours after exposure.

Materials of intermediate solubility such as chlorine, bromine, ozone and dimethyl sulphate affect both the upper respiratory tract and the pulmonary tissue. Once dissolved in the moisture bathing the lung surfaces the gas or vapour may be absorbed by or react with lung tissue or blood. Ethyl ether is an upper respiratory irritant, is of intermediate solubility, penetrates to the alveoli and is absorbed there without affecting the lungs. It is transported by the blood to the central nervous system, particularly the brain, where it exerts its anaesthetic action. Phosgene, that is carbonyl chloride, while slightly soluble and irritating to the upper respiratory tract, also has a profound effect on the alveolar walls and blood capillaries when in sufficient concentration for any to reach the alveoli. In high concentration it gives rise to lung oedema which seriously interferes with gas exchange and may prove fatal, albeit after a lag of several hours.

Gas taken up by the blood from alveolar air is carried in the arterial stream to the remote capillaries where exchange of gas takes place by diffusion between the blood and tissues. The blood in the venous stream, with the concentration of gas built up therein, leaves the tissues and is returned to the lungs where it is brought into equilibrium again with the gas in the alveoli. Absorption of gas from alveolar air eventually slows and ceases when the concentration of gas in venous blood attains equilibrium with the concentration of gas in alveolar air. Furthermore, when exposure ceases gas from venous blood enters alveolar air and is then eliminated by being exhaled. The principles are similar to those of absorption in reverse.

Gases with the least solubility, often wrongly termed 'insoluble', may penetrate to the alveoli, remain unabsorbed and be exhaled relatively unchanged. Enough dissolves to bring the pulmonary blood into equilibrium with the gas phase within a few breaths after which continued uptake from the lungs is independent of minute volume but depends upon the rate of transport and storage elsewhere in the body. It may then take hours, days or even longer to reach some sort of stability. Once equilibrium has been attained throughout the body no more is absorbed no matter what the minute volume nor how prolonged may be any further exposure.

The maximum amount of the inhaled gas which any organ or tissue can absorb at a given concentration is governed by the solubility of the gas in that tissue. The rate at which the gas accumulates in the particular tissue depends both upon the solubility of the gas in that tissue and also upon the amount of the gas brought to the tissue in any period of time. The latter factor is governed by the blood supply in relation to the bulk of the tissue. Thus a tissue with a comparatively low solubility for the gas and a large blood flow approaches saturation rapidly, whereas a tissue of high solubility and small blood flow approaches saturation much more slowly.

Bone and fat, which together constitute a large part of the tissue of the body, receive a comparatively small blood flow. The flow to the brain, heart, lungs, bowels and liver is large. This latter group of organs therefore approaches saturation at a much more rapid rate than does the bone and fat. The same relation holds true of elimination.

The rate of saturation of tissues with a large blood supply follows closely the concentration of a gas or vapour in the arterial blood. The concentration in those tissues with a large blood supply follows closely the concentration of a gas or vapour in the arterial blood. The concentration in these tissues therefore rises at first much more rapidly than in the tissues with a smaller blood supply, but the less rapidly saturating tissues act as a buffer to prevent immediate and complete equilibrium. This is the case both during absorption and elimination of the gas. During absorption the partial pressure of a gas in any tissue cannot rise above the partial pressure of gas in arterial blood, nor can it fall below this limit during elimination.

Certain gases and vapours, particularly the anaesthetic hydrocarbons, exert their acute action specifically upon one organ—the brain—and are largely inert in regard to others. The brain has an extremely large blood supply; the anaesthetic action therefore tends to follow closely the concentration in the arterial blood. The concentration in the peripheral venous blood indicates only the degree of saturation in the particular organ from which the blood is returning. Because of these relationships brief inhalation of a high concentration of an anaesthetic hydrocarbon, such as ethyl ether, may cause marked symptoms before any large amount has been absorbed into the other parts of the body. On breathing fresh air the effects of brief exposure pass off equally rapidly since, when the inhalation of the vapour ceases, there is an immediate fall in the concentration in the arterial blood and correspondingly in the brain. If the inhalation has been prolonged with the accumulation of a large amount of the vapour in the body, this rapid fall does not occur, since the concentration in the arterial blood is maintained by the vapour returned in the venous blood; a longer time is then required for recovery.

The partial pressure of any vapour in the alveolar air corresponds to that in the arterial blood leaving the lungs at the time the alveolar air is collected. The ratio of the concentrations in the air and blood is the partition coefficient for the particular vapour.

The concentration of the vapour in urine at the moment the urine is secreted corresponds closely to that in the arterial blood passing through the kidneys at the time. It must be borne in mind that the urine collected in the bladder is a composite of the urine secreted during the entire period in which it has collected.

Gases and vapours are eliminated largely through the lungs and to a much less extent through the urine. Elimination through the lungs follows definite principles which are similar to those of absorption in reverse. When elimination begins and part of the air in the lungs is replaced by fresh air, a part of the gas is carried away in the expired air and the concentration of the gas in the arterial blood leaving the lungs is reduced below that in the venous blood coming to the lungs.

The solubility of the vapour is a primary factor in determining the rate of elimination; the more soluble the vapour the slower the elimination. As with absorption, the rate of blood circulation is the predominant influence for the rate of elimination of vapours of low solubility, and respiratory minute volume has little influence. These relationships are reversed with soluble vapours; rate of blood circulation has little effect on rate of elimination, and volume flow rate of respiration has a marked influence. Notwithstanding these remarks a greater percentage of vapour of high solubility is eliminated in the urine than is one of low solubility.

Non-reactive gases which have very low solubility may thus seem to present little hazard, and, indeed, this is usually the case. However they can still become hazardous if present in enormous amounts. They can act as simple asphyxiants by excluding oxygen from the lungs. If foreign gases reach levels of atmospheric concentration of 25% the result could be fatal in a matter of minutes due to suffocation alone. Examples of simple asphyxiants of this kind are hydrogen, acetylene, methane, ethane, propane and the rare gases; argon, helium and krypton. Since these gases are often odourless there may be no warning before the victim passes out from lack of oxygen. Even nitrogen acts as an asphyxiant when it is in higher than normal concentration in air. It must not be forgotten that some simple asphyxiants like hydrogen, acetylene, methane, ethane and propane also present risk of ignition and explosion.

A special but very important case of asphyxiation is what is known as 'chemical' asphyxiation. A simple asphyxiant does not combine with the blood but simply physically replaces oxygen in the air. The best known chemical asphyxiant is carbon monoxide which has a low solubility but has many times greater affinity for haemoglobin than oxygen. Hydrogen cyanide is another chemical asphyxiant and is one of the most rapidly acting of all toxic substances. It is readily absorbed in the respiratory tract, diffuses to all parts of the body and combines with cytochrome oxidase, a catalyst in the cells which is essential for the respiratory process. The hydrogen cyanide blocks the function of the catalyst, the haemoglobin returns to the lungs with its oxygen and unless exposure ceases promptly, death intervenes. However, if exposure is not too severe recovery may be complete (Dixon and Webb, 1958).

Deposition, retention and clearance of inhaled aerosols

Coarse particles are most easily caught by impaction in the convoluted, moist passages in the nose and at the back of the mouth. However, with decreasing particle size efficiency of impaction decreases sharply. The nose ceases to be important as a mechanism for capturing particles by impaction when they are smaller than 10–20 microns in diameter.

A considerable quantity of research work has been undertaken on the deposition of aerosols in different parts of the respiratory tract. Early in the twentieth century, and stimulated by the need to understand how silicosis is caused, studies of the size distribution were made of dust in the lungs of South African gold miners and others. This was done by examining the residues after burning to ash portions of lung obtained at post mortem (McCrae, 1913; Higgins, Lanza, Laney and Rice, 1917; Mavrogordato, 1926; McCrae, 1939). Gold is generally in rocks high in quartz content and the miners suffer from silicosis from the airborne dust. It was found that dust in the lungs at post mortem was mostly smaller than one micron in diameter, was only a small fraction of that originally inhaled but gives little idea of exactly where in the lungs it was originally deposited nor how much of it had been exhaled or swallowed.

A better approximation to the passage of aerosols into the deeper parts of the lungs was given by theoretical analysis of the amount of deposition due to impaction, settlement and diffusion, first elaborated in detail in a renowned paper by the German, Findeisen (1935). Much of this was confirmed and extended by experimental work in USA in which exhaled air was partitioned by an arrangement of valves into sequential fractions from different segments of the airways. It was possible to determine where the dust was lost through deposition by studying the differences in size distribution in the residual airborne dust in each segment (Brown, Cook, Ney and Hatch, 1950). It was found that the inhaled particles which penetrate to the alveoli are principally ones smaller than 5 microns in diameter, larger particles being deposited in the upper respiratory tract.

There are many different measures of size which could be used. The size could be measured in terms of particle length, breadth, surface area, volume, weight and so on, each of which parameters would put particles in a somewhat different order. The most relevant parameter in the present context is the aerodynamic diameter. This refers to the size of an airborne entity in standardised terms which indicates how and where it will deposit from suspension in air. It is the diameter of the unit density sphere which would fall at the same terminal velocity in air as the airborne entity in question. The airborne entity may be a single particle of any shape or form or a clump of two or more which are held together by surface forces as a single aggregate in suspension. Solid aerosols flocculate under electrostatic forces generated in their formation and by collision one with another. Liquid aerosols coalesce.

Terminal velocity is directly related to the aerodynamic properties of a particle suspended in air and takes into account its size, shape and density. It will be apparent that it is the aerodynamic behaviour of particles which governs their deposition by centrifugal force or gravity in various parts of the respiratory system and not their chemical composition. The aerodynamic properties of a coherent airborne entity, of whatever shape, whether it be a single large particle or a clump of many smaller ones, are best expressed by its 'aerodynamic' diameter.

The principle mechanism for deposition of coarse aerosols in the respiratory tract is impaction. Deposition from a moving air stream by impaction is caused by the centrifugal force acting on the particles in the direction away from the centre of rotation. The main factors influencing the magnitude of this force are shown in the formula for spherical particles:

$$F = \frac{\pi d^3(\sigma - \rho)r\omega^2}{6}, \tag{1}$$

where d is the diameter of the particle, σ is the density of the particle, ρ is the density of air, r is the radius of curvature of the path and ω is the angular velocity. Under the action of centrifugal force the particles move radially with increasing velocity. This motion is opposed by increasing drag or air resistance which, for spherical particles, is given by

$$R = C\pi\rho d^2u^2, \tag{2}$$

where C is a coefficient of resistance and u is the radial velocity of the particles through air. Over the range of conditions of present interest the drag coefficient is approximately inversely proportional to d and u for particles up to about 75 microns in diameter and decreases more slowly thereafter. So that for particles up to 75 microns in diameter the drag resistance is approximately proportional to particle diameter and velocity through air.

For spheres,

$$R = 3\pi\mu du, \tag{3}$$

where μ is the viscosity of the air. This is known as Stokes law.

The net radial force on particles in a rotating air stream is thus strongly favoured by high angular velocity and large particle diameter.

The mass median aerodynamic diameter of airborne dust in coal mining, typical of dusty trades, is 2.5–25 microns, with a geometric standard deviation of 2.0–4.0 (Mark, Vincent, Gibson, Aitken and Lynch, 1984; Treaftis, Gero, Kacsmar and Tomb, 1984). Exceptionally, when the liberated dust is coarse and released at high velocity the airborne dust can be coarser, as in the furniture trade, where the mass median diameter has been reported at 50–200 microns, with a geometric standard deviation of 4.0–6.0 (Whitehead, Freund and Hahn, 1985).

Particles of irregular shape have a higher drag than spherical particles of the same volume and they fall slower because of this. They have a diminished aerodynamic diameter. Fibrous particles, in the extreme are long, thin and straight and these fall at a speed little higher than spheres which have a diameter equal to twice the width of the fibres. This accounts for their ability to penetrate deep into the lungs. On the other hand, because of their length, the very longest fibres tend to deposit more easily through interception with the walls of the respiratory tract and at a bifurcation (Timbrell, 1965).

When hygroscopic particles are inhaled in relatively dry air they grow by condensation of water from the moist air as they penetrate the lungs to high humidity regions. Salt and cigarette smoke are notable in this regard (Hicks, Pritchard, Black and Megaw, 1989). For this reason, hygroscopic particles generally, such as sulphuric acid mist, are deposited to a higher degree than those of the same size which are not hygroscopic.

As a consequence of the branching of the respiratory tract there is increased opportunity for dust to deposit and adhere to the walls as the inhaled air changes direction. At the same time there is a rapid increase in total cross sectional area and a corresponding fall in velocity. At the terminal bronchioles the forward air velocity has fallen 40-fold from its value in the trachea and fallen a further 4-fold passing through the respiratory bronchioles. In the alveoli the air velocity falls to zero. The inhaled particles which deposit, by gravity, in the alveoli are principally smaller than 5 microns in aerodynamic diameter, although the very finest, smaller than 0.5 micron in diameter do not fall fast enough to deposit in large quantity and are mostly exhaled (Hatch and Gross, 1964). Electric charge causes significant deposition of particles of this size (John and Vincent, 1985; Vincent, 1986). Particles which are so fine that they have significant Brownian motion in air are even smaller, less than 0.1 micron in diameter. They move almost like large molecules of gas and are deposited on nearby surfaces mainly by diffusion. The surfaces may be of any orientation. Unlike gas molecules, however, once deposited they all stick on the surface and do not bounce off.

Methods have been developed for the direct measurement of the amount of dust deposited on the airways of volunteers by having them inhale an aerosol of artificial, spherical particles tagged with radionuclide tracers and by using collimated radiation detectors to locate the deposited particles from outside the body (Lippmann and Albert, 1969; Chan and Lippmann, 1980; Stahlhofen, Gebhart and Heyder, 1980, 1981). Another technique which has been used to give further understanding of the deposition mechanism is to make hollow casts of the respiratory tract from lungs obtained at post mortem and use these as experimental models through which dust laden air is first drawn and then examined in detail.

The regional deposition of particles in the respiratory tract according to aerodynamic diameter is shown in Figure 1.2. The values plotted are average values and the deposition in the lungs of a given individual may be anywhere between 10% higher or lower (Albert and Lippmann, 1972; Lapp, Hankinson, Amandus and Palmes, 1975; Tarroni, Melandri, Prodi, DeZaiacomo, Formignani and Bassi, 1980). Furthermore at high minute volume coarser dust tends to be deposited by impaction higher in the respiratory tract due to the higher air velocities involved.

Further down the respiratory tract the trachea, bronchi and bronchioles are lined with small, waving, lash-like appendages, cilia, whose action is to sweep a covering layer of mucus upwards towards the mouth, to be swallowed or spat out. The mucociliary blanket carries undissolved particles which were de-

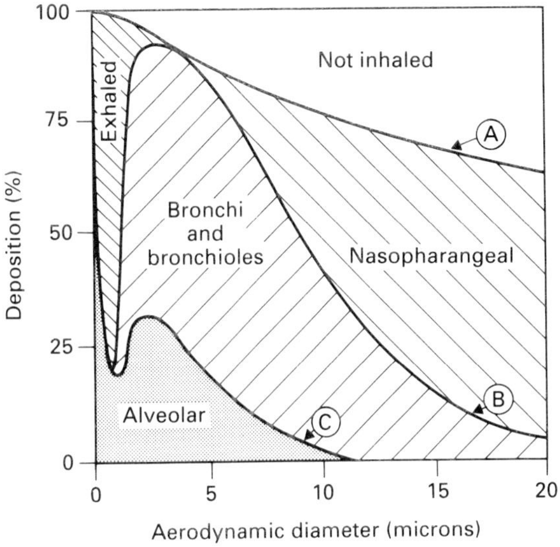

FIG. 1.2. Regional deposition of aerosols in the respiratory tract. Curve A—Total deposition. Curve B—Deposition beyond the pharynx. Curve C—Deposition in alveolar regions.

posited from the inspired air by impaction at bifurcations and by sedimentation elsewhere in the bronchi and bronchioles. These are particles smaller than 10–20 microns in diameter but not so small as to remain suspended and be carried onward by the inspired air into the alveoli. The time it takes for dust to travel carried by mucus from the terminal bronchioles to the larynx varies between different individuals from about 2.5 to 20 hours (Albert, Lippmann, Peterson, Berger, Sanborn and Bohning 1973).

The mucus carries particles which have been deposited in the bronchi and bronchioles and others, mostly smaller, which are engulfed in phagocyte cells and came originally from the alveoli. These cells first appear in the alveoli in large numbers under the stimulus of foreign bodies such as dust particles and microorganisms which have deposited there by gravity or other means and have not been dissolved. Phagocytes engulf most of the remaining particles in the alveolar regions and then migrate out of the alveoli to the mucociliary region. Precisely how these cells manage to move to the mucociliary is poorly understood although the lung surfactant system and the regular expansion-contraction of alveoli are believed to be intimately involved (Cosmi and Scarpelli, 1983; Podgorski and Gradon, 1990). It is estimated that clearance is over 98% effective for most of the deposited particles (Gross and de Treville, 1972). Rounded particles are more readily cleared than long, fibrous particles (Morgan, Talbot and Holmes, 1978). The remainder are retained in interstitial tissue and around the lymph nodes, where, unless slowly dissolved, dust particles may remain indefinitely.

When a spherical particle is dropped from rest it follows from Stokes law that its falling velocity after time t is

$$u = \tau g[1 - \exp(-t/\tau)], \tag{4}$$

where τ, called the particle relaxation time, is given by

$$\tau = \frac{d^3\sigma}{18\mu} \tag{5}$$

(Vincent, 1986). Irregularly shaped particles fall slower than spheres of the same volume because of their increased aerodynamic drag. The falling velocity of a dropped particle very soon ceases to increase and the particle has reached its terminal velocity,

$$u_t = \tau g. \tag{6}$$

The acceleration phase of fine particles falling under gravity is so short that it can be ignored for most purposes. However their terminal velocity is appreciable. In terms of aerodynamic diameter, over the range of streamline motion,

$$u_t = 0.003D^2 \tag{7}$$

where u_t = terminal velocity (cm $s^{-1)}$
D = aerodynamic diameter in microns ($0.5 < D < 100$).

The fluid motion around a falling particle ceases to be purely streamline for those above about 100 microns in aerodynamic diameter and this causes more air resistance. The terminal velocity of particles in suspension in air for this range of sizes is given by

$$u_t = 0.34D \quad \text{(approx.)}, \tag{8}$$

where u_t = terminal velocity (cm $s^{-1)}$
D = aerodynamic diameter in microns ($D > 100$).

Thus, a 5-micron diameter unit density sphere reaches 95% of its terminal velocity in air, 0.075 cm s^{-1}, in 0.0002 sec, by which time, calculation shows, it will have fallen only 1.5 microns. Mindful of the fact that alveoli are 100–200 microns deep, the same 5-micron diameter particle has time to fall to the bottom of an alveolus in between breaths, 2–4 seconds. On the other hand, a 0.5-micron diameter unit density sphere has a terminal velocity of 0.00075 cm s^{-1}, that is 7.5 microns per second. It can be appreciated that there will not be enough time between breaths for all of these to deposit. Many of the particles smaller than 0.5-micron diameter fail to deposit under gravity and are exhaled. The exhalation of such fine particles is exemplified by exhaled tobacco smoke which is patently visible but consists mostly of particles smaller than 1 micron in diameter.

The very finest particles, smaller than 0.5 micron in diameter are increasingly affected by Brownian motion to the extent that for particles smaller than 0.1 micron in diameter this is the dominant mechanism affecting their deposition in the alveoli. There is a fivefold increase in settling velocity over that given by purely streamline motion, equation (7), for particles 0.05 micron in diameter. They do not normally contribute much of the mass of industrial dust deposited in the alveoli but they have a greater potential for interaction with irritant gases, a fact which is of importance when considering the possibility of synergism with certain mixtures.

Ingestion of foreign substances

The principal contamination by hazardous materials which is ingested is in powder form, although liquids are also sometimes involved. In favourable circumstances solution from the acidic environment of the stomach, and absorption from the gastro-intestinal tract generally, may far exceed that absorbed from the lungs. Ingestion is believed to occur primarily from six main sources:

(1) Eating foodstuffs, sandwiches, fruit, tobacco, chewing gum and so on, which have been contaminated from the hands or from surfaces where they have been put down.

(2) Cigarettes, cigars, pipes and tobacco which have become contaminated.

(3) The hands, arms or clothing rubbed across the face.

(4) Contaminated drinking vessels.

(5) Lung clearance by ciliary action and swallowing of inhaled particles which have been inhaled and deposited in the lungs but have failed to be absorbed there.

(6) Aerosols deposited on the lips.

Absorption through exposure of the skin

Liquids, solids, gases and vapours in contact with the skin may be absorbed through the skin and enter the capillary blood, thereby becoming circulated around the body. Some of the many factors which influence the rate of absorption through the skin include the following properties of the substance and of the skin: physical form of the substance, its solubility in water and lipids, its molecular weight, the area of skin exposed, skin thickness, and its perfusion by blood (Scheuplein, 1977; Schaefer, Zesch and Stuttgen, 1982; Dugard, 1988).

Physical form of substance; gaseous, liquid or solid

There is only a limited amount of published experimental evidence on humans of absorption of vapours through the skin (Riihimaki and Pfaffli, 1978; Wieczorek, 1985).

Indirect evidence and extrapolation from absorption studies on animals indicates that absorption of common organic solvent vapours through the skin of exposed people is 3.5–7.5% of pulmonary uptake (Hefner, Watanabe and Gehering, 1975; McDougal, Jepson, Clewell and Anderson, 1985; McDougal, Jepson and Clewell, 1986; Tsuruta, 1989). However if the employees wear respiratory protection skin absorption would be correspondingly a much greater percentage of total intake.

A liquid spill over the clothing can result in a high concentration of vapour in contact with the skin. Gases and vapours diffuse through the skin. Absorption of solids through the skin relies upon deposition from the air or direct contact with the skin followed by solution in sweat. Liquids make more ready contact with the skin than do solids and the rate of absorption of a substance from the liquid phase is proportional to the concentration of the substance in solution, in a similar way to rate of absorption from the gaseous phase being proportional to atmospheric concentration (Tregear, 1966).

Solubility of the substance in water and lipids

The greater the solubility the more that is absorbed. By the same token, a substance insoluble in water and lipids is not absorbed through the skin.

Molecular weight of substance

Substances pass through the skin by diffusion. The higher the molecular weight the smaller is the diffusion coefficient and the less readily do substances penetrate the skin.

Area of skin exposed

Skin over the whole body is exposed to atmospheric gases and vapours as air readily penetrates most normal clothing. Aerosols are deposited on a more limited area of the body and liquids generally make contact with a small but variable area of the skin.

Skin thickness

The thinner the skin the more readily will the substance penetrate. Damaged skin is the most readily penetrated. Prolonged contact with liquids is one of the ways by which the skin is damaged.

Duration of exposure

When skin exposure is constant, in the first phase (Figure 1.3) there is a build-up of material in the skin itself, the lag phase. Transfer of the chemical from the dermis into capillary blood takes place in the second phase and the rate of transfer levels off as the substance accumulates in the blood. In the third, prolonged phase the substance accumulates in the rest of the body until the rate of elimination by all routes equals the rate of intake by all routes. In the fourth phase input and output of substance are in equilibrium.

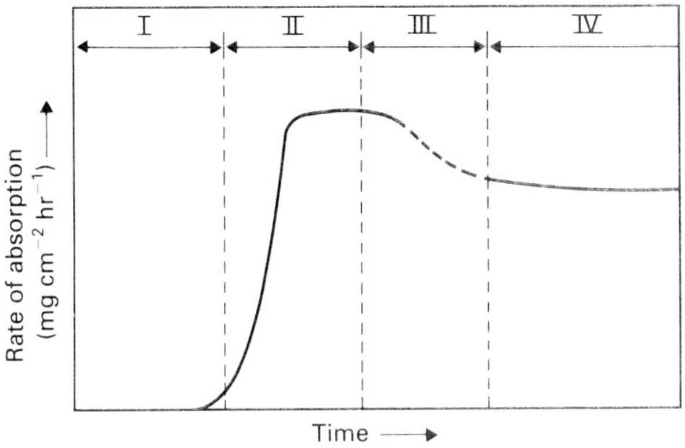

FIG. 1.3. Absorption of substances through the skin. Phase I. This is the lag period or break-through time before the 'front' of the contaminating substance penetrates through the skin. Phase II. The dominant action in this phase is transfer of the contaminating substance into the blood. Phase III. A variable length of time during which there is steady accumulation in the body. Phase IV. During this phase input to the body and output from the body come into equilibrium.

Perfusion of the skin by blood

The higher the energy expenditure and the higher the air temperature the higher will be the blood perfusion through the skin and the more rapidly is the substance taken away from the skin and circulated in the blood (Sheuplin and Blank, 1971; Bronaugh and Maibach, 1985).

In practice there is considerable variability in absorption rate from time to time and from one place to another over the surface of the body (Fiserova-Bergerova and Pierce, 1989). Furthermore exposure is not constant and the rate of absorption through the skin varies in a corresponding way, as does absorption through the respiratory tract from inhaled materials.

References

Albert R E and M Lippmann. Ann. N.Y. Acad. Sci. **200** 37 (1972).

Albert R E, M Lippmann, H T Peterson, J M Berger, K. Sanborn and D E Bohning. Arch. Intern. Med. **131** 115 (1973).

American Conference of Governmental Industrial Hygienists. Documentation of the Threshold Limit Values and Biological Exposure Indices. ACGIH, Cincinnati, Ohio, USA (1988).

American Conference of Governmental Industrial Hygienists Committee on Industrial Ventilation. Industrial ventilation. ACGIH PO Box 453, Lansing, Michigan, USA (1986).

American Conference of Governmental Industrial Hygienists. Threshold Limit Values and Biological Exposure Indices for 1988–1989. ACGIH, PO Box 1937, Cincinnati, Ohio 45201, USA (1988).

American Society for Testing and Materials. Test method for effectiveness of dedusting agents for powdered chemicals, 04331–84. American Society for Testing and Materials, 1916 Race St, Philadelphia, USA (1984).

Andreason A H M, N Hofman-Bang and N H Rasmussen. Kolloidzeitschrift 86 70 (1939).

Belden E A and L F Garber. J. Ind. Hyg. Toxicol. **31** 347 (1949).
Brain J D. Ann. Otol., Rhinol. and Laryngol. **79** 529 (1970).
British Occupational Hygiene Society. Technical Guide No. 4— Dustiness estimation methods for dry materials; their uses and standardisation towards a standard method. Science Reviews Ltd in association with H&H Scientific Consultants Ltd, Leeds, UK (1985).
Brown J H, K M Cook, F G Ney and T Hatch. Am. J. Public Health **40** 450 (1950).
Chan T L and M Lippmann. Am. Ind. Hyg. Assoc. J. **41** 399 (1980).
Commission for the Investigation of Health Hazards of Chemical Compounds in the Work Area. Maximum concentrations of the workplace and biological tolerance values for working materials 1988. Report XXIV. VCH Publishers, D-6940 Weinheim, FRG and Deerfield Beech, Fl, USA (1988).
Cowherd C, M A Grelinger, P J Englehart, R F Kent and K F Wong. Am. Ind. Hyg. Assoc. J. **50** 123 (1989).
Davies C N. Ann. Occup. Hyg. **29** 13 (1985).
Davies K M, C M Hammond, R W Higman and A B Wells. Ann. Occup. Hyg. **32** 535 (1988)
Dugard P H. Skin permeability theory in relation to measurements of percutaneous absorption in toxicology. In: Dermatotoxicology (Edited by F N Marzulli and H I Maibach). Hemisphere Publishing Corp., Washington DC, USA (1988).
Findeisen W. Arch. Ges. Physiol. **236** 367 (1935).
Fiserova-Bergerova V and J T Pierce. Appl. Ind. Hyg. **(4)**8 14 (1989).
Garlanda T. Dangerous substances. In: Encyclopaedia of occupational health and safety (Edited by L Parmegianni), pp. 583–586, Vol 1, 3rd Edition. International Labour Office, Geneva (1983).
Gross P and R T P de Treville. Am. Rev. Respir. Dis. **106** 684 (1972).
Health and Safety Commission. Information approved for the classification, packaging and labelling of dangerous substances for supply and conveyance by road, ISBN 0 11 885542 5. H M Stationery Office, London (Latest edition).
Health & Safety Executive. A guide to the notification of New Substances Regulations. HS(4)14 (REV), ISBN 0 11 885454 2 (1989).
Health & Safety Executive. Occupational exposure limits 1991. Guidance Note EH 40/91. H M Stationery Office, London (1991).
Hefner R E, P G Watanabe and P J Gehering. Toxicol. Appl. Pharmacol. **34** 529 (1975).
Hicks J F, J N Pritchard, A Black and W J Megaw. J. Aerosol Sci. **11** 289 (1989).
Higgins E, A J Lanza, F B Laney and G S Rice. Siliceous dust in relation to pulmonary disease among miners in the Joplin District, Missouri. Bulletin 132, United States Department of the Interior, Bureau of Mines, Washington DC, USA (1917).
Hoy K L. Tables of solubility parameters. Union Carbide Corporation, Chemicals and Plastics Research and Development Dept., South Charleston, W. Va., USA (1969).
International Commission on Radiological Protection. Report of the Task Group on Reference Man, Report No. 23. Pergamon Press, New York (1974).
John W and Vincent J H. Ann. Occup. Hyg. **29** 285 (1985).
Lapp N L, J L Hankinson, H Amandus and E D Palmes. Thorax **30** 293 (1975).
Lippmann M and R E Albert. Am. Ind. Hyg. Assoc. J. **30** 257 (1969).
Mark D, J H Vincent, H Gibson, R J Aitken and G Lynch. Ann. Occup. Hyg. **28** 125 (1984).
McDougal J N, G W Jepson, H J Clewell III and M E Anderson. Toxicol. Appl. Pharmacol. **79** 150 (1985).
McDougal J N, G W Jepson and H J Clewell III. Toxicol. Appl. Pharmacol. 85 286 (1986).
Morgan A, R J Talbot and A Holmes. Brit. J. Ind. Med. **35** 146 (1978).
Oliver G J A. Opportunities for using fewer animals in acute toxicity studies. In: Proceedings, Chemicals Testing and Animal Welfare, p. 119. The National Chemicals Inspectorate, PO Box 1384, S-117 27 Solna, Sweden (1986).
Pedley T J, R C Schroter and M F Sudlow. Gas flow and mixing in the airways. In: Bioengineering aspects of the lung (Edited by J B West). Marcel Dekker Inc., New York (1977).
Podgorski A and L Gradon. Ann. Occup. Hyg. **34** 137 (1990).
Riihimaki V and P Pfaffli. Scand. J. Work Environ. Health 4 73 (1978).

Scheuplein R J. Permeability of the skin. In: Handbook of physiology—Reactions to environmental agents (Edited by D H K Lee). American Physiology Society, Bethesda, Md., USA.
Scheuplein R J and I H Blank. Physiol. Rev. **51** 702 (1971).
Schroter R C and M J Lever. The deposition of materials in the respiratory tract. In: Occupational hygiene—An introductory text (Edited by H A Waldron and J M Harrington). Blackwell Scientific Publications (1980).
SI 1982 No. 1496 Health and Safety. The Notification of New Substances Regulations 1982. H M Stationery Office, London (1982).
SI 1984 No. 1244 Health and Safety. The Classification, Packaging and Labelling of Dangerous Substances Regulations 1983. H M Stationery Office, London (1984).
SI 1986 No. 1333 Health and Safety. The Ionising Radiation Regulations 1985. H M Stationery Office (1986).
Stahlhofen W, J Gebhart and J Heyder. Am. Ind. Hyg. Assoc. J. **41** 399 (1980).
Stahlhofen W, J Gebhart and J Heyder. Am. Ind. Hyg. Assoc. J. **42** 348 (1981).
Tarroni G, C Melandri, V Prodi, T DeZaiacomo, M Formignani and P Bassi. Amer. Ind. Hyg. Assoc. J. **42** 348 (1980).
Timbrell V. Ann. N. Y. Acad. Sci. **132** 255 (1965).
Treaftis H N, A J Gero, P M Kacsmar and T F Tomb. Am. Ind. Hyg. Assoc. J. **45** 826 (1984).
Tsuruta H. Ind. Health 27 37 (1989).
Vincent J H. The physics of gases and aerosols. In: Occupational hygiene—An introductory text (Edited by H A Waldron and J M Harrington). Blackwell Scientific Publications (1980).
Vincent J H. J. Electrostat. **18** 113 (1986).
Wells A B and D J Alexander. Powder Tech. **9** 271 (1978).
Whitehead L W, T Freund and L L Hahn. Am. Ind. Hyg. Assoc. J. 42 127 (1985).
Wieczorek H. Int. Arch. Occup. Environ. Health 57 71 (1985).
Wright G W. The influence of industrial contaminants on the respiratory system. In: The industrial environment—Its evaluation and control. U.S. Department of Health, Education and Welfare, US Government Printing Office, Washington 20402 USA (1973).

Bibliography

Bird R B, R D Stewart and E N Lightfoot. Transport phenomena. Wiley and Sons, New York (1960).
Bretherick L (Editor). Hazards in the chemical laboratory. The Royal Society of Chemistry, London (1981).
Bronaugh R I and H I Maibach (Editors). Percutaneous absorption. Marcel Dekker, New York (1985).
Cosmi E V and E M Scarpelli (Editors). Pulmonary surfactant system. Elsevier, Amsterdam (1983).
Dixon M and E C Webb. Enzymes. Academic Press, New York (1958).
Drinker P and Hatch T. Industrial dust. McGraw-Hill Book Company, Inc., New York (1954).
Gibbs W E. Clouds and smokes. The Blakiston Company, New York (1924).
Grant W M. Toxicology of the eye. Charles C. Thomas, Springfield, Illinois, USA (1962).
Hatch T F and P Gross. Pulmonary deposition and retention of inhaled aerosols. Academic Press, New York (1964).
Mavrogordato A. Contributions to the study of miner's pthisis. South African Institute of Medical Research, Johannesburg (1926).
McCrae J. The ash of silicote lungs. South African Institute of Medical Research, Johannesburg (1913).
McCrae J. The ash of silicotic lungs. Publs. S. African Inst. Med. Res. (1939).
Organisation for Economic Cooperation and Development. Labelling of chemicals—Overview of international labelling practices. OECD, Paris (1984).
Schaefer H, A Zesch and G Stuttgen. Skin permeability. Springer-Verlag, New York (1982).
Tregear H T. Physical functions of skin. Academic Press, Inc., New York (1966).
Walton A J. Three phases of matter. McGraw-Hill, New York (1976).
Weibel E R. Morphometry of the human lung. Springer Verlag, Berlin (1963).

CHAPTER 2

Diseases from Hazardous Substances

Historically, the occurrence of at least one occupational disease, silicosis, has been traced back several thousand years in those who work with flint-stone. Indeed, silicosis in Neolithic flint miners may possibly have been the first occupational disease in history (Collis, 1915). There is also little doubt that silicosis was prevalent in Egyptian miners, 2000 BC (Rosen, 1943). Another occupational disease, lead poisoning, was recorded by Hippocrates in the fourth century BC.

Coming closer to modern times, in the sixteenth century AD a remarkable German scholar, Georgis Agricola, described the hazards of the mining industry in some detail in 'De Re Metallica'. This famous old book was completed in 1553, in Latin, and was published in twelve volumes, in 1556. Interestingly, in US a mining engineer, Herbert C. Hoover, later to become President, and his wife Lou translated De Re Metallica into English in their spare time (English edition, 1912, 1950). Agricola reported that some women in the Carpathian mining district married seven husbands who were carried off to an early death after working in the silver mines. This was probably from lung cancer caused by radio-active air contamination. A hundred years later, in 1665 at Idria, Yugoslavia, tremors affecting the hands of cinnabar miners caused by mercurial dust and vapour had become notorious. This led to what is believed to be the first legislative measure of industrial hygiene; a decree shortening the working day to six hours (Hamilton, 1922; Rosen, 1958). Also in the seventeenth century, a renowned physician, an Italian, Bernadino Ramazzini (1633–1714) wrote a treatise on occupational diseases, 'De Morbis Artificium', which was published at the turn of the century, in 1700 (English edition, 1964). This classic treatise included descriptions of diseases from occupations as varied as miners, soap makers, cesspit cleaners and potters. Ramazzini is nowadays regarded as the founder of modern occupational medicine.

After Ramazzini the next major episode in the development of occupational medicine was marked by the Industrial Revolution which produced, among other things, an abundance of occupational injuries and diseases from exposure to hazardous substances at work (Thackrah, 1832). The health of employees and their families was badly affected (Rose, 1971). The British parliament, reflecting a growing concern for the safety and health of employees, passed the

Factories Act of 1833 for the protection of children and young persons, soon followed by the Mines and Collieries Act of 1842 to regulate conditions of employment in British mines. Occupational diseases could no longer be ignored. In the United States similar wide ranging legislation came later, at the beginning of the 20th century. The first state compensation laws were passed there in 1911 and, closely linked to this development, the Federal Government created the Office of Industrial Hygiene in the US Public Health Service in 1914. Occupational hygiene began to grow as a distinct discipline.

The diseases caused by excessive exposure to hazardous substances at work are very varied, from substances like the vapour of ethyl mercaptan, which is nauseatingly odorous and used as an odorizer for natural gas, to ones like benzidine crystals used in making dyestuffs. Benzidine is a potent bladder carcinogen. Each substance has a unique set of biological properties and excessive exposure to it at work produces a disease which is different in one or more respects from the diseases produced by other hazardous substances. The severity of the effects may cover the complete range: transitory annoyance—psychological and subjective disturbance—allergy—psychophysiological impairment—biochemical changes—decrement of physiological function—microscopic pathology—gross pathology and frank illness—death. Any of the organs or organ systems may be affected (Hunter, 1974).

Surprisingly, perhaps, dermatosis is by far the most common of the occupational diseases, accounting for approximately one-half of all reported occupational illnesses (Dionne, 1984). But the lungs provide the most common portal of entry into the body taken by hazardous substances in air and may themselves be affected. The blood, liver, kidneys or nervous system may also be harmed. Several forms of cancer are produced by exposure to particular hazardous substances and reproductive effects are not unknown.

A sample of the different adverse effects which have been influential in the past in setting occupational exposure limits for various hazardous substances is summarised in Table 2.1, constructed from the Documentation of ACGIH TLVs (American Conference of Governmental Industrial Hygienists, 1986).

Cancer

Most of the numerous cell types in the body share in the process of division and multiplication to replace dying cells or, under intermittent stress, to augment specific cell types such as the phagocyte. Some cells may completely replace themselves in a few weeks. Usually, cell division proceeds in an orderly fashion with the continuous development of identical, normally formed and constituted cells. However, on long continued exposure to certain hazardous substances it appears that the substance itself or a metabolite begins to influence cell growth, injury or repair, the cells markedly change their character and undergo a fundamental transformation. The cells seem to lose their customary organisation and orderly multiplication and are subject to increasingly disorderly growth, that is, a malignant transformation. This is the beginning of a tumour, carcinoma or cancer, the most feared of industrial diseases.

TABLE 2.1. *Examples of various methods employed to detect the presence of early changes due to exposure to hazardous substances. Least adverse effect described in documentation of published (ACGIH) exposure limits*

Adverse effect	Example substance	Reference (first author, year)
Sensory irritation	Acetone	Di Vincenzo, 1973
	Allyl alcohol	Dunlap, 1958
	Camphor	Gronka, 1969
	Ethyl alcohol	Lester, 1951
	Hexane	Nelson, 1943
Tooth erosion	Sulphuric acid	Malcolm, 1961
Garlic breath	Tellurium	Steinberg, 1942
Dermatitis	Ammonium picrate	Sunderman, 1945
Pruritus	Tetryl	Bergman, 1952
Acne	Chlorodiphenyl	Meigs, 1954
Basal rales	Asbestos	BOHS, 1968
Bronchial tightness	Acetonitrile	Pozzani, 1959
Bronchitis	Chlorine dioxide	Ferris, 1967
Reduced ventilatory capacity	Quartz	Theriault, 1974
Reduced airway conductance	Sulphur dioxide	Weir, 1972
Jaw necrosis	Phosphorus	Heimann, 1946
Osteosclerosis	Fluorides	Largent, 1961
Blood cell destruction	Benzene	Pagnotto, 1961
Fever	Magnesium oxide	Drinker, 1927
Nausea	Carbon tetrachloride	Kazantzis, 1960
Narcosis	Chloroform	Challen, 1958
	Toluene	von Oettingen, 1942
Headache	Nitroglycerine	Trainor, 1966
	Propylene glycol dinitrate	Stewart, 1974
Neurological signs	Carbon disulphide	Rubin, 1950
	Manganese oxide	Schuler, 1957
Tremor	Mercury	Bidstrup, 1951
Coproporphyrinuria	Lead	Tsuchiya, 1965

The exact manner and point in the carcinogenic process that the substance acts to initiate or enhance tumour formation probably differs from one substance to another. It is believed that in some instances mutations may be caused during cell division, possibly in direct response to the interaction of the hazardous substance with DNA. Other substances seem to change the cellular environment so as to enhance tumour formation rather than initiate it, or perhaps to change a tumour from benign to malignant. Whatever the mechanism by which the hazardous substance causes cancerous cells, in some individuals the cancerous cells survive, become established and propagate indefinitely to produce clinical, malignant tumours. Metastasis is common; that is, the transfer of cells of malignant tumours from one part of the body to another by the lymphatic channels or blood vessels. The well-being of the individual as a whole then suffers.

Carcinogens comprise a highly diverse collection of substances, including inorganic and organic chemicals, solids, liquids, vapours and gases. Any organ may be affected; skin, lungs, liver, blood, bladder and so on. It was Ludwig Rehn who in 1895 first reported three cases of bladder cancer from coal tar dyes (Gross, 1967). Exposure to benzidine or B-naphthylamine crystals also causes bladder tumours. Some hazardous substances foster cancer in several sites around the body, not necessarily where first deposited. Most forms of cancer carry a high risk of premature death, although new types of treatment have improved the prognosis. A risk of cancer mortality has long been known in the rubber industry (Sorahan, Parkes, Veys and Waterhouse, 1986).

There is commonly a long delay, sometimes decades, between first exposure and the appearance of cancer. Because of this latent period there is no short term indication that a particular employee exposed to a carcinogenic substance is being adversely affected. As the result of activities of the synthetic chemical industry new organic carcinogens are being continually introduced into the environment, many of which are not known to be carcinogenic to humans until long after they are introduced. Many substances have been identified as being carcinogenic to some degree in experimental animals (International Agency for Research on Cancer, 1987; NIOSH Registry of Toxic Effects of Chemical Substances, latest edition; Singh, 1988). For this reason, and because of the lag time inherent in the carcinogenic process, it is likely that in future years it will become necessary to learn, retrospectively, whether populations have been exposed to substances not now known to occur in the environment as human carcinogens (Commoner and Vithayathil, 1979).

The finding of cancer in an employee due to exposure to a hazardous substance is conditional upon three factors being operative. First, the substance must be carcinogenic to man. A substance carcinogenic to rodents, for example, is probably, although not necessarily carcinogenic to man. Second, the employee must have had sufficient exposure to the substance. The greater the intensity of exposure the more likely is the employee to get cancer. Third, it is necessary that a history of exposure has been in place for a sufficient length of time. The later in life exposure occurs the less the likelihood that cancer will develop in the employee's life time.

The respiratory system

The basic role of the respiratory system is to provide the conditions under which rapid exchange of oxygen and carbon dioxide can take place between the atmosphere and the blood in the pulmonary capillaries. Air inhaled through the nose and mouth passes into the pharynx and down the trachea or windpipe. The olfactory receptors are in the upper portion of the internal noses. The trachea divides into two bronchi leading to the left and right lungs respectively (Figure 2.1). Repeated sub-divisions of the airways, up to 23 times, lead to

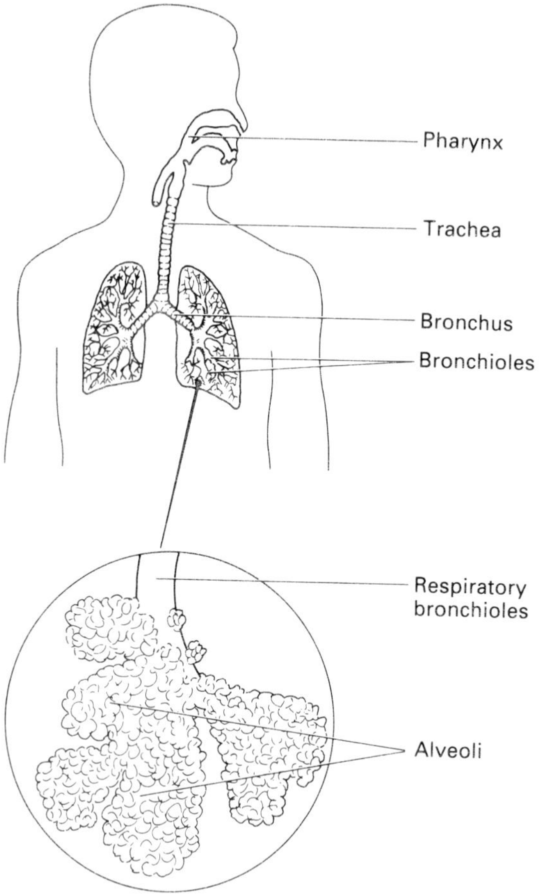

FIG. 2.1. Structure of the lungs.

progressively smaller bronchioles not unlike the branches of a tree, terminating in 250–350 million alveoli, crenellated air sacs, 0.1–0.3 mm diameter, covered inside with a thin, fluid layer of surfactant (Nagaishi, Nagasawa, Yamashita, Okada and Inaba, 1972). The path length from the beginning of the trachea to alveolus averages about 15 cm, varying from 10 to 25 cm. Oxygen is absorbed in the alveoli through membranes 0.35–2.5 microns thick, 90 m^2 in total area, into the blood contained in a corresponding capillary network (Montcastle, 1976).

The trachea, bronchi and bronchioles are lined with a ciliated epithelium containing cells which secrete a mucous fluid upon the walls of these airways. Mucus is constantly being moved up towards the larynx by the synchronous, rhythmic beating of myriads of cilia, which are tiny, lash-like projections from epithelial cells. The current of fluid induced by the cilia carries upward any foreign bodies that chance to touch the linings of the respiratory passages. The nasal passages likewise are bathed in mucus and lined with cilia. All the mucus moves towards the exits of the nose and mouth and is never stagnant. The mucus carries with it dust particles and other material deposited on the airways and phagocytes from the alveoli which have engulfed material scavenged there. The mucus is swallowed or spat out.

By the time air reaches the alveoli it has been warmed to 38C and is saturated with water vapour, a relative humidity of 100%. The alveoli have exceedingly thin walls almost completely filled with a network of capillary blood vessels, set in an elastic framework of connective tissue and covered only by an extremely thin membrane. The alveoli have, in total, a surface area of some 70–90 square metres. The exchange of oxygen and carbon dioxide between the blood in these vessels and the inhaled air all takes place in alveolar regions. The blood from the capillary bed surrounding the alveoli flows to the heart from where it is pumped to the tissues. The blood returns from the tissues to the heart, charged with carbon dioxide in combination as alkali bicarbonate, and is pumped to the alveoli, to complete the cycle.

Accompanying each inhalation and exhalation the volume of air in the lungs naturally increases and decreases. This regular change in air volume is caused by contraction and relaxation of the muscles of respiration. Nearly all the change in volume is accounted for by the expansion and contraction of alveoli. A trace of the volume changing with time illustrates the main features of the breathing cycle. These are shown in Figure 2.2. On taking a deep breath the volume of the lungs is at most about 6 litres for men and about 4.2 litres for women. It is not possible to fully collapse the lungs voluntarily and at maximum expiration the residual volume of the lungs is still a little over a litre. The difference in volume between the maximum inspiration and maximum

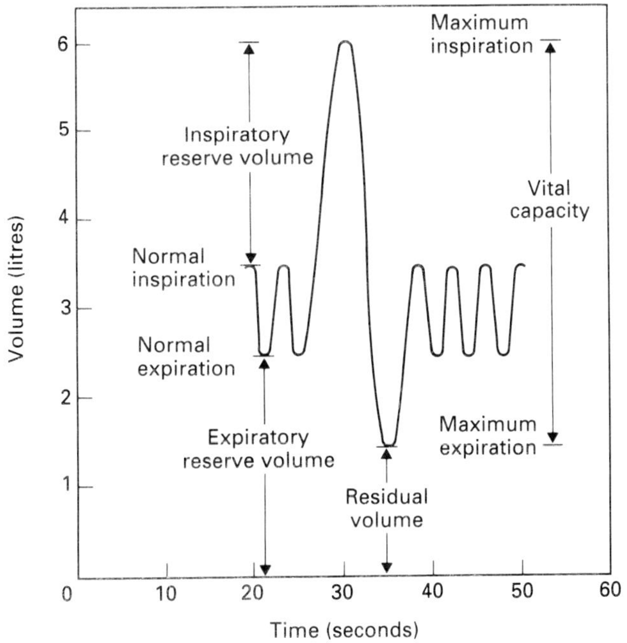

FIG. 2.2. Volume of the lungs and its partitions.

TABLE 2.2 *Minute volume of air breathed in different activities*

Activity	Air breathed (l min^{-1})
Resting in bed	5
Sitting	7
Standing	8
Walking	15
Fast walking	25
Jogging	45
Maximum work rate	65–100

expiration, the vital capacity, is an important physiological parameter. In adults it falls with increasing age by about 20 ml per year. It is reduced in individuals with pneumoconiosis.

The difference in volume between normal inspiration and expiration, the tidal volume, is 0.5–2.0 litres depending on how hard the person is working. Of this tidal air only about two thirds reaches the alveoli, where it is mixed instantly and completely with the expiratory reserve volume, about 1.5–2.5 litres, adding to the oxygen content and diluting the carbon dioxide. The other third of the tidal air stops momentarily in the so-called respiratory dead space—the mouth, nose , trachea and bronchi—and takes virtually no part in the exchange of oxygen and carbon dioxide. It is swept out of the dead space by the next expiration. Expired air is therefore a mixture of about two-thirds alveolar air and one-third air from the dead space from which little or no oxygen has been absorbed, and only a little carbon dioxide added, plus water vapour.

The respiratory rate varies from 10 to 30 respirations per minute depending on work rate. The product of the tidal volume and respiratory rate, the minute volume, thus varies from 5 to 60 litres depending on the work rate (West, 1974). A grasp of the amount of air breathed when undertaking a given task of work is given by studying Table 2.2. An untrained employee is apt to work at a higher rate than a trained employee undertaking a given task.

Lung diseases affect the functioning of the lungs in various ways. The highest achievable minute volume, the maximum breathing capacity, is a good, overall measure of an individual's respiratory function. A reduction in maximum breathing capacity could be produced either by a reduction in vital capacity or by increased resistance to breathing such as may come about by a narrowing of the airways. However the direct measurement of maximum breathing capacity is a very strenuous test to perform for several minutes on end as may be required. Consequently a rapid, even though indirect test of respiratory function is much preferred. The most common of these used for measuring impairment of lung function is the maximum volume flow-rate averaged over 1 second from a maximum inspiration, the forced expiratory volume in 1 second, or FEV_1. An increasing variety of other pulmonary function tests may be undertaken to examine different aspects; diffusion capacity, respiratory resistance, lung stiffness/compliance, and so on (Bouhuys, 1974).

Lung diseases

The most important occupational diseases in history have been those affecting the lungs. They range from mild irritation of a temporary nature to severe, chronic diseases which continue to progress even in the absence of further exposure.

Respiratory irritation

The first effect that is noticed when inhaling hazardous substances is often irritation of the eyes, nose and throat. Sensory irritants stimulate trigeminal nerve endings in the cornea and nasal mucosa and evoke a stinging or burning sensation. Coughing is excited by even slight irritation of the larynx from stimulation of laryngeal nerve endings (Alarie, 1973). The protective reflexes consist in coughing, constriction of the larynx and bronchi, closure of the glottis and inhibition of respiration. This does not prevent the passage of irritants but it warns of the presence of these substances, especially those which irritate the upper respiratory tract. The common property of respiratory irritants is their ability to cause inflammation of mucous membranes with which they come in contact.

On continued exposure to some respiratory irritants they may cease to seem irritating. There is a marked irritation threshold shift. This threshold shift may otherwise be called a development of tolerance, desensitisation, adaptation or inurement. It takes place rapidly with sulphur dioxide, for example, more slowly with formaldehyde or ammonia and not at all with the vapour of acrolein, which is a chemical intermediate, or benzoquinone (Anderson, Lundquist, Jensen and Proctor, 1974; Alarie, Kane and Barrown, 1980; Alarie, 1981). Benzoquinone is a yellow crystalline solid at room temperature but it has a significant vapour pressure (0.1 mm mercury at 25C) and its vapour is particularly irritating to the eyes. One of the other features of some irritants which renders more hazardous any exposure to those which primarily act on the lungs is the very fact that they do not greatly irritate the upper respiratory tract and go unnoticed for short while. Phosgene, arsenic trichloride and nitrogen dioxide are notable in this respect.

It is conceivable that death could result from an exceedingly high exposure to an irritant but this will only occur when it is not possible to escape from the environment, as may nevertheless happen in emergency situations or in confined spaces. With certain important exceptions irritants have sufficient warning properties to deter employees from staying very long in concentrations so high as to be dangerous to life. Temporary discomfort and annoyance is the primary consideration. More than one third of published exposure limits are ones set at such levels as to minimise irritation (Stokinger, 1970). This does not mean that other, more severe effects cannot occur with these substances but that, by inference, they occur at a higher concentration than that necessary to cause irritation.

The immediate action of irritant gases, vapours, mists and dusts is to cause inflammation at the site where they are absorbed or deposited. Most act on the upper respiratory tract and these characteristically cause irritation within a few minutes of the start of exposure. Irritants are sometimes subdivided into primary and secondary irritants. A primary irritant is a substance whose irritant action is far in excess of any other adverse effect and a secondary irritant is one whose irritant effects, while present, are exceeded by its other, systemic effects.

A wide variety of industrial operations liberate irritants into work-room air. Ammonia is liberated in the production of fertilisers, sulphur dioxide in paper production and zinc chloride in battery manufacture. Ozone is liberated in argon-shielded welding, and vanadium pentoxide occurs in cleaning boiler soot.

One conspicuous example is chromic acid mist from anodising operations, as is explained by the nature of the process. Anodising is an electrolytic process whereby non-corroding oxide film is deposited on aluminum and light alloys generally. The aluminum is made the anode in a large electrolytic cell containing chromic or sulphuric acid. Hydrogen is evolved in the process. The bubbles of hydrogen carry a fine acid mist. The process of pickling has a similar effect. Pickling is removing a coating of scale, oxide, tarnish and other impurities from metal objects so as to obtain a chemically clean surface prior to plating or other surface treatment. Pickling is effected by immersing the objects in an acid bath. Sulphuric acid is commonly used, although hydrochloric, phosphoric, nitric, chromic or hydrofluoric acid are sometimes used according to the metal being pickled. The acid attacks the metal, giving off bubbles of hydrogen which carry a fine acid mist. The primary health hazards in anodising and pickling operations are from splashing the acid on the skin and from inhaling the acid mist, which causes upper respiratory tract irritation.

Some, typically less soluble substances act chiefly on the lungs and cause pulmonary oedema and pneumonia, such as the war gas, chlorine and nitrogen dioxide (Vedder, 1925; Flury and Zernik, 1931). Chlorine has a wide range of industrial uses as an oxidising agent, in the manufacture of chlorinated hydrocarbons, bleaches and disinfectants. After moderate exposure, as in an industrial setting, oedema reaches its height in from 4 to 24 hours after initial exposure, but tends to regress within 48 hours.

Irritants are associated with effects on pulmonary function; namely increased flow resistance and decreased compliance. Besides temporary irritation there may be a progressive disease, chronic bronchitis, associated with prolonged exposure to a wide range of mild respiratory irritants; dusts, fumes and gases (Vigdortschik, Andreeva, Matussevitsch, Nikulina, Frumina and Striter, 1937; Flechsig, 1989). Irritant gases include the ones already mentioned and acid gases generally. Acid mists are well known respiratory tract irritants and additionally cause acid-etched teeth (Sappington, 1943). One of the very lowest exposure limits is for the highly irritant vapour of osmium tetroxide, a solid with an appreciable vapour pressure at room temperature (11 mm mercury at

27C). It is used by pathologists for staining fat. The odour of osmium tetroxide is described as 'having the kick of a mule' (Hamilton and Hardy, 1974). Typical dusts which are particularly irritating to the upper respiratory tract include vanadium pentoxide, used extensively as a catalyst in the chemical industry and as a yellow colorant in glass manufacture. Vanadium pentoxide has the dubious merit of being recognisable as the upper respiratory irritant which also gives rise to a green tongue (Hudson, 1964).

Pneumoconiosis

Historically the most important of all industrial diseases have been the lung diseases silicosis, coal workers pneumoconiosis, asbestosis and other pneumoconioses caused by mineral dusts accumulated in the lungs. High airborne dust concentrations in mining are a product of restricted ventilation as the mines get deeper, and blasting and rock drilling which became particularly widespread at the end of the last century. In the first half of the twentieth century this was compounded by increasing mechanisation underground, one product of which was copious quantities of fine dust. Byssinosis, from the Greek word for flax, byssos, is also a modern disease which has been widespread since ancient times but increasingly prevalent since the industrial revolution. It is caused by inhaling high concentrations of flax dust and cotton dust in work-room air in the textile industry.

Silicosis

Silicosis is caused by excessive exposure to airborne dust of free crystalline silica, and in modern times occurs most commonly in gold mine workers and pottery workers. It is also found in craftsmen working in the granite industry. The common crystalline form of silica in the extractive industries is quartz. When heated to 870C quartz converts to another variety of silica, tridymite and when further heated to 1470C it converts to cristobalite, which is a crystalline form of free silica, extremely hard and chemically inert, used extensively in precision casting by the hot wax process and certain speciality ceramics.

Quartz in refractory bricks and amorphous silica in diatomaceous earth are altered to cristobalite when exposed to temperatures in excess of 1470C. The development of silicosis from silica is intimately associated with the crystalline form of the silica. Cristobalite dust is more fibrogenic than quartz dust, which in turn is much more fibrogenic than amorphous silica (King, Mohanty, Harrison and Nagelschmidt, 1953).

The presence of silica dust in the lungs stimulates the connective tissue of the lung to produce fibrous tissue. This leads to the formation of small nodules of fibrotic tissue which slowly increase in size and coalesce as the disease progresses. These nodules are seen on a chest radiograph as small, round opacities scattered throughout the lung fields. As the disease continues to progress the employee's respiratory function diminishes and symptoms of breathlessness appear. Once established the disease continues to progress even in the absence of further exposure, there is increasing difficulty in breathing and death frequently results from combined heart and lung failure. Silicosis appears to render the individual susceptible to tuberculosis, which manifestly increases the suffering. Fortunately, since the introduction of antibiotics the tubercle bacillus is of vanishing importance.

Natural diatomaceous earth is 90% amorphous silica, which is, curiously, relatively innocuous (Cooper and Cralley, 1958). Diatomaceous earth is made up of the compacted remains of prehistoric diatoms, which are elementary aquatic plants.

Coal workers pneumoconiosis

Coal workers pneumoconiosis is of two kinds; simple pneumoconiosis and complicated pneumoconiosis. The one being a precursor of the other. Simple pneumoconiosis is not disabling in any marked degree and is recognised by the appearance of small, discrete rounded opacities in chest radiographs associated with the accumulation of coal dust in the lungs. The condition does not as a rule appear to progress on cessation of exposure to coal dust. If it does progress, the progression is so slight as not to be discernible in chest radiographs. However with increasing exposure there is an increasing profusion of rounded opacities in chest radiographs. The individual is then liable to develop a second condition, complicated pneumoconiosis, in which there is progressive massive fibrosis recognised by the presence of large opacities of irregular shape in chest radiographs and increasing shortness of breath (Davis, Chapman, Collins, Douglas, Fernie, Lamb and Ruckley, 1983; Seaton, 1983). This second condition progresses even on cessation of exposure and can be severely disabling.

Asbestosis

Asbestos has been known at least since Roman times. Its versatile heat resisting properties were described by Marco Polo in the late 13th century (Latham, 1958). However it was not until the latter part of the 19th century that quarrying and preparation of asbestos products became widespread.

The first published report on adverse health effects associated with asbestos exposure is believed to be one drawn up by Murray in 1906 (Murray, 1907). Since that time the prevalence of lung diseases from asbestos exposure has seemed to increase exponentially. The increase closely parallels the increase in production of asbestos products some years previously (Gilson, 1965; Murray, 1990).

Asbestosis is the pneumoconiosis which results from excessive exposure to airborne asbestos dust. After several years of exposure to the airborne dust fibrosis of the lungs develops in a characteristic linear form in the lower parts of the lung as observed on the chest radiograph. This causes increasing shortness of breath. The development of asbestosis is believed to be intimately connected with the fibrous shape of the airborne particles (Timbrell, 1972). Smokers who are exposed to airborne asbestos dust are at very much greater risk of developing lung cancer than those who do not smoke and, indeed, greater risk than those who smoke but are not exposed to asbestos (Doll, 1955; McDonald, Liddell, Gibbs, Eyssen and McDonald, 1980; Selikoff, Seidman and Hammond, 1980; Sluis-Cremer and Bezuidenhout, 1989). The two factors are synergistic.

All the many kinds of asbestos minerals have one thing in common; they have a crystalline structure and exhibit a pronounced fibrous cleavage. The associated airborne dust contains many thin fibres. The most common form of asbestos is chrysotile or white asbestos. Some of the larger fibres have a distinct curl. Other important kinds of asbestos are crocidolite and amosite. Dust from crocidolite and amosite asbestos contains mostly very straight and thin fibres. Dust from crocidolite or 'blue' asbestos causes an otherwise rare cancer, mesothelioma of the pleura. The tumour spreads from the pleura into the underlying tissues and is inevitably fatal (Wagner, Sleggs and Marchand, 1960; Becklake, 1976). The adverse effects of asbestos dust are commonly believed to be caused in part by its fibrous form and other, artificial fibrous materials have therefore also been suspected of being capable of producing similar biological effects (Pott, Ziem and Mohr, 1984; Wagner, Skidmore, Hill and Griffiths, 1985).

Byssinosis

Byssinosis is a dust disease of workers in the cotton and flax industries. It is not caused by the cotton or flax fibres themselves which are cellulose and innocuous. However in the cleaning and preparation of the material dust from plant debris is also liberated, which is a complex mixture of organic substances and micro-organisms. The precise causative agent and its mechanism is not certain but the symptoms of byssinosis are characteristic. Occasional tightness of the chest is first experienced on Monday mornings or after a period of absence, known in the industry as 'Monday tightness'. There may be a cough, chest tightness or difficulty in breathing. The affected worker may first notice symptoms after annual holidays but later they usually occur after weekends.

Early effects may be noticed during the first year of exposure to dust and at this stage the first and only complaint may be cough or chest tightness after the work shift immediately following the weekend break (Mondays in Western countries). The cough, the feeling of chest tightness or difficulty in breathing may disappear shortly after leaving the work place. On Tuesdays there are no symptoms. As the disease progresses, the symptoms worsen and are accompanied by breathlessness. A fall in forced expiratory volume can be observed over the shift. Over a period of years the chest tightness gradually becomes more pronounced and occurs on more days of the week, although at this stage of the disease there is still improvement as the week goes on. Difficulty in breathing is noticed and eventually there is evidence of permanent respiratory impairment which does not materially diminish even on leaving the industry. At this stage, the effects of the dust cannot be distinguished from chronic bronchitis, except that the past history of chest symptoms, characteristically worse at the beginning of the week, may suggest the etiology.

Benign pneumoconiosis

The term 'pneumoconiosis' on its own means simply dust in the lungs, without necessarily inferring adverse effects. The condition caused by dusts which accumulate in the lungs without producing significant adverse effects is benign pneumoconiosis. Two such dusts are tin oxide and barium oxide. Both are highly radio-opaque as is evidenced by their high atomic weight. Consequently chest radiographs of employees with moderate exposure to these dusts typically show dense opacities, without any associated disability.

Metal fume fever

Dusts do not only affect the lungs. Many substances which dissolve from dust in the lungs are subsequently carried around the body and cause systemic effects. The metals and their compounds: such as lead, manganese and mercury, are obvious examples. However, another compound, zinc oxide, in high concentration causes 'metal fume fever', a peculiar transient disturbance characterised by chills followed by fever a few hours later. Repeated exposure day after day appears to impart some immunity from further attacks. Magnesium and copper oxides may also produce similar fevers (Drinker and Hatch, 1954).

Sensitising agents

Allergic rhinitis from exposure to soluble salts of platinum was first reported in 1804 (Harris, 1975). Since then it has been well recognised that exposure to a wide variety of quite different organic and inorganic hazardous substances, including certain platinum salts, can give rise to occupational asthma in which the foreign material acts as an antigen (Hunter, Milton and Perry, 1945; Chan-Yeung and Lam, 1986). The time taken for sensitisation to occur varies from a few days exposure to many years (Newman Taylor, 1988). The symptoms can be distressing. The sensitised person will normally respond with rapid and severe bronchial narrowing even to an extremely small exposure; much smaller than the exposure which caused the original sensitisation.

Good examples of such sensitising agents are toluene-2,4-diisocyanate, which is used principally in the manufacture of polyurethane foams (Wegman, Pagnotto, Fine and Peters, 1974; Butcher, Jones, O'Neill, Glindmeyer, Diem, Dharmarajan, Weill and Savaggio, 1977), polyisocyanates used in paint sprays (Sequin, Allard, Cartier and Malo, 1987; Ferguson, Schaper and Alarie, 1987), airborne dust when working with flour or cereal grains (Darby, 1982) and dust containing proteolytic enzymes, which were at one time used extensively in the manufacture of biological washing powders (Flindt, 1969; Greenberg, Milne and Watt, 1970).

Another allergy, extrinsic allergic alveolitis, which affects the alveoli, as the name implies, is known by various colourful names indicating an occupational association: Farmer's lung from mouldy hay, Bird Fancier's lung from bird droppings, Bagassosis from crushed sugar cane, Malt Worker's lung from mouldy malt, Mushroom Worker's lung from mushroom compost, to name but a few. The most common causal factor is exposure to small diameter fungal spores (Pepys, 1969). The disease is characterised by shortness of breath, which generally disappears on removal from exposure for a few hours, but with repeated exposure may lead to pulmonary fibrosis and permanent impairment of lung function.

Lung cancer

Lung cancer is usually produced only by excessive exposure for many years to the carcinogenic substance (Conning, Magee, Oesch and Clemmesen, 1980). Well known carcinogenic substances which affect the lungs include bis(-chloromethyl) ether, nickel carbonyl vapour, certain chromium compounds, various forms of asbestos dust and uranium compounds, coal-tars and other agents (Parkes, 1982; Cotes, Steel and Leathart, 1987). Cigarette smoking appears to considerably increase the risk of lung cancer from exposure to such hazardous substances. Exposure to the minerals asbestos and silica appears to increase the risk of lung cancer markedly in the presence of asbestosis or silicosis respectively (Mastrangelo, Zambon, Simonato and Rizzi, 1988). There is a hazard of lung cancer in the rock and slag wool industry (Doll, 1987).

Skin complaints—the visible marks of occupation

Work involving friction or pressure on the skin may result in an abrasion or a callus caused mechanically. Viable organisms: bacteria, viruses, fungi and parasites, may attack the skin directly or gain entry to the rest of the body through lesions caused by chemicals. Burns can, of course, result from any substance which is hot enough. Absorption of substances through the skin is faster through skin which is abraded or inflamed.

Examples of substances which cause occupational skin disorders of one sort or another include

—acids and alkalis
—adhesives
—alcohols
—chlorinated naphthalenes
—detergents
—dyes and dye intermediates
—explosives: tetryl, TNT, amitol
—ketones
—metals and their compounds: beryllium, chromium, cobalt, mercury, nickel
—paint, varnish and lacquers
—pesticides and fungicides: organo-phosphates
—petroleum products: oils, lubricants, coolants, degreasing agents
—resins

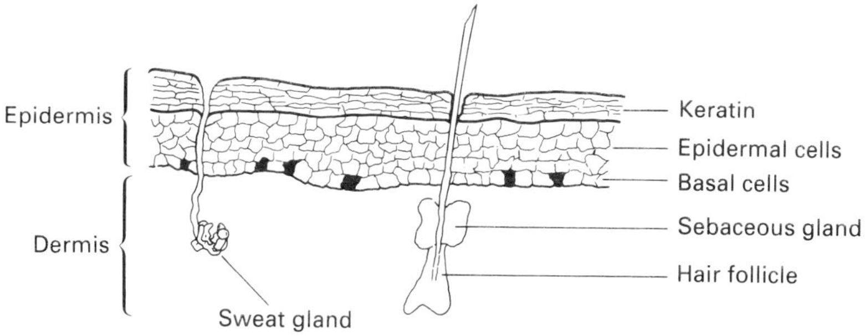

FIG. 2.3. Diagrammatic section through skin.

—synthetic rubber, accelerants, antioxidants, plasticisers
—tar products

The effects on the skin from handling hazardous substances are usually mild but can be quite severe. Descriptive terms commonly used are: dermatitis, for an inflammation of the surface of the skin; dermatosis, for an affection of the deeper layers of the skin; and eczema, for an itching, inflammatory condition occurring characteristically as a reaction to irritants. Chemicals spilled on the skin or eyes may cause severe chemical burns or dermatitis. Those which are absorbed through the skin can enter the body and cause damage to internal organs (Stokinger, 1977).

Structurally, skin is composed of two layers: the epidermis and the dermis, each layer having a composite structure. The principal elements are indicated in Figure 2.3. The epidermis has an outermost protective layer normally about 10 microns thick, 500 microns in the soles and palms, consisting of dead, flattened epidermal cells or horn cells and a thicker, inner layer about 50 microns thick, consisting of living epidermal cells which are continually being reproduced and which eventually die and become horn cells. The horn cells take the wear and are thickest on the palms of the hands and soles of the feet. The basal layer of the epidermis, where new cells are continually being formed, includes melanocytes. These are the cells which produce pigment which absorbs ultraviolet light. Below the epidermis is the thicker dermis, about 1.25 mm thick, composed of elastic and collagen tissue which gives the skin its resilience. The sweat glands and hair follicles are in the dermis. The sebaceous or oil glands are connected to the hair follicles. Capillaries are invested in the dermis and nerve fibres terminate there. However, there are many free nerve endings which branch repeatedly, reaching into the epidermis.

Most occupational dermatoses are caused by primary irritants which on contact attack the skin chemically, starting with the outermost, horn cell layer (Griffiths and Wilkinson, 1985). Strong irritants generally produce their effects immediately they make contact with the skin. Some common examples are sulphuric acid, hydrofluoric acid, hydrochloric acid, phosphoric acid, chromic acid, nitric acid, caustic soda and bleaching powder. The skin can develop deep sores or become badly ulcerated in this way. Other irritants take a little longer but the damage is essentially established within a few hours. Dilute hydrofluoric acid gives no immediate warning of injury, but the skin is penetrated and after several hours painful ulcers may develop. Most acids and alkalis are skin irritants. They attack or at least soften the keratin layer and thereby remove part of it.

Most of the problems which are produced by contact with chemicals occur as a result of either primary irritation or sensitisation. There are other ways in which the protective layers can be destroyed by direct, if more subtle chemical action. For example, organic solvents in contact with the skin dissolve the surface lipids and disturb the outermost layers this way. Repeated exposure gives rise to a very dry and cracked skin. The salts of arsenic, chromium, mercury and zinc precipitate protein in the skin and thereby denature it. Salicylic acid, oxalic acid, urea and other substances act as reducing agents with the keratin of the horn cells.

Chloracne, known as 'blackhead itch' or 'cable rash', is a distinctive form of acne caused by chronic exposure to chlorinated naphthalenes or chlorodiphenyl. It is also caused by 2,3,7,8-tetrachlorodibenzodioxine, a contamination of certain pesticides. Chloracne is a characteristically widespread acne which is quite slow to disappear on redeployment of employees away from direct exposure, taking months rather than weeks (Good and Pensky, 1943).

Epoxy resins and formaldehyde solutions are perhaps the most well documented examples of the group of chemicals which cause allergic contact dermatitis. Acrylic and epoxy resins have also been implicated. Other examples are the metal nickel, its salts, benzoyl peroxide, turpentine and phthalic anhydride. This disease is rarely seen before 5 days from initial exposure and it may take weeks or months of repeated exposure to establish the sensitivity. A sensitiser does not cause any visible alteration to the skin immediately following first contact but by the time several contacts have been made specific changes in the skin have been caused so that further contact on the same or, indeed, other parts of the body will induce a dermatitis. Not everyone exposed becomes sensitised in this way but the sensitivity, once acquired, is long lasting (Adams, 1984).

Several chemicals affect the skin in such a way that it undergoes peculiar growth patterns and leads to the formation of a tumour or cancer. Occupational cancer of the skin was first recorded by the English physician Sir Percival Pott in 1775. He described scrotal cancer, then known as soot wart, which was prevalent in chimney sweeps. The same condition was later found in mule spinners in cotton mills (Henry and Irvine, 1936) and more recently in workers exposed to mineral oils in the engineering industry (Southam and Wilson, 1922; Cruikshank and Squire, 1950; Desoille, Philbert, Ripault, Cavigneaux and Rossignolli, 1973). It is believed that the carcinogenic activity of mineral oils is intimately connected with the 4–6 ring polycyclic aromatic hydrocarbons they contain (Medical Research Council, 1968; Grimmer, Dettbarn, Brune, Deutsche-Wenzel and Misfeld, 1982). Occupational exposure to certain petroleum products, a number of the coal tar based materials, asphalt fumes, creosote, arsenic and some of the chlorinated hydrocarbons can also cause skin cancer (Birmingham, 1971).

Blood diseases caused by exposure to certain substances

The main function of the blood is to transport oxygen and carbon dioxide between the lungs and body cells. The blood circulates to the capillaries throughout the body where, as a result of respiratory activities, the oxygen is given up and carbon dioxide is liberated and carried back to the lungs in the blood. The higher the energy expenditure the higher is the rate of respiration and the higher is the rate of circulation of the blood (Burton, 1968). The blood also transports the nutrients that cells need and carries waste products to the organs of excretion. It plays an important role in maintaining a uniform temperature in the body.

Blood consists of a fluid plasma in which are suspended cells of several different types, the blood corpuscles. The plasma consists of a complex mixture of inorganic salts in solution and blood proteins in colloidal sol. The corpuscles normally present in blood fall into two well-defined groups; red corpuscles which contain the respiratory pigment, haemoglobin, and white corpuscles. Haemoglobin in the red cells, under conditions of high oxygen concentration, as at the respiratory surface, combines with this gas to form the bright red oxyhaemoglobin. Red corpuscles are biconcave, circular, disc-like cells, similar in size, 7.5 microns in diameter and 2 microns thick, formed in the bone marrow. They have a short life and after a few weeks circulation in the blood stream they are normally destroyed by phagocytes.

Possibly the most notorious blood disease from a hazardous substance is aplastic anaemia from atmospheric exposure to benzene vapour in low concentration for several years or more (Infante, 1987). Brief exposure to a high enough concentration, in excess of 1,000 ppm in air, can cause narcosis very rapidly, but in low concentration it is damage to the blood-forming tissue in the bone marrow which is the greatest cause for concern. Extended benzene exposure reduces the production rate of red blood cells. It may also be a cause of leukaemia, by which there is an excessive growth of the tissues producing white cells. However, whether benzene truly does cause leukaemia is still open to question in the minds of some authorities.

Arsine (arsenic hydride), a gas used in the manufacture of semi-conductors and given off from wet ores and metals containing arsenic when attacked by dilute acid also causes anaemia when inhaled. The causation in this instance is by combination of arsine with the haemoglobin, resulting in destruction of the red blood cells themselves.

Nitrobenzene and aniline, formed by the reduction of nitrobenzene, are both oily liquids whose vapours represent two of a series of compounds which act specifically on the haemoglobin in blood to form methaemoglobin, a stable oxide in which the oxygen is so firmly bound it ceases to be available for tissue exchange. Nitrobenzene and aniline are used in the manufacture of many chemical intermediates. They are liquid at room temperature and may be absorbed through the contaminated skin as well as by absorption through the lungs on inhalation of the vapour. The consequences of excessive exposure are, progressively, headache, dyspnoea on exertion and very marked cyanosis, giving rise to typical 'blue lip' or 'huckleberry pie' faces. Relief is possible with oxygen and recovery within 24 hours is usual.

Another substance which affects the haemoglobin in a different way but with similar consequences is the gas, carbon monoxide. Poisoning by carbon monoxide was the subject of some classic studies in the early part of the century (Douglas, Haldane and Haldane, 1912; Henderson, Haggard, Teague, Prince and Wunderlich, 1921; Sayers and Yant, 1923; Drinker, 1938). The gas has low solubility and penetrates to the alveoli. Once there it does not affect the lungs as such but rather their function. Carbon monoxide is known as a 'chemical' asphyxiant. It has a stronger affinity for haemoglobin than does oxygen by 200–300 times and thereby replaces oxygen in the blood to form carboxy-haemoglobin. Interestingly, this reaction is completely reversible and when exposure ceases the carbon monoxide is steadily eliminated through the lungs. These reversible reactions of oxygen and carbon monoxide with haemoglobin are expressed by the following equation, in which haemoglobin is designated by the symbol **Hb**:

$$HbO_2 + CO \rightleftharpoons HbCO + O_2$$

The red corpuscles are not injured. Repeated exposure to low but significant amounts of carbon monoxide in air does not ordinarily cause permanent ill effects but higher exposure, sufficient to cause unconsciousness, may cause permanent injury to the brain and nervous system from anoxaemia. Exposure to the very highest concentrations, in excess of 4,000 ppm of carbon monoxide in air, leads to death.

Liver disorder of occupational origin

The liver is a large, spongy organ weighing about 1.7 kg as compared, for example, with the lungs which weigh about 1.0 kg. It is situated on the right side of the upper part of the abdomen. It has many important functions including regulating the amino acids in the blood, forming and secreting bile which aids in absorption and digestion of fats and transforming glucose into glycogen. Many toxic materials are metabolised in the liver before they or their metabolites are excreted in the urine. The kidney can only excrete molecules which are water soluble.

Symptoms of liver dysfunction include nausea, anorexia, flatulence, vomiting, jaundice and enlarged and tender liver. The diagnosis of liver disease of occupational origin is especially difficult because the incidence of non-occupational liver disease from drinking alcohol or viral infection is relatively high and commonly used laboratory tests of liver function have low specificity. One of the best clues to cause in the individual case is a clear temporal relationship between exposure and the onset of clinical symptoms in a previously healthy person (Proctor, Hughes and Fischman, 1988).

Excessive accumulation of toxic chemicals in the liver can interfere with its normal function. Examples of hepatotoxins include carbon tetrachloride, once widely used as a solvent (Kazantzis and Bomford, 1960), toluene, still widely used as a solvent (Greenburg, Mayers, Heinmann and Moskowitz, 1942),

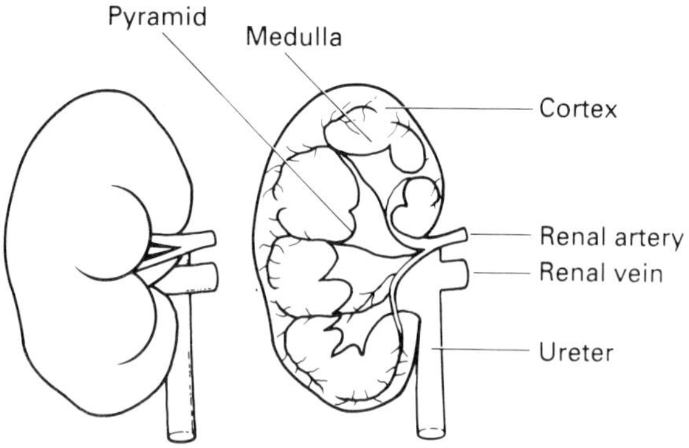

FIG. 2.4. Structure of the kidneys—external view and cross section.

chloronaphthalenes, tetrachloroethane (von Oettingen, 1955) and yellow phosphorus (Fischler, 1941). A rare cancer, angiosarcoma of the liver, is found in some individuals exposed to very high concentrations of vinyl chloride monomer, used in the manufacture of PVC (polyvinylchloride) (Fox and Collier, 1977; Maltoni and Lefemine, 1974).

The kidneys may be the target organ

The kidneys are situated high up at the back of the abdominal cavity, one each side of the vertebral column (Figure 2.4). The kidneys perform two major functions: first, they expel most of the end-products of metabolism and second, they control the concentrations of most of the constituents of the body fluids. Blood brought to the kidneys by the renal arteries contains the waste products which are removed in the kidneys by a delicate filtration/separation process. Urine is formed in the kidneys from where it passes through the ureter into the bladder before excretion.

Kidney damage is caused by excessive exposure to the heavy metals, especially uranium, and to organic solvents such as carbon tetrachloride, which are typical nephrotoxins. The damage is made evident by the appearance in the urine of substances normally held back by the kidney: protein, albumin, glucose and so on. Severe damage can give rise to complete cessation of kidney function, as may happen with high exposure to carbon tetrachloride. But the specialised cells in the kidney regenerate themselves and recovery of kidney function is possible.

Nervous system

Nerves consist of numerous fibres united in bundles, collected together in a sheath. There are numerous types of fibre, the two main ones being those which carry sensory impulses from the receptor organs to the central nervous system and those which carry impulses from the central nervous system to the various effector organs.

The nervous system comprises two different kinds: the voluntary system and the autonomic system. In humans the central nervous system, which includes the brain and spinal chord, coordinates and controls the activities. The peripheral system, which includes the cranial nerves, spinal nerves and autonomic nervous system, provides the connecting links between the organs and tissues and the central nervous system.

Narcosis is a general depression of the central nervous system and may be caused by a wide variety of organic solvents, including chloroform, trichloroethylene, benzene and carbon tetrachloride. Carbon monoxide, a gas, may also be classed as a narcotic. The symptoms of narcosis include, progressively, sleepiness, dizziness, headache, unconsciousness and, in the extreme case, death. On cessation of exposure mild narcosis gradually reduces as the causative substance is slowly eliminated, and recovery is generally complete.

Neurotoxins include a wide variety of gases, vapours and dusts. Substances which affect the central nervous system and cause permanent mental changes have been found to include carbon disulphide, manganese, thallium and others, besides inorganic mercury whose adverse effects have been well known since Roman times. Carbon disulphide may cause progressively; dizziness, headache, some paralysis of all four limbs, through to depression, melancholy and acute mania (Paluch, 1948). Manganese was discovered by Gahn in 1774. Poisoning by manganese, like carbon disulphide, is of fairly recent discovery, being first reported in France in the 19th century (Couper, 1837). Airborne dust from manganese dioxide ore is inhaled, deposited in the lungs and ultimately affects the brain, specifically the extrapyramidal motor system. Employees with manganese poisoning exhibit tremor, Parkinsonism, shuffling gait, monotonous speech and uncontrollable spasmodic laughter (Berry and Bidwai, 1959). The insecticides dieldrin and the closely related aldrin also predominantly affect the central nervous system causing irritability, muscle spasms, convulsions and in extreme cases coma (Jager, 1970). Organic solvents in paints and lacquers are associated with a characteristic neurotoxicity known as 'solvent syndrome' leading to chronic toxic encephalopathy (Axelson, Hane and Hogstedt, 1976; Arlien-Soborg, Bruhn, Gyldensted and Melgaard, 1979; Cranmer and Goldberg, 1986).

Mercury was known to ancient Chinese and Hindus. It has been found in Egyptian tombs of 1500 BC. The adverse effects on health of mercury in mining the ore and refining the metal were known to the Romans. They used slave labour in the Spanish mercury mines at Almaden, and a sentence to work there was considered equivalent to a death sentence. In history, whenever mercury or its compounds have been handled the risk of mercurialism has been notorious. The most well known example is in the hatting trade. Top hats used to be made most commonly with the fur of rabbits or hares. In order to make the fur lay flat French hatters of the 17th century had discovered a treatment with an acid solution of mercuric nitrate, a process called carroting. The hatting trade in top hats was flourishing in the last century. At the workshops where the complete felting process was undertaken, comprising mixing, blowing, wetting-down and sizing, the inhaled air became contaminated with inorganic mercury salts (Neal, Fliner, Edwards, Reinhart, Hough, Dallavalle,

Goldram, Armstrong, Gray, Coleman and Poyman, 1941). One of the earliest signs of mercurialism, whether from the vapour of the metal or dust from inorganic mercury compounds, is a fine tremor in the writing. Advanced cases of chronic mercurialism have a characteristic tremor known as the hatters' shakes and those affected often display an excessive irritability. Those severely affected may also experience depression, insomnia, hallucinations, delusions and mania. Hence the expression *as mad as a hatter* personified by the Mad Hatter in Lewis Carroll's tale of Alice's Adventures in Wonderland, published in 1865.

Organic (alkyl) mercury compounds affect the central and peripheral nervous systems (Hill, 1943; Ahlmark, 1948). These compounds are employed in seed dressings as a fungicide, although the notorious Minimata Bay episode in Japan originated from inorganic industrial waste methylated by micro-organisms on the sea bed.

Materials which affect the peripheral nervous system generally do so by damaging the nerve cells, thereby producing defects in motor function, manifested by a decrease in muscular activity and/or a loss of sensation. Examples of substances which can affect the peripheral nervous system, causing polyneuropathy, include n-hexane vapour (Cavigneaux, 1972), inorganic lead compounds (NIOSH, 1978) and acrylamide dust (Kuperman, 1958; Waldron, 1979).

Effects on reproduction

The genetic effects of hazardous substances pose a difficult problem of recent concern in the hazard evaluation process. There is, as yet, little hard evidence to go on from industrial surveys, although this problem has the potential for even more emotional interplay than does chemical carcinogenesis. Three distinct kinds of adverse effect on reproduction are probable:

(1) Decreased reproductive activity.

(2) Effects on male and female germ cells, giving rise to decreased fertility.

(3) Effects from exposure during pregnancy on the development of the unborn foetus. An example of a substance with teratogenic potential is carbon monoxide, whose ability to cause foetal death, toxic effects on the nervous system and anatomical malformations is established in non-occupational exposure (Bankl and Jellinger, 1967; Caravati, Adams, Joyce and Schafer, 1988; Norman and Halton, 1990).

Harbison has reported on the possible types of deleterious effects on the human foetus for 200 different chemicals (1980). Investigation of the reproductive effects of exposure to specific substances at work is in its infancy but will undoubtedly reveal at least some teratogens as yet not known to exist. It is not possible to consider exposure limits to prevent adverse effects on reproduction until the possibility of such effects has been examined in human or animal studies. Hopefully, the disciplines of industrial hygiene, occupational medicine and toxicology are scientifically better prepared to handle the associated employee concern than they were during the era of the 'carcinogen of the week' (Guidoffi, 1963; Carter, 1982; Owen, 1985).

References

Ahlmark A. Brit. J. Ind. Med. **8** 480 (1948).
Alarie Y. CRC Crit. Rev. Toxicol. **2** 299 (1973).
Alarie Y. Fd. Cosmet. Toxicol. **19** 623 (1981).
Alarie, Y, L Kane and C S Barrown. Sensory irritation: the use of an animal model to establish acceptable exposure to airborne chemical irritants. In: Toxicology: Principles and practice (Edited by A L Reeves), **1** 48. John Wiley and Sons, New York. (1980).
American Conference of Governmental Industrial Hygienists. Documentation of the Threshold Limit Values and Biological Exposure Indices. 5th Edition. American Conference of Governmental Industrial Hygienists, Cincinnati, Ohio USA (1986).
Andersen I, G R Lundquist, P L Jensen and D F Proctor. Arch. Envir. Health **28** 31 (1974).
Arlien-Soborg P, P Bruhn, C Gyldensted and B Melgaard. Acta Neurol. Scand. **60** 149 (1979).
Axelson O, M Hane and C Hogstedt. Scand. J. Work Environ. Health **2** 14 (1976).
Bankl H and K Jellinger. Beitr. path. Anat. **135** 350 (1967).
Becklake M R. Am. Rev. Respir. Dis. **114** 187 (1976).
Bergman B B. Arch. Ind. Hyg. & Occ. Med. **5** 10 (1952).
Berry J N and P S Bidwai. Neurology, India **7** 34 (1959).
Bidstrup P L, J A Bonnell, O G Harvey and S Lockets. Lancet **2** 856 (1951).
Birmingham D J. Occupational dermatoses. In: Dermatology in general medicine (Edited by T B Fitzpatrick). McGraw-Hill. Inc., New York (1971).
British Occupational Hygiene Society—Committee on Hygiene Standards. Ann. Occup. Hyg. **11** (1968).
Butcher B T, R N Jones, C E O'Neil, H W Glindmeyer, T E Diem, V Dharmarajan, H Weill and J E Savaggio. Am. Rev. Resp. Dis. **116** 411 (1977).
Caravati E M, C J Adams, S M Joyce and N C Schafer. Ann. Emerg. Med. **17** 714 (1988).
Cavigneaux A. Securite et Hygiene du Travail **67** 199 (1972).
Carter V l. Ann. Am. Conf. Govt. Ind. Hyg. **3** 113 (1982).
Challen P J R, D E Hickish and J Bedford. Brit. J. Ind. Med. **15** 243 (1958).
Chan-Yeung M and S Lam. Am. Rev. Resp. Dis. **133** 686 (1986).
Commoner B and A Vihayathil. The analysis of urine samples as a means of detecting exposure to environmental organic carcinogens. In: The use of biological specimens for the assessment of human exposure to environmental pollutants (Edited by A Berlin, A H Wolff and Y Hasegawa), p. 121. Martinus Nijhoff Publishers, The Hague/Boston/London (1979).
Conning D M, P Magee, F Oesch and J Clemmessen (Editors). Quantitative aspects of risk assessment in chemical carcinogenesis. Arch. Toxicol. Suppl. 3: 1–330 (1980).
Cooper W C and L J Cralley. Pneumoconiosis in diatomite mining and processing. Public Health Service Publication No. 601, USPGO, Washington, DC, USA (1958).
Couper J. Brit. Ann. Med. Pharm. **1** 41 (1837).
Cranmer J and L Goldberg. Neurotoxicol. **7** 1 (1986).
Cruikshank C N D and J R Squire. Brit. J. Ind. Med. **7** 1 (1950).
Davis J M G, J Chapman, P Collins, A N Douglas, J Fernie, D Lamb and V A Ruckley. Am. Rev. Resp. Dis. **128** 118 (1983).
Darby F J. Clinical notes on occupational asthma. H M Stationery Office, London (1982).
Desoille H, M Philbert, G Ripault A Cavigneaux and H Rossignoli. Arch. Mal. Prof. Med. Trav. Sec. Soc. (Paris) **34** 669 (1973).
Dionne E D. Natl. Sof. News **129** 17 (1984)
Di Vincenzo G D, F J Yanno and B D Astill. Am. Ind. Hyg. Assoc. J. **34** 329 (1973).
Doll R. Brit. J. Ind. Med. **12** 81 (1955).
Doll R. Ann Occup Hyg **31** 805 (1987).
Douglas C G, J S Haldane and J B S Haldane. J. Physiol. **44** 275 (1912).
Drinker P, R M Thompson and J L Flinn. J. Ind. Hyg. **9** 187 (1927).

Dunlap M K, J K Kodama, M D Wellington, M D Anderson and C H Hine. A.M.A. Arch. Ind. Health **18** 303 (1958).
Ferguson J S, M Schaper and Y Alarie. Toxicol. Appl. Pharmacol. **89** 332 (1987).
Fischler F. Munch. med. Wochschr. 621 (1941).
Ferris B G, W A Burgess and J Worcester. Brit. J. Ind. Med. **24** 26 (1967).
Flechsig R. Ind. Health **27** 27 (1989).
Flindt M L H. The Lancet, p. 177 (June 14, 1969).
Fox A J and P F Collier. Brit. J. Ind. Med. **34** 1 (1977).
Gilson J C. Trans. Soc. Occup. Med. **16** 62 (1965).
Greenberg M, J F Milne and A Watt. Brit. Med. J. **2** 629 (1970).
Greenburg L, M R Mayers, H Heinmann and S Moskowitz. J. A. M. A. **118** 573 (1942).
Grimmer G, G Dettbarn, H Brune, R Deutsch-Wenzel and J Misfeld. Int. Arch. Occup. Environ. Health **50** 95 (1982).
Good C K and N Pensky. Arch. Dermatol. and Syphilol. **48** 251 (1943).
Gronka P A, R L Bobkoskie, G J Tomchick and A B Rakow. Am. Ind. Hyg. Assoc. J. **30** 276 (1969).
Guidoffi T L. J. Occup. Med. **25** 9 (1963).
Hamilton A J. Ind. Hyg. **4** 219 (1922).
Harbison R D. In: Toxicology—the basic science of poisons (Edited by Casarett and Doull). Macmillan, New York (1980).
Harris S. J. Soc. Occup. Med. **25** 133 (1975).
Heimann H. J. Ind. Hyg. & Tox. **28** 142 (1946).
Henderson Y, H W Haggard, M C Teague, A L Prince and R M Wunderlich. J. Ind. Hyg. **3** 79,137 (1921).
Henry S A and Irvine E D. J. Hyg. **36** 310 (1936).
Hill W H. Can. J. Public Health **34** 158 (1943).
Hunter D, R Milton and K M A Perry. Brit. J. Ind. Med. **2** 92 (1945).
Infante P F. Am. J. Ind. Med. **11** 599 (1987).
International Agency for Research on Cancer. Overall evaluations of carcinogenicity: An updating of IARC Monographs Volumes 1 to 42, Supplement 7. World Health Organisation, 1211 Geneva 27, Switzerland (1987)
Kazantzis G and R R Bomford. Lancet **1** 360 (1960).
King E J, G P Mohanty, C V Harrison and G. Nagelschmidt. Brit. J. Ind. Med. **10** 9 (1953).
Kuperman A S. J. Pharm. Exptl. Therap. **123** 180 (1958).
Lester D and L A Greenberg. Q. J. Stud. Alc. **12** 167 (1951).
Malcolm D and E Paul. Brit. J. Ind. Med. **18** 63 (1961).
Maltoni C and G Lefemine. Lincei-Rendiconte Della Classe di Science, Tesiche, Mathmatische **56** 1 (1974).
Mastrangelo G, P Zambon, L Simonato and P Rizzi. Int. Arch. Occup. Environ. Health. **60** 299 (1988).
McDonald J C, F D K Liddell, G W Gibbs, G E Eyssen and A D McDonald. Brit. J. Ind. Med. **37** 11 (1980).
Medical Research Council. The carcinogenic action of mineral oils: A chemical and biological study. Medical Research Council Special Report Series No. 306. H M Stationery Office, London (1968).
Meigs J W, J J Albom and B L Kartin. J. Am. Med. Assoc. **154** 1417 (1954).
Murray H M. Report of Departmental Committee on Compensation for Industrial Disease, c.d. 3495/6, H M Stationery Office, London (1907).
Murray R. Brit. J. Ind. Med. **47** 361 (1990).
Neal P A, R H Fliner, T I Edwards, W H Reinhart, J W Hough, J M Dallavalle, F H Goldram, D W Armstrong, A S Gray, A L Coleman and B F Posyman. Mercurialism and its control in the felt-hat industry. US Publ. Hlth. Bull. No 263 (1941)
Newman Taylor A J. Postgrad. Med. J. **64** 505 (1988).
NIOSH. Revised criteria for a recommended standard—Occupational exposure to inorganic lead. Department of Health Education and Welfare (NIOSH), Pub. No. 78–158, US Government Printing Office, Washington DC, USA (1978).

Norman C A and D M Halton. Ann. Occup. Hyg. 34 335 (1990).
von Oettingen W F. The halogenated hydrocarbons, their toxicity and potential dangers. Public Health Service Pub. No. 414 (1955).
von Oettingen W F, P A Neal D D Donahue, J L Svirbely, H D Baerstein, A R Monarco, P J Valaer and J L Mitchell. The toxicity and potential dangers of toluene, with special reference to its maximal permissible concentration. Pub. Health Bull. 279. Government Printing Office, Washington DC, USA. (1942).
Owen R. Occupational exposure: How the workers see it. CEFIC symposium on occupational exposure limits and harmonisation in the setting and control of OELs for the protection of workers, European Council of Chemical Manufacturers' Federations, Avenue Louise 250, Brussels, Belgium (1985).
Pagnotto L D, H B Elkins, H G Brugsch and J E Walkley. Am. Ind. Hyg. Assoc. J. **22** 417 (1961).
Paluch E J. Ind. Hyg. **30** 37 (1948).
Pott F, U Ziem and U Mohr. Lung carcinomas and mesotheliomas following intra-tracheal instillation of glass fibres and asbestos. Presented at VI International Pneumoconiosis Conference. Conference report published by Bergbau-Berufsgenossenschaft, Bochum (1984).
Pozzani U C, C P Carpenter, P E Palm, C S Weil and J H Nair. J Occup. Med. **1** 634 (1959).
Rose M E. Brit. J. Ind. Med. **28** 22 (1971).
Rubin H H, A J Arieff and F W Tauber. Arch. Ind. Hyg. & Occup. Med. **2** 529 (1950).
Sayers R R and Yant W P. US Bur. Mines Rept. Invest. No. 2476 (1923).
Schuler P, H Oyanguren, V Maturana, A Valenzuela, E Cruz, V Plaza, E Schmidt and R Haddad. Ind. Med. & Surg. **26** 167 (1957).
Seaton A. Thorax **38** 241 (1983).
Seguin P, A Allard, A Carter and J L Malo. J. Occup. Med. **29** 340 (1987).
Selikoff I J, H Seidman and E C Hammond. J. Natl. Cancer Inst. **65**(3) 507 (1980).
Singh J B. Appl. Ind. Hyg. **3** 58 (1988).
Sluis-Cremer G K and B N Bezuidenhout. Brit. J. Ind. Med. **46** 537 (1989).
Sorahan T, H G Parkes, C A Veys and J A H Waterhouse. Brit. J. Ind. Med. **43** 363 (1986).
Southam A H and S R Wilson. Brit. Med. J. **2** 971 (1922).
Steinberg H H, S C Massari, A C Miner and R Rink. J. Ind. Hyg. & Tox. **24** 183 (1942).
Stewart R D, J E Peterson, P E Newton, C L Hake, M J Hosko, A J Lebrun and G M Lawton. Tox. Appl. Pharm. **30** 377 (1974).
Stokinger H E. Criteria and procedures for assuming the toxic responses to industrial chemicals. In: Permissible levels of toxic substances in the working environment. International Labour Office, Geneva, Switzerland (1970).
Stokinger H E. Routes of entry and modes of action. In: Occupational diseases—a guide to their recognition (Edited by M M Key). US Government Printing Office, Washington DC, USA (1977).
Sunderman F W, F D Weidman and O V Batson. J. Ind. Hyg. & Tox. **27** 241 (1945).
Theriault G P, W A Burgess, L J Di Berardinis and J M Peters. Arch. Environ. Health. **28** 12 (1974).
Theriault G P, J M Peters and W M Johnson. Arch. Environ. Health **28** 23 (1974).
Timbrell V. Inhalation and biological effects of asbestos. In: Assessment of airborne particles (Edited by T T Mercer, P E Morrow and W Stober). C C Thomas Publisher, Springfield, Illinois, USA (1972).
Trainor D C and R C Jones. Arch. Environ. Health **12** 231 (1966).
Tsuchiya K and S Harashima. Brit. J. Ind. Med. **22** 181 (1965).
Vigdortschik NA, E C Andreeva, I Z Matussevitsch, M M Nikulina, L M Frumina and V A Striter. J. Ind. Hyg. & Tox. **19** 469 (1937).
Wagner J C, J W Skidmore, R J Hill and D M Griffiths. Brit. J. Cancer **51** 727 (1985).
Wagner J C, C A Sleggs and P Marchand. Brit. J. Ind. Med. **17** 260 (1960).
Weir F W, D H Stevens and P A Bromberg. Tox. Appl. Pharm. **22** 319 (1972).
Wegman D H, L D Pagnotto, L J Fine and J M Peters. J. Occup. Med. **16** 258 (1974).

Bibliography

Adams R M. Occupational contact dermatitis. J. B. Lippincott Co., Philadelphia, USA (1984).

Agricola G. De Re Metallica, Basel (1556). Transl. by H C Hoover and L H Hoover, The Mining Magazine, London (1912). Dover Publications, New York (1974).

Bouhuys A. Breathing, physiology, environment and lung disease. Grune & Stratton, New York (1974).

Burton A C. Physiology and biophysics of the circulation. Year Book Medical Publishers Inc., Chicago, Il., USA (1968).

Collis E L. Industrial pneumoconioses, with special reference to dust phthisis, Milroy Lectures, 1915. H M Stationery Office, London (1919).

Cotes J E, I Steel and G L Leathart. Work-related lung disorders. Blackwell Scientific Publications, Oxford, UK (1987).

Drinker C K. Carbon monoxide asphyxia. Oxford University Press, New York (1938).

Drinker P and T Hatch. Industrial dust. McGraw-Hill Book Company, Inc., New York (1954).

Flury F and F Zernik. Schadliche Gase. Springer, Berlin (1931).

Griffiths W A D and D J Wilkinson. (Editors). Essentials of industrial dermatology. Blackwell Scientific Publications, Oxford, UK (1985).

Gross E. Berufskrebs. Bericht uber die fruhere Kommission fur Berrufskrebs der Deutschen Forschungsgemeinschaft. Boppard: Boldt, Germany (1967).

Hamilton A and H I H Hardy. Industrial toxicology. Publishing Sciences Group, Acton (1974).

Hudson T G F. Vanadium, toxicology and biological significance. Elsevier, Amsterdam (1964).

Hunter D. The diseases of occupation. Little Brown and Co., Boston, USA (1974).

Jager K W. Aldrin, dieldrin, endrin and telodrin. Elsevier Publishing Company, Amsterdam/London/New York (1970).

Largent E J. Fluorosis. Ohio State University Press, Columbus, USA (1961).

Latham R E (Editor). The travels of Marco Polo. Penguin Books, London (1958).

Montcastle V B. Medical physiology (Editor). C V Mosby Co., St. Louis, MO, USA (1976).

Nagaishi C, N Nagasawa, M Yamashita, Y Okada and N Inaba. Functional anatomy and histology of the lung. Igaku Shoin Ltd, Tokyo (1972).

NIOSH. Registry of toxic effects of chemical substances. US Government Printing Office, Washington 20402 USA (latest edition).

Parkes W R. Occupational lung disorders. Butterworths, London (1982).

Pepys J. Hypersensitivity diseases of the lungs due to fungi and organic dusts. Karger, Basle, Switzerland (1969).

Pott P. Chirurgical observations relative to cancer of the scrotum. Hawes, London (1775).

Proctor N H, J P Hughes and M L Fischman. Chemical hazards of the workplace. J B Lippincott Company, Philadelphia, USA (1988).

Ramazzini B. Diseases of workers (1713). (English Edition). New York Academy of Medicine. Hafner, New York (1964).

Rosen G. The history of miners' diseases. Schuman's, New York (1943).

Rosen G. The history of public health. M. D. Publications, New York (1958).

Sappington C O. Essentials of industrial health. J B Lippincott Co., Philadelphia, USA (1943).

Thackrah C T. The effects of arts, trades and professions and of civic states and habits of living on health and longevity (& others), 2nd Edition. Longman, London (1832).

Vedder E B. The medical aspects of chemical warfare. Williams & Wilkins, Baltimore, USA (1925).

West J B. Respiratory physiology—the essentials. The Williams & Wilkins Co., Baltimore, USA (1974).

Waldron H A. Lecture notes on occupational medicine. Blackwell Scientific Publications, Oxford, UK (1979).

CHAPTER 3

Understanding Thresholds

In the 16th century Paracelsus published the truism which, in translation, reads 'All things are poisons for there is nothing without poisonous qualities; it is the dose which makes a thing nonpoisonous' (Paracelsus, 1538). Today the issue is much the same, although it is nowadays taken up concerning the notion of 'threshold'. Everything, some proclaim, has a threshold concentration, save for carcinogens.

The modern use of the term 'threshold' in relation to exposure limits probably stems from its use in physiology to describe the intensity of a stimulus causing the least perceptible sensation, be it odour, touch, sight, noise or pain. Use of the terms 'threshold concentration' and 'threshold dose' of substances other than, perhaps, carcinogens creates the impression that there must be a certain concentration or dose of these substances below which there is no effect on the human body whatsoever and above which there is a prompt and adverse reaction. The basic concept, that there is a level of exposure below which there is no health risk to anyone, has particular allure, since by keeping below this level one could avoid all argument over just what level of risk would be tolerable. In point of fact the body's reaction to increasing exposure to non-carcinogens is probably more remarkable for its imperceptible gradation of response from near-zero exposure upwards, whatever the kind of response. By contrast, with carcinogens it is often held that there is no hope of handling them with complete safety; the health risk does not have a threshold, or if it does it is at zero concentration and zero dose. In reality there is growing evidence that there are thresholds with at least some carcinogens in much the same sense as there are with non-carcinogens.

The issue of thresholds is clarified by distinguishing between the reaction of one particular person to a dose of toxic substance and the reactions of members of a large group or population. Some individuals may be susceptible to a particular substance and others seemingly very resistant (Festing, 1987). Next, it helps to make a further distinction where possible between, on the one hand, the accumulation of body burden, that is, more generally, the toxico-kinetics of hazardous substances and on the other hand the reaction of the body to that accumulation, that is, the toxico-dynamics of the matter. With some substances the accumulated body burden at any instant may largely be the result of exposure over the previous few minutes, whereas with others it may be longer, possibly the result of a lifetime of exposure. The way the body burden may accumulate in various organs when under exposure to a varying concentration of air contamination can be studied in simplified models which have similar accumulation and elimination properties. Consideration of these developments in toxico-kinetics will be taken up in the next chapter.

The presence of a substance in a particular organ does not necessarily imply injury at that site. A significant quantity of lead in the bones is not especially harmful to the bones whereas, in contrast, even a relatively small quantity of cadmium in kidney may cause adverse renal effects. The reaction of the body to the accumulated burden of hazardous substance may be immediate, or may take hours, months or many years to express itself. The reaction, once started, may be rapid or slow, steadily increasing or decreasing. Study of these reaction rate processes is firmly in the realm of toxico-dynamics. There is an infinite variety of kinds of reaction within the body, as well as non-linear rate processes under way. The reactions may be broadly classified:

(1) intrinsically progressive; that is, the reactions progress even in the absence of further exposure,

(2) stable; that is, once established the reactions neither progress nor regress in the absence of further exposure,

(3) intrinsically regressive; that is, in the absence of or on reduction of exposure the body returns to normal.

Aside from frank occupational disease and overlaying these rate problems there is genuine cause for disagreement about how much and what kind of reaction should be cause for concern. The kind of reaction and least amount of reaction which constitute an adverse effect are largely a matter of subjective judgement. The concern aroused by a particular adverse effect is influenced by three main considerations:

(1) The nature of the effect.
A substance which produces long lasting impairment of health, for example, is of greater concern than one which is merely malodorous or mildly irritating.

(2) The extent of the effect.
A severe effect on a whole organ system causes more concern than a mild, local effect.

(3) The likelihood of the effect being progressive.
An adverse effect which gets progressively worse without further exposure is of more concern than one which is stable or temporary.

An assessment of the hazard of exposure includes due account being taken of all the above aspects, insofar as they are known or suspected.

Reaction within the body of an individual employee

Gases, vapours and aerosols differ widely in their inherent physical and chemical nature and this influences the amount and kind of injury which occurs when they are inhaled. There are some gases and dusts which are regarded as being very nearly inert in terms of the tissue reaction to their presence, as for example nitrogen, carbon and many silicates. Some dust particles, when deposited in the lungs, are engulfed in macrophages and ultimately come to reside in the tissues or in lymph nodes where reaction is either non-existent or at most is a mild, foreign body, inflammatory process. It is tempting to regard such particles as being inert. However, under unusual circumstances of exceedingly high concentration, as for example nitrogen under several atmospheres of pressure or carbon dust in extraordinarily excessive amounts a cell reaction of greater significance may occur. Moreover, both nitrogen and carbon dust in sufficiently high concentration are asphyxiants. In contrast there are gases such as phosgene and dusts such as quartz which, because of their inherent nature, are biologically very active and can give rise to a biological reaction of important magnitude when inhaled in even the lowest concentration.

Clinical symptoms are experienced late in the disease process. The first cellular reactions occur a great deal earlier. The conceptual framework for relating effects to dose is pictured in Figure 3.1. On the basis of practical evidence it is generally held that for any hazardous substance there is a finite dose, albeit small, that can be tolerated by the body without producing any overt evidence of tissue reaction. In terms of an important reaction which threatens health this is undoubtedly the case. However, cellular reactions which do not threaten health in any way could also occur, such as macrophage accumulation in the lung or subtle changes only made visible under the electron microscope. There may be minor disturbances of cell structure or function which are completely reversible and do not shorten the life of the cell. Such changes permit the system to experience a transient perturbation while it

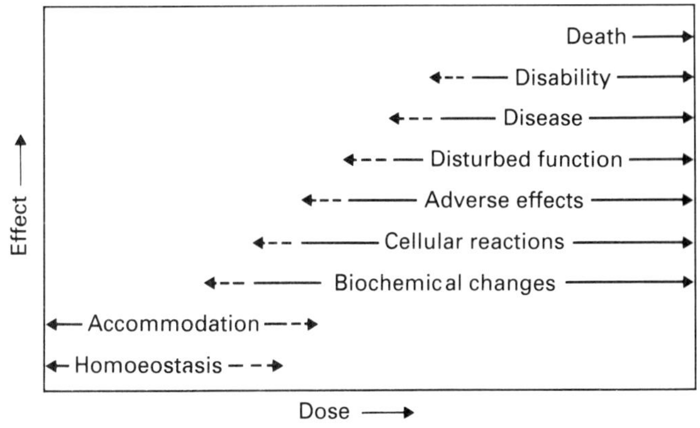

FIG. 3.1. Progressive effects of increasing dose of a hazardous substance.

carries out the process of accommodation, a response to the demands imposed by a dynamic environment (Dinman, 1984).

It is also possible that a cell reaction to the substance may be an appropriate one and quite properly considered to be, by itself, a normal reaction without any adverse connotation. Premature death of a cell and its replacement by a new cell is a normal body mechanism for tolerating exogenous agents and is not necessarily of itself threatening to health. In the same sense, phagocytic action, including storage of inert particles therein, is a normal body function and can scarcely be considered an injury (Wright, 1973).

Threshold of disease

The term 'threshold dose' may have any of several different definitions. It is generally held that for every substance there is some small dose which may be completely accommodated. The substance has mass and occupies volume but otherwise at this dose it is without any effect whatsoever on the body. Next there are small changes to cells and organ systems which are homoeostatic, completely reversible and below the level which requires adaptation. At some higher but still small dose the reaction of the body may be one of adaptation but no more, and totally within the adaptive capacity of the body. The reserve of the body to adapt to a further dose of the substance may be diminished but there is no effect other than to reduce this reserve adaptive capacity. The highest of these doses is a kind of threshold. But it is still not the threshold of disease. Even so, a sufficient presence of a substance in the body serves at least to reduce the reserve adaptive capacity with respect to the substance and this reduction of reserve adaptive capacity is considered by some to be an adverse effect although not necessarily constituting disease.

The concept of disease in society has quantitative aspects when viewed against the local socio-cultural background. There must be a dose which may cause small but recognisable alterations to cell structure and function without being quite great enough to be regarded as disease by society (Hatch, 1962). The highest of such doses causes a reaction which is at the threshold of disease for the individual. A scar, for example, is an effect which is undoubtedly real but may be minor in extent and in no way interfere with normal function nor shorten the life of the person. Fibrosis in the lungs may cause a reduction in respiratory reserve below the level of detectability by the individual concerned but detectable by special machines for measuring respiratory ability. This is not to deny that progressive increments in scar tissue and disability could be the harbinger of disease, sickness and possibly even premature death.

Another good example but of a different kind is exposure to inorganic lead, the consequences of which are known in great detail. With increasing exposure the first evidence of lead absorption is a raised concentration of lead in the blood and in the urine. In this first stage, however, no recognisable biological change is detected. The mere presence of a substance in the body is not an adverse health effect. The second stage is a biochemical change in the enzyme system for porphyrins; this might be regarded as one of the earliest biological changes produced by lead but not one which, by itself, many authorities believe indicates injury nor, for that matter, impending loss of health. The next stage, basophilic stippled red cells, typically follows the increased coproporphyrin excretion, while lead anaemia is likely to appear if the exposure increases even further (Lane, 1931). This may be followed, in the next stage, by various symptoms, indicative of frank lead poisoning, typically including colic and palsy. The stages do not necessarily follow the same sequence in every single case and at any given time there is often more than one stage present (Tsuchiya and Harashima, 1965). The definition of just which of these stages or combination shall constitute an adverse effect or, for that matter, a disease from lead is an arbitrary decision and hence inherently contentious as with hazardous substances generally.

Thresholds of adverse effect

Study of the adverse effects of chemical agents on biological systems is embraced within the science of toxicology (Klaassen, Amdur and Doull, 1986). It takes different threshold doses of a toxic substance to cause different kinds of effect in one particular organ or system of the body. Furthermore, in different organs or systems it takes different doses to cause the same minimal adverse effects. Consequently, a threshold dose with respect to one organ may be quite different from the threshold dose with respect to another.

A World Health Organisation Study Group considered the following types of effect as adverse:

—effects that indicate early stages of clinical disease;

—effects that are not readily reversible, and indicate a decrease in the body's ability to maintain homoeostasis;

—effects that enhance the susceptibility of the individual to deleterious effects of other environmental influences;

—effects that cause relevant measurements to be outside the 'normal' range, if they are considered as an early indication of decreased functional capacity;

—effects that indicate important metabolic and biochemical changes

(World Health Organisation, 1980).

For our purposes, from the threshold dose upwards, however low that may be defined, all adverse effects of a meaningful sort increase as dose of toxic substance is increased. This appears to be the case, in the minds of most if not all students of the problem, even with respect to carcinogenic and teratogenic effects (Harbison and Baker, 1969).

Dose is, by definition, an amount of administered substance; concentration of hazardous substance in inhaled air is in its essence an implied time rate of dose. Such a concentration lasting for zero time is of no medical consequence whatsoever. It is concentration of hazardous substance in air coupled with a significant period of exposure to it which causes the hazard to health. The

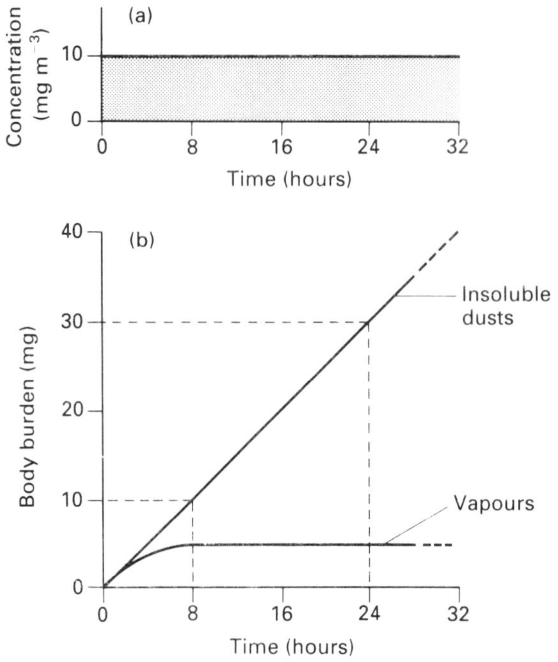

FIG. 3.2. Accumulation of body burden on continuous exposure to air contamination for 24 hours per day. (a) Constant concentration of air contamination. (b) Corresponding accumulation of body burden.

period or periods of exposure have to be specified as well as the level of concentration of hazardous substance in air.

There is a small, finite, steady concentration of a hazardous substance which could conceivably persist in inhaled air for a lifetime and cause no adverse effect on the individual. There is a corresponding finite body burden of the hazardous substance and its metabolites. The body burden of substances which accumulate steadily, day after day, such as insoluble dusts, continues to increase until exposure ceases. Gases and vapours which are absorbed in the blood are exhaled in increasing amounts as the concentration in the blood builds up until there is an equilibrium between that inhaled and that exhaled (Figure 3.2). There exists a range of constant concentrations from zero upwards, any one of these producing no adverse effects from continual exposure, 24 hours per day for a lifetime. The highest of these concentrations is a threshold concentration. It might be appropriate for use in laboratory studies of 24 hours per day exposures or in studies of community air pollution, but it is not commonly used in occupational hygiene. There also exists a range of constant concentrations of a hazardous substance in air exposure to which for just 8 hours each day, 5 days each week and 50 weeks each year for a working life time produces no adverse effects in the individual. The highest of these is a threshold concentration which might be called the threshold 8 hours per day steady concentration.

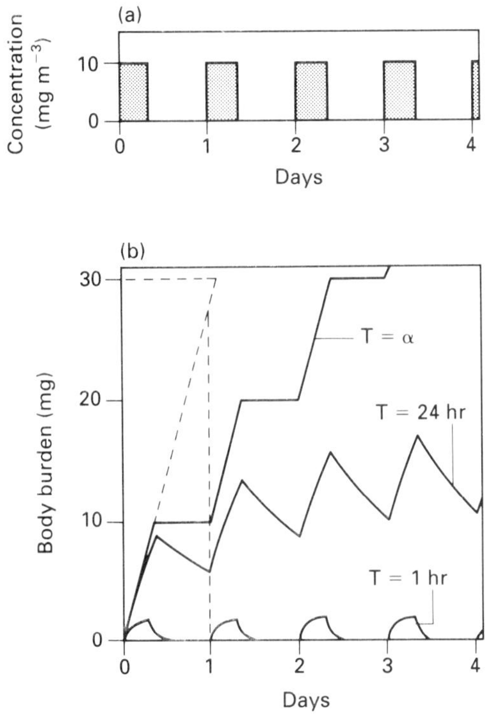

FIG. 3.3. Accumulation of body burden on exposure to air contamination for 8 hours per day. (a) Constant concentration of air contamination for 8 hours per day. (b) Corresponding accumulation of body burden.

With many gases and vapours that rapidly accumulate to the equilibrium body burden, a threshold for exposure 8 hours each day to a steady concentration would differ little from the threshold for exposure twenty four hours each day. This is illustrated by the lowest curve in Figure 3.3, which shows how the body burden levels off before the end of each shift. On the other hand, with many dusts which slowly accumulate, the body burden from exposure 24 hours per day may be up to three times the body burden from exposure 8 hours per day, and the threshold concentration for exposure 24 hours per day is correspondingly lower.

Suppose now that the individual was exposed to the air contamination for only 8 hours each day but that within the 8 hours its concentration in air is constant for a short time at a value 50% higher than the average and then changes to a value 50% lower than the average, and this pattern is repeated indefinitely. There exists a maximum 8-hours time weighted average concentration in inhaled air with such a variation about the mean which the individual can work in each day without being adversely affected and spend the remaining time in an atmosphere free of contamination. This is yet another threshold concentration. For the substances with a very short biological half time this third threshold concentration may be lower than the other two by up to 50% of its value, since the body burden closely follows the momentary concen-

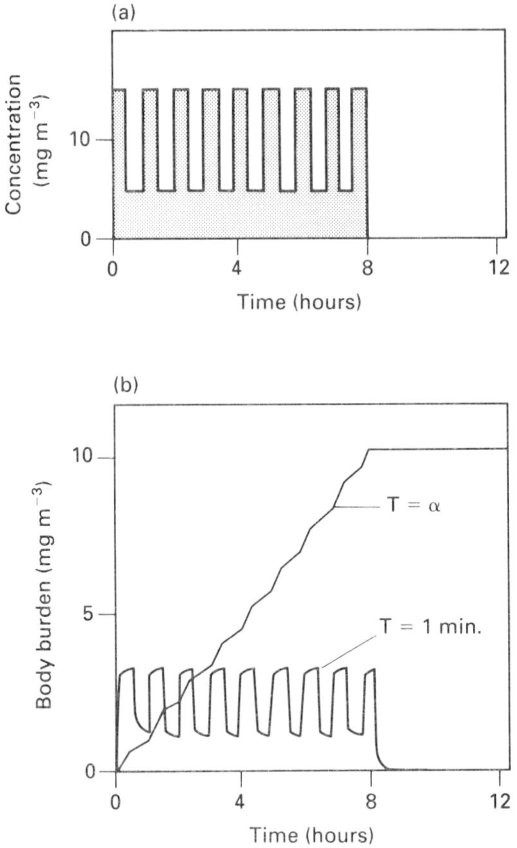

FIG. 3.4. Accumulation of body burden on exposure to air contamination for 8 hours per day. (a) Square wave variation about the mean concentration of air contamination for 8 hours per day. (b) Corresponding accumulation of body burden.

tration. But for the substances with a long biological half time this third threshold concentration will be the same as the second threshold concentration and three times the first threshold concentration, since the accumulated body burden depends merely on the total quantity inhaled per day (Figure 3.4).

An individual's biological response to the accumulated body burden of a particular foreign substance can be expressed quantitatively by a line (Figure 3.5). There is a body burden which produces no response whatsoever. This is marked by the threshold A. The position of threshold A is somewhere along the horizontal axis. In theory, at least, for some substances threshold A may be at zero body burden. For others the position of threshold A depends on the precise definition of the least 'response'. In any event if the individual accumulates progressively higher body burden the response progresses correspondingly. The least positive response may be real enough but merely undesirable. A higher response may have mild sensory manifestations. A still

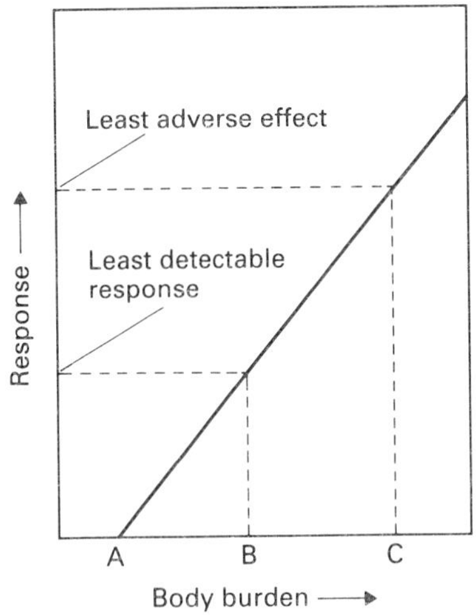

FIG. 3.5. Response to body burden in the individual.

higher response might be associated with observable physiological changes and overt clinical effects. The highest responses of all would perhaps be signified by loss of consciousness, or even death.

The degree of response in an individual may be conceived to be measurable objectively on a continuous scale starting at zero. The response in an individual is sometimes scaled in a quantitative or at least semi-quantitative fashion. In the case of a respiratory disease, for example, this might be degree of irritation, increasing symptoms of chronic bronchitis, a physiological test of lung function or the appearance of opacities in a chest radiograph. The body burden/response curve may be naturally linear or the scales of body burden and response could be chosen to make the relationship between them linear, as in Figure 3.5. The body burden/response relationship could then be described by the slope of the line and the point where the line strikes the body burden axis, point A. The slope of the line expresses the rapidity with which unit increases in body burden cause increases in response in this individual. The steeper the slope, for example, the greater the safety factor necessary in practice to allow for the inevitable occasional excursions of exposure above the maximum desirable.

Body burdens which cause zero response can never, in point of fact, be known with certainty since the sensitivity of measurements of response is limited (Hatch, 1972). The least response which can just be detected in an individual corresponds to another and higher body burden, shown as threshold B on Figure 3.5. Since threshold A occurs at zero response, that is a response below the level of detection, the location of A on the horizontal axis can only be inferred by extrapolation beyond the range of possible experimental observations. It is essentially *hypothetical* in nature. To suffer a response which is so small as to be incapable of being detected is hardly likely to be unacceptable and threshold B is of more practical interest. There may be a detectable response but this response may still be so mild in itself as to be considered insignificant, immaterial or not sufficient to be called an 'adverse' effect or 'disease'. This response is a higher one, shown as threshold C on Figure 3.5.

Threshold C is the one which has played the leading role in the history of exposure limits. Most published data on human health experience of individuals exposed to hazardous substances is in the form of a diagnosis of the presence or absence of a defined adverse effect. Diagnosis of the presence or absence of a particular effect reduces potentially quantitative effect data into a quantal, all-or-none type. This is not necessarily a disadvantage but it is important to appreciate its limitations.

Reactions in a defined population of employees

The reaction of primary interest to the occupational hygienist intent on prevention is the threshold of adverse effect. The two different aspects to consider are firstly the threshold for this reaction in a single identified individual and secondly the amount the threshold may differ from one individual to another within a defined population. In any real population the threshold body burden for an adverse effect to occur will vary from one individual to another about an average value, as in other medical norms. Biological variation in susceptibility to exposure to hazardous substances is in reality quite large.

There is ample evidence of considerable variation from individual to individual in their activity, metabolism, cellular responses to a foreign substance and in their organ responses. More serious tissue injury may develop in one individual than another even though the dose administered to both individuals is the same. A natural variability exists among any group of normal, healthy individuals in their response to a given magnitude of exposure to a particular agent. The fact that employees are not equally affected by exposure to hazardous substances is realised only with difficulty by the average layman. Yet everyday experience with infections, to say nothing of differences in strength, skill and intelligence is ample evidence of innate differences between people.

The response of each individual in a given population will be characteristic of that particular individual; each will have a slightly different threshold body burden and a different slope to the line relating amount of response to body burden (Figure 3.6). A particular population of employees under study would have to be represented by a family of lines centred around the line for the average employee.

It is to be expected that the frequency distribution of threshold body burdens would be a hump-shaped curve tailing away on either side of the central value, as shown in Figure 3.7 (Koch, 1966). It is common to find that the distribution of biological tolerances is approximately lognormal over the observed range.

The importance of the lognormal distribution in statistical analysis has been established for a century or more. It fits the tolerance distribution of humans and experimental animals in the centre part of its range. It is also used in

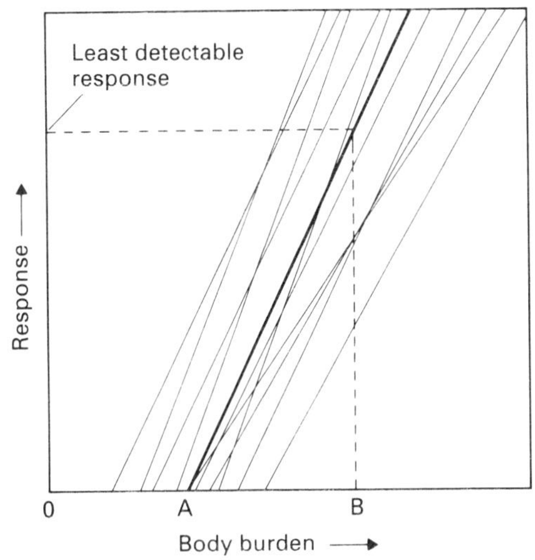

FIG. 3.6. Response to body burden in a group of individuals. Each employee has a different threshold adverse effect and a different slope to the curve relating quantity of response to body burden. A population of employees under study would be represented by a family of body burden vs response lines, centred about the curve for the average employee.

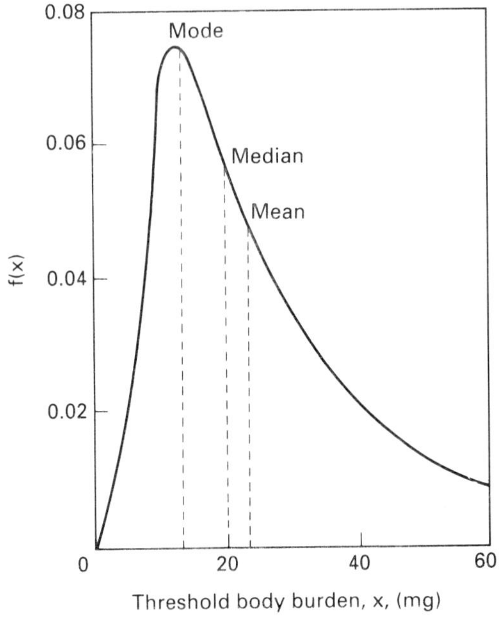

FIG. 3.7. Lognormal frequency distribution of thresholds.

describing temporal variation of the concentration of air contamination by hazardous substances at work and for describing particle size distribution of aerosols in industry. The two principle parameters of a simple lognormal distribution are the geometric mean and geometric deviation, corresponding to the mean and standard deviation of a Normal distribution.

In 1879 Galton showed that in certain cases the geometric mean is to be preferred to the arithmetic mean as a measure of the location of a distribution (Galton, 1879). In point of fact Galton derived his ideas from a consideration of the Weber-Fechner law relating biological response to the magnitude of the stimulus. This law asserts that the response is proportional to the logarithm of the stimulus (Weber, 1834; Fechner, 1897).

Besides the effect of varying tolerance the relationship between exposure and body burden will also vary to some extent from one individual to another because of differences in energy expenditure, biological half time and so on. Consequently the threshold exposure will vary more from person to person than will the threshold body burden.

In an industry study, once 'adverse effect' has been defined, it may be established whether an employee has or has not been adversely affected by the exposure. The apparent variation in exposure threshold is enhanced by the inevitable experimental errors in determining exposure and response in the field. For the purpose of the study the response is quantal, that is, all-or-none. Exactly when the employee first developed the characteristic response is usually not known. What is known is the proportion of employees with a certain measured exposure who appear to exhibit the response. A graph showing exposure against the proportion of employees adversely affected by it would be a cumulative sigmoid curve, illustrated in Figure 3.8. In a finite population the curve would not be smooth and there would be a specific individual who has the least exposure before responding and, at the other extreme, some other individual who has the greatest exposure before responding. The larger the population the further apart would the minimum and maximum exposure to produce a response be expected to be. In an infinite

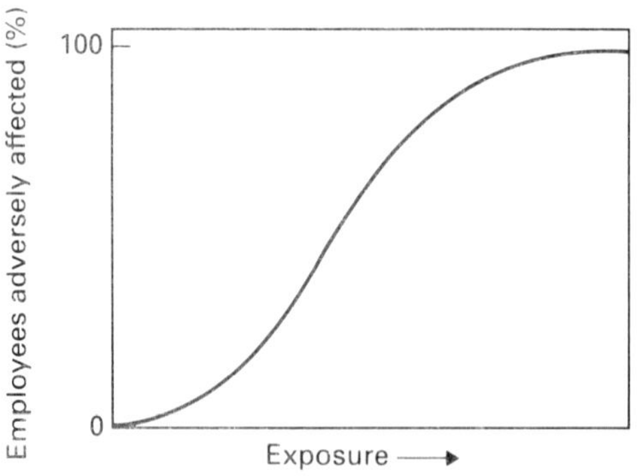

FIG. 3.8. Relationship between employees adversely affected and their exposure.

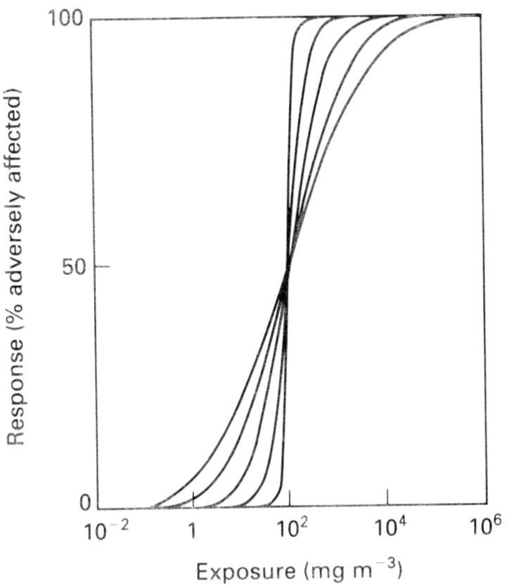

FIG. 3.9. Exposure vs response curves based upon lognormal distributions of tolerance—hypothetical data.

population the exposure-response curve would be smooth and would be expected to tail away down to zero response at zero exposure whatever the slope of the line relating body burden with response on one individual and wherever such lines intercept the axes. If the lognormal tolerance distribution holds true down to zero exposure this would also mean that complete absence of risk for all individuals could never be guaranteed. Lognormal curves of exposure versus response are shown in Figure 3.9 to illustrate the case when exposure is plotted on a logarithmic scale. The same curves are shown in Figure 3.10 when plotting on an arithmetic scale, to illustrate the distortion this causes.

A disease such as pneumoconiosis has a recognised pattern of increasing response indicated by progressively increasing size and multiplicity of opacities on the chest radiograph, decreasing lung function measured in a variety of ways and increasing clinical symptoms. Each level of response can be examined for the whole population of exposed employees (Hatch, 1968). An example of this is taken from a survey at a tin smelter, which included detailed measurements of dust exposure to respirable tin oxide dust conducted by the author, occupational histories of employees to establish their years of exposure and, amongst other tests, readings of chest radiographs of the employees. Exposure–response curves are plotted in Figure 3.11 for the population of employees at the smelter exposed to tin oxide. The semi-quantitative response which was thereby examined was the grading of radiographs according to the increasing size and multiplicity of small opacities visible on the radiographs: Categories I, II and III. The fitted curves were parallel in the sense that they were straight and parallel when plotted on logarithmic probability paper. They must be parallel since, as a little thought will show, they cannot truly cross. Tin oxide is highly

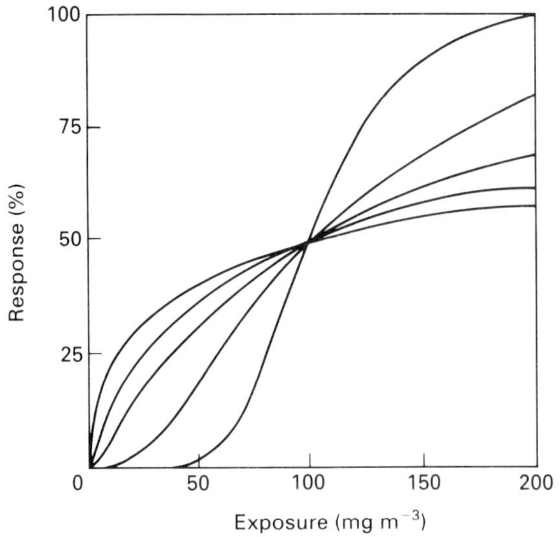

FIG. 3.10. Exposure vs response curves based upon lognormal distributions of tolerance—same data as Figure 3.9. Exposure plotted on arithmetic scale.

radio-opaque, as expected from the high atomic weight of tin, consequently the results are less alarming than might be first thought. The amount of respiratory disability was minimal.

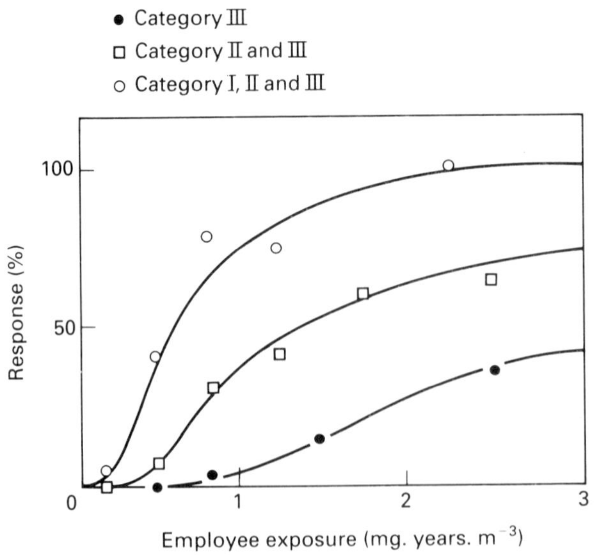

FIG. 3.11. Exposure vs response to respirable tin oxide dust. Exposure = [Mean dust concentration of employees in common job-exposure group] × [Years of exposure]. Response = Percent of employees with at least Category I, II or III chest radiograph.

The conclusions from these considerations of response in a population are twofold and far reaching:

(1) For each and every human there is a determinate threshold concentration, albeit ever so small, for every substance and adverse effect.

(2) Due to natural variation between individuals and inevitable measurement errors, amongst any large population the lowest threshold for any substance or adverse effect is inherently indeterminate. There is no non-zero level of concentration which can be guaranteed safe for everyone, whatever the substance or adverse effect.

Common pitfalls

There are, unfortunately, many traps and pitfalls for the unwary in studying biological data, whether it be human experience in industry or animal experiments in the laboratory.

Association does not prove cause

There are innumerable examples of a medical condition being wrongly attributed to exposure to a hazardous substance because they were associated in a few cases. Epidemiological support should be sought. A thorough listing of pertinent questions has been given by Bradford Hill (1965) and elaborated by Doll in connection with suspect carcinogenicity (1985).

(1) Strength of association—is the disease more common in a particular group of workers? If so, by how much?

(2) Consistency—has the association been described by more than one researcher in a different locality and preferably using different methods of enquiry?

(3) Specificity—is the disease isolated to certain groups of employees and to certain sites? The materials to which these employees are exposed and/or the intensity of exposure may be unique to those employees.

(4) Time—does the suspect cause (excessive exposure to some substance or combination) always precede the disease and is the time interval constant?

(5) Biological gradient—is there a good dose/response relationship? To answer this question it is necessary that dose or exposure and response both be recorded.

(6) Biological plausibility—does the association seem reasonable or is it absurd?

(7) Coherence—do all aspects of the causality hang together in a logical and feasible way?

(8) Experimental evidence—can the causality be tested experimentally? The best test is to remove the suspect cause and study the outcome.

(9) Analogy—has a similar suspect cause been shown for related causes or effects?

A positive answer to every one of the nine questions is not essential, but the more positive answers there are the more likely it is that there is a causal relationship (Harrington, 1980). In particular, where the disease is severe and irreversible, as it is with most forms of occupational cancer, for example, even weak association may be sufficient to justify prompt action (Doll, 1984).

Present day cases and present day conditions

Irritants produce their effects during exposure or shortly thereafter but other substances produce adverse effects which advance imperceptibly day by day over many years. Thus present day cases, in the latter instance, may have been caused by conditions which no longer exist.

Non-occupational causes

Occupational diseases such as dermatitis and asthma have a range of causes only some of which are related to exposure to hazardous substances. Other diseases such as lead poisoning and asbestosis may be caused by exposure to the substance away from work. These possibilities make attribution difficult because an isolated case of a common occupational disease cannot necessarily be put down to work. Association does not prove cause.

Graphic extrapolation

It is important to note that the exposure response curves in Figure 3.10 are almost straight over a substantial range of exposures. This arises from the underlying lognormal distribution of tolerances. This phenomenon can lead to mistaken interpretation of exposure data. Thus, after having fitted a straight line to a limited range of data it is tempting to extrapolate the line to a 'no effect' level or through zero (Figure 3.12). The solid lines shown are, in point of fact, the relatively straight parts of lognormal distributions shown by the broken lines.

Pitfalls inherent in animal models may be no less a threat to sound conclusions than is the association-does-not-prove-cause trap of human epidemiology. Possible pitfalls lie both in the design of experiments and in interpretation of their results (Smyth, 1984). Results from random samples from a lognormal distribution are plotted in Figure 3.13. They show the response to exposure from large sample sizes of 100 animals drawn at random from a population with median 10 g per kg body weight and geometric standard deviation 2.0. Many would conclude from such data that there is a 'no effect' level at 2.0–2.5 g per kg body weight. Only if sufficiently large samples had been taken over a larger range of dose would this have been shown not to be the case.

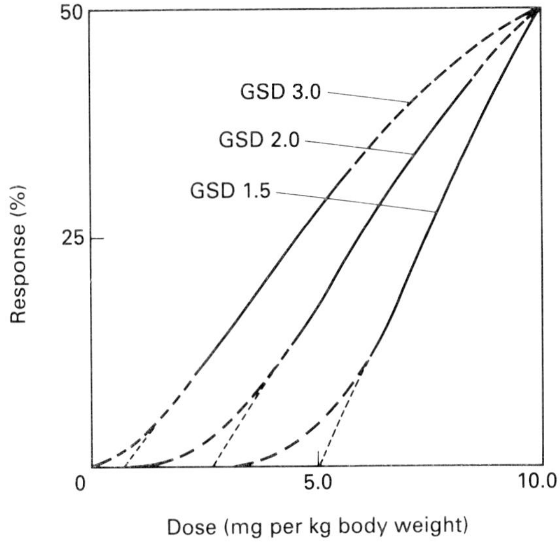

FIG. 3.12. Graphic extrapolation of data from lognormal distributions.

Zero risk

An exposure limit is still associated with a small but non-zero probability of employees being affected by the substance in question. Zero risk is unattainable unless the substance is banned or kept in hermetically sealed containers at all times. The risk of employees being affected can be made as small as technical

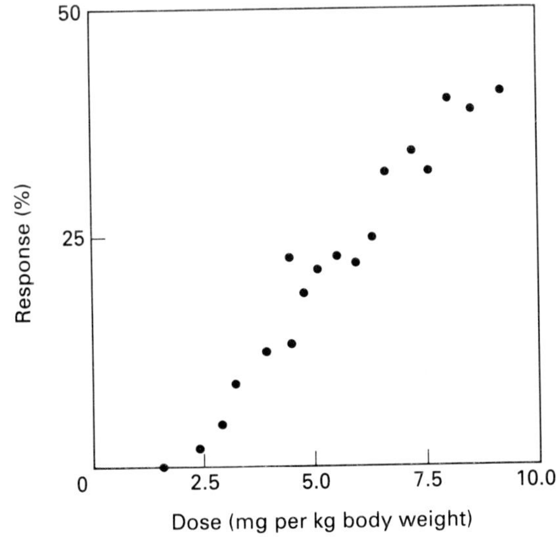

FIG. 3.13. Response in random samples of 100. Lognormal distribution, having a median of 10 mg per kg body weight and geometric standard deviation of 2.0.

feasibility will permit. A judgement is made explicitly or implicitly about the kind and greatest magnitude of risk which will be tolerated in the conditions of use. This may be as high as 10^{-1} for mild irritation at work to as low as 10^{-8} lifetime risk for carcinogenic risks in the general population (Mantel and Bryan, 1961). Zero risk remains the ultimate aim. A practical exposure limit becomes a compromise between costs of one sort or another, not necessarily financial; the costs of bearing a health risk and the costs of engineering and personnel control (Roach, 1970).

Thresholds of cancer induction

There is continuing debate about thresholds of cancer induction by hazardous substances. Studies on ionising radiations are indirectly instructive about carcinogenic substances. They show that more is known about the potential of ionising radiations to cause cancer than about all other environmental carcinogens. There is a wealth of epidemiological information about irradiated human populations, spanning from the survivors of the atom bombs dropped at Hiroshima and Nagasaki, the Chernobyl accident, to innumerable patients exposed to medical X-rays, as well as an extensive literature about radiation on experimental animals and living cells. Yet the cancer risks from low doses of radiation remain a subject of great dispute.

There are two fundamental difficulties. First, it is not possible to obtain clear, direct evidence of cancer caused by radiation at the very low levels characteristic of current medical practice or emanating from nuclear power reactors. Compared with the natural or spontaneous levels in the human the additional incidence of cancer induced by radiation acceptable in the general population is very small. So small that the sample sizes that would need to be studied to obtain statistically accurate information are impractically large (Hall, 1984). Consequently, the only practical approach is to observe much smaller groups of people who have been exposed to far higher levels of radiation: patients treated with medical X-rays or the Japanese atom bomb survivors, for example, and extrapolate downwards (Los Alamos Scientific Laboratory, 1950). But then risk analysis encounters the second fundamental difficulty: extrapolation from high doses well above the contemplated use levels down to the very lowest doses of radiation.

Many ingenious statistical distribution models for low dose extrapolation of tolerance distributions have been devised in recent years, principally for radioactive substances, and all give rise to different answers. The variety of possible extrapolations is so extensive as to make the conclusions seem little more than a wild guess (FDA, 1971; Chand and Hoel, 1974). The most lasting of these models have been the one-hit model, the k-hit model and the Armitage-Doll multi-stage model (Crump, Hoel, Langley and Peto, 1976). The applicability of the results is still by no means automatic or immediate. It is tempting to draw a straight line through zero. However, as with non-carcinogenic hazardous substances the response to increasing dose of carcinogen in an individual is unlikely to be a linear one except over a very short range (Watanabe, Young and Gehring, 1977). Moreover, plausible mathematical models for cancer induction by ionising radiation and by radioactive substances cease to be so plausible when advanced as models for chemical carcinogens (US Office of Science and Technology Policy, 1985).

The very earliest reaction to a chemical carcinogen also merits mention. In theory, at least, since the cancer cell is self replicating the disease could be the result of a process which begins with a minute quantity of substance causing a change in only one molecule of DNA in a cell nucleus. But at least some carcinogenic substances can be taken in low dose without producing cancer, as is well known experimentally (Shimkin and Stoner, 1975). Small amounts of ingredients that are harmful in high doses are known to be present in many natural foods such as cabbage which contains goitrogens, roast meat which contains the carcinogen 3–4 benzopyrene, and potatoes which contain the mitotic poison solanine (Goldberg, 1967). Any of several mechanisms could prevent cancer developing from a sufficiently small dose of carcinogen:

(a) the substance may be metabolised before reaching the target (Stokinger, 1977, 1981),

(b) altered DNA may be repaired before the process becomes irreversible (Zapp, 1977; Williams and Weisburger, 1986),

(c) the cancer cell does not find conditions favourable for its development (Becker, 1982),

(d) the process may not reach fulfilment before the death of the individual from other causes (Yanysheva and Antomonov, 1976; Truhaut, 1980). One such other cause might be occupational disease other than cancer. An example of this is carbon tetrachloride, which is a liver carcinogen in rodents. However, that tumour only occurs at concentrations which result in substantial liver destruction, and may even be a consequence of that destruction. For the purpose of setting exposure limits the most relevant aspect of the long term toxicity of carbon tetrachloride may be the earlier effect, liver necrosis.

A different aspect of the same problem arises when extrapolating from animal studies. A massive dose may overwhelm mechanisms which might otherwise prevent cancer development. Thus, the ACGIH Threshold Limits Sub-Committee has recommended absolute upper limits to the animal dosages of carcinogens that are used for extrapolation to exposure limits for employees (American Conference of Governmental Industrial Hygienists, 1986). These limits are:

Respiratory route

1 g m^{-3} for the mouse and 2 g m^{-3} for the rat.

Dermal route

1.5 g per kg body weight for the mouse and 3 g per kg body weight for the rat.

Gastro-intestinal route

500 mg per kg body weight per day, equivalent to 10 g T.D. for the mouse and 100 g T.D. for the rat.

References

American Conference of Governmental Industrial Hygienists. Sub-committee on threshold limits. Threshold Limit Values and Biological Exposure Indices. American Conference of Governmental Industrial Hygienists, 6500 Glenway Ave., Bldg. D-7, Cincinnati, Ohio, USA (Latest edition).

Chand N and D G Hoel. A comparison of models for determining safe levels of environmental agents. In: Reliability and biometry statistical analysis of lifelength (Edited by F Proschan and R J Serfling). SIAM, Philadelphia, Pa., USA (1974).

Crump K S, D G Hoel, C H Langley and R Peto. Cancer Res. **36** 2973 (1976).

Dinman B D. The present state of occupational medicine. In: Occupational and industrial hygiene: Concepts and methods (Edited by N A Esmen and M A Mehlman), p. 165. Princeton Scientific Publishers Inc, Princeton, New Jersey 08540, USA (1984).

Doll R. Ann. Occup. Hyg. **28** 291 (1984).

Doll R. Int. J. Epidem. **14** 22 (1985).

FDA. Food and Drug Administration Advisory Committee on Protocols for Safety Evaluation—Panel on Carcinogenesis. On cancer testing in the safety evaluation of food additives and pesticides. Toxicol. Appl. Pharmacol. **20** 419 (1971).

Festing M F W. CRC Crit. Rev. Toxic. 18 1 (1987).

Galton F. Proc. Roy. Soc. **29** 365 (1879).

Goldberg L. J. Roy. College of Physicians of London 1 385 (1967).

Harbison R D and B A Baker. Teratology 2 305 (1969).

Harrington J M. Epidemiology. In: Occupational hygiene (Edited by H A Waldren and J M Harrington). Blackwell Scientific Publications, Oxford, London, Edinburgh, Boston, Melbourne (1980).

Hatch T F. Am. Ind. Hyg. Assoc. J. **23** 1 (1962).

Hatch T F. Arch. Environ. Health. **16** 214 (1968).

Hatch T F. J. Occup. Med. **14** 134 (1972).

Hill A Bradford. Proc. Roy. Soc. Med. **58** 295 (1965).

Koch A L. J, Theoret. Biol. 12 276 (1966).

Lane R E. J. Ind. Hyg. **13** 276 (1931).

Mantel N and W R Bryan. J. Nat. Cancer Inst. **27** 455 (1961).

Roach S A. Ann Occup. Hyg. **13** 7 (1970).

Shimkin M B and G D Stoner. Adv. Cancer Res. **21** 5 (1975).

Smyth H F. Ann. Am. Conf. Ind. Hyg., **9** (1984).

Stokinger H E. Occupational carcinogens. In: Patty's 'Industrial hygiene and toxicology'. 3rd Ed **2B**. Toxicology, Chapter 39, p. 2879. John Wiley & Sons, New York (1981).

Stokinger H E. The case for carcinogen TLV's continues strong. In: Workplace control of carcinogens—Proceedings of a topical symposium, p. 54. American Conference of Governmental Industrial Hygienists, Cincinnati, Ohio, USA (1977).

Truhaut R. Am. Ind. Hyg. Assoc. J. **41** 685 (1980).

Tsuchiya K and S Harashima. Brit. J. Ind. Med. **22** 181 (1965).

US Office of Science and Technology Policy. Chemical carcinogens; a review of the science and its associated principles, February 1985. Federal Register 10371–10442 (14 March 1985).

Watanabe P G, J D Young and P J Gehring. J Env. Path. & Toxicol. **1** 47 (1977).

Williams GM and J H Weisburger. Chemical carcinogens. In: Casarett and Doull's Toxicology—The basic science of poisons (Edited by C D Klaassen, M O Amdur and J Doull). Macmillan Publishing Company, New York (1986).

World Health Organisation. Recommended health-based limits in occupational exposure to

heavy metals: Report of a WHO Study Group. WHO Techn. Rep. Ser. No 647. Geneva, Switzerland (1980).
Wright G W. The influence of industrial contaminants on the respiratory system. In: The industrial environment—its evaluation and control. U.S. Department of Health, Education and Welfare, Public Health Service, Center for Disease Control, National Institute for Occupational Safety and Health (1973).
Yanysheva N Y and Y G Antomonov. Environ. Health Perspect. **13** 95 (1976).
Zapp J A. Am. Ind. Hyg. Assoc. J. **38** 425 (1977).

Bibliography

Becker F F (Editor). Cancer: A comprehensive treatise, Vol 1, 2nd ed. Plenum Publishing Corp, New York (1982).
Fechner G T. Kollektivmasslehre. Engelmann, Leipzig (1897).
Hall J H. Radiation and life. Pergamon Press (1984).
Klaassen C O, M O Amdur and J Doull (Editors). Cassarett and Doull's Toxicology—The basic science of poisons. Macmillan Publishing Company, New York (1986).
Los Alamos Scientific Laboratory. The effects of atomic weapons. McGraw-Hill Book Company Inc (1950).
Paracelsus Th. B. von Hohenheim. Epistola dedicatora St. Viet Karnten: Sieben defensionen oder sieben schutz-, schirmund trutzreden dritte defension (August 24, 1538).
Weber H. De pulsa resorptione auditu et tactu. Annotationes anatomicae et physiologicae. Leipzig. Koehler (1834).

CHAPTER 4

Washout Curves—Toxico-kinetic Data on Hazardous Substances

In life, the total body burden of a hazardous substance and its distribution in different organs is not measurable directly. The type of information which can be made available most easily is concentration level in samples of urine and exhaled air either of the substance itself or of a metabolite. When necessary, the blood of employees may also be sampled occasionally and analysed. Sequential analysis of samples of urine, exhaled air or blood is used to indicate the changes which take place from time to time within the body.

The study of the time course of toxic substance concentration levels in excreta, blood, other fluids and tissues of the body resulting from exposure to the substance in the gaseous, liquid or solid state comes under toxico-kinetics. It includes study of the mathematical relationships required to develop models to interpret such data. The use of mathematical models to simulate these processes allows predictions about the body burden of substances, their residence time in the body and other information which may aid in assessing the risks to employees from a peculiar pattern of exposure (Himmelstein and Lutz, 1979; Andersen, 1982).

The importance or otherwise of exposure at work to a sudden burst of air contamination may be judged in terms of its likely influence on the body burden of a hazardous substance or its metabolites. At one extreme, a substance which is rapidly excreted, metabolised to a non-toxic substance or otherwise rapidly eliminated is, by definition, one which does not accumulate. At the other extreme, a substance or metabolite which is not eliminated produces its adverse effects on the body through the sum result of steadily accumulating body burden from successive exposures to the substance since birth. In the latter instance the increment of body burden due to a burst of air contamination represents a much smaller, possibly almost negligible percentage of the total accumulated body burden. The problems in assessing the risk to health by means of measurements of the environmental conditions may thus appear to be very different from one substance to another. They have more in common when viewed in relation to the length of time the substance resides in the body. Materials rapidly eliminated by the body have a short residence time, so that for these the body burden is a reflection of exposure over the recent past. Others which are eliminated more slowly have a longer residence time.

Single compartment models have long been employed by some authorities to simulate the absorption and elimination of gaseous contamination in inhaled air (Haggard, 1924; Henderson and Haggard, 1943; Kety, 1951; Fiserova-Bergorova, Vlach and Singhal, 1974). Less attention has been paid to the relationships between exposure to aerosols and the accumulating burden in the critical organ, except, perhaps, for dusts causing pneumoconiosis (Vincent, Johnston, Jones, Bolton and Addison, 1985). More quantitative recognition has been given to these interrelationships in radiobiology and health physics than in older areas of toxicology and industrial hygiene (International Commission on Radiological Protection, 1968, 1971). This is probably accounted for by the strong theoretical mathematical-physical base upon which radiobiology has been built from the start.

The use of differential equations to describe the observed behaviour of body burden leads naturally to the development of interest in their compartmental form. The body burden of absorbed substances is divided unevenly between various 'compartments' such as lungs, blood, soft tissues, muscles, skin, hair, bone and fat, which each have somewhat different toxico-kinetic parameters. Multiple compartment models are now increasingly used to simulate these systems in toxico-kinetics (Piotrowski, 1971; Clewell and Andersen, 1985). Most biological data on employees exposed to hazardous substances may be sufficiently well represented by the behaviour of models having one, two or at most three compartments between which the transfer of material is governed by linear differential equations (Papper and Kitz, 1963). For some purposes the need is for a mathematical description of concentration-time curves of biological determinants analysed into a number of exponential washout curves, without necessarily identifying their precise physiological basis (Salmowa, Piotrowski and Neuhorn, 1963; Fernandez, Humbert, Droz and Caperos, 1975). For others the analogy between connected compartments and specific organs illuminates the physiology governing the fate of foreign substances in the body (Dengler, 1970).

The physiological and anatomical parameters of these models help to provide a basis for extrapolating toxico-kinetic data from test animals to man and from laboratory volunteers to full time employees. The model parameters may be identified with measurable physiological parameters and tissue solubility as between test animals and man, thereby allowing a prediction of the comparative body burden and its disposition in different organs or tissues in different species (Dedrick, 1973; Dedrick and Bischoff, 1980; Andersen, 1981a; Baselt, 1983; Fiserova-Bergerova, 1983; Ramsey and Andersen, 1984; Andersen, Clewell, Gargas, MacNaughton, Reitz, Nolan and McKenna, 1991).

Conclusions about the behaviour of compartmental models are embodied in a comprehensive and concise form by mathematical results and formulae. Those who find the simplest mathematics baffling will perhaps prefer to know only the main results, expressed in words. Nevertheless the mathematics employed are not very difficult, and are very rewarding for the effort required for their comprehension. Notes to remind the reader of the elementary principles are given in Appendix 1.

Exposure at work is notoriously variable both from time to time and place to place. The importance of short term and long term temporal fluctuations in exposure is a central problem in occupational hygiene of hazardous substances. The issues may be quantified in a reasoned and consistent manner with the aid of a knowledge of the general toxico-kinetic properties of the substance.

In this book the underlying concept which is employed to interpret such properties is that in life there is a critical body burden of a hazardous substance in the critical organ which should not be exceeded. This does not mean that peak body burden is the only parameter governing toxicity. Adverse health effects may indubitably appear long after the body burden which caused them has disappeared but they cannot occur before it has arrived. It is this preventive principle which is at the heart of the matter.

There is a handful of hazardous substances whose toxico-kinetics is known in some detail but for most hazardous substances evidence about the residence time in the critical organ is limited. Fortunately, for the purpose of controlling health risks by restricting exposure peaks at work, such shortcomings can be accommodated by choosing a minimum likely value of the residence time and accepting the safety factor thereby indirectly incorporated in the intervention strategy. The choice of averaging period for air sampling data taken on such a basis is preferable to an arbitrary choice of 15 minutes, 8 hours or a period chosen solely for the convenience of the sampling instrument, analytical procedures or measuring personnel.

It is not supposed that the simple accumulation and decay models discussed here are an exact reproduction of behaviour in the body. The complexity of the human body cannot possibly be imitated exactly by a few well-mixed, interconnected compartments. It would be unduly arrogant to suppose otherwise. However, for comparing the relative merits of alternative atmospheric sampling strategies to trigger environmental controls the primary properties are exhibited by the simplest compartmental models. There is insufficient evidence about most hazardous substances to warrant wider use of models having a more sophisticated network of compartments with multiple exponential or like type of accumulation and decay curves.

At the close of these introductory remarks it should be emphasised that the present chapter is taken up with the average rise and fall of body burden in the critical organ resulting from exposure to hazardous substances at work of the average employee. The equations employed represent functional relationships; but there is no suggestion that they are mathematically exact representations of an employee, and still less of every employee (Droz, 1989; Opdam, 1989).

Rate of accumulation and elimination

The primary exposure to hazardous substances at work is usually air contamination. A substance may enter the respiratory tract in the form of a gas, liquid droplets or solid particles. Skin contamination by the liquid and solid phase of a substance is often contributory but skin absorption of gases and vapours as such is rarely significant (Guy, Hadcraft and Maibach, 1985).

The period of accumulation of body burden is generally much shorter in the case of foreign gases and vapours than it is for dusts. The time spent away from work each day is more than adequate to permit complete elimination of many absorbed gases and vapours, consequently none of these may be present in the body at the beginning of the next daily exposure. This is mainly because significant elimination takes place by exhalation. Absorption from the lungs by the blood continues until the concentration in the blood has reached an equilibrium concentration with gas and vapour in the lungs. Accumulation in the rest of the body, when it takes place to any significant extent, is dependent on the concentration in the blood. When inhaling a vapour such as benzene, for example, equilibrium between the vapour and the blood is reached within a few minutes; and equilibrium between benzene in the blood and the fatty tissues is reached in two or three days (Schrenk, Yant, Pearce, Patty and Sayers, 1941). On reversing the process, immediately after exposure ceases the concentration in exhaled air falls rapidly as the benzene in blood comes into equilibrium with the benzene in alveolar air, but the rate of change lessens later as the limiting factor becomes the rate of removal from fatty tissue (Teisinger, Fiserova-Bergerova and Kudrna, 1952; Hunter, 1968; Hunter and Blair, 1972).

Solid particles and liquid droplets are deposited mainly on the upper respiratory tract, the coarser particles being deposited first and finer particles sedimenting out in the deeper parts of the tract. If they are not immediately dissolved the coarser particles are removed by ciliary action. Most of the insoluble particles deposited in the alveoli are removed by phagocytes which can engulf particles and migrate to the bronchioles from which they too are removed by ciliary action. However, a proportion of the finer particles which have been deposited in the alveoli remain in the lungs indefinitely if they are insoluble. Such material remains in the lungs more or less permanently, even after exposure ceases so that at any one time the amount of dust in the lungs is roughly proportional to the product of a person's age and the average concentration of the dust to which the person has been exposed since birth.

It has long been known that in contrast to relatively insoluble dusts like coal and silica, all of a more soluble substance such as lead in the body is eventually excreted when exposure ceases, albeit slowly and the lead content even of the skeleton usually reaches normal values within about 18 months (Kehoe, Thamann and Chalk, 1933; Task Group on Metal Accumulation, 1972). Thus the total amount of lead in the body is governed by the level of exposure over the previous 18 months and, to a great extent by that over the previous few weeks.

A relatively soluble dust such as lead oxide is absorbed in the lungs so that lead fairly rapidly accumulates in the blood, soft tissues and finally the skeleton. But it is also eventually excreted. The rate of solution of lead oxide increases with the amount of material deposited in the lungs and the rate of excretion of lead increases with the amount in the body, so that under exposure to a constant concentration a level of equilibrium will eventually be reached

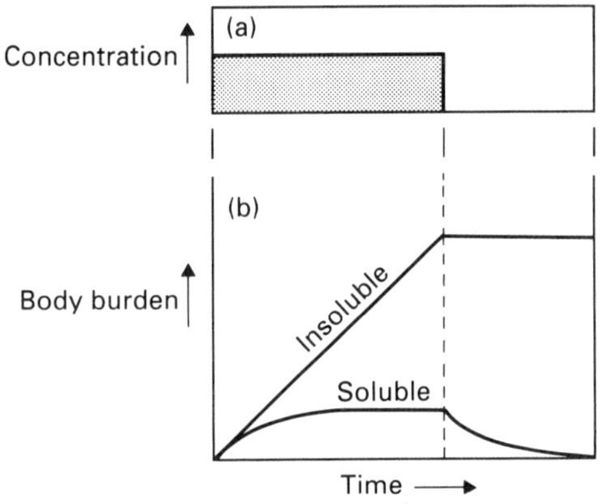

FIG. 4.1. Accumulation of body burden on exposure to air contamination. (a) Graph of atmospheric concentration of hazardous substance against time. (b) Corresponding body burden against time—insoluble materials and soluble materials.

when the amount of lead in the body is such that on the average the rate of excretion just equals the rate of deposition in the lungs.

These points are illustrated in Figure 4.1. The top half of the diagram indicates the situation in which an employee is exposed to a constant atmospheric concentration of the material for a given length of time. The bottom half of the diagram illustrates how, as a result of this exposure, the material accumulates in the body to an extent dependent upon the solubility of the material deposited in the lungs. With all gases and those solids which are removed from the lungs in solution and are eventually excreted or removed in some other way, the amount accumulated in the lungs, tissues and other organs approaches a maximum asymptotically at a value governed by the atmospheric concentration and toxico-kinetic properties of the material. At the other extreme, with solids which are for all practical purposes insoluble, the amount accumulated in the body increases steadily at a rate broadly proportional to the atmospheric concentration.

Notwithstanding the above remarks, transient concentrations of a material having even a short biological half time could, if high enough, lead to a temporary accumulation in the body sufficiently great to be harmful. Hydrogen sulphide is an example of a substance which is normally metabolised in the blood plasma to non-toxic substances but is rapidly lethal at high concentrations as evidenced by the many accidental deaths it has caused. In high enough concentration, over 500 ppm, the metabolism is not sufficiently rapid to counter the intake of the gas and it may produce respiratory paralysis leading to consequent death from asphyxiation unless the individual is promptly removed to fresh air and given artificial respiration. As another example from occupational hygiene, it is generally accepted that exposure to an average concentration of carbon monoxide of 50 ppm for a 40-hour week is perfectly tolerable, even with moderate fluctuations about the mean and no harm could be detected. However, were the week's dose inhaled in 30 minutes, that is with a concentration of 4,000 ppm, then almost certainly the exposure would be fatal (Henderson, Haggard, Teague, Prince and Wunderlich, 1921; Coburn, Forster and Kane, 1965). In this case the magnitude of fluctuations of concentration from hour to hour are of great importance.

These considerations have an important bearing on the choice of sampling period of investigative instruments used to measure concentration. In the case of most hazardous gases and vapours an instrument which integrates the exposure over a week would certainly fail to respond to short term fluctuations in concentration which nevertheless are of sufficient duration to give rise to serious effects on the employees. An instrument which samples for an hour or even less might then be required to demonstrate these short period fluctuations. However, in the case of airborne dust of quartz, for example, a sampling instrument into which air had to be drawn for a week to collect sufficient dust to weigh and analyse accurately would not be rejected on this account. A low 3- or even 6-month moving average level of concentration would not hide a danger from silicosis. In other words, the ideal time weighted averaging period, the reference period, is determined from a knowledge of the residence time of the substance in the body in relation to the time scale of the fluctuations in concentration in work place air.

Biotransformation

Toxico-kinetics deals mainly with the behaviour of hazardous substances in their passage through the body and toxico-dynamics deals mainly with the complex, consequent cellular changes, tissue changes and functional changes in the body. Straddling the interface between the two disciplines lies biotransformation, which is concerned specifically with the conversion of molecules from one sort to another. The new molecules may be more complex or simpler, more toxic or less so. Biotransformation rate processes may be linear with concentration, accelerative or exhibit saturation at high concentrations (Andersen, Clewell and Gargas, 1987). There is evidence that in rats the metabolism of styrene, for example, increases during exposure (Andersen, Gargas and Ramsey, 1984).

Biotransformation directly influences both toxico-kinetics and toxico-dynamics. Thus metabolism often results in the conversion of fat-soluble substances into ones which are more water-soluble and thereby promotes their excretion. Other things being equal those hazardous substances which are especially fat-soluble are generally more readily absorbed through the lungs, skin or gastro-intestinal tract and are poorly excreted. Water-soluble substances, on the other hand, are readily excreted. Systems which catalyse such biotransformation are found mainly although not exclusively in the liver (Andersen, 1981b).

The more rapidly a substance is eliminated from the body the lower will be the concentration of it in a particular organ. However, the rate at which a substance is excreted is intimately bound up with its distribution and biotransformation. If the substance is distributed to and stored in bones or fat, for example, it is likely to be eliminated slowly because it is not so readily available for excretion. Furthermore, many commercial chemicals are very fat-soluble and cannot be excreted until they are biotransformed into a water-soluble product. Thus the processes of absorption, distribution, biotransformation and excretion are closely inter-related and integral factors in determining whether a chemical will be toxic (Andersen, 1991).

Body burden matters

Air contamination may adversely affect an employee by being inhaled and deposited in the lungs or absorbed through the respiratory tract or absorbed directly through the skin or eyes. Hazardous substances may also produce their ill effects by contaminating food, hands and objects handled so that they are subsequently absorbed through the alimentary tract. In any case, by whatever route it enters, the body burden of the substance would generally be a more accurate indication of whether a worker is likely to be adversely affected by it than would the level of concentration in the air of the workshop. It is the substance or its metabolites in the body which produces the response in the individual. Further, if the substance produces its harmful effect in particular organs or in particular parts of the body, the burden of the substance in these organs will be a more accurate index of adverse effects than the whole body burden. Thus exposure limits for a changing environment should be based in part on the varied but intimate connection between body burden and concentration in inhaled air of the substance in question (Rappaport, 1985). An understanding of the toxico-kinetics is essential for settling three major issues in occupational hygiene:

(1) Exposure limits. That combination of maximum concentration and period(s) of exposure which is most likely to result in prevention of health risk.

(2) The most suitable averaging period of exposure measurements for detecting the presence of hazardous conditions.

(3) The scheduling of sampling for biological specimens.

The biological half time of a substance in the body varies from one individual to another about an average value. Furthermore the biological half time of an individual will vary to some extent from time to time with variation in respiratory minute volume, cardiac output, liquid intake, diet and other parameters. A consolidated list of selected substances according to average biological half time is in Table 4.1. The published information is generally indicative only and conservative estimates have been employed in this tabulation for use in assessing health risk from exposure data (Roach, 1966; American Conference of Governmental Industrial Hygienists 1987; Leung and Paustenbach, 1988; Saltzman, 1988). The arrangement of the gases and vapours in a separate column from the aerosols makes it very apparent that as a group they tend to have a short biological half time. They tend to be more easily expelled from the body. This factor is also one of the reasons why gases and vapours tend to have higher exposure limits. In the compilation in Table 4.1 one third

TABLE 4.1. *Average biological half time of selected inhaled substances in humans*

	Substances	
Biological half time	Gases and vapours	Aerosols
Less than 10 minutes	Ammonia	Sulphuric acid
	Amyl acetate	
	Chlorine	
	Chloroform	
	Dichlorodifluoromethane	
	Hydrogen chloride	
	Hydrogen sulphide	
	Osmium tetroxide	
	Sulphur dioxide	
10 minutes–1 hour	Carbon disulphide	
	Dioxane	
	Ethyl alcohol	
	Formaldehyde	
	Nitrogen dioxide	
	Styrene	
	Trichlorofluoroethane	
1 hour–8 hours	Acetone	Benzidine
	Aniline	
	Benzene	
	Carbon monoxide	
	Carbon tetrachloride	
	Dimethyl formamide	
	Ethyl acetate	
	Ethyl benzene	
	Furfural	
	Hexane isomers	
	Methanol	
	Methyl cellosolve	
	Methylene chloride	
	Phenol	
	Toluene	
	1,1,1-Trichloroethane	
	Vinyl chloride	
8 hours–1 month	Halothane	Arsenic (As_2O_3)
	Methyl chloride	Chromium cpds
	Nitrobenzene	Iron oxide fume
	Tetrachloroethylene	Nickel
	Trichloroethylene	Pentachlorophenol
		Thallium
		Vanadium (V_2O_5)
More than 1 month	Mercury	Asbestos
		Cadmium
		Coal
		DDT
		Dieldrin
		Lead
		Mica
		Silica

of the substances possess a biological half time not dissimilar to the daily duration of exposure (1 hour to 8 hours), one third possess a longer biological half time and the remainder a shorter biological half time.

Single compartment models

On being inhaled, a gas or vapour is absorbed in the respiratory tract and solid particles are deposited in it at a rate directly proportional to the concentration in inspired air. Also it is reasonable to suppose that, the greater the amount of a foreign substance in the body, the greater will be the rate at which it is excreted, exhaled, detoxified or otherwise eliminated.

In the basic scheme it is assumed that the substance is excreted, exhaled, detoxified or otherwise eliminated by the body at a rate directly proportional to its concentration in the body or that part of the body containing the substance. The rate of accumulation and decay is assimilated to that in a single, well mixed compartment of fixed volume with one input and one output. The maximum concentration in the compartment is governed by the input and output rate constants in effect at the time. The half time is governed by the output rate constant and compartment volume.

Constant exposure

Suppose a substance enters a single, well mixed compartment. The consequences of exposure to a constant concentration are fairly easy to predict. Where the concentration in the inspired air is C mg m^{-3}, it is assumed that the substance is taken in from the air at a rate proportional to C. The constant of proportionality is the respiratory minute volume multiplied by the percent retention. The percent retention varies with the substance. A highly soluble vapour may be wholly retained at the outset whereas others are at least partly exhaled. Coarse particles may all deposit in the upper respiratory tract whereas particles of 0.5 microns diameter may be inhaled into the alveoli and be mostly exhaled again without being deposited anywhere.

Suppose, then, a substance enters a model, which is a single, well mixed compartment, at a rate kC mg h^{-1}, k being a constant which is characteristic of the substance and respiratory minute volume (Figure 4.2a). Where the concentration of hazardous substance in the compartment after a time t hours from the start of exposure is x mg m^{-3}, the instantaneous rate of elimination is ax mg h^{-1}, a being another characteristic constant, and

$$\frac{dx}{dt} = \frac{kC - ax}{V}, \tag{1}$$

where V is the volume of the compartment. Discrete compartments may be identified with a specific organ or organ system whose physiological parameters are documented. The effective volume of the organ of interest in humans and experimental animals is often known. In any event the rate of elimination of many substances is measurable by sampling and analysis of their urine and exhaled air. Blood may also be sampled for a direct measure of the concen-

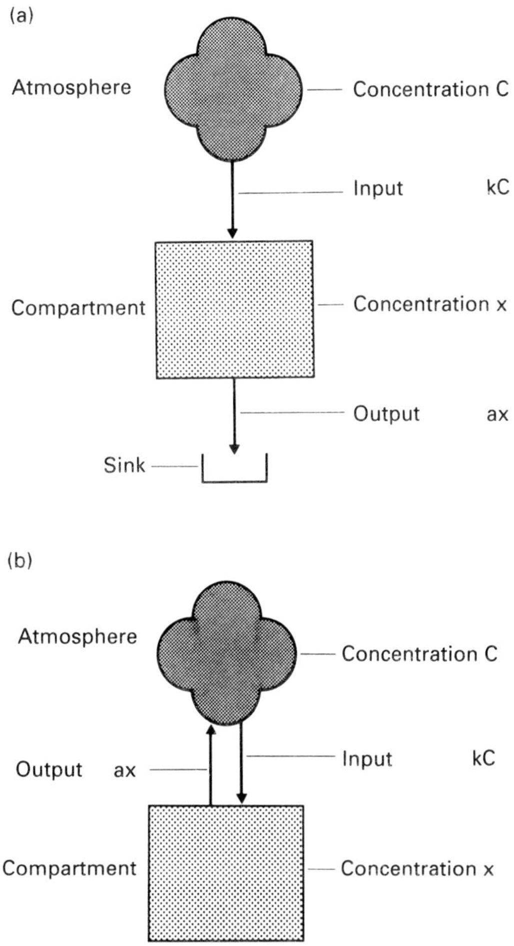

FIG. 4.2. Single compartment, with single input and single output. Compartment volume = V. (a) Output by excretion. (b) Output by exhalation.

tration of substance present there. These are the primary parameters employed in biological monitoring of exposed employees at work and experimental animals in the laboratory. A model in which elimination is by excretion is in Figure 4.2a and one in which elimination is by exhalation is in Figure 4.2b. In the latter instance the volume of air inhaled equals the volume exhaled and when the system is in equilibrium the concentration of contamination in the inhaled air equals the concentration exhaled.

Solving the differential equation (1), as explained in Appendix 1, the general solution is

$$x = (kC/a) + A \exp(-at/V), \tag{2}$$

where A is an arbitrary constant.

By substituting the initial conditions, $x = x_o$ when $t = 0$, and solving for A it follows that $A = [x_o—(kC/a)]$.

Thus, on exposure to a constant atmospheric concentration it follows that the concentration of substance in the model compartment is a simple exponential function:

$$x = (kC/a) + [x_o - (kC/a)]\exp(-at/V), \tag{3}$$

where C = concentration in air (mg m^{-3})
x = concentration in compartment (mg m^{-3})
k = minute volume × percent retention (m^3 min^{-1})
t = time (min)
a = rate of removal from compartment (m^3 min^{-1})
V = volume of compartment (m^3)
t_o = time when investigation begins (min)
x_o = the concentration of the substance in the compartment at time t_o, when the investigation begins (mg m^{-3}).

It is readily apparent from equation (3) that, when the duration of exposure, t, is very large, $\exp(-at/V)$ tends to zero, and $x \rightarrow kC/a$. At this stage the rate of elimination is keeping pace with the rate of intake. The equilibrium value—the full saturation of the body for any one concentration of air contamination—is of great practical importance: it expresses completion of absorption/retention and therefore does away with all further consideration of time of exposure in relation to concentration of contamination in the air. If the total amount of hazardous substance that can have been taken up by the body at any given concentration in air causes no harm, then the concentration may be inhaled indefinitely, since at equilibrium no more of the substance can accumulate in the body. That is:

- The mass of substance in the compartment, the equivalent of the body burden, would tend asymptotically to an equilibrium value of kCV/a mg.
- At equilibrium the ratio of the concentration in the compartment to the concentration in air is, $x/C = k/a$. Experimentally, given three of these variables the fourth follows. In the case of exposure to a gas or vapour, when elimination is solely by exhalation (Figure 4.2b) the ratio of concentration in the blood to concentration in air is equal to the solubility of the gas or vapour in the blood.

Where the initial concentration, $x_o = 0$, equation (3) simplifies to:

$$x = (kC/a)[1 - \exp(-at/V)]. \tag{4}$$

Equation (4) can be arranged in a more handy form by making use of the fact that the equilibrium value of substance mass in the compartment is kC/a. Thus,

$$\text{Percentage of equilibrium mass} = 100[1 - \exp(-at/V)],$$

where t = time from start (min)
a = rate of removal from compartment (m^3 min^{-1})
V = volume of compartment (m^3).

Now, in order to find out when a certain percentage of the equilibrium concentration will be reached, say, 99%, denoted t_{99}, 99 is substituted in the left hand side and the equation solved (MacFarland, 1976). Thus,

$$t_{99} = \frac{4.605 \times V}{a}.$$

Similarly,

$$t_{50} = \frac{0.693 \times V}{a}.$$

It is apparent that t_{99}, t_{50} or the time of any definite percentage of the equilibrium concentration is dependent on only two variables; the volume of the compartment, V, and the characteristic constant, a. It is independent of the value of the atmospheric concentration and independent of the equilibrium concentration in the compartment. The value of t_{99} may alternatively be expressed as a multiple of the half time, t_{50}, or T:

$$t_{99} = 6.645 \times T.$$

Values of the multiple for various percentages of the equilibrium concentration are:

$$\begin{aligned} t_1 &= 0.015 \times T \\ t_5 &= 0.074 \times T \\ t_{10} &= 0.152 \times T \\ t_{20} &= 0.322 \times T \\ t_{25} &= 0.416 \times T \\ t_{50} &= T \\ t_{75} &= 2.000 \times T \\ t_{80} &= 2.322 \times T \\ t_{90} &= 3.323 \times T \\ t_{95} &= 4.323 \times T \\ t_{99} &= 6.645 \times T \end{aligned}$$

Obviously, once t_{90} has been passed, the concentration will increase by less than 10% no matter how long the system is in operation. Similarly, once t_{99} has been passed the level of concentration will increase by less than 1% no matter how long the system is in operation.

Exponential decay

When an employee comes out of a contaminated environment, the inhaled air should be free of contamination; C is then zero. Under these compartmental conditions, it follows from equation (3) that the concentration in the compartment would then be given by:

$$x = x_o \exp(-at/V) \qquad (5)$$

Thus, the concentration of substance in the model compartment falls from the value x_o at the start of the period of zero input, tending to a concentration of zero as t increases.

Another property of this decay curve is that the slope of the curve at the point (x,t) is, from equation (1):

$$\frac{dx}{dt} = -\frac{ax}{V}.$$

Hence the time it would take to reach zero compartment concentration at this rate of decay is:

$$x\frac{dt}{dx} = -\frac{V}{a} \quad \text{minutes,}$$

which is a constant with the dimensions of time, independent of the values of x and t. This is the 'time constant' of the system.

The time it takes for the compartment concentration to fall to half its initial value is the half time, T, which bears a simple relationship to the time constant of the system, V/a. The biological half time

$$T = \frac{V \ln 2}{a} = \frac{0.693 \times V}{a}. \tag{6}$$

The half time may be read off directly from a graph of experimental data of exponential decay. However, in practice a more accurate method, taking account of all the data, is to measure the area under the curve, which, by integrating equation (4), is directly related to the half time:

$$T = \frac{0.693 \times \text{total area under the curve}}{x_o}.$$

Fractions of the area under the curve are obtained from successive urine samples and integrated exhaled air samples.

Accumulation and decay

The rise and fall of the concentration of substance in the compartment on exposure to a constant air concentration for two hours only is shown in Figure 4.3. To crystallise ideas the illustration is for exposure to a substance which has a biological half time of 30 minutes and to an air concentration which, if it persisted indefinitely, would result in a maximum compartment burden of 40 mg. From the commencement of exposure the concentration rises rapidly at first but on continued exposure slowly approaches the equilibrium value asymptotically. The difference between the mass of substance in the compartment and its maximum, equilibrium value diminishes in an exponential fashion with a half time T.

On cessation of exposure the mass of substance in the model compartment falls exponentially to zero with a half time T. The form of the decay curve of concentration in the model compartment is the same whatever the magnitude or duration of atmospheric exposure. This is not a general result applying to all systems and, as will be shown shortly, strictly applies only to a single compartment system.

An example of a common hazardous substance which exhibits exponential clearance from the body is carbon monoxide. In humans carboxyhaemoglobin

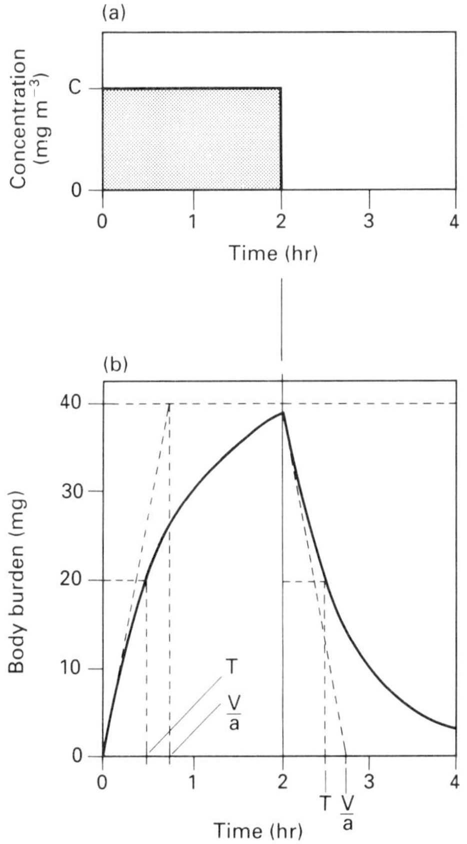

FIG. 4.3. Exponential accumulation and decay. (a) Exposure to a constant concentration for 2 hours. (b) Compartment burden resulting from 2-hour exposure as given by equation (2). Half time, $T = 0.5$ hr. Maximum possible compartment burden, $kC/a = 40$ mg. System time constant, $V/a = T/(\ln 2) = 0.722$ hr.

in blood resulting from up to 24 hours of exposure to carbon monoxide behaves as a substance accumulating in a single compartment system (Hatch, 1952; Coburn, Forster and Kane, 1965; Peterson and Stewart, 1975; Stewart, 1975; Stewart, Stewart and Stam, 1976). It decays exponentially on cessation of exposure with a half time of about 320 minutes. Under any physical exertion which increases the minute volume the half time is inversely decreased. Similarly, small children with more active metabolism and relatively larger minute volume for their size also have a reduced biological half time for carbon monoxide. A small animal, such as a canary, has a very much shorter half time than humans, which accounted for its use in mines to detect vitiated air. When the canary falls off its perch men breathing the same atmosphere have time to escape because of their inherently longer biological half time.

When the biological half time of a substance is very short, in terms of the model compartment a/V is large, $\exp(-at/V) \rightarrow 0$, and equation (3) becomes

$x = kC/a$. Thus with such substances the mass of substance in the model compartment would be directly proportional to C, the concentration in inspired air, and would be largely independent of the duration of exposure.

The way the body burden would be expected to accumulate on exposure to a substance with a half time of very long duration can best be found by considering what happens to the quantity of material in the model single compartment when its initial concentration is zero, equation (3).

Since the biological half time under consideration is a very long one, in terms of the single compartment T is very large, V/a is near zero and $x = kCt/V$. Thus, when the biological residence time is a very long one, provided the body burden of a substance behaves as would the mass in a single compartment system, the body burden should be directly proportional to time as well as concentration; that is, to the total quantity of substance inhaled.

Residence time of substances in the body

Biological residence times for substances commonly encountered vary from a few seconds to at least several months and possibly much longer before they are eliminated. Depending on the agent, residence times in each compartment may differ greatly. The effective residence time of an upper respiratory irritant such as sulphuric acid may be measured in seconds, whereas a substance which is more or less fixed in bone, such as lead, obviously has a relatively long residence time in that compartment. The shortest periods of medical importance for time weighted averages differ proportionately. The key residence time as regards health risk is the residence time in the critical organ.

The mean length of time the mass of substance entering a single compartment resides in the model can be calculated. For example, consider a compartment to which has been input a unit dose, say 1 mg. The fraction of it remaining after time t is $[\exp(-at/V)]dt$. This quantity has resided in the compartment t hours (Figure 4.4). The total area under the curve is

$$\int_0^{\infty} \exp(-at/V)dt = V/a \text{ mg hours} \qquad (7)$$

The mean residence time is found by weighting the amount remaining from the input dose by its residence time t, Figure 4.5, summing to find the area under the curve and dividing by V/a mg hours, the area under the previous curve, Figure 4.4. Therefore,

$$\text{Mean residence time} = \int_0^{\infty} \frac{at}{V} [\exp(-at/V)]dt,$$

$$= \frac{V}{a} \text{ hours.}$$

The mean residence time, V/a hours, when expressed in terms of the compartment half time, is $1.44T$ hours. The mean residence time of material in a single compartment model is thus equal to the time constant of the system and is directly proportional to the half time, the constant of proportionality being 1.44.

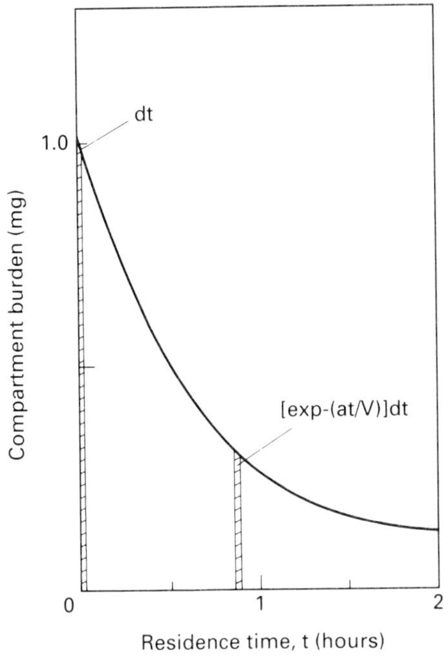

FIG. 4.4. Fate of a unit element of dose.

Saturation effects

At low exposures human uptake of hazardous substances in air is at least approximately proportional to its atmospheric concentration, whether the substance be gas, vapour, liquid or dust. At work, employee exposure would normally be regarded as low in these terms. However at a sufficiently high concentration saturation of uptake mechanisms will be approached and the assumption of strict proportionality will possibly be in error. Saturable mechanisms also exist in secondary compartments and can themselves induce

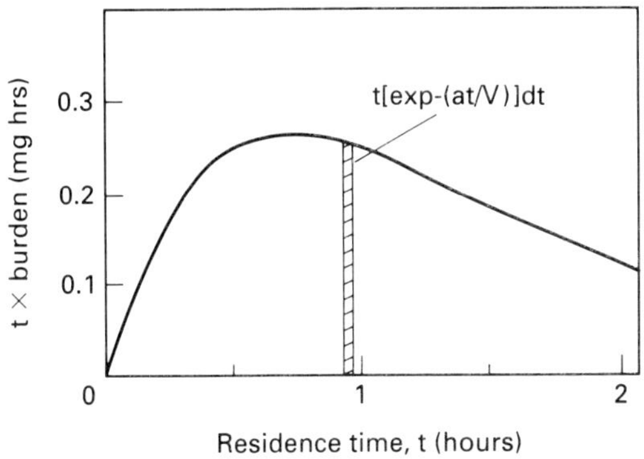

FIG. 4.5. Compartment burden remaining from an element of dose, weighted by residence time. Mean residence time = V/a hr.

undesirable biochemical or functional changes, so that the consequences in toxico-kinetic models is not just of theoretical interest (Andersen, 1981c; Kjellstrom, Elinder and Friberg, 1984; Kennedy, 1990; Fiserova-Bergorova, 1990).

Simulation of a saturable process in a model compartment is instructive. For example a simple formulation of a saturable process is to suppose that uptake is at a rate given by:

$$kC\left[1 - \frac{x}{x_{max}}\right],$$

where x_{max} is the maximum possible concentration of hazardous substance in the compartment. This formula has the property of tending to kC when x is low or x_{max} is high and tending to zero as $x \rightarrow x_{max}$.

On solving this system, as before, it is found that the compartment concentration still follows a simple exponential accumulation and decay, but the time constant is now longer, being

$$\frac{a}{V}\left[1 + \frac{kC}{ax_{max}}\right].$$

Furthermore, the equilibrium compartment concentration on exposure to a constant concentration is reduced, being

$$\frac{kCx_{max}}{kC + ax_{max}}. \quad (8)$$

A property of this function is that for large values of C, $x \rightarrow x_{max}$. Consequently, the equilibrium compartment burden is no longer sensitive to changes in compartment exposure, nor is the rate of elimination of substance from the compartment. As saturation of the compartment is approached the rate of excretion of the substance or metabolite ceases to be of value as an indication of exposure.

Haber's rule?

A much quoted rule of toxicology, known as Haber's rule, states that the product of administered concentration and time produces a constant effect. When c is the concentration of a substance in the medium of interest and t is time, the toxic effect E would be given by

$$E = Act, \quad (9)$$

where A is a constant. The rule was based on studies of the mortality among experimental animals exposed to phosgene (Flury, 1921; Haber, 1924). The plot of c against t, keeping E constant, would, by Haber's rule, be a rectangular hyperbola whose asymptotes are at right angles, coincident with the axes (Figure 4.6). By Haber's rule the effect would be the same if the concentration is doubled and the time halved or vice versa.

It is of interest to compare Haber's rule with single compartment toxicokinetics discussed previously. Under exposure to a constant concentration the

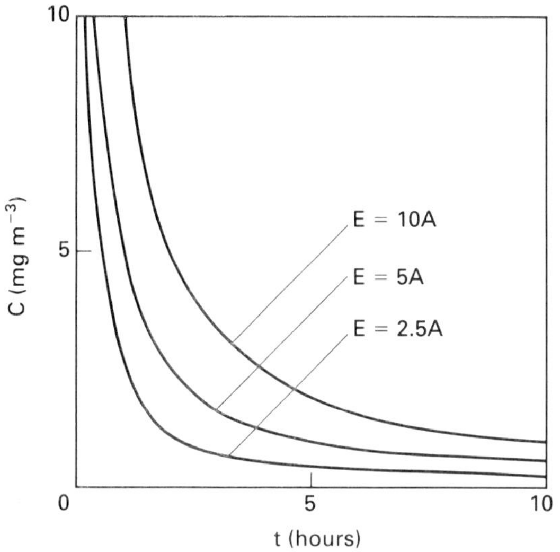

FIG. 4.6. Haber's rule of constant effect. By this rule a curve showing concentration against time for constant effect is a rectangular hyperbola with its axes coincident with the horizontal and vertical axes. Formula: $E = Act$, where E = toxic effect, c = atmospheric concentration, t = duration of exposure, A = constant.

formula for the consequential concentration of substance in the model compartment given by equation (4) may be expanded in powers of t:

i.e.
$$x = A_1 c(t—A_2 t^2 + A_3 t^3—. . .), \tag{10}$$

where A_1, A_2, A_3 . . . are constants.

For small values of t or for large values of the half time this reduces to

$$x = A_1 ct, \tag{11}$$

where $A_1 = k/V$.

Were concentration in the body proportionate to toxic effect, equation (11) would be identical with Haber's rule, equation (9). A series of relevant experiments, employing airborne dust of alumina and of coal dust, has been conducted on animals exposed for 2 hours per day compared with animals exposed for 20 hours per day at one-tenth the concentration (Wright, 1957). After 3 months' exposure and after 18 months' exposure those animals exposed 2 hours per day to the higher concentration had accumulated the same amount of dust in their lungs or somewhat less than those exposed for 20 hours per day at one-tenth the concentration. Where toxic effect in the body is measured by the LD50 or LC50 for experimental animals it is found empirically that Haber's rule is obeyed by most substances when t is small, less than 30 minutes, or the biological half time is long, more than 24 hours. An example for which the rule has been studied in detail and broadly confirmed is by exposure to the highly toxic gas, hydrogen selenide (Dudley and Miller, 1941; Elkins, 1959).

Haber's rule, unmodified, does not predict effect thresholds at low body burdens and is increasingly in error for substances with small values of biological half time. Combinations of high exposure for short periods with long periods of low exposure may produce effects which are different from those following a constant level of exposure, although the integrated dose may be the same in both cases.

The fact that there may be a noninjurious concentration was accommodated in early work by a simple modification of Haber's rule. The modified rule is:

$$E = A(c—e)t, \quad (12)$$

where e, in English translation, is 'the sum of all events that stand against toxifying effect'. This modification was based upon the observed mortality among experimental animals exposed to hydrogen cyanide, whereby below a certain concentration none died even after very long exposure (Flury and Heubner, 1919). A plot of c against t for a constant effect would again be a rectangular hyperbola, this time with the horizontal asymptote coincident with $c = e$ (Figure 4.7). This is easier to reconcile with compartmental toxicokinetics as may be shown by rearranging equation (4), thus:

$$ax/k = C[1 - \exp(-at/V)]. \quad (13)$$

Again, provided concentration in the body could be equated with toxic effect the plot of atmospheric concentration against time for constant toxic effect is not so very different from a rectangular hyperbola with the horizontal axis

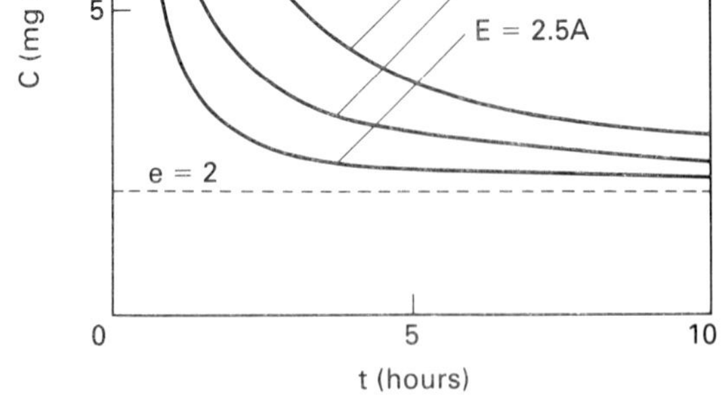

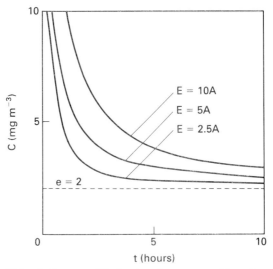

FIG. 4.7. Haber's first modified rule of constant effect. By this rule a curve of concentration against time for constant effect is a rectangular hyperbola with its horizontal asymptote coincident with $c = e$. In this example $e = 2$ mg m^{-3}. Formula: $E = A[c - e]t$, where E = toxic effect, c = atmospheric concentration, e = noninjurious atmospheric concentration, t = duration of exposure, A = constant.

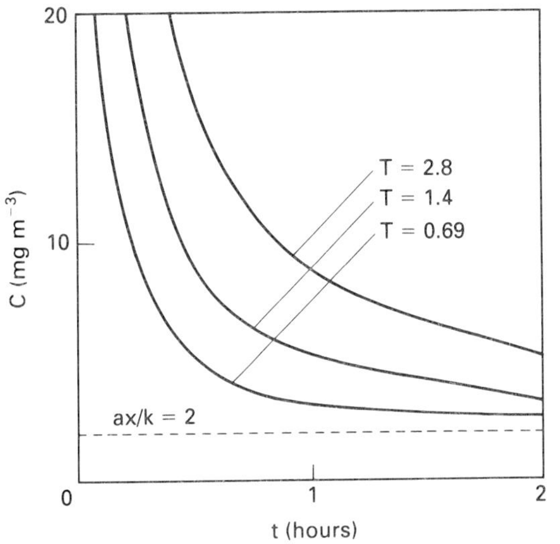

FIG. 4.8. Concentration and time of exposure for constant compartment burden according to single compartment toxico-kinetics, equation (11). Example shown is for $ax/k = 2$.

coincident with $C = ax/k$ (Figure 4.8). The approximation is a good one for duration of exposure less than the biological half time.

Finally, it is known that a better fit with experimental data is sometimes obtained by a further modified form of Haber's rule by which

$$E = (c - e)t^b, \tag{14}$$

where b is a constant (Lazarev and Brusilovskaya, 1934; Atherley, 1985). A good example is acute intoxication of rats by nitrogen dioxide (Gray, Patton, Goldberg and Kaplan, 1954; Carson, Rosenholtz, Wilinski and Weeks, 1962). This allows an even better fit with the simple exponential curve by a judicious choice of value for b for the range of the observations.

This modification can also make some allowance for cases where there is progression or regression of a biological effect with time when concentration is held constant. Biological reaction to a foreign substance in the body is never instantaneous even though the time which elapses before a reaction is noticed may be very brief indeed, as with many irritants. Sometimes, as with pneumoconiosis, it takes many years of exposure before there is a noticeable response. This is also true with carcinogens, for which it is known as the 'latent' period. With these substances many studies show that the larger the dose of carcinogen the shorter is the delay before tumours appear. In studies on experimental animals Druckrey found that

$$dt^n = k, \tag{15}$$

where d is one day's dose of carcinogen, t is the subsequent elapsed time in days before the appearance of tumours, the power n is a constant >1 and k is a constant representing carcinogenic activity (Druckrey and Kupfmuller, 1948;

Druckrey, 1967). Such an expression is in accord with the notion of a threshold. When the dose of carcinogen is low enough the time before appearance of tumours will exceed the normal life span of the animals. Where n is near unity, as with liver cancer from paradimethylaminoazobenzene (PDAB) the carcinogenic activity is dependent on total dose administered. When n is higher than unity, as with diethylnitrosamine (DENA) the carcinogenic activity is influenced more by time increments than by daily dose increments. High daily doses for a short time are less effective than low daily doses for a long time. Whatever the value of n, when the dose of carcinogen is made low enough, the time before appearance of tumours can exceed the normal life-span of the animal. Thus, for each individual and each carcinogen there appears to be a certain threshold dose although, as with non-carcinogenic hazardous substances, the least of the thresholds amongst a large population is indeterminate (World Health Organisation, 1974). Nevertheless, even in its modified forms, Haber's rule is still only a rough approximation over a limited range of concentration and time.

Temporal fluctuations

A common problem in industrial hygiene is how to interpret the set of results from a number of short period or grab samples taken in one work shift. This arises from the hygienist's desire to find out as much as possible about the causes of exposure and to assess the conditions as accurately as possible during a short visit. The variability and its correlation with activities in the vicinity provide information about the control of sources of contamination not available with a single time weighted average over the whole shift. Also, many sampling instruments take only grab samples, and a single grab sample is hardly likely to be sufficient. At the same time there is little published information from industrial surveys about the health hazard from random variability as such. Animal inhalation experiments are usually designed so that the concentration stays constant during the exposure period. There are only occasional reports about varying exposure schemes (Coffin, Gardner, Sidorenko and Pinigin, 1977; David, Frantik, Holusa and Novakova, 1981; Savolainen, Kurppa, Pfaffli and Kivisto, 1981; Van Stee, Boorman, Moorman and Sloane, 1982).

In assessing the effect on risk to health of fluctuations in concentration of air contamination the primary difference to appreciate between different substances lies in the residence time of the body burden of the substances. The importance of these temporal fluctuations of concentration is determined by the position of the harmful material on the scale of biological half time. In the case of material which becomes excreted, exhaled or removed in some other way, the body can tolerate materially higher atmospheric concentrations almost indefinitely. An equilibrium is soon reached when the material is being eliminated from the body as fast as it is being accumulated in the lungs. On the other hand if the amount of material in the body depends upon lifetime exposure short-term fluctuations of concentration are of no importance except in so far as they render the average level of concentration difficult to determine. This will be the case with the pneumoconioses, in which the rate of accumulation is proportional to the long term average concentration. There is a progressive accumulation of the material in the lungs under exposure to even the lowest concentrations.

In environments that vary in a random fashion the body burden is related to the magnitude of the variability and the half time of the substance. Suppose that air samples of t minutes duration were taken one after another and the average concentration over eight hours is C_{av}, with individual sample values varying at random about this average with a variance Var(C).

Algebraically,

$$\text{Var}(C) = \frac{(C - C_{av})^2}{n}, \tag{16}$$

where n is the number of samples. The assumption of random variations implies that the observed variance would increase as the inverse of t, the duration of the samples.

The body burden will vary in harmony with this varying concentration, tending to rise at first but eventually varying around an average value when the average rate of elimination equals the average rate of accumulation. That is, in terms of a single compartment model,

$$x_{av} = \frac{kC_{av}}{a}. \tag{17}$$

Furthermore, it may also be shown that when analysing the results of a varying compartment mass caused by input varying in a random fashion,

$$\text{Coeff. var. } x = 0.59.(\text{Coeff. var. } C).(t/T)^{1/2} \tag{18}$$

(Roach, 1966). Interestingly, only the sampling period and biological half time would need to be known.

Some of the intriguing implications from this result are:

- In order for a sequence of sampling results of exposure to hazardous substances to be directly comparable from one substance to another the averaging period would be made proportionate to the biological half time.
- If the duration of the samples, t, were made $2.88T$ the observed coefficient of variation would match the coefficient of variation of the body burden. The output of a machine producing a $2.88T$ hours moving average would parallel the variation in body burden. If the duration of each sampling could be made about three times the biological half time the results from sampling conducted at a constant flow rate should vary in harmony with the body burden.
- In the event that the concentration is highly autocorrelated the constant of proportionality would be made somewhat less than 2.88 and thereby provide a safety factor, in the sense that the variability would assuredly be as high or higher than the variability of the body burden (Rappaport and Spear, 1988). A factor of one half the biological half time would probably be low enough for all conceivable conditions (Roach, 1977).

Moving averages

The body burden at any moment in time is the sum of residuals of increments made at earlier times. It is a type of moving average and as such comparable with the common arithmetic moving average. Single compartment toxico-kinetics can be readily adapted to contrast and compare the two types.

Suppose that observations were made at equal, very small intervals of time, dt. An arithmetic moving average scheme would weight all the observations made over a predetermined prior period of time equally, and all earlier observations would have zero weight. Contrast this with the way the substance mass in a toxico-kinetic model having a single compartment is accumulated. The present substance mass in the compartment resulting from exposure at concentration C, for a very small interval of time dt, at some prior time t hours ago would be $kC[\exp(-at/V)]dt$. The weight given to the increment in dose, $kCdt$, is thus $\exp(-at/V)$, a weight which would be greatest for the recent past and diminishes exponentially with length of elapsed time since an observation was made (Figure 4.9). The process is an example of exponential 'smoothing' and the result may be called an 'exponential' average.

Now, the average age of the incremental doses weighted exponentially is $1.44T$, as worked out previously, and a moving arithmetic average employing an averaging period of $2.88T$ would also have the same coefficient of variation when input with random noise data. Data from an averaging period of $2.88T$ and weighted equally would have an average age of $1.44T$. The arithmetic and exponential averages would be closely correlated, although insofar as they are not exactly parallel with one another the exponential average would be

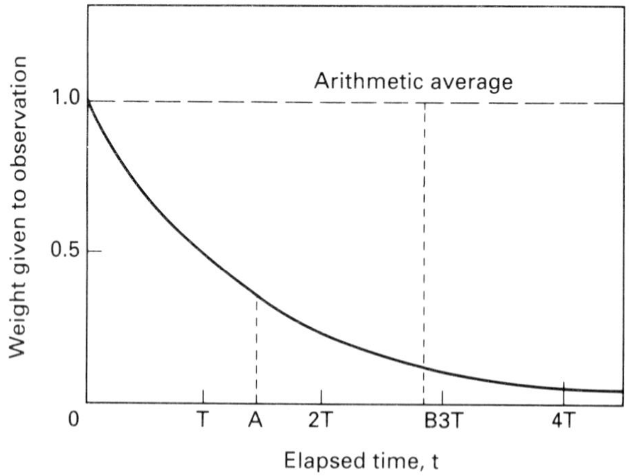

FIG. 4.9. Exponential smoothing. In an arithmetic moving average equal weight is given to all past observations made over the previous period of one hour, or week, or year, etc. In exponential smoothing a weight, $\exp(-0.693t/T)$, is given to observations made t hours previously (see text). A $= 1.44T =$ Average age of compartment burden resulting from exponential smoothing with half time T. B $= 2.88T$ = Period of arithmetic moving average of data having an average age of $1.44T$.

expected to be more closely in tune with health risk as it is more closely mimicking the body burden.

Exponential smoothing is a common sort of averaging in many fields. In the field of systems engineering, this may be recognised as being the simplest case of proportional control. The estimate is corrected with each new observation in proportion to the difference between the previous estimate and the new observation. In this way, some manufacturers set a new production plan each month. They consider the inventory they have now and the inventory they would like to have in the future. This inventory difference, together with a forecast of orders, determines the ideal rate of production to be followed in the next month. But it is not desirable to change production rates abruptly, so the actual plan is equal to the present production rate plus a fraction of the difference between the current production rate and the ideal rate. The computation is exactly that of exponential smoothing (Brown, 1967).

The moving exponential average of sequential results is relatively easy to work out stepwise, even simpler than a moving arithmetic average. In terms of model compartment concentration the essence of the process is illustrated by rearranging equation (2):

$$x = x_o \exp(-at/V) + (kC/a)[1 - \exp(-at/V)]. \tag{19}$$

At each step the new value of the compartment concentration, x, is equal, in sequence, to the sum of two terms; the first term is proportional to the previous value, x_o, and the second is proportional to the new value of input atmospheric concentration, C. Given the two constants of proportionality a sequence of values of atmospheric concentration is rapidly converted term by term into a moving exponential average in accordance with the chosen biological half time. This may be done step by step with a hand calculator, or, if necessary, more quickly with a programmable calculator or desk top computer.

Problem 4.1

(a) Set up a work sheet for calculation of the body burden from exposure to an atmospheric concentration of hazardous substance measured by successive 10-minute samples, given it behaves as if the body was a single compartment in which $k = 10$ lpm, $a = 1$ lpm, $V = 10$ l.

(b) Calculate the moving average employing an averaging period of $2.88T$.

Solution

Data: Atmospheric concentration: a sequence of 10-minute TWA levels.

Model: Single compartment: $k = 10$

$a = 1$

$V = 10$

$t = 10$

Formula: $x_n = x_{n-1} \exp(-at/V) + (kC_n/a)[1 - \exp(-at/V)]$

$= 0.368x_{n-1} - 6.32C_n$

$$T = \frac{V \ln 2}{a} = \frac{0.693 \times V}{a} = 6.93 \text{ minutes}$$

$2.88T = 20$ minutes

20-minute moving average $= (C_n + C_{n-1})/2$

Time t	Data C_n	Moving average atmospheric concentration $(C_n + C_{n-1})/2$	Body burden concentration x_n
8.00	0	0	0
8.00–8.10	12	6	76
8.10–8.20	46	29	319
8.20–8.30	49	42	427
8.30–8.40	37	43	391
8.40–8.50	20	29	270
8.50–9.00	0	10	99
9.00–9.10	0	0	37
9.10–9.20	0	0	13
9.20–9.30	0	0	5
.	.	.	.
.	.	.	.
.	.	.	.

The last two columns are closely correlated and, as expected, during the exposure period they are broadly in the ratio 1:10, equal to the ratio $a:k$.

Real models

Consideration has so far been given to the reasoning by which the sampling period of instruments would ideally be adjusted so that the results reflect changes proportionate to changes in body burden. It is of interest to consider the properties of a sampling instrument which, rather than meet a specified sampling period, would imitate the way in which substances accumulate in the body.

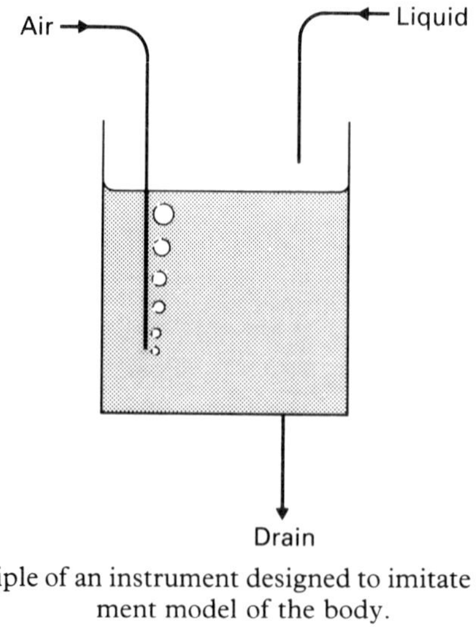

FIG. 4.10. Principle of an instrument designed to imitate a single compartment model of the body.

The principle of such an instrument is illustrated schematically in Figure 4.10. Contaminated air drawn from the work place is bubbled at a constant flow rate through an absorber in which the liquid is being drained at a constant flow rate and being replaced at the same flow rate from a reservoir. This system has the properties ascribed to the human body. The bubbling corresponds to respiration, the absorber corresponds to the body, the drain corresponds to excretion and the replacement liquid corresponds to liquid intake. When the air is no longer contaminated, the amount of contamination in the absorber liquid decays exponentially due to the flushing from the reservoir to the drain via the absorber.

When contaminated air starts to come through for the first time the concentration of contamination in the absorber rises rapidly at first but on continued exposure slowly approaches an equilibrium value asymptotically, at which time the rate contamination is leaving by the drain equals the rate it is entering with the air. The difference between the concentration in the absorber liquid and its maximum equilibrium value diminishes exponentially. First the flushing rate is adjusted so that the half time in the absorber equals the biological half time for the substance in humans.

Where the volume of the liquid in the absorber is v ml and the half time required is T minutes, the flushing rate, R ml min^{-1}, from the reservoir and to the drain is given by

$$R = \frac{0.693v}{T}. \tag{20}$$

Provided that the air entering the absorber was held at a constant concentration of contamination at a level just avoiding significant adverse effects in a group of employees the equilibrium amount of contamination accumulated in the absorber would then be W mg, say. Now, an appropriate indicator is attached to the absorber to show the amount of the contamination in the absorber liquid. Whenever the amount of contamination accumulated in the absorber exceeds W mg, this corresponds to the critical organ concentration in humans. Thus this continuous system simulates the human system and when applied to fluctuating exposure conditions it shows the precise moment when the compartment burden exceeds its critical concentration, while taking due account of all peak concentrations of contamination in air. It will also be observed that this system is very flexible. The absorber air flow rate, the volume of liquid and the flushing rate can be adjusted to accommodate a considerable range of half times and atmospheric concentration levels.

A variety of chemical principles can be drawn upon to record automatically the quantity of air contamination absorbed by the liquid. A colorimetric reaction may be suitable or conductivity or pH may be continuously indicated with appropriate meters to give a warning when the critical compartment concentration is exceeded.

Existing continuous read-out meters can also be modified to read changes in compartment burden rather than changes in concentration of contamination in air. Given the biological half time of the substance under study the meter has to be matched to it. This may be done internally electronically or even more

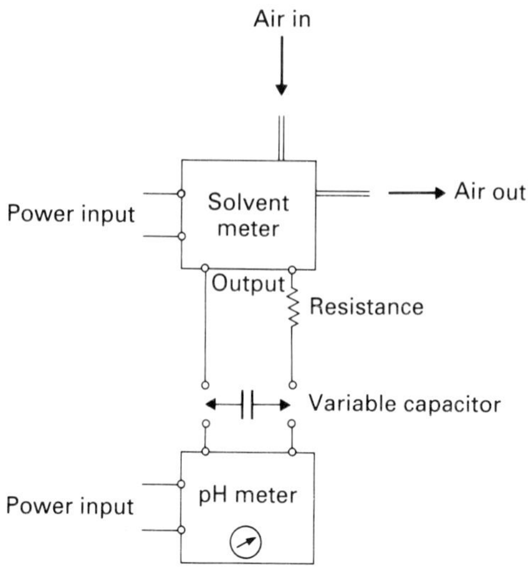

FIG. 4.11. Modification to a continuous read-out meter so as to provide an output proportional to a single compartment model of the body.

simply by connecting an additional integrating circuit to the recorder which alters the time constant of the instrument to the value $1.44T$. This makes the half time of the meter equal to that of the body compartment and the recorder reading is then directly proportional to the model compartment burden (Mapleson, 1963).

For a practical illustration, the reading given by a particular commercial continuous read-out meter was known to be approximately proportional to the concentration of various solvent vapours in air. A circuit was required which can accumulate electricity from the meter and which has a time constant of up to several hours. There are numerous ways this can be done and the simple circuit illustrated in Figure 4.11 was made up from available components. The current from the meter charges up a variable capacitor and the charge leaks through a resistance. The charge is proportional to the model compartment burden and can be read off by connecting to a voltmeter of relatively small capacity but high resistance (Roach, 1963). Such a read-out meter is found in common pH meters used in chemical laboratories. The reading on the pH meter is proportional to the model compartment burden and is conveniently scaled 0–100 such that 100 corresponds to the exposure limit.

The same object may also be achieved by interposing a mixing chamber between the sampling probe and the instrument. The volume of the chamber is such that the outlet from it has a concentration proportional to the body burden. The volume of the chamber required, V_c litres, where f is the flow rate through the recorder in litres per minute, is given by

$$V_c = 1.44\, f\, T. \qquad (21)$$

Models with two compartments

There can be few substances which are accumulated in only one part of the body and the average residence time may be very different in one part than another. Evidence to support this idea is found by analysing washout curves, which, not infrequently, are apparently made up of two or more exponential decay curves. Consequently, to assess the importance of fluctuations in concentration their periodicity should be compared against the residence time in the particular organ which is most easily affected by the substance, the so-called 'critical' organ. Analysis of models with more than one compartment gives valuable insights into the working of organ systems.

A model with several linked compartments is created whose behaviour matches the known behaviour of biological determinants (Fernandez, Droz, Humbert and Caperos, 1977; Droz and Guillemin, 1983; Clewell, Andersen, MacNaughton and Stuart, 1988; Johanson and Naslund 1988). The compartments may be assimilated to different parts of the body or different organs. Otherwise the identity may be less specific as with a central compartment and a peripheral compartment. An elementary treatment is given in Gibaldi and Perrier (1975). A thorough mathematical analysis of complex compartmental systems is given by Godfrey (1983).

In order to enter gently into the mathematics of multi-compartment models the cases considered here are ones in which the single lumped compartment considered previously is divided into two, so that there are just two homogeneous compartments through which the substance passes before elimination. The compartments may be in parallel or in series. The mathematics of parallel systems is the simplest. Appropriate rate constants and compartment volumes may be calculable from published data or determined experimentally (Sato and Nakajima, 1979). On cessation of exposure the decline of compartment burden from such systems is generally biphasic.

Examples of reported multi-phasic urinary excretion in humans are the washout of metabolites, such as phenol following atmospheric exposure to benzene vapour and mandelic acid following atmospheric exposure to styrene vapour (Teisinger and Fiserova-Bergerova, 1965; Sherwood, 1972; Guillemin and Bauer, 1979). An example from the field of exhaled breath analysis is exponential pulmonary excretion of tetrachloroethylene vapour and 1,1,1-trichloroethane with which a high percentage is excreted during the first 24 hours following exposure, followed by a protracted period of relatively slow removal caused by the solvent's affinity for fat tissue (Stewart, Gay, Schaffer, Erley and Rowe, 1969; Stewart, Baretta, Dodd and Torkelson, 1970). Multiphasic clearance of mineral dusts deposited in the respiratory tract is well known, coarser particles deposited in the upper respiratory tract being cleared first.

CASE 1. The first case considered is of compartments in parallel (Figure 4.12). The atmospheric concentration of the substance in question is C mg m^{-3}.

Suppose the compartments, numbered 1 and 2, have volumes V_1 and V_2, and that the substance enters compartments 1 and 2 at rate k_1C mg h^{-1} and k_1C mg

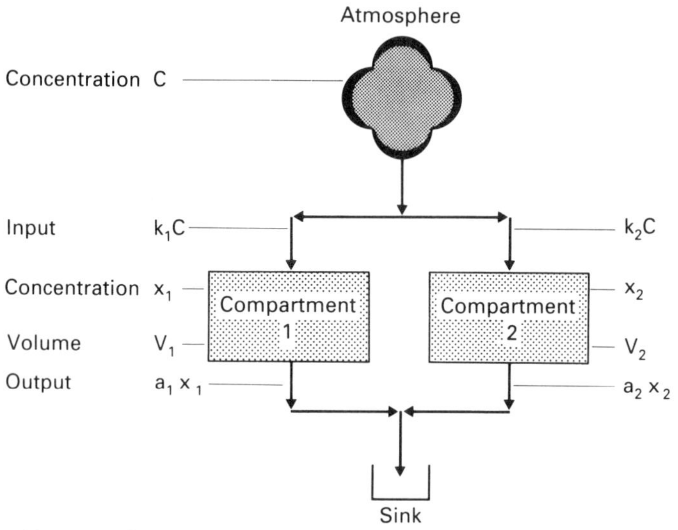

FIG. 4.12. Toxico-kinetic two-compartment, parallel system.

h^{-1}, respectively, where k_1, k_2 are constants characteristic of the substance, having the dimensions $m^3 h^{-1}$. The concentration of substance in the compartments is at x_1 mg m^{-3} and x_2 mg m^{-3} in compartments 1 and 2 respectively. The substance is finally eliminated by removal from the compartments at rates a_1x_1 mg h^{-1}, a_2x_2 mg h^{-1}, respectively, where a_1, a_2 are constants characteristic of the substance.

Then,

$$\frac{dx_1}{dt} = \frac{k_1 C - a_1 x_1}{V_1},$$
$$\frac{dx_2}{dt} = \frac{k_2 C - a_2 x_2}{V_2}. \qquad (22)$$

The differential equations, (22), are independent of one another. Solving as explained in Appendix 1, the general solutions are

$$x_1 = (k_1 C/a_1) + A_1 \exp(-a_1 t/V_1),$$
$$x_2 = (k_2 C/a_2) + A_2 \exp(-a_2 t/V_2), \qquad (23)$$

where A_1, A_2 are arbitrary constants.

By substituting the initial conditions, $x_1 = x'_1, x_2 = x'_2$ when $t = 0$, and solving for A_1, A_2, it follows that $A_1 = [x'_1 - (k_1 C/a_2)]$, $A_2 = [x'_2 - (k_2 C/a_2)]$.

Thus, on exposure to a constant atmospheric concentration it follows that the respective concentrations of substance in the model compartments are simple exponential functions:

$$x_1 = (k_1 C/a_1) + [x'_1 - (k_1 C/a_1)] \exp(-a_1 t/V_1),$$
$$x_2 = (k_2 C/a_2) + [x'_2 - (k_2 C/a_2)] \exp(-a_2 t/V_2). \qquad (24)$$

It is readily apparent from the form of equations (24) that, when the duration of exposure, t, is very large, $\exp(-a_1 t/V)$, $\exp(-a_2 t/V)$ tend to zero, and the equilibrium concentrations, $x_1 \rightarrow k_1 C/a_1$, $x_2 \rightarrow k_2 C/a_2$.

Problem 4.2

Given the two-compartment system, above, in which

$$k_1, k_2 = 1.0 \text{ m}^3 \text{ h}^{-1}$$
$$V_1, V_2 = 100 \text{ cm}^3$$
$$a_1 = 500 \text{ cm}^3 \text{ h}^{-1}$$
$$a_2 = 0.5 \text{ cm}^3 \text{ h}^{-1}.$$

Find solutions for x_1, x_2. Evaluate the system.

Solution

Inserting the given values for the constants in equations (24):

$$x_1 = 2{,}000C + [x_1' - 2{,}000C]\exp(-5t),$$
$$x_2 = 2{,}000{,}000C + [x_2' - 2{,}000{,}000C]\exp(-0.005t). \qquad (25)$$

By substituting the initial conditions, $x_1' = x_2' = 0$, it follows that, if initially the substance mass in the two compartments is zero, on exposure to a constant concentration,

$$x_1 = 2{,}000C[1 - \exp(-5t)]$$
$$x_2 = 2{,}000{,}000C[1 - \exp(-0.005t)] \qquad (26)$$

When the duration of exposure, t, is very large, both $\exp(-5t)$ and $\exp(-0.005t)$ tend to zero, $x_1 \rightarrow 2000C$ and $x_2 \rightarrow 2{,}000{,}000C$. That is, the substance mass in the first compartment tends to an equilibrium value of $0.2C$ mg and the substance mass in the second compartment tends to an equilibrium value of 200C mg. At this stage the rate of elimination from the system, the rate of transfer between the two compartments and the rate of intake are all equal.

When exposure ceases the concentration of substance in the two compartments subsequently is given by substituting the initial conditions. Suppose the initial conditions were:

$$x_1 = 0.2,\ x_2 = 200, \quad \text{when} \quad t = 0.$$

Then, on substituting in equation (25),

$$x_1 = 0.2[\exp(-5t)]$$
$$x_2 = 200[\exp(-0.005t)] \qquad (27)$$

The biological half times are, respectively,

$$T_1 = 0.139 \text{ hours}, \quad T_2 = 139 \text{ hours}.$$

The consequences of exposure of the system to a constant concentration of 1 mg m^{-3} for just 1 hour is illustrated in Figure 4.13.

In the above case it will be observed that the total substance mass in the system is given by the sum of two exponential decay curves with half times 0.139 hours and 139 hours, respectively, and the washout curve would be biphasic. Were the system to consist of 3 compartments in parallel the washout curve would be tri-phasic, and so on. The half time time for decay of total substance mass in the system is a function of the initial conditions; specifically the relative mass in the two or more compartments. There is no unique half time for the system as a whole as there is in a single compartment model.

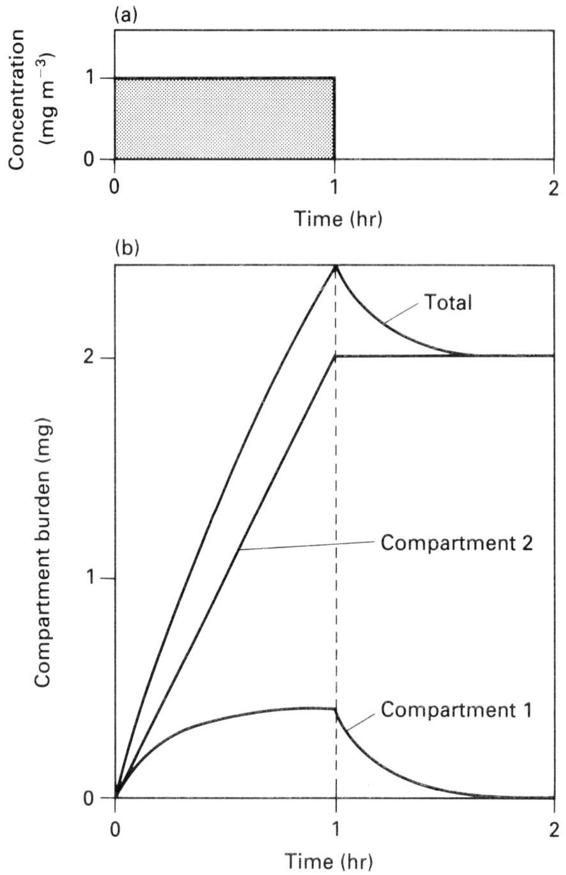

FIG. 4.13. Two-compartment parallel system, given in Figure 4.12—and Problem 4.2, exposure for 1 hour. (a) Graph of atmospheric concentration of hazardous substance against time. (b) Corresponding compartment burdens against time.

CASE 2. The second case considered is of two compartments in series. A substance enters the system at the first compartment, leaves by the second compartment and the passage of material through the connection between the two compartments is governed by a simple but plausible mathematical function, Figure 4.14.

The concentration of substance in the inhaled air is C mg m^{-3}. Suppose the two compartments are numbered 1 and 2, have volumes V_1 and V_2, and that the substance enters compartment 1 at rate kC mg h^{-1}, where k is a constant characteristic of the substance, having the dimensions m^3 h^{-1}. The concentration of substance in the compartments is at x_1 mg m^{-3} and x_2 mg m^{-3} in compartments 1 and 2 respectively. Suppose the rate of transfer of substance through the connection between compartments is directly proportional to the difference in concentration. Thus the transfer from compartment 1 to compartment 2 proceeds at rate $a_{12}(x_1—x_2)$ mg h^{-1}, where a_{12} is a constant. The

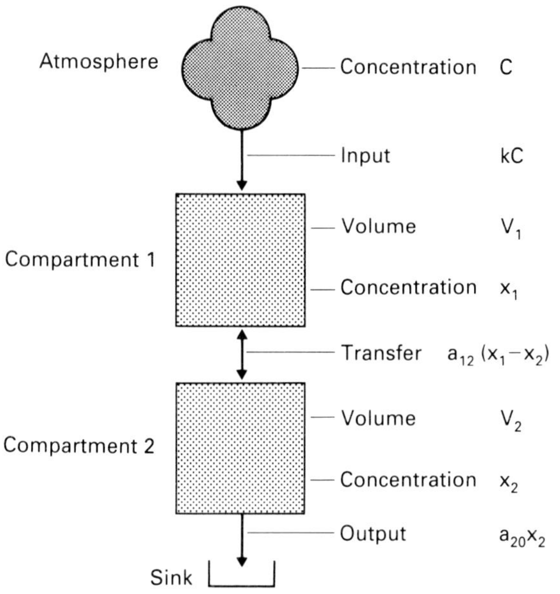

FIG. 4.14. Toxico-kinetic two-compartment, series system.

substance is finally eliminated by removal from the second compartment at $a_{20}x_2$ mg h^{-1}, where a_{20} is another constant.

Then,

$$\frac{dx_1}{dt} = \frac{kC - a_{12}(x_1 - x_2)}{V_1},$$

$$\frac{dx_2}{dt} = \frac{a_{12}(x_1 - x_2) - a_{20}x_2}{V_2}. \tag{28}$$

Such a pair of simultaneous linear differential equations is solved by standard methods (see Appendix 1). The general solution contains two arbitrary constants which are evaluated from the initial conditions. Analysis of various models in which the parameters are given numerical values is instructive.

Problem 4.3

Given the two-compartment system, above, in which

$k = 0.5$ m^3 h^{-1}

$V_1 = 0.2$ m^3

$V_2 = 0.1$ m^3

$a_{12} = 0.2$ m^3 h^{-1}

$a_{20} = 0.1$ m^3 h^{-1}.

Find solutions for x_1, x_2. Evaluate the system.

Solution

Inserting the given values for the constants in equations (22) and solving by standard methods, the general solution is

$$x_1 = 7.5C - 0.365A_1[\exp(-3.73t)] + 1.37A_2[\exp(-0.268t)]$$
$$x_2 = 5.0C + A_1[\exp(-3.73t)] + A_2[\exp(-0.268)], \quad (29)$$

where C = the inhaled air concentration in mg m^{-3} and A_1, A_2 are arbitrary constants.

By substituting the initial conditions, $x_1 = x_2 = 0$, and solving for A_1, A_2 it follows that $A_1 = 0.37C$, $A_2 = -5.37C$. Thus, if initially the substance mass in the two compartments is zero, on exposure to a constant concentration,

$$x_1 = 7.5C - 0.14C[\exp(-3.73t)] - 7.36C[\exp(-0.268t)]$$
$$x_2 = 5.0C + 0.37C[\exp(-3.73t)] - 5.37C[\exp(-0.268t)]. \quad (30)$$

When the duration of exposure, t, is very large, both $\exp(-3.73t)$ and $\exp(-0.268t)$ tend to zero, $x_1 \rightarrow 7.5C$ and $x_2 \rightarrow 5.0C$. That is, the substance mass in the first compartment tends to an equilibrium value of $1.5C$ mg and the substance mass in the second compartment tends to an equilibrium value of $0.5C$ mg. At this stage the rate of elimination from the system, the rate of transfer between the two compartments and the rate of intake are all equal.

When exposure ceases the concentration of substance in the two compartments subsequently is given by substituting the initial conditions. Suppose the initial conditions were:

$$x_1 = 7.5, x_2 = 5.0, \quad \text{when} \quad t = 0.$$

Solving for A_1, A_2; $A_1 = -0.37$, $A_2 = 5.37$. Then, on substituting in equation (27),

$$x_1 = 0.14[\exp(-3.73t)] + 7.36[\exp(-0.268t)]$$
$$x_2 = -0.37[\exp(-3.73t)] + 5.37[\exp(-0.268t)] \quad (31)$$

Biological monitoring of employees is sometimes feasible and the rate of excretion in the urine or exhaled air may be measured, as appropriate. The rate of reduction in substance mass in the two model compartments is given by

$$-V_1\frac{dx_1}{dt}, \quad -V_2\frac{dx_2}{dt},$$

respectively. From (31),

$$-V_1\frac{dx_1}{dt} = 0.52[\exp(-3.73t)] + 1.97[\exp(-0.268t)],$$
$$-V_2\frac{dx_2}{dt} = -1.38[\exp(-3.73t)] + 1.44[\exp(-0.268t)] \quad (32)$$

The consequences of exposure of the system to a constant concentration of 1 mg m^{-3} for just 1 hour is illustrated in Figure 4.15.

In the above case it will be observed that after cessation of exposure the substance mass in compartment 1 is given by the weighted sum of two exponential decay curves with half times 0.19 hours and 2.59 hours, respectively, whereas the substance mass in compartment 2 is given by the difference between two decay curves with those half times. The half time time for decay of substance mass in the two compartments is a function of the initial conditions; specifically the relative mass in the two compartments. There is no unique half time for either compartment as there is in a single compartment model or in the individual compartments of a system of parallel compartments.

When investigating biological determinants in humans, such as concentration in blood, urine or exhaled air the results are commonly plotted against

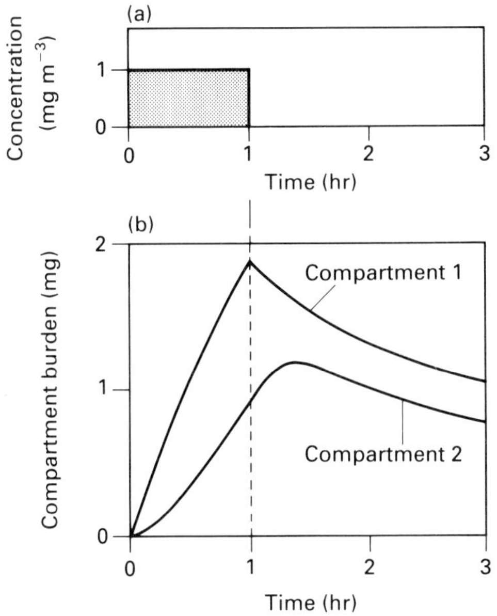

FIG. 4.15. Two-compartment series system, Figure 4.14—Problem 4.3, exposure for 1 hour. (a) Graph of atmospheric concentration of hazardous substance against time. (b) Corresponding compartment burdens against time.

time and exponential washout curves fitted empirically. In general, when there are two compartments in series, if the substance has a longer half time in the second compartment the maximum burden in this compartment may occur some time after exposure has ceased. This may be the underlying explanation for there being a noticeable lag between exposure and effect with some substances (Roach, 1977).

CASE 3. The values of the parameters in Case 2 were chosen to demonstrate the behaviour of a two compartment series system as contrasted with a single compartment system. In practice, very often either the compartments behave very similarly or the behaviour of only one is of significant interest.

Problem 4.4

As Problem 4.2, save for the value of the rate constant between the two compartments, a_{12}, which is much higher than in Case 2.

Given, then, the following values:

$k = 0.5\ \mathrm{m^3\ h^{-1}}$

$V_1 = 0.2\ \mathrm{m^3}$

$V_2 = 0.1\ \mathrm{m^3}$

$a_{12} = 10.0\ \mathrm{m^3\ h^{-1}}$

$a_{20} = 0.1\ \mathrm{m^3\ h^{-1}}$.

Find solutions for x_1, x_2. Evaluate the system.

Solution

Employing the methods for solving simultaneous differential equations, as described in Appendix 1, the general solution of this example is

$$x_1 = 5.05C - 0.5A_1[\exp(-151t)] + 1.007A_2[\exp(-0.332t)]$$
$$x_2 = 5C + A_1[\exp(-151t)] + A_2[\exp(-0.332t)] \qquad (33)$$

where C = the inhaled air concentration in mg m^{-3}. A_1, A_2 are arbitrary constants.

It is evident that in both equations the second term is for all practical purposes negligible compared with the third and the substance mass in the two compartments will follow each other step for step.

Problem 4.5

As Problem 4.2, save for the value of the rate constant between the two compartments, a_{12}, which is much lower than in Case 2.

Given, then, the following values:

$k = 0.5$ m^3 h^{-1}

$V_1 = 0.2$ m^3

$V_2 = 0.1$ m^3

$a_{12} = 0.01$ m^3 h^{-1}

$a_{20} = 0.1$ m^3 h^{-1}.

Find solutions for x_1, x_2. Evaluate the system.

Solution

The general solution of this example is

$$x_1 = 55C - 0.0475A_1[\exp(-1.105t)] - 10.55A_2[\exp(-0.0452t)]$$
$$x_2 = 5C + A_1[\exp(-1.105t)] + A_2[\exp(-0.0452t)], \qquad (34)$$

where C = the inhaled air concentration in mg m^{-3}. A_1, A_2 are arbitrary constants.

By substituting the initial conditions, $x_1 = x_2 = 0$, and solving for A_1, A_2 it follows that $A_1 = 0.214C$, $A_2 = -5.21C$. Thus, if initially the substance mass in the two compartments is zero, on exposure to a constant concentration,

$$x_1 = 55C - 0.0102C[\exp(-1.105t)] - 55C[\exp(-0.0452t)]$$
$$x_2 = 5C - 0.214C[\exp(-1.105t)] + 5.21C[\exp(-0.0452t)] \qquad (35)$$

When the duration of exposure, t, is very large, both $\exp(-1.105t)$ and $\exp(-0.0452t)$ tend to zero, $x_1 \rightarrow 55C$ and $x_2 \rightarrow 5C$. That is, the mass of substance in the first compartment tends to an equilibrium value of $11C$ mg and the mass of substance in the second compartment tends to an equilibrium value of $0.5C$ mg. At this stage the rate of elimination from the system, the rate of transfer between the two compartments and the rate of intake are all equal. The concentration ratios are then:

$$\frac{x_1}{C} = \frac{k}{a_{12}} + \frac{k}{a_{20}} = 55, \quad \frac{x_2}{C} = \frac{k}{a_{20}} = 5.0.$$

It is evident that in this system the mass of substance in the second compartment is small compared with the mass in the first. Also, in these equations the second term is generally negligible. Thus, with such a combination of parameters interest regarding the major site of the body burden is focussed on the first compartment and the system behaves essentially as two single compartments. On the other hand, the first compartment may represent an organ which is relatively insensitive to the substance. The primary focus of attention regarding the major hazard should be the compartment representing the critical organ, even though its burden may be the lesser of the two.

CASE 4. Gases and vapours absorbed in the lungs are also eliminated through the lungs at the termination of exposure and carried away by the exhaled air. An example of observed multi-phasic pulmonary elimination is n-hexane vapour, which is eliminated from body tissues and blood by being removed in the exhaled air, in which there are two detectable half times of 14 minutes and 2.5 hours, respectively (Nomiyama and Nomiyama, 1974).

Elimination through the lungs follows definite principles which are similar to those of absorption in reverse. This prompts the question of what is the behaviour of a compartmental system with similar properties. For example, in the case of unreactive gases the constants k, a may be equated to the respiratory volume flow rate which ventilates the alveoli; about two thirds the minute volume. The gas is carried by the blood to and from the body tissues, organs, bones and fat. The subsequent disposition between different compartments is governed by their associated blood flow, volume and partition coefficient.

In the compartmental model in Figure 4.16, after exposure for a very long time to atmospheric concentration C mg m^{-3} there is equilibrium between the atmospheric concentration and the concentration in the compartments. The ratio x/C would equal the solubility of the gas in blood. The two compartments are numbered 1, 2 (Figure 4.16). The concentration of substance is x_1 mg m^{-3}, x_2 mg m^{-3} in compartments 1 and 2 respectively. Suppose the solubility of the air contamination in compartment 1 is N in the sense that when the contaminated air and compartment are in equilibrium the concentration of substance in the compartment is NC mg m^{-3}. Initially the air inhaled into the alveoli comes almost instantly into equilibrium with the pulmonary blood and the blood:air partition coefficient is N. The substance, then, enters compartment 1 from the air at net rate $k[C - (x_1/N)]$ mg h^{-1}. The transfer from compartment 1 to

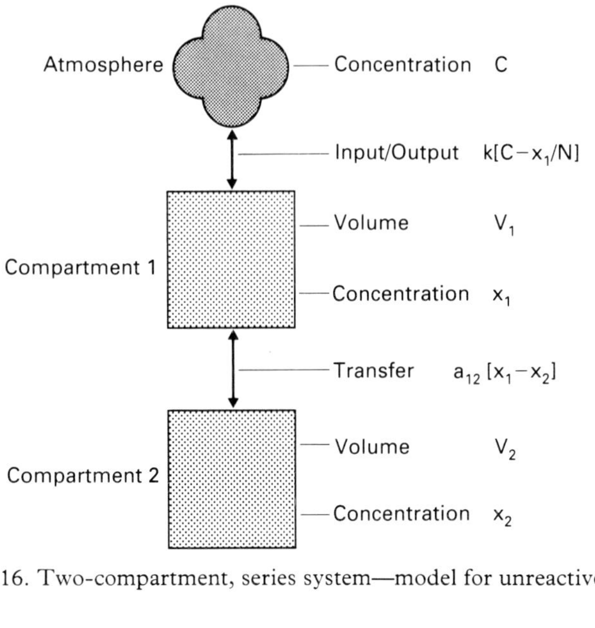

FIG. 4.16. Two-compartment, series system—model for unreactive gases.

compartment 2 proceeds at rate $a_{12}(x_2 - x_1)$ mg h^{-1}. The volume of the compartments is V_1 m^3, V_2 m^3, respectively.

$$\frac{dx_1}{dt} = \frac{kC}{V_1} - \frac{kx_1}{V_1 N} - \frac{a_{12}(x_1 - x_2)}{V_1}$$
$$\frac{dx_2}{dt} = \frac{a_{12}(x_1 - x_2)}{V_2} \qquad (36)$$

These simultaneous linear differential equations are solved by standard methods and the arbitrary constants determined from the initial conditions.

Problem 4.6

Given the two compartment system, above, in which:

$k = 0.5\ m^3\ h^{-1}$
$V_1 = 0.2\ m^3$
$V_2 = 0.1\ m^3$
$a_{12} = 0.2\ m^3\ h^{-1}$
$N = 2.5$

Find solutions for x_1, x_2. Evaluate the system.

Solution

The general solution is found by solving the system of differential equations as in Solution 4.2 (see Appendix 1),

$$x_1 = 2.5C - 0.71A_1[\exp(-3.41t)] + 0.71A_2[\exp(-0.59t)]$$
$$x_2 = 2.5C + A_1[\exp(-3.41t)] + A_2[\exp(-0.59t)], \qquad (37)$$

where C = the inhaled air concentration in mg m^{-3}. A_1, A_2 are arbitrary constants.

By substituting the initial conditions, $x_1 = x_2 = 0$, and solving for A_1, A_2 it follows that $A_1 = 0.51C$, $A_2 = -3.01C$. Thus, if initially the substance mass in the two compartments is zero, on exposure to a constant concentration,

$$x_1 = 2.5C - 0.36C[\exp(-3.41t)] — 2.14C[\exp(-0.59t)]$$
$$x_2 = 2.5C + 0.51C[\exp(-3.41t)] — 3.01C[\exp(-0.59t)] \qquad (38)$$

When the duration of exposure, t, is very large, both $\exp(-3.41t)$ and $\exp(-0.59t)$ tend to zero, the concentration in the first compartment is in equilibrium with the concentration in air, $x_1 \to 2.5C$ and $x_2 \to 2.5C$. That is, the mass of substance in the first compartment tends to an equilibrium value of $0.5C$ mg and the mass in the second compartment tends to an equilibrium value of $0.25C$ mg. At this stage no more gas is being absorbed, the rate of elimination from the body and the rate of intake are equal and the rate of transfer between the two organs is zero. The concentration ratios are then:

$$\frac{x_1}{C} = \frac{k}{a_{12}} = 2.5, \quad \frac{x_2}{C} = \frac{k}{a_{12}} = 2.5.$$

When exposure subsequently ceases the concentration of substance in the two compartments is given by substituting the initial conditions. Suppose the initial conditions were $x_1 = 2.5$, $x_2 = 2.5$ when $t = 0$. Solving for A_1, A_2; $A_1 = -0.51$, $A_2 = 3.01$. Then, on substituting in equation (28):

$$x_1 = 0.36[\exp(-3.41t)] + 2.14[\exp(-0.59t)]$$
$$x_2 = -0.51[\exp(-3.41t)] + 3.01[\exp(-0.59t)] \qquad (39)$$

Under these conditions, elimination from compartment 1 to atmosphere is biphasic with half times of $0.693/3.41 = 0.2$ hours and $0.693/0.59 = 1.2$ hours, respectively. The consequences of

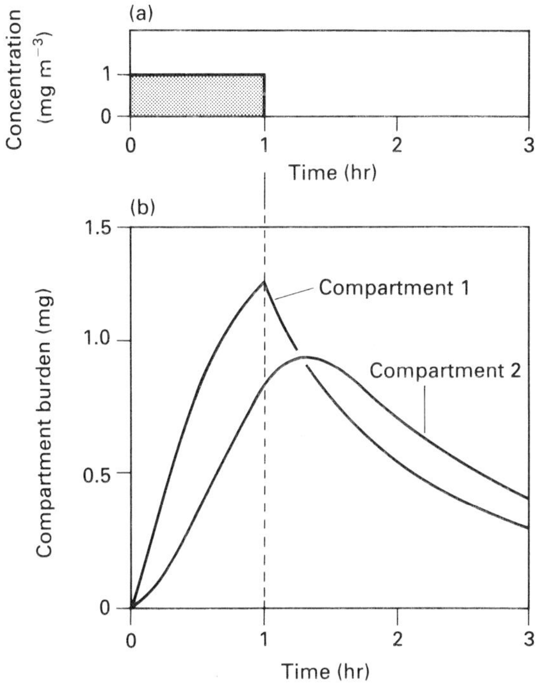

FIG. 4.17. Two-compartment, series system—model for unreactive gases, Figure 4.16—Problem 4.6, exposure for one hour. (a) Graph of atmospheric concentration of hazardous substance against time. (b) Corresponding compartment burdens against time.

exposure of the system to a constant concentration of 1 mg m^{-3} for just 1 hour is illustrated in Figure 4.17. It will be noted again that, as in Case 1, in the first compartment the concentration of substance is the sum of two exponential washout curves and in the second it is the difference between two exponential washout curves. The delay in reaching levels approaching the equilibrium value of concentration in compartment 1 is due to the limited rate of entry from the air and the high solubility of the gas in the compartment. The lag of accumulation in compartment 2 behind that in compartment 1 is due to the limited rate of transfer between the two compartments and the magnitude of the volume of compartment 2.

Inhaled air contamination is taken up in part in the upper respiratory tract. The concentration of contamination in air entering alveolar regions is diminished by this factor. In the case of an unreactive gas or vapour the upper respiratory tract acts as another compartment which takes up the gaseous substance according to its solubility, until there is equilibrium between inhaled air and liquid bathing the respiratory tract. Reactive gases and aerosols are taken up as by a sink.

The modelling of a two-compartment system for unreactive gases and vapours is also illustrated by Figure 4.16. Absorption from the upper respiratory tract could be represented by intake to compartment 1, alveolar ventilation effects the transfer between compartments and compartment 2 may represent the critical organ. On cessation of exposure the substance returns to the lungs in the blood, is taken up by alveolar air and exhaled.

Models with three or more compartments

Whilst a surfeit of information is unlikely, in particular cases there may be sufficient knowledge about the distribution of body burden between several different compartments, about toxicity mechanisms and the toxico-dynamics so that a closer estimate may be made of the exposure regimes yielding equivalent risk than is calculated from the single compartment approximation. Such instances may justify detailed analysis on an individual basis.

An example of hazardous substance is mercury vapour, for which three distinct compartments have been identified as being operative; namely, blood, brain and kidney (Hursh, Clarkson, Cherian, Vostal and Mallie, 1976). Another example is inorganic lead, for which three different half times of elimination of lead from blood have been identified, associated with soft tissues (half time 35–40 days) and bone (half times of 6 months and 20 years), respectively (Piotrowski, 1971; Marcus, 1979). Three phases with different half times are distinguishable in data from exhaled air analysis after exposure to toluene (Sato, Nakajima, Fujiwara and Hirosawa, 1974). Exhaled air analysis also shows apparent triphasic elimination of trichloroethylene (Fernandez, Humbert, Droz and Caperos, 1975; Sato, Nakajima, Fujiwara and Hirosawa, 1974). Dust deposited in the lungs has been compartmentalised into three or four linked compartments to explain the clearance phenomena of relatively insoluble dust deposited there (Vincent, Johnston, Jones, Bolton and Addison, 1985).

The necessary number of compartments in a model is commonly found as follows. A continuous input of substance is made for a long time until the system is in a steady state. This may not be feasible when one or more of the elements has an extraordinarily long half time. However, assuming the steady state has been reached, or nearly so the input is then discontinued and the washout curve is plotted. The last portion of the curve is dominated by the contribution with the longest decay constant. Successive portions are 'peeled off' with progressively smaller decay constants. Experimental data are of limited accuracy. The number of distinct exponential washout curves identifiable in a real concentration-time curve is the number of compartments needed in the model (Fiserova-Bergerova, Vlach and Cassady, 1980; Ramsey and Andersen, 1984). The amount of 'noise' in experimental data is such that it is rarely possible in practice to distinguish more than three exponential washout curves. In the great majority of instances fitting one or at most two should be adequate for the immediate practical purposes of setting the parameters of appropriate exposure limits.

The procedure for solving compartmental models with three or more compartments is an extension of the procedures considered so far. The toxico-kinetic problems reduce to finding the roots of a cubic or higher order equation. Formulae for the roots of a cubic are readily evaluated with a hand calculator. An understanding of Laplace transforms may be helpful for finding solutions to the differential equations. Higher orders can be handled by numerical methods or, where the facilities are available, by standard computer programmes (Johanson,1986; Paterson and Mackay, 1986; Perbellini, Mozzo, Brugnone and Zedde, 1986; Gradon and Podgorski, 1991). The identification of the model constants with specific organs is often uncertain and new biological data may be needed to resolve any uncertainties before proceeding with compartmental analysis (Bellman and Astrom, 1970; Cobelli and DiStefano, 1980).

References

American Conference of Governmental Industrial Hygienists. Documentation of the Threshold Limit Values. ACGIH, 6500 Glenway Ave., Bldg. D-7, Cincinnati, OH 45211, USA (1987).

Andersen M E. Neuro-behavioural Toxicol. Teratol. **3** 383 (1981a).

Andersen M E. Toxicol. Appl. Pharmacol. **60** 509 (1981b).

Andersen M E. CRC Crit. Rev. Toxicol. **9** 105 (1981c).

Andersen M E. Drug. Metab. Rev. **13** 799 (1982).

Andersen M E. Ann. Occup. Hyg. **35** 309 (1991).

Andersen M E, H J Clewell and M L Gargas. Toxicol. Appl. Pharmacol. **87** 185 (1987).

Andersen M E, H J Clewell, M L Gargas, M G MacNaughton, R G Reitz, R J Nolan and M J McKenna. Toxicol. Appl. Pharmacol. 108 14 (1991).

Andersen M E, M L Gargas and J C Ramsey. Toxicol. Appl. Pharmacol. **73** 176 (1984).

Atherley G. Am. Ind. Hyg. Assoc. J. **46** 481 (1985).

Bellman R and K J Astrom. Math. Biosci. 7 329 (1970).

Bolt H M, J G Filser and A Buchter. Arch. Toxicol. **48** 213 (1981).

Carson T R, M S Rosenholtz, F T Wilinski and M H Weeks. Am. Ind. Hyg. Ass. J. **23** 457 (1962).

Clewell H J and M E Andersen. Toxicol. Ind. Health. **1** 111 (1985).

Clewell H J, M E Andersen, M G MacNaughton and B O Stuart. Aviat. Space Environ. Med. **59** A125 (1988).

Cobelli C and J J DiStefano. Am. J. Physiol. **239** R7 (1980).

Coburn R F, R E Forster and P B Kane. J. Clin. Invest. **66** 1899 (1965).

Coffin D L, D E Gardner, G I Sidorenko and M A Pinigin. J. Toxicol. Environ. Health **3** 811 (1977).

Cope R, B Pancamo, W E Rinehart and G L Ter Haar. Am. Ind. Hyg. Ass. J. **40** 47 (1979).

David A, E Frantik, R Holusa and O Novakova. Int. Arch. Environ. Health **48** 49 (1981).

Dedrick R L. J. Pharmacokin. Biopharm. **1** 435 (1973).

Dedrick R L and K B Bischoff. Fed. Proc. **39** 54 (1980).

Droz P O. App. Ind. Hyg. **4** F-20 (1989).

Droz P O and M P Guillemin. Int. Arch. Occup. Environ. Hlth. **53** 19 (1983).

Droz P O and M Yu. Biological monitoring strategies. In: Exposure assessment for occupational epidemiology and hazard control (Edited by S M Rappaport and T Smith). Lewis Publishing Co., Chelsea, Michigan, USA (1990).

Druckrey H. Potential carcinogenic hazards from drugs. UICC Monograph Series, 7 60. Springer-Verlag, New York (1967).

Druckrey H and K Kupfmuller. Z. Naturforsch **3** 254 (1948).

Dudley H C and J W Miller. J. Ind. Hyg. Toxic. **23** 470 (1941).

Ehrenberg L, E Moustacchi and S Osterman-Golkar. Mut. Res. **123** 121 (1983).

Fernandez J G, P O Droz, B E Humbert and J R Caperos. Brit. J. Ind. Med. **34** 43 (1977).

Fernandez J G, B E Humbert, P O Droz and J R Caperos. Arch. mal. prof. **36** 397 (1975).

Fernandez J G, B E Humbert, P O Droz and J R Caperos. Arch. Mal. Prof. Med. Trav. **35** 397 (1975).

Fiserova-Bergerova V. Ann. Occup. Hyg. **34** 639 (1990).

Fiserova-Bergerova V, J Vlach and J C Cassady. Brit. J. Ind. Med. **37** 42 (1980).
Fiserova-Bergerova V, J Vlach and K Singhal. Brit. J. Ind. Med. **31** 159 (1974).
Flury F. Z. ges. exp. Med. **13** 1 (1921).
Flury F and W Heubner. Biochem. Z. **95** 249 (1919).
Gray E L, F M Patton, S B Goldberg and E Kaplan. AMA Arch. Ind. Hyg. & Occup. Med. **10** 418 (1954).
Gradon L and A Podgorski. Ann. Occup. Hyg. **35** 249 (1991).
Guillemin M P and D Bauer. Int. Arch. Occup. Environ. Health **44** 249 (1979).
Guy R H, J Hadgraft and H I Maibach. Toxicol. Appl. Pharmacol. **78** 123 (1985).
Hatch T F. Arch. Ind. Hyg. Occup. Med. **6** 1 (1952).
Haggard H W. J. Biol. Chem. **59** 737 (1924).
Henderson Y, H W Haggard, M C Teague, A L Prince and R M Wunderlich. J. Ind. Hyg. **3** 79 (1921).
Himmelstein K J and R J Lutz. J Pharmacokinet. Biopharm. **7** 127 (1979).
Hoel D G, N L Kaplan and M W Anderson. Science **219** 1032 (1983).
Hunter C G. Proc. R. Soc. Med. **61** 913 (1968).
Hunter C G and D Blair. Ann. Occup. Hyg. **15** 193 (1972).
Hursh J B, T Clarkson, M G Cherian, M G Vostal and R V Mallie. Arch. Environ. Health. **31** 302 (1976).
International Commission on Radiological Protection. Report of Committee 4 on evaluation of radiation doses to body tissues from internal contamination due to occupational exposure. ICRP Publication 10, Pergamon Press, Oxford UK (1968).
International Commission on Radiological Protection. The assessment of internal contamination resulting from recurrent or prolonged uptakes. A report of ICRP Committee 4. ICRP Publication 10A, Pergamon Press, Oxford UK (1971).
Johanson G. Toxicol. Lett. **34** 23 (1986).
Johanson G and P H Naslund. Toxicol. Lett. **41** 115 (1988).
Kehoe R A, F Thamann and J Cholak. J. Ind. Hyg. **15** 320 (1933).
Kennedy G L. Biological monitoring in the American chemical industry. In: Biological monitoring of exposure to industrial chemicals (Edited by V Fiserova-Bergerova and M Ogata). Proceedings of the United States—Japan Co-operative seminar on biological monitoring, Honolulu, 1989. American Conference of Governmental Industrial Hygienists, Cincinnati, Ohio, USA (1990).
Kety S S. Pharmacol. Rev. **3** 1 (1951).
Kjellstrom T, C G Elinder and L Friberg. Environ. Res. **33** 284 (1984).
Koizumi A, T Sekiguchi, M Konno and M Ikeda. Am. Ind. Hyg. Assoc. J. **41** 693 (1980).
Lazarev N V and A I Brusilovskaya. J. Physiol. USSR **XVII** 611 (1934).
Leung H-W and D J Paustenbach. Am. Ind. Hyg. Assoc. J. **49** 445 (1988).
Lindstedt G, I Gottberg, B Holmgren, T Jonsson and G Karlsson. Scand. J. Work Environ. Health. **5** 59 (1979).
Mapleson W W. J. appl. Physiol. **18** 197 (1963).
Marcus A H. Environ. Res. **19** 79 (1979).
Nomiyama K and H Nomiyama. Int. Arch. Arbeitsmed. **32** 85 (1974).
Opdam J J D. Brit. J. Ind. Med. **46** 831 (1989).
Paterson S and D Mackay. Toxicol. Appl. Pharmacol. **82** 444 (1986).
Paxman D and S M Rappaport. J. Reg. Tox. Pharmacol. **11** 275 (1990).
Peterson J E and R D Stewart. J. Appl. Physiol. **39** 633 (1975).
Perbellini L, P Mozzo, F Brugnone and A Zedde. Brit. J. Ind. Med. **43** 760 (1986).
Piotrowski J. More complex kinetic problems. In: The application of metabolic and excretion kinetics to problems in industrial toxicology, p. 119. US Department of Health, Education and Welfare, Government Printing Office, Washington DC (1971).
Ramsey J C and M E Andersen. Toxicol. Appl. Pharmacol. **73** 159 (1984).
Rappaport S M. Ann. Occup. Hyg. **29** 201 (1985).
Rappaport S M. Biological considerations for designing sampling strategies. In: Advances in air sampling (Edited by W John), pp. 337–352. Lewis Publishers, Chelsea, Michigan, USA (1988).

Rappaport S M and R C Spear. Ann. Occup. Hyg. 32 21 (1988).
Roach S A. The interpretation of hygienic standards. In: Proceedings of the 2nd International Symposium on Maximum Allowable Concentrations of toxic substances in industrial environments. Institut National de Securite, Paris (1963).
Roach S A. Am. Ind. Hyg. Assoc. J. 27 1 (1966).
Roach S A. Ann. Occup. Hyg. 20 65 (1977).
Salmowa J, J Piotrowski and U Neuhorn. Brit. J. Ind. Med. 20 41 (1963).
Saltzmann B E. J. Air Pollut. Control Assoc. 20 660 (1970).
Saltzman B E. Am. Ind. Hyg. Assoc. J. **49** 213 (1988).
Saltzmann B E and S H Fox. Environ. Sci. Technol. 20 916 (1986).
Sato A, T Nakajima, Y Fujiwara and K Hirosawa. Int. Arch. Arbeitsmed. 33 169 (1974).
Sato A, T Nakajima, Y Fujiwara and N Murayama. Brit. J. Ind. Med. 34 56 (1977).
Sato A and T Nakajima. Brit. J. Ind. Med. 36 231 (1979).
Savolainen H, K Kurppa, P Pfaffli and H Kivisto. Chem. Biol. Interact. 34 315 (1981).
Schrenk H H, W P Yant, S J Pearce, F A Patty and R R Sayers. J. Ind. Hyg. 23 20 (1941).
Sherwood R J. Ann. Occup. Hyg. 15 409 (1972).
Stewart R D. Ann. Rev. Pharmacol. 15 409 (1975).
Stewart R D, E D Baretta, H C Dodd and T R Torkelson. Arch. Environ. Health 20 224 (1970).
Stewart R D, H H Gay, A W Schaffer, D S Erley and V K Rowe. Arch. Environ. Health 19 467 (1969).
Stewart R D, R S Stewart and W Stam. J. Am. Med. Assoc. 235 390 (1976).
Task Group on Metal Accumulation. Environ. Physiol. Biochem. 3 65 (1972).
Teisinger J and V Fiserova-Bergerova. Archives des Maladies Professionnelles 16 221 (1965).
Teisinger J, V Fiserova-Bergerova and J Kudrna. Prac. Lek. 4 175 (1952).
Van Stee E W, G A Boorman, M P Moorman and R A Sloane. J. Toxicol. Environ. Health 10 785 (1982).
Vincent J H, A M Johnston, A D Jones, R E Bolton and J Addison. Brit. J. Ind. Med. 42 707 (1985).
Watanabe P G, J D Young and P J Gehring. J. Environ. Path. Tox. 1 147 (1977).
World Health Organisation. Report of WHO Scientific Group on the assessment of the carcinogenicity and mutagenicity of chemicals. WHO Tech. Rep. Ser. No 546. Geneva, Switzerland (1974).
Wright B M. Brit. J. Ind. Med. 14 219 (1957).

Bibliography

Baselt R C. Disposition of toxic drugs and chemicals in man. Biomedical Publications, Davis, CA, USA (1983).
Brown R G. Smoothing, forecasting and prediction of discrete time series. Prentice-Hall International, London (1967).
Dengler H J (Editor). Pharmacological and clinical significance of pharmacokinetics. Schattauer, New York (1970).
Elkins H B. The chemistry of industrial toxicology. John Wiley & Sons, Inc., New York (1959).
Fiserova-Bergerova V (Editor). Modeling of inhalation exposure to vapors: uptake, distribution and elimination. CRC Press, Boca Raton, Fl, USA (1983).
Gibaldi M and D. Perrier. Pharmacokinetics. Marcel Dekker, New York (1975).
Godfrey K. Compartmental models and their application. Academic Press, London (1983).
Haber F. Funf vortrage aus den jahren 1920–1923. Springer, Berlin and New York (1924).
Henderson Y and H W Haggard. Noxious gases and the principles of respiration influencing their action. Reinhold, New York (1943).
MacFarland H N. Essays in toxicology. Academic Press, New York (1976)
Papper E M and R J Kitz (Editors). Uptake and distribution of anaesthetic agents. McGraw-Hill, New York (1963).

PART II

OCCUPATIONAL EXPOSURE LIMITS

Introduction

Data from occupational hygiene surveys in industry indicate the level of control of the substance reasonably achievable in practice. A simple 'as low as reasonably achievable' approach to standard setting, has the attraction of convenience. But implicit in the acceptance of conditions which are no more than reasonably achievable is the possible realisation of a significant health risk. Maintaining conditions no better than is generally achieved in the industry is, regrettably, no guarantee of freedom from risk.

The assessment of health risks from exposure to many thousands of hazardous substances is made feasible by a procedure for determining satisfactory health-based atmospheric exposure limits. These limits are pivotal in the practice of occupational hygiene. First, the substances involved must be identified and second, appropriate exposure limits must be adopted or set which, if respected, would ensure no significant risk to health of those exposed.

There are published exposure limits for air contamination by one or other of about 2,000 different substances. These limits specify exposure both by the atmospheric concentration and by a period of exposure. An upper limit is recommended to the time weighted average concentration in air of the substance and the reference period over which the concentration is to be averaged. Lists of published limits are described in Chapter 5. In so far as these limits must be conditioned in part by value judgements about our tolerance of different health risks, about the technical feasibility of complying with exposure limits, economic considerations, social factors and so on, they embrace both scientific risk assessment and the social function of risk management (National Academy of Sciences, 1983).

In order to proceed with the same certainty when handling all the hazardous substances without published limits it is necessary somehow to devise exposure limits which are comparable with those which have public recognition. There are tens of thousands of inorganic and organic chemicals produced for use in chemical and process applications, for example. These chemicals are not generally sold as such to the ultimate consumer, but rather to other manufacturers or formulators. They range from relatively simple basic commodities to extremely complex dyes and bulk pharmaceuticals. Other industrial chemicals are used as solvents, as intermediates in further chemical manufacture and as the basis for making synthetic rubber, plastics and man-made fibres.

Exposure limits for all these substances need to be chosen with care. Too high will engender a false sense of security, too low could result in the use of a substance being discontinued and its benefits lost. Published limits and their documentation provide the source material for deciding on acceptable criteria for in-house limits. The development of in-house limits does not necessarily mean that special steps need be taken to reduce exposure or even monitor the atmospheric concentration of every substance, but does enable those substances to be identified where such steps are advisable.

Some published limits are based on extensive human experience. Procedures for developing exposure limits from epidemiological data from industry, from volunteer studies and other human experience are given in Chapter 6. Sources of information sought when setting an in-house occupational exposure limit for a substance include one or more of the following:

(a) Epidemiological studies of exposed populations. These should be evenly balanced, with comparable effort having been put into health assessment and into exposure measurement.

(b) Data from health surveillance of persons routinely or accidentally exposed to the substance. For such data to be of real value the health surveillance should have been of all persons in one or more defined job-exposure groups, their exposure should be known and all should have been medically examined with equal thoroughness.

(c) Results of a literature search for human volunteer studies which have been undertaken and published on the substance of interest. The exposure of the volunteers must be known.

Overlapping the studies of substances exposed to volunteers is the whole range of data on dose and effect of therapeutic agents. These data are directly or indirectly useful for setting exposure limits for employees exposed to these substances at work.

(d) Chemical and physical properties of the study substance compared against other substances with similar chemical and physical properties which have established limits. This is known as the 'analogy' approach. Unfortunately, there are several examples in the literature of substances with very similar chemical and physical properties but very different biological properties. Chemical and physical analogy should only be used in a supporting role.

Data from human experience is sparse and very often reliance has to be placed upon experimental animal toxicology. It may be argued that animal toxicology is a poor substitute for human experience. This may be true, but it took 5,000 years of human experience before silicosis was controlled, 3,000 years before lead poisoning was well enough understood to bring it satisfactorily under control. It took a hundred years of human suffering to eliminate scrotal cancer in chimney sweeps and mule spinners. From the first observation of bladder cancer in 1895 it took 71 years until control regulations were in place in Britain, and it took 80 years of asbestosis and lung cancer for regulations for asbestos control. It only took 35 years from introducing vinyl chloride until effective measures were taken to control the hazard and this was largely due to the evidence from animal toxicology. It is no longer necessary to wait for human injury before preventive measures can be put in place. The means are available through experimental animal toxicology and occupational hygiene based on exposure limits derived from them.

In the documentation of published limits of common chemicals it is quite usual to find that there has not been sufficient information available about the effects of exposure on humans directly and the best data available has been on animal toxicology. A similar situation is found when setting in-house limits for new substances and most others which have been in substantial use for only a few years.

Data from quantitative animal toxicology studies provides a good means of developing exposure limits for humans. Studies are conducted to study the nature and degree of toxicity of the substance, its speed of absorption and elimination, the metabolic pathways traversed and the 'target' organ. Ideally these studies are expanded to include atmospheric exposure at several different levels of concentration, over the lifetime of the animals and which relate toxic, carcinogenic and genetic responses to exposure in a quantitative manner. Comprehensive studies in great detail are expensive and consequently exceptional. Procedures for developing exposure limits from comprehensive toxicity studies, from more limited data or just from acute toxicity data are given in Chapter 7. The number of substances which have been the subject of fully comprehensive studies is limited but there are many thousands of substances whose acute toxicity in rodents has been determined and published.

Most published exposure limits are set for exposure to single substances. A few are set for exposure to a common industrial material which comprises a complex mixture of substances, such as asphalt, mineral oil or welding fumes. However, in reality employees are not exposed to just one substance or a material with a unique composition. Pure materials are not usually met in industry. Grades are more likely to be the so-called technical type with a percentage of process impurities present. Factory air is rarely contaminated with only one substance. There is an infinite variety of possible mixtures to contend with, from adventitious emissions as well as processes which are deliberately engaged in manufacturing, preparing or using mixtures. Air entering a work-room may be contaminated by large or small quantities of substances from different processes in other work-rooms. Even 'fresh' air contains small quantities of contamination from neighbouring factories and sometimes these quantities are not as small as might be desired. Another kind of mixed exposure comes about through employees being exposed to different substances sequentially by the nature of the processes, job assignments or their movements during the work shift.

A comprehensive basis for the adaptation of the exposure limits of single substances to encompass exposure limits of mixtures is hard to find. Are the effects additive in a simple arithmetic or proportional way? Do two together potentiate? Does the effect of one perhaps work in opposition to the effect of another? The major published lists have appended guidance and formulae covering some aspects. None deals with sequential mixtures. Exposure limits for concurrent and sequential mixtures are worked out in Chapter 8.

Exposure limits are usually set for exposure on five consecutive work days each week followed by two days off work, and a work shift of 8 hours beginning the same time each work day. However, the work schedule in some occupations is very different from this. A work shift may be 12 hours long and fewer days worked per month or the work pattern may be more complicated. Even the standard work week can be repeatedly disturbed by overtime and weekend working. The question then arises of what adjustments, if any, should be made to exposure limits when work schedules or exposures differ very markedly from the standard work schedule.

In batch processes and maintenance operations exposures lasting for only a brief period of the work shift are common. There is also a need for special exposure limits for use in advance planning for possible emergencies such as a massive spill or vessel rupture. Closely related to the issue of exposure to a high concentration for a short period is the problem of short term peaks inserted on a backgound of lesser concentrations. This merges into the most common situation of more or less random excursions from the mean. In Chapter 9 principles for extrapolating from 8-hour time weighted average exposure limits to much shorter periods or much longer periods are developed from toxicokinetic considerations.

Reference

National Academy of Sciences. Risk Assessment in the Federal Government: Managing the Process. Academy Press, Washington, DC (1983).

CHAPTER 5

Published Exposure Limits

The history of atmospheric exposure limits barely covers 100 years. The origins are strongly rooted in the German chemical industry. Gruber is believed to be the first to report a limit; 500 ppm for carbon monoxide (Gruber, 1883; Lehman, 1886). The first comprehensive list, for 20 substances, was prepared in Germany by Kobert (1912). This was followed, 15 years later, by publication in USA of a list of exposure limits for 27 substances, prepared by Sayers (1927).

Lists of occupational exposure limits promulgated by organisations began to appear in the 1940s and onwards and are now found in most industrialised countries throughout the world. In USA in 1942 what was then known as the Subcommittee on Threshold Limits of the National Conference of Governmental Industrial Hygienists compiled a list of atmospheric exposure limits for 63 substances (National Conference of Governmental Industrial Hygienists, 1942; Lippmann, 1983). In 1945 that list was extended to 144 substances (Subcommittee on Threshold Limits, 1946). Furthermore the list was accompanied by a brief commentary on the references from which the limits were extracted (Cook, 1945). This was the precursor of ACGIH Documentation of Threshold Limit Values, which was first published under that name in 1962 for 257 substances (Committee on Threshold Limit Values, 1962).

In Britain a list of exposure limits for 151 substances in Imperial Chemical Industries Ltd was reported by Goldblatt in 1955 (Goldblatt, 1955). In Germany the regular production of a list of MAK-values began in 1968 (Henschler, 1984).

New chemicals have been developed and brought to the market at an ever increasing pace since 1856, when William Henry Perkin at the Royal College of Chemistry in London created the first truly man-made industrial chemical, the coal tar dye mauveine (Osteroth, 1985). New ones are now being developed at an incredible pace. Each year about 300,000 new chemicals are discovered and of these, each year 500–1,000 new chemicals come into use industrially (American Chemical Society, 1978). There are about 7 million known chemicals at the present time and the number of industrial chemicals in practical use has been estimated to be of the order of 70,000. Over 100,000 substances are traded in the European Economic Community (EINECS, 1987). Exposure limits have been published so far for a tiny minority of all these chemicals, amounting in total to 1,200 or so different substances. Substances for which there are published exposure limits are naturally the most important ones commercially and ones which are known to be particularly hazardous to health (Chemical Industries Center, 1977). These are the ones considered in this chapter.

The published limits are guides to the maximum concentration of the substance there should be in work-place air when averaged over a specified period of time. It is generally believed by those who promulgate limits that these exposure limits could be tolerated at work indefinitely without causing significant adverse effects on health of employees there (Rantanen, Aitio and Hemminki, 1982; Paull, 1984). Enlightenment about exactly what constitutes the adverse effect and how small or large the risk for those exposed to a particular substances at its exposure limit must be sought in the documentation associated with that limit. In setting exposure limits some account may also be included of socio-economic, socio-cultural, and technological constraints (Zielhuis and Notten, 1979; World Health Organisation, 1980).

Most atmospheric exposure limits refer to individual chemical substances. A few refer to materials as a whole rather than one of the constituent substances; like, for example, the exposure limits of welding fumes, liquefied petroleum gas, coal tar pitch volatiles, mineral oil mist, coal dust and emery.

Units of concentration

The atmospheric concentration of gases and vapours in published lists of exposure limits is usually expressed by volume/volume and the concentration of aerosols is usually expressed by mass/volume. Specifically, the volume/volume concentration of gases and vapours is expressed as parts of gas or vapour per million parts of air by volume, abbreviated to ppm. The mass/volume concentration of aerosols is expressed as milligrams of dust or mist per cubic metre of air, abbreviated to mg m^{-3}.

It is custom and practice in occupational hygiene when converting the mass/volume atmospheric concentration of gases and vapours into a volume/volume concentration to assume the contamination follows the ideal gas laws. Furthermore the standard conditions normally adopted for the conversion are 25*C* and 760 mm mercury pressure. Under these conditions the gram molecular weight of an ideal gas or vapour occupies 24.45 litres. From which,

$$\frac{\text{Concentration in ppm}}{\text{Concentration in mg m}^{-3}} = \frac{24.45}{\text{molecular weight}}.$$

Except for mineral dusts, particles of all sizes are included in the published exposure limits for aerosols, although this policy may change (ACGIH, 1984). For many years exposure limits of mineral dusts were expressed in terms of number of particles in unit volume of air, but this practice has largely fallen into disuse as practical devices for measuring the mass concentration of respirable particles reliably have become readily available. The size range of particles included in exposure limits for several mineral dusts is limited to defined respirable sizes. Exposure limits for asbestos dust are the last to be expressed in number of particles in unit volume. Limits for asbestos are in terms of the number of fibres per cubic centimetre. The sampling methods for asbestos, the microscopy, and laboratory procedures are all defined in great detail in an attempt to achieve reliable results (Health and Safety Executive, 1988a, 1988b).

Reference period

An important consideration is the reference period of an atmospheric exposure limit. This is the averaging period to which the limit refers. The ideal period from the toxicological point of view is unlikely to be the best period for practical engineering control purposes or for air monitoring. Sometimes an upper limit is given to the reading of a continuous read-out meter (ceiling value) which could be used to trigger an audible alarm or warning light. Otherwise the averaging period for comparing exposure against a limit might be 10 minutes, 1 hour, 4 hours, a shift (8 hours), a week (40 hours), a month or even a year. In continuous operations maxima corresponding to different averaging periods are to some extent correlated and the averaging period used for the limit might be one chosen mainly for its practical convenience (Rappaport, Selvin and Roach 1988). Published lists of exposure limits usually include ones for the time weighted average concentration in inhaled air over a work shift (8-hour TWA). This owes as much to the practical convenience of a unit work shift sampling period as to toxicological considerations. It is for this reason that the limit with respect to an 8-hour reference period is the principal one considered in this chapter. Nevertheless it should be noted that some substances in some published lists have multiple limits, that is a limit or limits for reference periods shorter and/or longer than a work shift, in addition to the limit for a work shift. Maintaining and testing for compliance with several such limits is obviously more demanding than when there is only one.

Time weighted averages

Most exposure limits are expressed in terms of the time weighted average concentration over a specified period: 10 minutes, 15 minutes, 1 hour, 8 hours and so on. The quantity of contamination inhaled in a single breath or over a given period is the product of the volume of air inhaled and the average concentration in it. The time weighted average concentration over the period of exposure is equal to the area under the concentration-time graph divided by the time from start to finish. Assuming the respiratory minute volume is constant over the period of exposure the average concentration in the air which was inhaled over that period would be equal to the *time weighted* average concentration over the period in question. This accounts for the importance of the time weighted average in occupational hygiene.

In order to determine the time weighted average exposure of an individual over a work shift time studies might first be made. At each location occupied by the employee measurements could be made of the time spent there and the

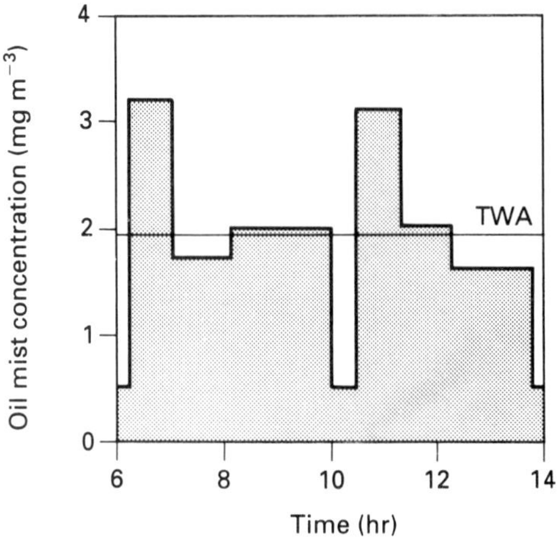

FIG. 5.1. Time weighted average (TWA)—Problem 5.1, oil mist exposure. The TWA is the area under the concentration vs time graph divided by the time from start to finish.

concentration of contamination present in the air. The sum of the products of the duration of exposure multiplied by the level of air contamination of a specified substance during the time interval, divided by the total shift time, is the time weighted average (TWA) concentration of the employee's exposure to this substance. Algebraically, where t_i is the period of time spent at the ith location and C_i is the concentration there:

$$\mathrm{TWA} = \frac{t_1 C_1 + t_2 C_2 + \cdots}{t_1 + t_2 + \cdots} = \frac{\Sigma t_i C_i}{\Sigma t_i}. \qquad (1)$$

Problem 5.1

An operator tending an automatic lathe is exposed to average levels of mineral oil mist as indicated below.

Time	Concentration of oil mist (mg m^{-3})
6:00– 6:15	0.5
6:15– 7:05	3.2
7:05– 8:10	1.7
8:10–10:00	2.0
10:00–10:30	0.5
10:30–11:20	3.1
11:20–12:20	2.0
12:20–13:50	1.6
13:50–14:00	0.5

Calculate the time weighted average concentration of oil mist to which the operator is exposed. The concentration-time graph is in Figure 5.1.

Solution

The time weighted average concentration is best found by setting out the calculation as in Table 5.1.

The time weighted average concentration over the work shift of 8 hours is thus 1.95 mg m^{-3}.

TABLE 5.1. *Calculation of the time weighted average concentration in Example 5.1*

	Period, t_i (mins)	Concentration, C_i (mg m^{-3})	t_iC_i
	15	0.5	7.5
	50	3.2	160.0
	65	1.7	110.5
	110	2.0	220.0
	30	0.5	15.0
	50	3.1	155.0
	60	2.0	120.0
	90	1.6	144.0
	10	0.5	5.0
Total	480		937.0

$$\text{TWA} = \frac{937.0}{480} = 1.95 \text{ mg m}^{-3}.$$

Problem 5.2

An attendant in a mercury cell control room leaves the control room to take readings of meters twice during a work shift. The concentration of mercury vapour in the vicinity of the meters is measured when the attendant is taking readings.

Time	Concentration of mercury vapour (μg m^{-3})
10:20–10:30	220
15:20–15:35	170

(a) Calculate the time weighted average concentration of mercury vapour to which the attendant is exposed when reading meters.

(b) For the remainder of the work shift the attendant is in the control room, where the concentration of mercury vapour is virtually zero. Calculate the time weighted average concentration of mercury vapour to which the attendant is exposed over the work shift. The duration of the work shift is 8 hours.

Solution

(a) During the period spent reading meters:

$$\text{TWA} = \frac{(10 \times 220) + (15 \times 170)}{(10 + 15)} = 190 \ \mu\text{g m}^{-3}.$$

The time weighted average concentration of mercury vapour to which the attendant is exposed over a total of 25 minutes when reading meters is thus 190 μg m^{-3}.

(b) During the work shift, lasting 8 hours, or 480 minutes:

$$\text{TWA} = \frac{(25 \times 190) + (455 \times 0)}{480} = 9.9 \ \mu\text{g m}^{-3}.$$

As a general rule, when measuring time weighted average concentrations of air contamination the shorter the averaging period the higher will be the observed maximum concentration and the lower will be the observed minimum concentration. Some find this puzzling at first. The data in Table 5.2 was constructed to show how this occurs. The 32 values in the first column were chosen at random from a lognormal distribution to make the exercise seem more realistic,

TABLE 5.2. *Time weighted averaging of 32 successive concentration values*

Sample duration: 15 minutes
Units: parts per million (ppm)
Grand mean concentration: 6.1 ppm
Distribution: lognormal

	Reference period					
	15 min	30 min	1 hr	2 hr	4 hr	8 hr
	2.9	3.2	6.4	7.9	5.8	6.1
	3.6					
	9.0	9.5				
	10.0					
	25.0	14.6	9.4			
	4.2					
	7.5	4.2				
	1.0					
	2.8	2.5	4.0	3.8		
	2.2					
	7.0	5.5				
	4.0					
	3.8	4.1	3.5			
	4.5					
	2.0	3.0				
	3.9					
	1.8	2.2	6.2	6.7	6.7	
	2.6					
	5.6	10.3				
	15.0					
	8.0	12.5	7.2			
	17.0					
	0.6	2.0				
	3.3					
	1.5	6.7	6.1	6.6		
	12.0					
	6.2	5.5				
	4.8					
	3.2	4.1	7.1			
	5.1					
	6.0	10.0				
	14.0					
Max.	25.0	14.6	9.4	7.9	6.7	6.1
Min.	0.6	2.0	3.5	3.8	5.8	6.1

although it may be shown that the exact form of the distribution is not important in the present context. Suppose the results were from successive samples, each taken over 15 minutes. Time weighted averages over 30 minutes are found by averaging successive pairs of values from thc first column, time weighted averages over 1 hour are found by averaging successive pairs of values from the second column, and so on. It can be seen that in successive columns the maximum value is lower and the minimum value is higher. It is a matter of arithmetic.

Lists of limits

Exposure limits are not physical or chemical constants like the boiling point of water or the molecular weight of sodium. A closer analogy might be speed limits for road vehicles which, like exposure limits, may differ from one country to another, from one locality to another within a country and even from time to time. Exposure limits are value judgements about the degree of safety protection afforded and the importance attached locally to the adverse effect against which the protection is aimed (Parmegianni, 1983). Furthermore, if one country sets a low exposure limit for a particular substance it does not follow that the prevailing conditions in industry in that country are necessarily any different.

Methods of deriving limits differ from one country to another (World Health Organisation, 1977). Much depends on local attitudes; the vigour with which Regulations are enforced, whether they embrace exposure limits and the regard in which the pronouncements of voluntary organisations are held (Vigliani, 1977). The legal status of limits varies according to the authority responsible for them and changes from time to time. Nevertheless in every compilation of exposure limits it will be found that different hazardous substances have very different limits, spanning a spectrum of at least a million-fold in atmospheric concentration. Accordingly, a difference of two-fold or even ten-fold between limits for certain substances promulgated in different countries is not necessarily very disturbing, even though difficult to explain to lay people (Fairchild, 1982).

The list of exposure limits published by the American Conference of Governmental Industrial Hygienists is the best known in the English speaking world and has been used as the basis of many other lists. The American Conference of Governmental Industrial Hygienists (ACGIH) is a society of industrial hygienists in government employ or in research/teaching. The ACGIH still publish a revised list of exposure limits annually (American Conference of Governmental Industrial Hygienists, 1990). They describe their recommended exposure limits as Threshold Limit Values or TLVs. These limits are designed to protect 'nearly all' workers repeatedly exposed to airborne substances on a daily, 40-hour week basis for a working lifetime. There are three types of TLV; shift time weighted average (TWA), short term exposure limit (STEL) and ceiling value (TLV-C). The shift time weighted average is self explanatory. The TLV-TWAs for aerosols (median, 0.16 mg m^{-3}) tend to be lower than those for gases and vapours (median, 2.5 mg m^{-3}). The frequency distribution of values on the 1990 list are compared in Figure 5.2 for aerosols as compared with gases and vapours, employing a log scale. Short term exposure limit (TLV-STEL) refers to time weighted averages over periods of 15 minutes. Exposures at the STEL should not be repeated more than four times a day and there should be at least 60 minutes between successive exposures at the STEL. The TLV-STELs for aerosols (median, 3 mg m^{-3}) also tend to be lower than those for gases and vapours (median, 100 mg m^{-3}). The frequency distribution of values on the 1990 list are compared in Figure 5.3. Ceiling value refers to the maximum concentration at

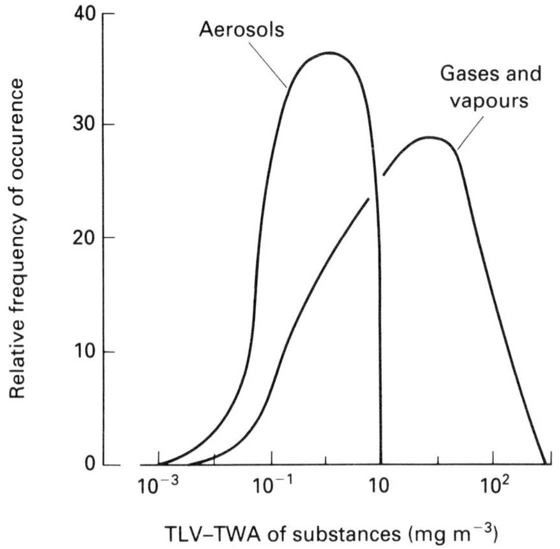

FIG. 5.2. Frequency distribution of ACGIH TLV-TWAs (1990).

any moment. The values are of the same order as TLV-STELs but in this instance should not be exceeded even once during a work shift.

As a general rule, in the context of exposure assessment a current national limit, if any, would be given precedence over other limits for a substance. In Britain this would normally be the Health and Safety Executive Occupational Exposure Standard (OES) which aims to be set at a level which is unlikely to be injurious and can reasonably be complied with. In some instances a Health and Safety Commission Maximum Exposure Limit (MEL) is promulgated in cases where a numerically higher value than the OES is necessary if certain uses are to be reasonably practicable (Carter, 1989). MELs are predominantly but

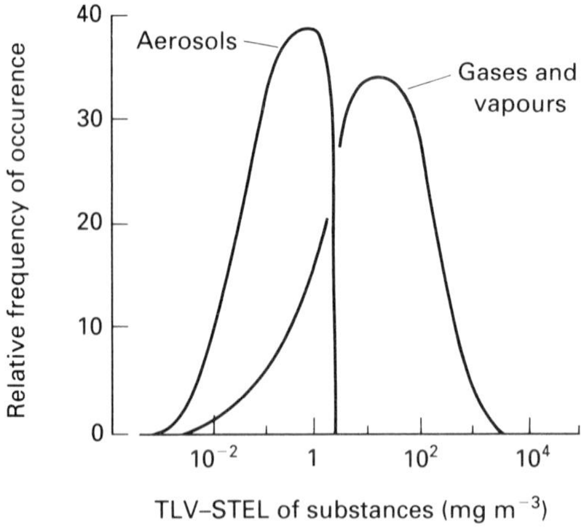

FIG. 5.3. Frequency distribution of ACGIH TLV-STELs (1990).

not exclusively carcinogenic substances. OESs and MELs are in UK Health and Safety Executive Guidance Note EH 40 (Health and Safety Executive, latest edition). Most substances are given two exposure limits: a 'long-term' limit, which is for an 8-hour reference period and a 'short-term' limit, which is for a 10-minute reference period. Short-term samples are those taken specifically to cover short-term tasks or periods of work where peaks of exposure are expected (Burns and Beaumont, 1989).

In USA there is no one list. Several different national bodies produce rival lists of exposure limits. An appropriate limit might be the Occupational Safety and Health Administration Permissible Exposure Limit (PEL), National Institute for Occupational Safety and Health Recommended Limit (REL), the American Conference of Governmental Industrial Hygienists Threshold Limit Value (TLV), or American Industrial Hygiene Association Work-place Environmental Exposure Level (WEEL) (Occupational Safety and Health Act, 1970; OSHA, 1987; American Conference of Governmental Industrial Hygienists, 1990; American Industrial Hygiene Association, 1988). In Canada the Federal Department of Labour recommends the TLVs of the ACGIH as amended from time to time. There are also various other lists promulgated by different States in USA and in different Provinces in Canada.

In USSR Maximum Permissible Concentrations (MPCs) were determined by the Ministry of Public Health (Sanotsky, 1974; Roschin, 1974; INRS, 1987). They are listed together with other data on toxicity of the substances by Izmerov, Sanotsky and Siderov (1982).

In Germany the primary limit would be one promulgated by the Senate Commission for the Testing of Industrial Materials— Maximum Work-place Concentration (MAK Value or TRK) (Henschler, 1979; Commission for the Investigation of Health Hazards of Chemical Compounds in the Work Area, 1988). A substance is assigned a MAK Value where possible or, failing that, a TRK. The MAK value is an 8-hour time weighted average defined as the maximum permissible concentration of a chemical compound present in the air within a working area which, according to current knowledge, generally does not impair the health of the employee nor cause undue annoyance. Excursions above the MAK value are limited according to the classification of the substance into one of five categories determined by the nature of the adverse effects: irritation, systemic effects or intensive odour and according to the biological half time of the substance as shown in Table 5.3 (Commission for the Investigation of Health Hazards of Chemical Compounds in the Work Area, 1989).

The technical guidance concentration (TRK) of a hazardous working material defines that concentration of gas, vapour, or airborne particulates which serves as a directive for necessary protective measures and surveillance by measuring techniques. TRK values are assigned only for hazardous working materials for which MAK values confirmed by toxicological or industrial medical experiences cannot be established at the present time. MAK values are established by a commission of scientists, whereas TRK values are established by a tripartite committee (Henschler, 1985a). Adherence to TRK values is meant to reduce the risk of a health hazard but cannot completely eliminate it

TABLE 5.3. *Limitation of exposure peaks for substances in the MAK Values List (Commission for the Investigation of Health Hazards of Chemical Compounds in the Work Area, 1989)*

TABLE 5.3.1. *Substance categories*

Category	
Cat. I	Local irritants
Cat. II	Substances with systemic effects. Onset of effect <2 h Cat. II,1: half-life <2 h Cat. II,2: half-life 2 h shift length
Cat. III	Substances with systemic effects. Onset of effect >2 h Half-life $>$ shift length (strongly cumulative)
Cat. IV	Substances eliciting very weak effects. MAK $>$ 500 ppm
Cat. V	Substances having intensive odour

TABLE 5.3.2. *Limitation of exposure peaks*

Category	Short-term level peak	Short-term level duration	Frequency per Shift
Cat. I	2 × MAK	5 min, momentary value*	8
Cat. II,1	2 × MAK	30 min, average value	4
Cat. II,2	5 × MAK	30 min, average value	2
Cat. III	10 × MAK	30 min, average value	1
Cat. IV	2 × MAK	60 min, momentary value*	3
Cat. V	2 × MAK	10 min, momentary value*	4

*The momentary value is a level which the concentration should never exceed. It represents a limit to be observed in work area technical planning; the analytical testing can then be carried out by use of sampling procedures designed for recording average values.

(Commission for the Investigation of Health Hazards of Chemical Compounds in the Work Area, 1988).

In Japan maximum permissible exposure levels (kyoyo nodo) are recommended by the Committee for Permissible Exposure Limits of the Japan Society for Industrial Health (1984). For a limited number of substances there are also control limit indices (CLIs) determined by a tripartite committee and promulgated by the Department of Safety and Health of the Ministry of Labour in Japan (1984; Toyama, 1985; Reich and Frumkin, 1988). Limits employed in 18 countries are listed by the ILO (1980).

Occupational exposure limits proposed by international bodies are relatively few in number. Various WHO Study Groups have proposed purely health-based occupational exposure limits for some two dozen substances (World Health Organisation, 1980, 1981, 1982, 1983, 1984). The Council of the European Communities has issued binding exposure limits in Directives on vinyl chloride monomer, lead and asbestos (1978, 1982, 1983). Subsequently, the Commission of the European Communities established an initial list of Indicative Limit Values for 27 other substances (Commission of the European Communities, 1991). These latter are limits of which Member States 'shall take account' when establishing their own limit values.

Sometimes limit values for a substance differ inexplicably from one authority to another. In such instances the lowest might be adopted so as to be sure to err on the side of caution as the documentation of some lists is hard to find. A review and comparison of national limits has been given by the Commission of the European Communities (1979). Another, rather more comprehensive review of limits, including many details has been prepared by Cook (1987).

Authoritative occupational exposure limits should not be regarded as target levels to be met some time in the indeterminate future. They are regarded as minimum standards of air cleanliness in work places, acceptable to industry generally and tolerable by most employees. Allegiance to the standards is engendered firstly by scientific authority as to the biological effects of the substances on individuals, secondly by socio/economic conditions and thirdly by their technical feasibility and practicability. They are not fixed in perpetuity but are meant to be subject to amendment as and when new information comes to light in any or all of these areas.

Carcinogens

Because of the gravity of the condition and the long latent period special consideration is usually given to carcinogens. Depending on their carcinogenic potency some are prohibited completely or prohibited for certain uses, whereas others may be permitted on condition that they are handled under completely enclosed processes or nearly so. Still others are treated little different from other hazardous substances and are given normal exposure limits.

The ACGIH, for example, identify substances which are confirmed human carcinogens and others which are merely 'suspected'. Both types may or may not be allotted a TLV. When employing confirmed human carcinogens the user is urged to eliminate exposure to them altogether.

In the German lists a subtle distinction is also made between carcinogenicity evidenced by animal experiments and that unequivocally proven in humans. It is also necessary to wrestle with the notion of a sub-group of substances whose carcinogenicity is merely 'justifiably suspect'. In the list of MAK values these 'justifiably suspect' substances are included. No MAK values are given for substances which, on the basis of pertinent experience, are recognised to be capable of inducing cancer in man, but TRK values are promulgated for these to eliminate potential hazards wherever possible.

In UK some carcinogens are prohibited and others are allotted exposure limits (MELs). The same is true in USSR and many other countries (Carnevale, Montesano, Partensky and Tomais, 1987; Cook, 1989). Biological thresholds have been determined in the USSR for a number of carcinogens, including benzo(a)pyrene (Yanysheva, 1972a, 1972b).

Skin protection

Atmospheric exposure limits, by themselves, place limits on the amount of hazardous substance which may be inhaled. But the rate of transport through other important routes of entry, namely skin absorption and ingestion may also be significant in particular circumstances. Many published exposure limits carry a 'skin' notation in order to caution the user that those limits are inextricably linked with the provision of due skin protection. The skin notation refers to the substance in the liquid or solid state which may contaminate the skin and thereby be absorbed into the body or affect the skin locally (Fiserova-Bergerova, Pierce and Droz, 1990). Thus the atmospheric exposure limit for these substances refers to atmospheric exposure when adequate skin protection against the liquid or solid is also employed. About one in four of the substances on American, British and West German lists have a skin notation although rather fewer of the substances on the USSR list do so (Scannsetti, Piolatto and Rubina, 1988).

Documentation

An exposure limit for air contamination is preferably founded on published quantitative data from thorough medical/environmental surveys in industry. When setting exposure limits for substances which cause irritation or mild narcosis it is sometimes feasible to employ results of human volunteer studies. Data from acute accidental exposure may also be of value for setting exposure limits. In the absence of fully documented human experience an exposure limit may have been arrived at from the results of studies in experimental animals. Finally, in the absence of data from medical-environmental surveys in industry, studies on volunteers or animal studies, an exposure limit has often been set for a substance on the somewhat tenuous basis of its chemical and/or physical similarity to other chemicals whose biological effects are thought to be related to those particular properties.

The background information is rarely, if ever completely satisfactory and is commonly very sparse and indirect. Exposure limits are best judgements at the time of their adoption and are believed to confer a minimum degree of protection which would command acceptance in the conditions of use at the time. They are under continuous review and subject to change.

In every system for setting or revising authoritative limits it is possible to detect three distinct strands at work which when bound together, support the limit:

(a) Collation and appraisal of the available biological information, which may include human and animal toxicology, epidemiology and exposure data.

(b) Development of the exposure/response relationships and preparation of a position paper laying out all the known chemical, physical and biological data.

(c) Review of the position paper plus input on costs, engineering and availability of methods for measuring the substance.

Value judgements are then made to recommend a realistic new or revised exposure limit which would sustain what is believed to be an appropriate degree of health protection. In point of fact current conditions of employee exposure may already be in keeping with such a new or revised exposure limit.

One reason for this is that the chemical industry is acutely conscious of the information flow about its products from whatever source. It adapts rapidly to any new knowledge, often long before it appears in published form. Published exposure limits tend to follow current best practice, not to create it (Roach, 1982).

Documentation of the highly respected ACGIH TLVs is regularly updated and revised (American Conference of Governmental Industrial Hygienists, 1986). A very full account of the history, development and status of ACGIH TLVs has been prepared by La Nier (1984). WEELs are published annually in the American Journal of Industrial Hygiene and include a brief commentary on the main references consulted.

The West German MAK values were originally taken from the American Conference of Governmental Industrial Hygienists list of TLVs and the ACGIH Documentation applied to those values. Since then additions and amendments to MAK values have been made by the MAK Commission. Critical evaluations of the present state of knowledge about MAK values originating from the Deutsche Forschungsgemeinschaft is updated annually (Henschler, 1981; Henschler, 1985b; Commission for the Investigation of Health Hazards of Chemical Compounds in the Work Area, 1989). As a matter of principle economic considerations and problems of practical realisation do not enter into the setting of MAK values. They are, however, specifically taken into account in the development of technical guidance concentrations for carcinogenic substances (TRKs) laid down by the West German Ministries of Labour and Social Affairs (Woitowitz, 1985). The following factors determine the TRK values:

—the analytical possibilities of determining the concentration of a harmful material in the range of the TRK value,

—the current state of industrial processing methods and ventilation technology, taking into account technical improvement which can be achieved in the near future,

—absence of available adverse industrial medical experience.

(DFG, 1984)

The recent British Occupational Exposure Standards (OESs) are supported by the documentation published by the American ACGIH or German MAK Commission and in some 20 or so instances by British HSE Toxicity Reviews prepared by the Health and Safety Executive (1991). The Japanese permissible exposure levels (PEL's) are recommended annually, with appropriate documentation in the Japanese Journal of Occupational Health (Toyama, 1985). An instructive compilation of USSR literature was prepared by Levine in the sixties (Levine, 1960–1983).

Possible limitations of the process

In every system for setting published exposure limits there are groups which lobby to have a major influence on the values set; chemical trade associations, organised labour, enforcing authorities and independent scientific experts (Duncan, 1981). Keeping the balance correct at all times is obviously not easy. Values which have had the most widespread use are those of the American Conference of Governmental Industrial Hygienists. Although originally intended as unofficial guides of acceptable exposure to chemical agents in the work place, these limits are widely applied as official limits by many states and countries. They formed the basis of the German list of maximum work place concentrations (MAKs) (Henschler, 1984), the British list of occupational exposure standards (OESs) (Health and Safety Executive, latest edition), the Japanese list of maximum permissible exposure limits (PELs) (Toyama, 1985), the Swedish list of hygienic limit values (HLVs) (Nordberg, Frostling, Lundberg and Westerholm, 1988) and many other lists.

The exposure limits for a given substance in these different lists are highly correlated and in most instances identical. The existence of a level of concentration for every hazardous substance, below which no adverse health effects will occur and above which injury will result, is accepted without question by the average layman. Exposure limits gain ever increasing prestige and authority through mere copying and repetition. But they are not without critics. Although the ACGIH membership excludes industry employees it is impossible to avoid industrial influence entirely, even if that were desired. In any event values of published exposure limits are usually accorded more respect than they deserve, in view of the paucity of data on which many such figures are based. Both the process by which the TLVs have been established and the data supporting the limits have been seriously called into question (Henschler, 1984; Castleman and Ziem, 1988; Zielhuis, 1988; Roach and Rappaport, 1990).

The process

Threshold Limit Values for Chemical Substances represent conditions under which it is believed by the American Conference of Governmental Industrial Hygienists that 'nearly all workers' may be repeatedly exposed without adverse effect. For a better understanding of the phrase 'nearly all' a detailed research has been completed by the author and others into the 6,000 published references cited in the 1986 edition of the ACGIH Documentation of TLVs and the 4,500 published references cited in the 1976 printing, 10 years earlier (Roach, 1982; Roach and Rappaport, 1990). When evaluating the evidence compiled in the documentation it becomes clear that the TLV Committee has traditionally placed greatest importance upon studies involving human experience. This has been confirmed unequivocally by a former Chairman of the Committee, Herbert Stokinger (Stokinger, 1984). Attention was therefore focussed on references in the documentation to data on 'industrial experience' and 'experimental human studies'. These references were then further sorted for ones which included data on both the incidence of adverse effects on humans and the corresponding exposure to the substance in question.

The ACGIH Threshold Limit Values were found to be poorly correlated with the incidence of adverse effects ($r^2 = 0.17$). It was found that in environments where the exposure was at or below the levels in the list of Threshold

Limit Values one in every five cited studies showed no adverse effects on the employees. However all the remainder, four in every five studies, indicated that up to 100% of exposed employees may be adversely affected at or below the corresponding Threshold Limit Values. Coincidentally, a strong correlation ($r^2 = 0.61$) and direct correspondence was found between Threshold Limit Values and mean exposure reported in the cited studies. The same was true of studies cited in the 1986 and the 1976 ACGIH Documentation.

These findings indicate that firstly, Threshold Limit Values which are justified by ACGIH on the basis of 'industrial experience' have not been based only upon considerations of health. Secondly, ACGIH Threshold Limit Values set on the basis of human experience have reflected the levels of atmospheric exposure which were perceived at the time to be achievable in industry. ACGIH Threshold Limit Values up to 1986 may or may not represent relevant guides for purposes of control but they are not the thresholds of adverse health effects.

It seems to be important that the exposure of employees in a particular process must be so controlled that during each work-shift the atmospheric exposure is kept well below the levels in the list of ACGIH Threshold Limit Values. This is in keeping with the sentiments expressed in the TLV booklet that '. . . the best practice is to maintain concentrations of all atmospheric contaminants as low as practical . . . ' (American Conference of Governmental Industrial Hygienists, 1988).

Since many ACGIH Threshold Limit Values for chemical substances appear to offer relatively limited health protection in themselves, occupational hygienists and other health professionals routinely examine the published references pertaining to particular air contamination rather than accepting on faith that every published exposure limit provides the protection proclaimed in the booklet listing them. As a general principle it would be prudent, wherever possible, to keep the average exposure of the employees below one-tenth of published exposure limits. If this is done, and assuming exposure distributions to be quasi-lognormal, then fewer than 1% of exposures would be expected to exceed the TLV (Rappaport, Selvin and Roach, 1988). Reasonable assurance might then be given that the health protection is adequate.

Skin notation

A separate critical examination of the ACGIH TLV Documentation has been conducted by Scannsetti, Piolatto and Rubino into the skin notation employed in the list of Threshold Limit Values to warn of the risk to health from absorption through the skin (1988). They found inconsistency in the use of the notation for different classes of compound, from one year to another and in comparison with other national lists. They urged that the skin notation should be the result of full scientific judgement based on past and present reported animal and/or human data on absorption through the skin.

Atmospheric exposure can be measured in many different ways, but it is difficult to quantify skin exposure and ingestion, with the possible exception of those substances where biological exposure indices are available and appropriate. Consequently for most substances regular and quantitative measurements of exposure are confined to air contamination. Control of other sources of exposure is hidden in the rules for wearing protective clothing and entrusted to labels on which is said 'avoid skin contact'. There is a danger that compliance with atmospheric exposure limits engenders a false sense of security for exposure to substances which:

(a) are readily absorbed through the skin,
(b) attack the skin,
(c) are ingested.

Health risk criteria

A hazardous substance commonly has several quite different effects on man. A limit level which protects against acute irritation may be inadequate to protect against narcosis. Such a limit may be insufficient to protect against long term bodily injury. A lower limit again may be necessary to protect the sensitised employee. Still lower may be necessary to protect against cancer. Also, the importance of brief exposures may be quite different for these different effects. Consequently, ideally both the limit levels and the appropriate time weighted averaging periods would differ according to the nature of the biological effect for which protection is given.

In the view of this author published exposure limits of hazardous substances should include the following five specifications as a minimum:

(1) Atmospheric concentration of the specified substance or mixture.

(2) Duration of exposure to which the limit refers.

(3) Method of measuring the concentration in (1).

(4) Minimum adverse effect that is considered significant.

(5) Percentage of the population intended to be protected against the adverse effect in (4).

The equivalent five specifications are all commonly available for exposure limits of the main physical factors which may cause risk to health in industry; ionising radiations, noise and thermal comfort (Fanger, 1970; Atherley and Noble, 1971; Chrenko, 1974; Halton, 1988).

Published exposure limits for hazardous substances are generally satisfactory with respect to the specification of duration of exposure and concentration. With certain exceptions, the specification of the methods to be employed—where, when, how, with what—is usually deficient in detail, but in the hands of an experienced hygienist is arguably sufficient for the purpose. The adverse effect that is considered significant can usually be found by a close reading of the documentation, if that is available. But as to the degree of protection provided or to be provided, with rare exceptions the information is non-existent or at best vague or only qualitative.

References

American Chemical Society. CAS report. 7:2. (1978).

American Conference of Governmental Industrial Hygienists. Documentation of the Threshold Limit Values and Biological Exposure Indices. 5th Edition. ACGIH, Cincinnati, Ohio USA (1986).

American Conference of Governmental Industrial Hygienists. Threshold Limit Values and Biological Exposure Indices for 1990–1991. ACGIH, P O Box 1937, Cincinnati, Ohio 45201 USA (1990).

American Conference of Governmental Industrial Hygienists. Particle size-selective sampling in the workplace. American Conference of Governmental Industrial Hygienists, P O Box 1937, Cincinnati, Ohio 45201 USA (1984).

American Industrial Hygiene Association. WEEL Guides. American Industrial Hygiene Association, 475 Wolf Ledges Parkway, Akron, OH 44311–1087 USA (1988).

Atherley G R C and W G Noble. Occupational hearing loss (Edited by D W Robinson), Academic Press, London (1971).

Burns D K and Beaumont P L. Ann. Occup. Hyg. **33** 1 (1989).

Carnevale F, R Montesano, C Partensky and L Tomais. Am. J. Ind. Med. 11 453 (1987).

Carter J T. Ann. Occup. Hyg. **32** 653 (1989).

Castleman B I and G E Ziem. Am. J. Ind. Med. **13** 531 (1988).

Committee on Threshold Limit Values. Documentation of threshold limit values. American Conference of Governmental Industrial Hygienists, Cincinnati, P O Box 1937, Ohio 45201, USA (1962).

Commission for the Investigation of Health Hazards of Chemical Compounds in the Work Area. Maximum concentrations of the workplace and biological tolerance values for working materials 1988. Report XXIV. VCH Publishers, D-6940 Weinheim, FRG and Deerfield Beech, FL, USA (1988).

Commission for the Investigation of Health Hazards of Chemical Compounds in the Work Area. Toxikologisch-arbeitsmedizinische begrundungen von MAK-Werten. Verlag Chemie, D-6940 Weinheim Bergstr., Germany (1989).

Commission of the European Communities. Commission Directive of 29 May 1991 on establishing indicative limit values by implementing Council Directive 80/1107/EEC on the protection of workers from the risks related to exposure to chemical, physical and biological agents at work. Official Journal of the European Communities, No L 177, 5. 7. p. 22, Brussels (1991).

Cook W A. Ind. Med. Industrial Hygiene Supplement **6** 936 (1945).

Cook W A. Am Ind. Hyg. Assoc. J. **50** 680 (1989).

Council of the European Communities. Council Directive 78/610/EEC on asbestos. Official Journal of the European Communities, No L 263 24.9 (1983).

Council of the European Communities. Council Directive 78/610/EEC on lead. Official Journal of the European Communities, No L 247 23.8 (1982).

Council of the European Communities. Council Directive 78/610/EEC on vinyl chloride monomer. Official Journal of the European Communities, No L 197 22.7 (1978).

Department of Safety and Health of the Ministry of Labor of Japan. Control Limit Index for toxic substances in workplaces. Japan Industrial Safety and Health Association, 35 Shiba, Schome, Minato-ku, Tokyo 108, Japan.

Duncan K P. Exposure limits—whose responsibility? In: Recent advances in occupational health I (Edited by J C McDonald), pp. 247–255. Churchill Livingstone, Edinburgh, UK (1981).

EINECS. European inventory of existing commercial chemical substances, Vols I-VIII. Commission of the European Communities, EEC, Brussels, Belgium (1987).

Fairchild E J. Ann. Am. Conf. Govt. Ind. Hyg. **3** 83 (1982).

Fanger P O. Thermal comfort. In: Analysis and application in environmental engineering. McGraw Hill, Maidenhead, UK (1970).

Fiserova-Bergorova V, J T Pierce and P O Droz. Am. J. Ind. Med. **17** 617 (1990).

Goldblatt M W. Brit. J. Ind. Med. **12** 1 (1955).

Gruber M. Arch. Hyg. **1** 145 (1883).
Halton D H. Reg. Tox. Pharmacol. **8** 343 (1988).
Hart J W and Jensen N J. Regul. Toxic. Pharmac. 11 123 (1990).
Health and Safety Executive. Asbestos—control limits and measurement of airborne dust concentrations (rev). H M Stationery Office, London (1988a).
Health and Safety Executive. Asbestos fibres in air; Light microscope method for use with the Control of Asbestos at Work Regulations MDHS Series 39/2. H M Stationery Office, London (1988b).
Health and Safety Executive. HSE Toxicity Reviews, TR1—TR20. HSE Sales Point, Bootle, Merseyside, UK (1989).
Health and Safety Executive. Occupational Exposure Limits— Guidance Note EH40. H M Stationery Office, London (Latest edition).
Henschler D. Maximum Allowable Concentration Values in the Federal Republic of Germany. In: Trans. XLI Annual Meeting, American Conference of Governmental Industrial Hygienists, Chicago, IL, USA (1979).
Henschler D. Ann. Am. Conf. Ind. Hyg. **12** 37 (1985a).
Henschler D. Ann. Occup. Hyg. **28** 79 (1984).
Henschler D. Maximale Arbeitsplatzkonzentrationen—Grundlagen, entwicklung, beratungsmodell. In: Deutsche forschungsgemeinschaft: Wissenschaftliche grundlagen zum schutz von gesundheitsschaden durch chemikalien am arbeitsplatz: Boldt, pp. 29—40 (1981).
Henschler D. (Editor). Toxicologisch arbeitsmedizinische begrundung von MAK-Werten. Verlag Chemie, Weinheim, Germany (1985b).
INRS. Cah. Nat. Docum. **129** 581 (1987).
ILO. Occupational exposure limits for airborne toxic substances. Occupational Safety and Health Series, No 37 Second (Revised) edition. International Labour Office, Geneva (1980).
Japan Society for Industrial Health. Jpn. J. Ind. Health **26** 239 (1984).
Kobert R. Kopend. der prak. Toxicol., p. 45, Stuttgart (1912).
Lehmann K B. Arch. Hyg. **5** 1 (1886)
Lippmann M. Ann. Am. Conf. Gov. Ind. Hyg. **5** 5 (1983).
National Conference of Governmental Industrial Hygienists. Report of Subcommittee on Threshold Limits. Transactions of the Fifth Annual Meeting of the National Conference of Governmental Industrial Hygienists, Washington, DC, USA (1942).
Nordberg G F, H Frostling P Lundberg and P Westerholm. Am. J. Ind. Med. **14** 217 (1988).
Occupational Safety and Health Act of 1970. PL91–596, S.2193, Dec. 29. US Government Printing Office, Washington, DC, USA (1970).
OSHA. OSHA Safety and Health Standards, 29 Code of Federal Regulations 1900–1910, Subpart Z, Sect 1000–1047. US Government Printing Office, Washington, DC, USA (1987).
Parmegianni L. Exposure limits. In L Parmegianni (Ed) Encyclopedia of occupational health and safety, pp. 812–816. International Labour Office, Geneva, Switzerland (1983).
Paull J M. Am. J. Ind. Med. **5** 227 (1984).
Rantanen J, A Aitio and K Hemminki. Am. J. Ind. Med. **3** 363 (1982).
Rappaport S M, S Selvin and S A Roach. Appl. Ind. Hyg. **11** 310 (1988).
Reich M R and H Frumkin. Am. J. Public Health **78** 809 (1988).
Roach S A. In-plant industrial hygiene—the years ahead. In: Mens en Millieu 2000. IMG-TNO, postbus 214 2600 AE Delft, Netherlands (1982).
Roach S A and S M Rappaport. Am. J. Ind. Med. **17** 727 (1990).
Roschin A V. International Labour Review. **110** 235 (1974).
Sanotsky I V and D J Mendeleev. J. All-Union Chem. Society XIX 125 (1974).
Sayers R R. Toxicology of gases and vapors. In: International Critical Tables of Numerical Data, Physics, Chemistry and Toxicology, Volume 2, pp. 318–321. McGraw-Hill, New York (1927).
Scannsetti G, G Piolatto and G F Rubina. Am. J. Ind. Med. **14** 725 (1988).
Stokinger H E. Criteria and procedures for assessing the toxic responses to industrial chemicals. In: Threshold Limit Values: Discussion and thirty-five year index with recommendations (Edited by W E LaNier). Ann. Am. Conf. Ind. Hyg. **9** 155 (1984).

Subcommittee on Threshold Limits. Proc. 8th Ann. Mtg. Am. Conf. Gov. Ind. Hygienists, pp. 54–55 (1946).
Toyama T. Am. J. Ind. Med. **8** 87 (1985).
Woitowitz H-J. Am. J. Ind. Med. **8** 87 (1985).
World Health Organisation. Methods used in establishing permissible levels in occupational exposure to harmful agents. WHO Tech. Rep. Ser. No 601. World Health Organisation, Geneva, Switzerland (1977).
World Health Organisation. Recommended health-based limits in occupational exposure to heavy metals: Report of a WHO Study Group. WHO Tech. Rep. Ser. No. 647. World Health Organisation, Geneva, Switzerland (1980).
World Health Organisation. Recommended health-based limits in occupational exposure to selected organic solvents: Report of a WHO Study Group. WHO Tech. Rep. Ser. No. 664. World Health Organisation, Geneva, Switzerland (1981).
World Health Organisation. Recommended health-based limits in occupational exposure to pesticides: Report of a WHO Study Group. WHO Tech. Rep. Ser. No. 677. World Health Organisation, Geneva, Switzerland (1982).
World Health Organisation. Recommended health-based limits in occupational exposure to selected vegetable dusts: Report of a WHO Study Group. WHO Tech. Rep. Ser. No. 684. World Health Organisation, Geneva, Switzerland (1983).
World Health Organisation. Recommended health-based occupational exposure limits for respiratory irritants: Report of a WHO Study Group. WHO Tech. Rep. Ser. No 707. World Health Organisation, Geneva, Switzerland (1984).
Yanysheva N J. Gig. i Sanit. **1** 90 (1972a).
Yanysheva N J. Gig. i Sanit. **7** 87 (1972b).
Zielhuis R L. Occupational exposure limits for chemical agents. In: Occupational medicine—Principles and practical applications (Edited by C Zenz). Year Book Medical, Chicago Il, USA (1988).
Zeilhuis R L and W F R Notten. Int. Arch. Occup. Environ. Health 42 269 (1979).

Bibliography

Chemical Industries Center. Chemical origins and markets. Chemical Industries Center, Stanford Research Institute, Menlo Park, CA, USA (1977).
Chrenko F A. Bedford's basic principles of heating and ventilation. Lewis, UK (1974).
Commission of the European Communities. Comparative analysis of the principles and application of limit values in the Member States of the European Communities. PO Box 1907, Luxembourg (1979).
Cook W A. Occupational exposure limits—Worldwide. American Industrial Hygiene Association, 475 Wolf Ledges Parkway, Akron, OH 44311–1087, USA (1987).
Izmerov N F, I V Sanotsky and K K Siderov. Toxicometric parameters of industrial toxic chemicals under single exposure—USSR (English language edition). United Nations Environmental Program (1982).
La Nier M E (Ed). Threshold Limit Values—Discussion and thirty-five year index with recommendations. American Conference of Governmental Industrial Hygienists, Cincinnati, Ohio, USA (1984).
Levine B S. USSR Literature on air pollution and related occupational diseases (8 vols). US Department of Commerce, Washington DC, USA (1960–1963).
Osteroth D. Soda, teer und schwefelsaure. Der weg zur grobchemie. rororo-sachbuch. Deutsches Museum, Munchen (1985).
Vigliani E. Methods used in Western European Countries for establishing maximum permissible levels of harmful agents in the working environment. Fondazione Carlo Erba, Milano (1977).

CHAPTER 6

Guiding Concepts for Setting Exposure Limits—Human Experience

Cer-lap

There is little a risk assessor can conclude about the likelihood of adverse health effects from atmospheric exposure to a hazardous substance without knowing the exposure limit of the substance. The reason is that limits for different hazardous substances differ widely from one another over a range of a million-fold or more. The employee has little choice about being exposed to hazardous substances present at his work place. Be it only approximate it is necessary to ascertain the exposure limit for work place air contamination by the substance in question. To do this, it is not necessary to know exactly how a toxic substance in the body produces the diseases in question. Indeed there are many industrial diseases which have been overcome without knowing the precise mechanism by which the disease was produced. What is needed is to quantify the toxicity in terms of the maximum concentration that can be inhaled by employees every working day without significant risk to health. This is not to deny that to do this could still be daunting.

The primary manufacturer of a hazardous substance has a pressing responsibility for devising suitable exposure limits although this practice is far from universal (Morton, 1977). There might very well not be a suitable exposure limit for the substance in question in suppliers literature. Where the manufacturer or supplier does give a limit the customer should consider whether that limit is appropriate for the particular conditions of use. Exposure limits have also been published for some of the commercially important hazardous substances. The common industrial chemicals are the ones to be found in published lists of atmospheric exposure limits but it has to be recognised that numerically these are only a tiny fraction of all the chemicals in use. Consequently, to undertake a risk assessment of hazardous substances it will often become necessary to devise an appropriate exposure limit (Paustenbach and Langner, 1986).

Special situations may quite properly allow deviation from published exposure limits downwards or, sometimes, upwards. For example, in office jobs adjacent to works handling hazardous substances it is often feasible, practicable and eminently reasonable to adopt a lower exposure limit than would be appropriate in the works itself. Another example, the other way, is in developing countries where health risks at work are often of minor importance compared with risks from endemic disease and low living standards (Hatch, 1972). It is pertinent to note the priorities in Indonesia, a newly developing country, where it has been declared that levels should be selected to secure the continuity of existing production for the employment objective but minimise or, if possible, eradicate hazards of occupation (Suma'mur, 1976, 1977).

Good health can be bought

Industrial, economic and technological constraints seem to weigh heavily on the employer when trying to shape an opinion about realistic in-house exposure limits of hazardous substances. Many companies spend 10% of capital and research budgets to deal with such risk management problems (Barnard, 1987). If the available funds are not well spent the impact on the prosperity of the company could be serious. Nevertheless, the financial burden to industry to implement effective control promptly and meet strict exposure limits is unlikely to be seen by exposed employees as a good enough reason for taking a risk with their health. Aside from the use of personal protection provided by the employer, individual employees have little power to control or avoid the health risks in their day-to-day activities. They are fully aware that the cost of their health may well compete with company profit. A prosperous company is seen by its employees as one which has the means to spend huge amounts on keeping down health risks if it so chooses. However, even when there is a will to pay, expenditure on health cannot be open ended, and must be related in some way to what it is reasonable to pay to preserve it; that is, its worth.

Most health professionals believe profoundly that the costs in terms of time, trouble and money to prevent health risks at work pays off very handsomely indeed. An assumption that a certain risk would be an 'acceptable' one for employees cannot be made with the same equanimity. The exposure limit that is 'reasonable' for a company to adopt has to be established by reasoning which gives due precedence to prevention of adverse human health effects over comparative cost-benefit decision making or other trade-off methods. However a company's view of the importance of a particular kind of risk can often differ dramatically from the views of others. It is necessary for those involved in setting limits to try to strike a balance between opposing views.

The extra costs of the product being manufactured under strict environmental control should play little or no part in deciding which risks might be tolerable by employees. In the same sense, on a parochial level, the limitations of the sampling and analytical techniques which are feasible and 'reasonably achievable' in industry should not be allowed to impede the adoption of a strict exposure limit when one is needed on health grounds.

Once it is realised that a no-risk situation is truly unattainable, whatever the law may say, the limits devised for in-house use necessarily involve a value judgement about the comparative acceptability of a wide variety of possible effects (Corn, 1983). In this book there is one common guiding concept overriding all others, namely that conditions can and should be maintained so

.event significant risk to health. It is believed that in maintaining .nce to such a concept the exercise of judgement in setting a limit should ,ne in which a conscious endeavour is made to be consistent as far as ,ssible between one substance and another. But acceptance of this principle does not answer the question of what exactly makes a health risk significant.

Significant risk

It is seldom possible to foresee the exact nature, gradation and extent of all the biological effects that might occur as a result of inhaling excessive quantities of a particular substance. The symptomatology of occupational diseases differs strikingly from one hazardous substance to another. Different substances range from mild or severe irritants, on to simple asphyxiants, narcotics, systemic poisons and dusts causing fibrosis, through to the most highly toxic substances which could prove fatal after a single brief exposure. Each substance has unique biological properties just as it has unique chemical and physical properties. At first sight, therefore, there appears to be only limited room for general applicability between the foundations of exposure limits for different substances dispersed in air. An occupational exposure limit for a particular substance is unique to that substance.

The one common feature between exposure limits for different substances is the notion of significant risk to health. This feature is central to the schemes outlined below, although it has to be recognised that the implicit companion concept, that a health risk can be real but insignificant, is one which some find hard to accept. In the present context definition of the term 'significant' risk to health is in terms making reference to the reasonable employer. By 'significant' in this sense is not meant significant in a narrow, mathematical or statistical sense. Rather, it is suggested that the **risk to health is significant when it is known or foreseeable and a reasonable employer with full knowledge of the nature, extent and distribution of risk is more likely than not to take steps to reduce it**, for whatever weighting of reasons. The motivation may be medical, scientific, technological, ethical, social, economic, political, legal, statistical, emotional or anything else. Naturally the steps taken would be ones within the particular capabilities of the individual employer.

This does not exclude closing down an operation, process or even a whole company in extreme circumstances if this is what the informed, reasonable employer would do. Also, by 'reasonable' employer is not meant the most reasonable nor, for that matter, the least reasonable, but rather an employer of average reasonableness. The proverbial bus passenger on the top deck of a No. 19 omnibus would be expected to be neither more reasonable nor less reasonable than this employer.

Reference is made to the 'employer' rather than, say, 'citizen' or 'man in the street' or 'medical expert' since the responsibility for the continuing quality of the in-house environment rests principally with the employer (Illing, 1991). This is not to deny that minimum standards may be set or changed by external bodies with a wider view and enforced by the competent authority. Also the employer, in deciding on the exposure limit, would be expected to listen to the views of organised labour and technical experts insofar as they have anything to contribute.

The significance of a risk, in the above sense, is a value judgement and can, quite properly, vary according to circumstances both from place to place and from time to time. Exposure limits are not chemical constants, nor are they biological constants. The magnitude of a risk which is deemed just significant is dependent, in particular, upon the nature of the risk. For example, the least risk of mild irritation which is deemed significant would generally be higher than the least risk of systemic effects or of death which is deemed significant. Criteria of significant risk differ, commensurate with the potential seriousness or otherwise of the corresponding diverse effects. Published exposure limits attract a measure of public acceptance and may be studied for clues to what risks are tolerated. References in their current documentation and 'begrundungen' may be explored to shed light on what risks have been accepted and tolerated in the recent past.

Exposure limits by epidemiology

The experimental toxicologist who exposes animals to known amounts of hazardous substances is beset with the problem of extrapolating his results into the effects on humans. The epidemiologist suffers from none of this. Epidemiologists study the patterns of disease in groups of people directly. The patterns may be studied in relation to a host of related factors; be they human characteristics, environmental factors, historical factors, geography, economic conditions or otherwise (Lilienfeld and Lilienfeld, 1980; Monson, 1980). The classic example of the early use of epidemiology is by John Snow, anaesthetist to Queen Victoria, who, in 1854, mapped the distribution of the homes of cholera victims in London and thereby identified the source of an outbreak of the disease as being in the locality of the Broad Street communal water pump. By having the pump handle removed the epidemic was controlled, long before the micro-organism responsible for the disease had been discovered.

In the field of occupational hygiene epidemiologists are notable for their studies of the prevalence of diseases in relation to employee exposure to hazardous substances. Such studies are of immense value when the exposure of the employees has been measured directly. Semi-quantitative estimates such as 'low', 'medium' or 'high' are of only limited value. Atmospheric exposure limits which have been produced directly from the study of disease prevalence amongst employees in relation to their known exposure to hazardous substances are few in number but they are the most reliable limits. General guidelines for epidemiological studies have been discussed by the Epidemiology Work Group of the Inter-agency Regulatory Liaison Group (1981). The epidemiology of exposure to hazardous substances is, unfortunately, not simple (IPCS, 1983). The epidemiologist has, for example, no control over the work environment or the humans in it (Vanhoorne, Harrington, Parmeggiani, Hunter and Vuylsteek, 1985). The environment varies with the day, month, season and year; the exposure to hazardous substances varies with the raw materials, plant, process and output; the employees also vary with their eating, drinking and smoking habits, their recreational activities and interests (Goldsmith and Besser, 1984).

One type of study is known as a 'cross-sectional' study. As the name suggests, a group of employees is selected at a single point in time without regard to disease or exposure status. Each employee is then evaluated with regard to the presence and severity of disease and as to the level of his or her exposure to hazardous substances. Such a study would be appropriate for substances and levels of exposure causing mild, transient effects such as irritation or narcosis. Employees whose health is affected severely tend to leave.

The disease being studied may also be found in the general population or in other occupations. Then the parameter of major interest is the excess of prevalence over and above that which is likely in the absence of exposure to the particular hazardous substances at the work places under study. In such cases a major difficulty becomes that of collecting data on enough employees to yield a sufficient number of cases so as to be able tease out from all the other variables the effects of exposure.

Another type of study is one which is known as a 'longitudinal' study. These involve some time interval between when exposure begins and disease status is determined. A longitudinal study would be used for a substance which accumulates in the body over a long period of time or if there is a lengthy delay between exposure and effect as occurs with some diseases. The investigator usually begins by selecting groups of employees expected to be at risk, determines whether disease is present or absent and looks back in time to ascertain the level of exposure to which they have been exposed.

The prevalence may be grossly underestimated of some effects. This is especially true of effects which

(a) have long latent periods,
(b) are of such a severity that they alone can cause a person to change jobs or even cease work altogether,
(c) reduce life expectancy.

In order to overcome such difficulties a 'cohort' study may be made. In such a study a group of healthy employees who are exposed is followed up for a defined period of time and compared against a similar group who are not exposed. The most common type of cohort study is the 'retrospective' type. In such studies the investigation may have to be enlarged to include all persons who have left a company. A date is set, say, 20 years prior to the study and all persons who have left in the intervening period are traced, their disease status is determined and their exposure estimated. This is a very time consuming task and may not be feasible if the company has failed to keep records of all its past employees. In an endeavour to overcome such difficulties a 'prospective' cohort study may have to be undertaken. In a prospective cohort study the two groups of employees are followed into the future to determine the rate at which disease develops. Unfortunately for the epidemiologist, as soon as some disease begins to be found further exposure is inevitably reduced as far as possible. Published epidemiological studies of chronic effects should be examined for the care with which these factors have been taken into account.

Confounding factors such as smoking, drinking or obesity may influence the prevalence of the particular disease being studied so that elaborate provisions may have to be made to enable a suitable correction to be worked out. The human epidemiologist is never able to claim that his study can approach the rigour of that of the animal toxicologist (MacMahon and Pugh, 1974).

When one or several employees are found to have a particular disease the question springs to mind as to what other characteristic those individuals might have which distinguishes them from their fellows. A 'case control' study might therefore be undertaken. Here, the investigator studies the exposure of individuals with the disease in question in comparison with others closely matched for age, sex, place of residence, income group and so on but without the disease. Where the prevalence of disease is low it is found that case control studies have the particular attraction of being considerably less laborious than a comprehensive study of prevalence. The magnitude of exposure of employees would be established in a semi-quantitative form, usually by questionnaire designed to elucidate whether the cases have the largest exposure to the agent or agents in question. Such case control studies can establish that a particular type of exposure is or is not associated with the disease although it is not possible to define the reverse, namely how much disease is caused by how much exposure. Case control studies are thus of limited value by themselves for defining exposure limits, for which purposes prevalence studies are obligatory.

John Thomas Arlidge (1822–1899), a renowned epidemiologist working in the potteries in the last century around Stoke-on-Trent, was well aware that prevalence studies, when combined with semi-quantitative data about exposure, would be more informative about exposure limits than would case control studies (Arlidge, 1892). Among recent examples may be mentioned the incidence of adenocarcinoma of the nasal cavity related to dusty jobs in the furniture industry in Buckinghamshire, England (Acheson, 1968). This otherwise rare cancer was observed to be prevalent amongst those woodworkers exposed to the highest concentrations of hardwood dust for the longest time, as was subsequently confirmed in Belgium, France and Denmark (Debois, 1969; Gignoux and Bernard, 1969; Mosbech and Acheson, 1971). A lower exposure limit was recommended (Anderson, 1977).

Useful, but incomplete information related to exposure includes years of exposure, period each day spent heavily exposed, presence or absence of mechanical ventilation, type of respirator worn, if any, output and inventory of hazardous substances (Gregersen, Angelso and Nielsen, 1984; Baker, Letz and Fidler, 1985; Fidler, Baker and Letz, 1987). The analysis of this type of information in comparison with prevalence of adverse effects of various kinds can illuminate the nature of the disease, its time course, its toxico-kinetics and the means of prevention. This is a great advance, although still short of an exposure limit on the concentration of air contamination by hazardous substances. For this purpose actual measurements of concentration of contamination in air are needed at some stage.

Lack of adequate exposure data is a common cry from epidemiologists (Hernberg, 1980). It is possible to medically examine an employee, taking at most a couple of hours and establish whether or not that employee has the disease or adverse effect being studied. However, to find out what was the exposure which caused that effect the exposure needs often to have been monitored for the working lifetime of the employee. Furthermore, the exposure monitoring needs to have been of such a kind that the results are unbiased. Environmental sampling that has been purposively designed to focus the results on the times and places of highest exposure may be cost effective for keeping the exposure under control but is of limited use and quite possibly very misleading for epidemiological studies (Ulfvarson, 1983). Keeping the results for thirty years does not correct the errors.

Because exposure to hazardous substances at work is so varied the acquisition of reliable data on employee exposure is undoubtedly the major problem in epidemiological studies but it is not the only one. Keeping track of where employees have been, what they have been doing and their state of health also poses problems (Landrigan 1982). Records of each employee's exposure to hazardous substances covering their total working history would be invaluable. Unfortunately, this information is rarely available in industry due to frequent changes in work assignments within a plant, changes in materials and processes over the years as technology develops, limited air sampling data and the difficulty in assessing short periods of high concentration which occasionally occur due to failure of equipment. Employees join and leave a company of their own volition, normally. When they leave they rarely have a medical examination and are usually lost from a study. Occupational disease may be one of the reasons for leaving, or, they may leave in good health but succumb later. A way of correcting for this must be incorporated in a study or the prevalence will be underestimated.

A quantitative survey of exposure to a hazardous substance linked with medical examinations demonstrating the response, if any, should be well-proportioned and relevant for the purpose. Surveys are of diminished value if they have been based on only a handful of employees, a few atmospheric measurements taken for different reasons, disorganised medical examinations or examinations relying on the enthusiasm of a few 'volunteers'. Total satisfaction in all these areas is rare but the reliability of the resulting limit is critically influenced by the weakest aspect of a combined medical-environmental survey.

Care has to be taken in interpreting the exposures measured alongside medical examinations. Exposures vary from minute to minute, shift to shift and month to month. Employees may or may not wear respirators. The investigator has to consider the merit of short-term and long-term averages in setting exposure limits. The toxico-kinetics of the substance in question is of considerable importance in settling these issues. When comparing exposure limits for different substances, both the severity and prevalence of health effects found in surveys is considered. Exposure to a substance causing a low prevalence of a severe health effect may be awarded the same exposure limit as a substance causing a higher prevalence of a slight health effect. Furthermore, the toxicodynamics of the matter has to be taken into account. Health effects which are progressive in the absence of further exposure are of greatest concern and those which regress or disappear in the absence of further exposure are of less great concern.

Most published data on industrial medical-environmental surveys refer to agents which are cumulative in the body or produce effects which are cumulative and permanent in nature. There is a dearth of published data from surveys referring to agents which cause the milder effects such as irritation, narcosis and headaches. In coming years this may change. When people are affected and repeatedly so such effects cannot be ignored indefinitely.

The opportunity for cross-sectional surveys to yield useful data for setting exposure limits perversely diminishes as the environment is brought under tighter control, as present conditions become increasingly disconnected with past conditions. In future more reliance will have to be placed on long-term prospective studies of very large populations.

Exposure-response curves for employees

Reports of medical examinations of employees linked with comprehensive measurements of their atmospheric exposure in industry are of great value in the process of setting exposure limits for air contamination. Wherever possible, exposure limits are based on the findings from these industrial surveys. When it is known that a limit is based on such surveys this helps greatly in accepting the transfer of group experience from one situation to another.

The concept of deriving an exposure-response curve from industrial medical-environmental surveys was firmly established in the 1950's by the early work of Roach and of Hatch on dust diseases (Roach 1953, Hatch 1955). In the first example 'response' was the presence or absence of a defined degree of pneumoconiosis in coal-getters at a Lancashire coal mine as detected with the aid of chest radiographs (Roach 1953). In determining their dust exposure the duration of exposure had to be determined accurately as well as the average concentration to which they had been exposed. This was done to derive an index of the amount of dust retained in the lungs. The average concentration was measured at locations throughout the colliery employing Thermal Precipitators to collect samples of airborne dust and by counting the particles collected which lay in the size range 0.5–5.0 microns diameter. This size range was believed to be representative of sizes which deposit in the alveolar regions and become permanently retained in the lungs. Earlier workers had found by field studies that in all diseases due to accumulations of dust in the lungs the greatest incidence of pneumoconiosis occurs amongst employees who have been exposed for the longest time to the highest concentration (Sayers, Bloomfield, Dallavalle, Jones, Dreessen, Brundage and Britten, 1935; Dreessen, Dallavalle, Edwards, Miller, Sayers, Easom and Trice, 1938; Flinn, Dreessen, Edwards, Riley, Bloomfield, Sayers, Cadden and Rothman, 1939; Dreessen, Dallavalle, Edwards, Sayers, Easom and Trice, 1940; Flinn, Seifert, Brinton, Jones and Franks, 1941; Dreessen, Page, Hough, Trasko, Jones and Franks, 1942; Bedford and Warner, 1943). Since people with pneumoconiosis were known to have more dust in their lungs than others the logical conclusion was that the dust accumulates gradually in the lungs over a period of many years (Gough, 1947; King, Maguire and Nagelschmidt, 1956; Roach, 1959). The product of average concentration to which the miner had been exposed during his employment and the total length of this period in years can be thought of as an index of dose inhaled, assuming that the volume of air inhaled at work each year is approximately the same for each miner.

The mine surveyed was chosen after first spending much time searching for one where conditions had not changed for at least 20 years. Time spent on this preliminary step before embarking on a survey of a cumulative substance is time well spent. It greatly simplifies the subsequent analysis. Occupational histories were taken to establish where in the colliery the miners had worked and whether they had previously been employed in dusty occupations elsewhere. After the survey data from those miners with significant exposure in other dusty trades were also excluded from the main analysis. The lognormal exposure-response curve fitted to data from the remaining population was viewed in the same way as a biological assay in which the potency of the material is indicated by the position of the line relative to the axes. The shallow slope of the line represents a summation of the biological variation between individuals and inherent errors of measurement (Finney, 1964). Knowing the group response from a given average exposure then affords a means of setting appropriate exposure levels so as to limit the health risks in an orderly fashion. By focussing analysis on the earliest radiographic effects, which precede noticeable disability, any bias in the results through people having left the population because of ill health was much reduced. Extrapolation to low levels of risk below the range measurable with accuracy was also avoided.

In the US the procedure was taken up by Hatch (1955) to analyse the incidence of first stage silicosis, a more serious condition than 'category 1' coal workers pneumoconiosis. Hatch took as his criterion 1% silicosis in employees with 30 years' exposure.

In more recent times hygiene standards of the British Occupational Hygiene Society (BOHS) were typically also based on exposure-response analysis of industrial medical-environmental survey data on dusts hazardous to health. The exposure-response concept was employed by BOHS in its first hygiene standard, for chrysotile asbestos dust, in which the objective was to set an exposure limit which would reduce the risk of contracting the earliest demonstrable effects on the lung to less than 1% for those exposed to asbestos for a working life-time (BOHS 1968). An exposure-response curve was fitted to the data to seek an index of exposure corresponding to 1% basal rates, which, at that time was the earliest demonstrable effect (Roach 1970). No population of employees could be found who had worked in unchanging conditions but the next best thing, one was found where monitoring of exposure by one means or another had been carried out for 20 years.

An investigation by the British National Coal Board into coal workers pneumoconiosis by repeated examination of chest radiographs, lung function and clinical questionnaires of 50,000 miners whose exposure had been measured in terms of atmospheric concentration of respirable dust was the basis of coal dust exposure limits in many parts of the world (National Coal Board, 1969; Jacobsen, Rae, Walton and Rogan, 1970). An exposure-response curve was fitted to the data from 10 years of measurements, from which, by extrapolation it was concluded that after 35 years exposure at 4 mg m^{-3}

respirable dust the probability of developing 'category 1' pneumoconiosis would be 10% whereas at 1.6 mg m^{-3} the probability would be essentially zero. The exposure limit was accordingly set at 2 mg m^{-3} (American Conference of Governmental Industrial Hygienists, 1986).

The BOHS chrysotile asbestos dust hygiene standard was taken up by the US National Institute for Occupational Safety and Health when developing its first formal criteria for standards (NIOSH 1972). The Advisory Committee on Asbestos of the UK Health and Safety Commission later recommended an exposure limit of 1 fibre ml^{-1}. This corresponded to an upper estimate of about 1% (0.4% to 5%) excess mortality from lung cancer and asbestosis together caused by 50 years of exposure (paragraph 193 and Table 21, Health and Safety Commission, 1980).

The British Occupational Hygiene Society, by its hygiene standard for cotton dust and flax dust, took as its objective for these dusts the reduction to less than 4% in the prevalence of grade II byssinosis in working populations. This standard was based on industry-wide surveys of those showing grade II byssinosis, that is, chest tightness or difficulty in breathing each working day on three or more days each working week and evidence of permanent incapacity from diminished effort tolerance or reduced ventilatory capacity (BOHS 1972, 1980).

When the available data are trustworthy, though scanty, but suggestive of unduly high health risks it may be unavoidable to extrapolate downwards to estimate an exposure which will not be associated with significant risk. Extrapolation far beyond experimental points is quite properly frowned upon and should be resisted (Yardley-Jones, Anderson and Parkes, 1991). Recognition of a necessity to extrapolate does not make the outcome any the less uncertain (Berry, Gilson, Holmes, Lewinsohn and Roach 1979).

Extrapolation of quantitative risk assessments of carcinogenic agents has had much attention in recent years (Food Safety Council, 1978). The risk which will be accepted of adverse effects which are grave and possibly fatal in the general population, such as cancer risks, may very low indeed. For example, the US Food and Drug Administration adopted a concept of virtual safety for foods corresponding to a lifetime carcinogenic risk of 10^{-6} in the general population (Federal Register, 1977). The computation of the corresponding virtually safe dose of carcinogens for the general population, based on experimental observations involves extrapolation downwards through several orders of magnitude.

The only absolute and final proof of the effectiveness of a control system is one based on repeated, comprehensive and thorough medical examinations of those who have been exposed for a long time under that system. The size of the population of exposed employees that would be needed to measure risks as low as 10^{-6} with any accuracy is correspondingly great. Indeed, the numbers required would be so large there is no way of testing whether one extrapolation model is any better than another. For such levels of risk the theoreticians are free to create the most absurd possibilities. It may be argued that there exists no extrapolation model which relates such extremely low health risk to exposure concentrations based on solid experimental science or observation. It is simply not feasible to measure such low levels of risk directly either by epidemiological studies of employees or, for that matter, by animal experiments (Peto, 1985). The numbers required are too great.

However, a 'significant' risk in an industrial population, as defined earlier, is hardly likely to be so low as to be below the level of detectability. The level of risk deemed 'just significant' in a population of employees will almost certainly be well above 10^{-6}. For example, in the context of lifetime risk of leukaemia from industrial exposure to benzene the US Occupational Safety and Health Administration proposed 10^{-3} as being a significant risk in that instance (OSHA, 1987).

The lessons of experience in deriving occupational exposure limits with the aid of industrial medical-environmental surveys may be encapsulated briefly. Surveys in industry of particular value will probably include some of the following features:

(a) **Number medically examined.** Medical examinations should be conducted on at least 50–500 employees, in or from a defined occupational group working on a continuous process, exposed to the substance in question, at concentrations in air where early signs of adverse effects on health are observed in some employees. The lower extreme of 50 is applicable for studies of the less serious adverse effects such as irritation or mild narcosis, whereas the lower extreme is raised to 500 for studies of substances whose first detectable adverse effects are serious systemic effects. Batch processes are unlikely to yield satisfactory data because of their inherent variability and short term nature.

(b) **Number of exposure measurements.** Exposure to the substances in air inhaled by the employees should be measured over a period of at least two weeks and for at least 50 man-shifts. The duration of each sampling period should be chosen with regard to toxico-kinetic parameters. Substances whose effective biological half time is less than 30 minutes should be measured employing a unit sampling period of no more than 10 minutes. Job-exposure groups which don respiratory protection are unlikely to be useful because of the technical difficulty of measuring the concentration in their inhaled air. Where the survey period is extended over months or years and there have been changes in sampling or analytical methods over the period of the survey there has to be sufficient overlap between methods to permit interpretation in terms of the more modern methods.

(c) **Definition of adverse effects.** Adverse effects should be measured in a quantitative or at least semi-quantitative manner. There needs to be a clear definition of the criteria to be employed to decide the presence or absence of an adverse effect. A pilot study should be undertaken to test methods.

(d) **Relevance of exposure measurements.** Adverse effects, where recorded, should be reasonably attributable to exposure at the concentrations measured in air. When the adverse effects are cumulative in nature, or delayed, employees who have significant exposure in previous employments should, if possible, be excluded. When there is reason to believe the working conditions have improved (or deteriorated) in recent years medical data from employees who have been exposed in the earlier conditions should be excluded from analysis of cumulative effects. The reason for this is simple, but profound. Our knowledge of past exposure trends is so hazy. Each time the data is looked at afresh it seems to change. The constant revisions to historic data adjustments make it impossible to give much credence to estimates of accuracy. We still do it but hidden in the estimates is quite a lot of cynicism.

(e) **Specificity factors.** Some or all of the adverse effects recorded should be specific, that is, would not have occurred in the absence of exposure to the substance in question. If the adverse effects are non-specific, adequate provision must be made for measurement of their incidence in an appropriate control population. A non-exposed group of employees doing similar work in the same locality is ideal. Some types of disease such as chronic bronchitis and lung cancer are not uncommon in the population at large, and therefore an excess incidence among employees exposed to particular substances may fail to be detected. Comparison with national statistics may be misleading as there are often marked differences between urban and rural populations and between one geographical area and another (Veys, 1990). Other factors such as smoking, diet or exposure to other substances may act additively, independently or synergistically with occupational factors to affect the prevalence of disease.

(f) **Missing data.** Refusals or lost data in both the medical and environmental surveys should have been kept to less than 10%. The fittest people tend to be the most ready to volunteer for examination. Permission to survey environmental conditions is most readily given for jobs, operations, places and times where the conditions are most favourable.

(g) **Extrapolation.** Exposure of the employees should be within a factor of 10 either side of the relevant occupational exposure limits. Otherwise, extrapolation becomes too speculative.

In point of fact the prevalence of adverse effects amongst employees who have been exposed at the atmospheric exposure limit may be expected to be within a factor of 5 either side of the following values:

Adverse effect	Prevalence at exposure limit
Irritation	5×10^{-2}
Narcosis	2×10^{-2}
Impairment of health	10^{-2}

There are some circumstances which markedly reduce the value of medical-environmental surveys. Examples are:

(a) Occupational histories of employees are not taken or are taken inaccurately. This is a time consuming task but it should not be rushed or skimped.

Company personnel records are helpful but rarely adequate for this purpose.

(b) Studies of adverse effects of such a severity as to have led to some individuals leaving the occupational group may be of diminished value on this account. The remaining group of employees would be expected to have a reduced prevalence rate of these effects.

(c) An adverse health effect may not be revealed if it is one which is evident only when a significant passage of time has elapsed after exposure, that is when an effect has a so called 'latent period'.

(d) Observations are limited to small populations exposed to concentrations very much higher than (more than 10 times) the eventual exposure limit. Scientists do not know how to extrapolate from a high concentration to a low one when holding duration of exposure constant but varying concentration. The simplest solution is to adopt a linear concentration-response model, which assumes that the risk is proportional to concentration. This implies making an estimate of low concentration risk which is definitely on the high side and sometimes very much so. This should result in a built-in safety margin, although the resultant controls may be unnecessarily strict. This is illustrated in Figure 6.1. The extrapolation normally required is to estimate the concentration that would give rise to some stated reduction in risk. In this example several lognormal concentration-response curves are

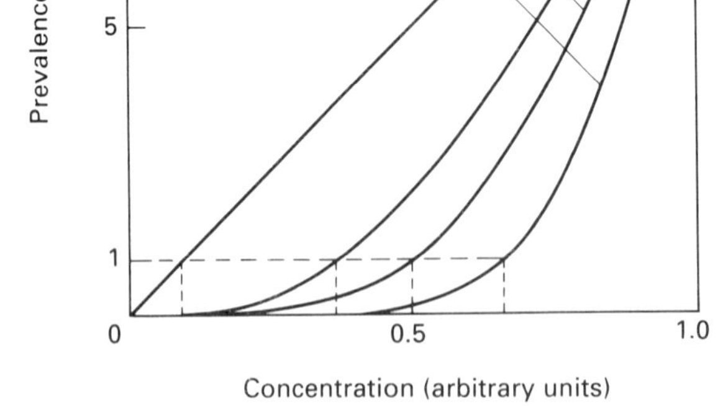

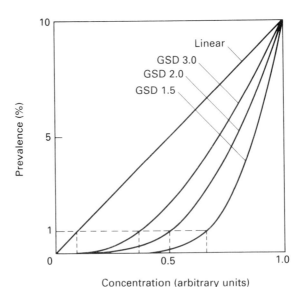

FIG. 6.1. Extrapolation of response from high to low concentration. Starting from the knowledge that the response is 10% after exposure to a certain concentration, linear or lognormal extrapolation could be employed to estimate the response at some lower concentration. Lognormal extrapolation is illustrated for curves assuming geometric standard deviation of 1.5, 2.0 and 3.0.

compared with linear extrapolation from a prevalence of 10^{-1} down to one of 10^{-2}. The contrast is more pronounced the greater the extrapolation.

(e) Industry wide *a posteriori* surveys in which exposure measurements have not been standardised. In industry measurements are not taken at random, nor are they taken systematically throughout every plant. The results are critically dependent upon the purpose of the measurements. The priorities in different companies can be very different. The sampling instruments employed should have been the same, as should analytical procedures, sampling strategy and intensity of monitoring (Enterline, Marsh and Esmen, 1983; Rockette and Arena, 1983; Checkoway, Mathew, Hickey, Shy, Harris, Hunt and Waldman, 1985; Blair, Stewart, O'Berg, Gaffey, Wairath, Ward, Bales, Kaplan and Cubit, 1986).

Human volunteer studies

Increasingly stringent regulations governing the use of humans in experimentation will continue to restrict this source of information for setting exposure limits. Nevertheless, human volunteer studies on groups of 10–20 volunteers have been conducted in the past on some of the airborne substances which give rise to mild irritation, offensive odour or headaches (American Conference of Governmental Industrial Hygienists, 1986; World Health Organisation, 1984). Such effects do not lend themselves readily to animal experimentation. Studies of this kind have been undertaken in the past where the effects are tolerable for a brief period, are transient and have no lasting sequelae. Information may also be obtained on effects on ability to perform simple tasks and the opportunity may be taken to establish toxico-kinetic parameters. Typically, 4 to 6 volunteers would enter a test room in which a known concentration of irritant has been raised and stay for 30 minutes to 4 hours. Their responses would be elicited by standard questionnaire about irritation of the eyes, nose, throat or lungs generally.

In the laboratory the concentration to which the volunteers are exposed is deliberately held constant. The interpretation of the results in terms of employee exposure in industry has to have regard to the variable concentration typical there. The time actually spent above a given atmospheric concentration of an acute irritant is more important than the time weighted average concentration over 8 hours. As a guide, it is suggested that a significant risk to the health of employees would reasonably be suspected if exposure to such agents is

(i) in excess of that level of exposure which is liable to give rise to 5% or more of exposed employees complaining of irritation of the eyes, nose or throat, and
(ii) in excess of that level for, say, 15 minutes or more per 8-hour work-shift.

In continuous processes the distribution of 10–30 minute time weighted average concentrations over a work-shift is quasi-lognormal and the geometric standard deviation is generally in the range 1.1–2.5 (Leidel, Busch and Crouse,

1975). To the extent that the distribution is lognormal it may be shown by mathematical analysis that there is a maximum to the frequency of values above a given multiple of the mean (Rappaport, Selvin and Roach, 1988). Accordingly, to assure the frequency of occurrence is less than 15 minutes in 8 hours the mean should be less than one fifth the maximum.

Human volunteer studies do not provide sufficient information about effects produced by long term exposure, so that their value is largely limited to providing a basis for exposure limits to protect against against irritation and other short term effects.

Setting limits for therapeutic agents

To calculate the least body burden that may be absorbed without harm there is a useful benchmark; the lowest therapeutic dose. This is, however, a benchmark which should be used circumspectly since, by definition, even a low therapeutic dose has some biological effect and some of the subtle side effects may be unknown. Furthermore, the usual route of administration of therapeutic agents to humans has been by ingestion, although for certain agents administration may have been by inhalation, by skin contact, intravenously or other means. Thus, the manufacture and handling of therapeutic agents may be associated with exposure through a different route from that employed in the experimental studies of the therapeutic properties. There may be adverse effects which occur on exposure many times less than would be equivalent to a therapeutic dose.

By their intended use, it is manifest that therapeutic agents are biologically active, and can therefore be hazardous if administered in excessive amounts or by an unusual route (Harrington, 1981; Teichman, Fleming Fallen and Brandt-Rauf, 1988). Setting exposure limits for therapeutic agents as contrasted to other substances is characterised by the relative abundance of data on the effects on experimental animals and humans produced by known doses. The therapeutic dose is usually, but by no means always administered orally. The interaction of these agents with specific receptors in the body is usually reversible. In principle an employee exposure would be acceptable which gives rise to a body burden in general and in the 'target' organ in particular of no more than one-tenth that obtained by the lowest therapeutic dose in humans (Paustenbach and Langner, 1986). On the other hand it may be argued in particular instances that the therapeutic effect effect is not 'adverse' and a less stringent safety factor might be acceptable. It would still remain to decide the relationship of employee exposure and the consequent body burden. Toxicokinetic data may have to be sought.

Additional safety factors could be added to remove any doubt about absorption calculations based on various means of exposure (Dollery, Davies and Conolly, 1971; Pepelko and Withey, 1985; National Research Council, 1986; Withey, 1987). In the same way, a dose administered orally may have little relevance to irritant effects on the eyes, upper respiratory tract or skin, for example.

There are two general facts in regard to the absorption of air contamination in inhaled air. First, no more of the substance can be absorbed than is taken into the lungs in the air breathed. Second, an equilibrium is eventually reached when the amount being eliminated from the body is equal to the amount being absorbed.

The amount of contamination brought to the lungs for absorption is the product of three factors:

(1) the volume flow rate of air inhaled (and exhaled) during exposure,
(2) the concentration of contamination in the inhaled air,
(3) the duration of exposure.

The volume flow rate of air inhaled by a man of average size performing work requiring moderate exertion may be taken as 10 litres per minute, or, in round numbers 5 cubic metres in an 8-hour working day. This, when multiplied by the concentration in the inhaled air, provides an effective maximum to the dose of substance per day by the inhalation route. The effects of this may well be considered comparable with the effects of a daily dose of therapeutic substance taken by the oral route, say.

Regular exposure at work, 5 days a week, with two days without exposure, is a different regimen than might be used in clinical trials of a drug and account should be taken of this in setting exposure limits. The consequences in terms of the body burden can be studied in relation to the toxico-kinetics of the substance. Data from a combination of different sources; animal experiments, clinical trials, toxico-kinetics and toxico-dynamics, may all be contributory in determining appropriate exposure limits for therapeutic substances (Sargent and Kirk, 1988). Specialist advice should be sought.

Physical/chemical analogy

When human experience is limited it is, of course, not unusual to make use of biological indices of toxicity in animals, such as LD50, LC50 and the like, to guide the choice of exposure limit. Very often, however, there are no quantitative biological test data available either for humans or experimental animals, particularly with the newer chemicals. In these instances an analogy might be considered with other substances having qualitatively similar physical, chemical and toxicological properties and thereby create a secondary exposure limit. Great care is needed with this approach. There are many notorious examples where it has failed, where the analogy rests solely upon physical and chemical properties. Because of its chemical similarity, toluene, for example, was once considered as toxic as benzene, despite human experience which is now overwhelmingly conclusive to the contrary.

The physical properties which are clearly of particular importance in seeking analogy are solubility, vapour pressure and, with aerosols, size distribution (Sato, 1987; Ulfvarson, 1987). Chemical analogy may be employed with fair confidence where a substance is one member in the midst of a large family or series of similar chemicals which exhibit a similar pattern of behaviour. A good example of this is found in the ACGIH Documentation of the Threshold Limit

Values for substances in workroom air for 1971 (American Conference of Governmental Industrial Hygienists, 1971). The example chosen is the documentation for Stoddard solvent at that time, from which the following is an extract:

'As a first approximation and for toxicological purpose, Stoddard solvent may be considered to consist of a mixture of 85% nonane and isodecane and about 15% trimethyl benzene.

'The Threshold Limit Value for n-heptane has been set at 500 ppm. While no comparable toxicity data are available for nonane and decane, the following figures have been given for pentane, hexane, heptane, and octane.

	CONCENTRATION NARCOTIC (MICE)	*CONCENTRATION LETHAL (MICE)*	*REFERENCE*
Hexane	30,000 ppm	37,000 ppm	Patty, 1963
Heptane	12,000 ppm	16,000 ppm	Patty, 1963
Octane	10,000 ppm		

	CONCENTRATION ISONARCOTIC (MICE)	*CONCENTRATION FATAL (MICE)*	*REFERENCE*
Pentane	130,000 ppm	130,000 ppm	Flury & Zernik
Hexane	42,000 ppm	52,000 ppm	Flury & Zernik
Heptane	16,000 ppm	16,000 ppm	Flury & Zernik
Octane	8,000 ppm	13,500 ppm	Flury & Zernik

'These data indicate that heptane is two and one-half to three times as toxic as hexane, while octane is one and one-quarter to two times as toxic as heptane. In the absence of data to the contrary, it is reasonable to predict that n-nonane and isodecane are considerably more toxic than heptane. A conservative assumption would be that they are twice as toxic, and that a TLV of 250 ppm, half that of heptane, would be appropriate.

'Threshold limits of 50 and 35 parts per million have been proposed for mesitylene -- one of the isomers of trimethyl benzene (Gerarde, 1960). For a mixture containing 15 mol percent mesitylene and 85 percent nonane and isodecane the threshold limit would be, assuming a TLV of 50 parts per million for mesitylene, as follows:

$$\frac{1}{\text{TLV}} = \frac{0.15}{50} + \frac{0.85}{250} = \frac{1.6}{250},$$

'If a TLV of 35 parts per million is assumed for trimethyl benzene, the Stoddard solvent value would be 131 parts per million.

'This reasoning is in general agreement with the statement of Flury and Zernik (1931) that heavy benzine (boiling range 102—160C) is one and one-half times as toxic as light benzine (boiling range 50—100C), although their data indicated a two-fold ratio of toxicity or even more if concentration is based on parts per million rather than milligrams per liter.'

(American Conference of Governmental Industrial Hygienists, 1971).

Chemical analogy is less satisfactory for chemicals towards the beginning of a homologous series.

The molecular structure of a chemical with an intricate structure may have particular features in common with others (Agius, Nee, McGovern and Robertson, 1991). It may be possible to identify key metabolites, metabolic pathways and mechanisms of toxicity (Silk and Hardy, 1983). A full evaluation can include molecular size, shape, structure and the presence or positioning of reactive functional groups (Stuper, Brugger and Jurs, 1979; Jurs, 1983;

Enslein, 1984). Quantitative structure activity relationships have been used to help predict respiratory irritation (Muller and Greff, 1984), organ toxicity (Schultz, Kier and Hall, 1982), carcinogenesis (Chou and Jurs, 1979) and mutagenicity (Tinker, 1981). With new substances some of the physical/chemical properties may be estimated from the molecular structure (Lyman, Reehl and Rosenblatt, 1984).

As a precautionary measure, a safety factor ranging from 2 to 10 might be employed according to the closeness of the physical and chemical analogy.

Generic limits

Exposure limits applying to a whole class of substances, generic limits, are applicable for similar substances without exposure limits. Examples are simple asphyxiants and nuisance dusts. Simple asphyxiants are gases and vapours with no known physiological effects other than to displace oxygen in inhaled air. An appropriate limit for these is one which reduces the oxygen content to 18% by volume. Argon, helium and neon are simple asphyxiants. Nuisance particulates are those which produce no significant adverse health effects below a concentration of 10 mg m^{-3} of dust of all sizes. The limit of 10 mg m^{-3} relates to the level commonly believed to cause general nuisance.

As regards acids and bases as a class, it is reasonable to suppose that the irritating potential of such substances is related in some way to their acidity/alkalinity. The exposure limits of organic acids and bases which have been based upon irritation are significantly correlated with their dissociation constants so that exposure limits for such substances may be estimated by analogy with organic acids/alkalis having similar dissociation constants. The regression equation for irritant organic acids on the 1986 list of ACGIH Threshold limit Values or 1985 WEEL Guides of the American Industrial Hygiene Association has been reported by Leung and Paustenbach (1988) and is

$$\log_{10} \text{OEL} = 0.43\ \text{p}K_a + 0.53$$

and for irritant organic bases is

$$\text{OEL} = -200\ \text{p}K_a + 2453,$$

where OEL = occupational exposure limit expressed as micro-mols per cubic metre,

$\text{p}K_a$ = equilibrium dissociation constant at 25C expressed as the negative decadic logarithm.

Limits calculated from these equations relate only to protection against irritation. They should be regarded as being essentially temporary limits and vigilance should be maintained against other effects. The majority of TLVs are based upon irritation but this does not mean that irritation is the only effect (Mackison and Stricoff, 1978). Other effects invariably occur at higher concentrations and with many substances adverse effects occur at a lower concentration than that which produces irritation.

Another kind of generic limit has been proposed by Gardner and Oldershaw for volatile organic substances which have not been allocated exposure limits by other means (1991). They recommend a limit based upon the labelling classification of the European Economic Community (EEC, 1987). The geometric mean of national 8-hour TWA occupational exposure limits for volatile organic substances labelled 'harmful by inhalation' (RP 20) was recommended for similar substances so labelled but which had not been allocated occupational exposure limits (approx. 25 ppm). The geometric mean of national 8-hour TWA occupational exposure limits for volatile organic substances labelled 'toxic by inhalation' (RP 23) or 'very toxic by inhalation' (RP 26) was recommended for substances labelled in either of these two ways (approx. 1 ppm). By the nature of the calculations approximately half the recommended occupational exposure limits would be too high, and half too low. As a precautionary measure a further safety factor of 10 might be employed so as to be reasonably sure the protection is adequate; that is, 2.5 ppm for substances labelled RP 20 and 0.1 ppm for substances labelled RP 23 or RP 26. In doing this it is recognised that these levels may be relaxed when there is subsequently accumulated sufficient toxicological information to set an occupational exposure limit directly.

The dermal factor

Many liquids and dusts may be absorbed through the skin in significant amounts unless appropriate skin protection is provided. In favourable circumstances absorption through the unprotected skin could be the major route of penetration of some substances into the body (Marzulli and Maibach, 1983). The physico-chemical properties of a material are the main determinants of whether or not a material will be absorbed through the skin. Among the important factors are pH, extent of ionization, water and lipid solubility and molecular size (Hodgson and Levi, 1987). Example substances where the major route of entry to the body could be skin penetration include phenol and phenolic derivatives, carbon tetrachloride, trichloroethylene, tetrachloroethylene, methylene chloride,1,1,1-trichloroethane, ethylbenzene, toluene, styrene and xylene (Dutkiewicz and Tyras, 1967, 1968; Stewart and Dodd, 1964; Sato and Nakajima, 1978). Skin protection should be provided when handling such substances. If the skin is damaged, the normal protective barrier to absorption of substances in contact with the skin is lessened and penetration may occur of substances which would otherwise fail to be absorbed to any significant extent. Skin protection should be provided wherever there is risk of physical injury to the skin.

Criteria for deciding whether a significant amount of a substance is likely to be absorbed through the skin have been proposed by Fiserova-Bergorova and Pierce (1989). They suggest firstly that the exposure limit of a chemical should carry a 'skin notation' if the rate of absorption through the skin is likely to be more than 30% that expected by inhalation at the atmospheric exposure limit.

That is, a skin notation is made when

$$\frac{F \times A}{0.3 \times \text{OEL} \times Q_a} > 1, \qquad (1)$$

where F = flux in mg cm^{-2} h^{-1}
A = exposed body surface area in cm^2
OEL = exposure limit in mg m^{-3}
Q_a = alveolar ventilation rate in m^3 h^{-1}

Further, as a guideline they suggest that the exposed area of the body be taken as either

(a) 2% of the body surface when exposure is to a liquid, that is, an area equal to the palm and fingers of both hands,

or

(b) the whole body surface when exposure is just to a vapour.

In the case of liquid contamination, when taking (male) body surface area as 18,000 cm^2 and alveolar ventilation as 0.9 m^3 h^{-1} the criterion for skin protection, expression (1), reduces to

$$\frac{1{,}333 \times F_1}{\text{OEL}} > 1, \qquad (2)$$

where F_l = flux of the liquid in question through skin, mg cm^{-2} h^{-1}
OEL = exposure limit of the vapour, mg m^{-3}

In the case of vapour absorption the criterion for skin protection, expression (1), reduces to

$$\frac{66{,}667 \times F_v}{\text{OEL}} > 1, \qquad (3)$$

where F_v = flux through the skin of the vapour, mg cm^{-2} h^{-1}
OEL = exposure limit of the vapour, mg m^{-3}

The flux of vapour will be approximately proportional to the atmospheric concentration. When respiratory protection is worn and employees are exposed to concentrations many times the exposure limit significant absorption of vapour through the skin is increasingly likely.

References

Acheson E D. Brit. Med. J. 2 587 (1968).

Agius R M, J Nee, B McGovern and A Robertson. Ann. Occup. Hyg. 35 129 (1991).

American Conference of Governmental Industrial Hygienists. Documentation of the Threshold Limit Values for Substances in Workroom Air. Publications Office, ACGIH, P O Box 1937, Cincinnati, Ohio 45201, USA (1971, 1986).

American Conference of Governmental Industrial Hygienists. Threshold Limit Values and Biological Exposure Indices for 1987–1988. ACGIH, P O Box 1937, Cincinnati, Ohio 45201, USA (1987).

American Industrial Hygiene Association. Documentation for WEEL Guides. AIHA, 475 Wolf Ledges Parkway, Akron, Ohio 44311–1087, USA (up-to-date).

Anderson H C. Brit. J. Ind. Med. **34** 201 (1977).
Baker E L, R E Letz and A T Fidler. J. Occup. Med. **27** 206 (1985).
Barnard R C. Am. Ind. Hyg. Assoc. J. **48** 798 (1987).
Bedford T and Warner C G. Chronic pulmonary disease in South Wales coal miners. II. Environmental studies, B. Physical studies of the dust hazard and of the thermal environment in certain coal mines. Spec. Rep. Ser. Med. Res. Coun. No. 244. London (1943).
Berry G, J C Gilson, S Holmes, H C Lewinsohn and S A Roach. Brit. J. Ind. Med. **36** 98 (1979).
Blair A, P Stewart, M O'Berg, W Gaffey, J Wairath, J Ward, R Bales, S Kaplan and D Cubit. J. Natl. Cancer Inst. **76** 195 (1986).
British Occupational Hygiene Society. Ann. Occup. Hyg. **11**, 47 (1968).
British Occupational Hygiene Society. Ann. Occup. Hyg. **15** 165 (1972).
British Occupational Hygiene Society. Ann. Occup. Hyg. **23** 1 (1980).
Checkoway H, R M Mathew, J L S Hickey, C M Shy, R Harris, E Hunt and G Waldman. J. Occup. Med. **27** 885 (1985).
Chou J T and P C Jurs. J. Med. Chem. **22** 792 (1979).
Corn M. Ann. Occup. Hyg. **27** 91 (1983).
Debois J M. Tidschrift voor Geneeskunde **25** 92 (1969).
Dollery C T, D S Davies and M E Conolly. Ann. N.Y. Acad. Sci. **179** 108 (1971).
Dreessen W C, J M Dallavalle , T I Edwards, J W Miller, R R Sayers, H F Easom and M F Trice. Pub. Health Bull. No. 241, Washington DC (1938).
Dreessen W C, J M Dallavalle, T I Edwards, R R Sayers, H F Easom and M F Trice. Pneumoconiosis among mica and pegmatite workers. Pub. Health Bull. No. 250, Washington DC (1940).
Dreessen W C, R T Page, J W Hough, V M Trasko, J L Jones and R W Franks. Pub. Health Bull. No. 277, Washington DC (1942).
Dutkiewicz T and H Tyras. Brit. J. Ind. Med. **24** 330 (1967).
Dutkiewicz T and H Tyras. Brit. J. Ind. Med. **25** 243 (1968).
EEC. Legislation on dangerous substances: Classification and Labelling in the European Communities: Consolidated Text of Council Directive 67/548/EEC. Volumes 1 and 2. Graham and Trotman Ltd, London.
Enslein K. Pharmacol. Rev. **36**:131S (1984).
Enterline P E, G M Marsh and N A Esmen. Am. Rev. Resp. Dis. **128** 1 (1983).
Epidemiology Work Group of the IRLG. Guidelines for documentation of epidemiological studies. Am. J. Epidem. **115** 609 (1951).
Federal Register. **42** (no. 35). 10412 (February 22, 1977).
Fidler A T, E L Baker and R E Letz. Brit. J. Ind. Med. **44** 133 (1987).
Fiserova-Bergerova V and J T Pierce. Appl. Ind. Hyg. **(4)8** 14 (1989).
Flinn R H, W C Dreessen, T I Edwards, E C Riley, J J Bloomfield, R R Sayers, J F Cadden and S C Rothman. Pub. Health Bull. No. 244, Washington DC (1939).
Flinn R H, H E Seifert, H P Brinton, J L Jones and R W Franks. Soft coal miners, health and working environment. Pub. Health Bull. No. 270, Washington DC (1941).
Food Safety Council. Quantitative risk analysis. Food Cosmet. Toxicol. **16** Suppl. 2, 109 (1978).
Gardner R J and P J Oldershaw. Ann. Occup. Hyg. **35** 51 (1991).
Gignoux M and P Bernard. Le Journal de Medecine de Lyon **50** 731 (1969).
Goldsmith J R and S Besser. Ann. Acad. Med. **13** 297 (1984).
Gough J Occup. Med. **4** 86 (1947).
Gregerson P, B Angelso and T E Nielsen. Am. J. Ind. Med. **5** 201 (1984).
Harrington J M. The health industry. In: Recent advances in occupational health (Edited by J C McDonald), pp. 75–84. Churchill Livingstone, Edinburgh, UK (1981).
Hatch T. Am. Ind. Hyg. Assoc. Quart. **16**, No 1, 30 (1955).
Hatch T F. Bull. World Health Org. **47** 151 (1972).
Hatch T F. J. Occup. Med. **14** 134 (1972).
Health and Safety Commission. Asbestos: Final Report of the Advisory Committee. Vol. 1 and 2. H M Stationery Office, London (1979/1980).
Hernberg S. Epidemiology—Principles and practical examples. In: Developments in occu-

pational medicine (Edited by P Zenz), pp. 3–40. Year Book Medical Publishers, Chicago, USA (1980).
Illing H P A. Human Expl. Toxic. 10 215 (1991).
IPCS. Guidelines on studies in environmental epidemiology. World Health Organisation, Geneva, Switzerland (1983).
Jacobsen M, S Rae, W H Walton and J M Rogan. Nature **227** 445 (1970).
Jurs P C. Studies of relationships between molecular structure and biological activity by pattern recognition methods. In: Structure-activity correlation as a predictive tool in toxicology—fundamentals, methods and applications (Edited by L Golberg) pp. 93–110. Hemisphere Publishing, London (1983).
King E J, B A Maguire and G Nagelschmidt. Brit J. Ind. Med. **13** 9 (1956).
Koch A L. J. Theoret. Biol. **12** 276 (1966).
Kroes R. Animal data. In: Proceedings of the International Conference on Environmental Carcinogenesis, Amsterdam (Edited by P Emmelot and E Kriek), p. 287. Elsevier/North-Holland Biomedical Press, Amsterdam—New York—Oxford (1979).
Landrigan P J. J. Environ. Sci. Health. **4**, 499 (1982).
Leidel N A, K A Busch and W E Crouse. Exposure measurement, action level and occupational environmental variability. U.S. Department of Health, Education and Welfare, HEW Publication No. (NIOSH) 76–131, Superintendent of Documents, U.S. Government Printing Office, Washington DC, 20402, USA (1975).
Leung H W and D Paustenbach. Appl. Ind. Hyg. **3** 115 (1988).
Morton W E. J. Occup. Med. **19** 258 (1977).
Mosbech J and E D Acheson. Danish Med. Bull. **18** 34 (1971).
Muller J and G Greff. Fundem. Chem. Toxic. **22** 661 (1984).
National Coal Board. Pneumoconiosis field research. NCB, London (1969).
National Institute of Occupational Safety and Health. Criteria for a recommended standard; Occupational exposure to asbestos (1972). NIOSH Publications Dissemination, 4676 Columbia Parkway, Cincinnati, Ohio 54226, USA (1972).
National Research Council. Dose route extrapolation using inhalation toxicity data to set drinking water limits. In: Drinking water and health, Vol. 6 (edited by R D Thomas). National Academy Press, Washington DC, USA (1986).
OSHA. Occupational exposure to benzene. Fed. Reg. **52** 34460–34579, USA (1987).
Patty F A. Industrial hygiene and toxicology, Vol II, 2nd Ed., p. 1198. Interscience, New York (1963).
Paustenbach D and R Langner. Am. Ind. Hyg. Assoc. J. **47** 809 (1986).
Peto R. Epidemiological reservations about risk assessment. In: Assessment of risk from low-level exposure to radiation and chemicals (Edited by A Woodhead, C Shellabarger, V Pond and A Holeander). Plenum Press, New York (1985).
Pepelko W E and J R Withey. Toxicol. Ind. Health **1** 153 (1985).
Peto R. Carcinogenic effects of chronic exposure to very low levels of toxic substances. Environmental Health Perspective **22** 155 (1978).
Rappaport S M, S Selvin and S A Roach. Appl. Ind. Hyg. **3** 310 (1988).
Roach S A. Brit. J. Ind. Med. **10** 220 (1953).
Roach S A. Brit. J. Ind. Med. **16** 104 (1959).
Roach S A. Ann. Occup. Hyg. **13** 7 (1970).
Rockette H E and V C Arena. J. Occup. Med. **25** 549 (1983).
Sargent E V and G D Kirk. Am. Ind. Hyg. Assoc. J. **49** 309 (1988).
Sato A. Scand. J. Work Environ. Health **13** 81 (1987).
Sato A and T Nakajima. Brit. J. Ind. Med. **35** 49 (1978).
Sayers R R, J J Bloomfield, J M Dallavalle, R R Jones, W C Dreessen, D K Brundage and R H Britten. Anthraco-silicosis among hard-coal miners. Pub. Health Bull. No. 221. Washington DC (1935).
Schultz T W, L B Kier and L H Hall. Bull. Environ. Toxic. **28** 373 (1982).
Silk S J and H L Hardy. Ann. Occup. Hyg. **27** 333 (1983).
Stewart R D and H C Dodd. Ind. Hyg. J. **25** 439 (1964).

Suma'mur P K. Indonesia J. Ind. Hyg. Occup. Health—Safety and social security **IX** 3 and 4, and **X** 1, 2, 3, 4, 53 (1976 and 1977).
Teichman R F, L Fleming Fallen and P W Brandt-Rauf. J. Soc. Occup. Med. **38** 55 (1988).
Tinker J. J. Chem. Info. Comput. Sci. **21** 3 (1981).
Ulfvarson U. Int. Arch. Occup. Environ. Health **52** 285 (1983).
Ulfvarson U. Scand. J. Work Environ. Health **13** 389 (1987).
Vanhoorne M, M Harrington, L Parmeggiani, W Hunter and K Vuylsteek. Int. Arch. Occup. Environ. Health **55** 337 (1985).
Veys C A. Ann. Occup. Hyg. **34** 349 (1990)
Withey J R. Approaches to route extrapolation. In: Toxic substances and human risk (Edited by R G Tardiff and J V Rodricks). Plenum Press, New York (1987).
World Health Organisation. Recommended health-based occupational exposure limits for respiratory irritants: Report of a WHO Study Group. WHO Tech. Rep. Ser. No. 707. World Health Organisation, Geneva, Switzerland (1984).
Wright G W. The influence of industrial contaminants on the respiratory system. In: The industrial environment—its evaluation and control. U.S. Department of Health, Education and Welfare, Public Health Service, Center for Disease Control, National Institute for Occupational Safety and Health, US Government Printing Office, Washington DC (1973).
Yardley-Jones A, D Anderson and D V Parkes. Brit. J. Ind. Med. 48 437 (1991).

Bibliography

Arlidge J T. The hygiene, diseases and mortality of occupations. Percival and Co., London (1892).
Finney D J. Statistical methods in biological assay. Hafner, New York (1964).
Flury F and F Zernik. Schadliche gase, pp 257–264. Springer, Berlin (1931).
Hodgson E and P E Levi. A textbook of modern toxicology. Elsevier, New York (1987).
Lilienfeld A M and D E Lilienfeld. Foundations of epidemiology. Oxford University Press (1980).
Lymon W J, W F Reehl and D H Rosenblatt. Handbook of chemical property estimation methods—Environmental behaviour of organic compounds. McGraw Hill, New York (1982).
Mackison F W and R S Stricoff. NIOSH/OSHA Pocket guide to chemical hazards. US Department of Health Education and Welfare, Public Health Service, Center for Disease Control, National Institute for Occupational Safety and Health. DHEW (NIOSH) Publication No. 78–210 (1978).
MacMahon B and T F Pugh. Epidemiology: Principles and methods. Little Brown & Co., Boston USA (1974).
Marzulli F M and H I Maibach (Editors). Dermatotoxicology. Hemisphere, New York (1983).
Monson R R. Occupational epidemiology. CRC Press, Boca Raton, Fl., USA (1980).
National Research Council. Toxicity testing: Strategies to determine needs and priorities. National Academy Press (1984).
Stuper A J, W E Brugger and P C Jurs. Computer assisted studies of chemical structure and biological function. John Wiley, Chichester, UK (1979).

CHAPTER 7

Guiding Concepts for Setting Exposure Limits—Animal Experiments

In the early and middle part of the 19th century the usual method of toxicity testing of new chemicals was to try the new chemical on fellow chemists or technicians. However, towards the end of the century manufacturers began occasionally to use laboratory animals in toxicity tests (Boyd, 1959). Thus in 1899 the original report on the new drug acetylsalicylic acid referred to tests on one small fish and two freshly captured frogs (Dreser, 1899). In the early part of the 20th century the concept of determining lethal doses of new drugs in animals took hold and predictive toxicology began to evolve (Boyd, 1972). A tremendous impetus was given to the discipline by the elegant mathematical analyses of animal death rates in a seminal report on biological assay by Gaddum (1933). This was quickly reinforced by developments in the use of probits in the statistical analysis (Bliss, 1934, 1935). The groundwork for extrapolation was in place.

Over the last 50 years it has become common practice to study the toxicity of hazardous substances in experimental animals, usually mammals, regarded as appropriate models or substitutes for humans. Experimental toxicology has an established role in assessing the hazards of new chemicals prior to use beyond exploratory stages. Extensive studies at pilot plant stage are commonplace. Information on the qualitative and quantitative behaviour of a hazardous substance in the animal body can be used to estimate exposure limits as well as to predict the signs and symptoms to be vigilant about in medical surveillance. For the foreseeable future exposure limits devised for making health risk assessments of work activities will depend in large measure on data from such animal studies. Unfortunately, equal concentrations of chemicals and their metabolites do not necessarily mean there will be equal toxic effects across all animal species and allowance has to be made for the consequent differences. Rate and extent of metabolic processes, cell permeability, tissue perfusion, enzymes, and receptors are by no means always the same in different animal species.

It has been found experimentally that the manifestation of adverse effects from a toxic substance may vary widely between one mammalian species and another because of major differences in the type and severity of toxicity exhibited. Consequently proper allowance has to be made for the considerable uncertainties in extrapolating from the results of animal exposures in the laboratory to possible effects on people exposed at work. Two procedures will be considered in detail here at opposite ends of a spectrum. The first procedure is possible where extensive studies on the substance have been made and published about the nature of the toxicity, particularly the early effects and possibly in several species. The more common situation, however, is at the other end of the spectrum when it is found that quantitative information is sorely lacking and procedures are needed which take into account the consequent huge uncertainties. Such a procedure is outlined second, designed for use where all that can be gleaned from the literature is an index of the minimal lethal dose in rodents. Many instances also occur, of course, somewhere intermediate between the two situations described.

For the first approach to be successful there must be sufficient information about comparative toxicology for the probable effects of the particular substance on humans to be inferred directly, that is without reference to exposure limits of other substances. Extensive experience guides the interpretation of toxicological experimental findings in rodents and, to a lesser extent in other animals, in terms of the probable effects on humans.

For the second approach to be successful there must be sufficient substances whose animal toxicity and exposure limits are both known sufficiently well for a reliable correlation to be established. Then extrapolation factors which most closely predict ACGIH or other exposure limits can be calculated from their correlation with acute toxicity indices of the substances (Ljublina and Filov, 1975).

Extrapolating from comprehensive animal toxicology

This procedure is used for interpreting comprehensive animal toxicology on experimental animals to which a substance has been administered in terms of adverse effects on health that may be caused by employee exposure to that substance at work (Feron, Van Bladeron and Hermus, 1990). It begins by assembling the information of both a quantitative and qualitative nature, including metabolic pathways followed by the hazardous substance and its metabolites, biotransformations involved, toxico-kinetics of the substance, cage-side observations, results of physiological function tests and pathological findings. It continues in a circumspect manner to define the gaps in knowledge about the behaviour of the substance in animals and man and concludes by extrapolation to an exposure limit for people at work that may be used with confidence.

The most accurate and rapid way of administering a dose to experimental animals is by injection and this is used to assess the intrinsic toxicity of an absorbed chemical. In animal experimentation intravenous injection may have advantages in particular cases as it introduces the material directly into the circulation with no delay, hence comparison of response by another route can provide information on the rate of uptake of the material by that other route. Administration of a substance by its introduction into the peritoneal cavity, that is the intraperitoneal route, is frequently preferred to intravenous injection as a base line since it usually provides 'the path of least resistance' for the chemical to reach the critical organ, resulting in steady but quite rapid absorption.

Oral administration by stomach tube or by adding the substance in food or water for an extended period of weeks, months or even the lifetime of the animal is the favoured method for investigation of toxic effects of ingestion of hazardous liquids and solids. But this leaves many uncertainties in route-to-route extrapolation.

When considering occupational exposure to air contamination the most important route of input to the body is by inhalation. Animal experiments employing administration of a substance by the inhalation route are easier to interpret in terms of atmospheric exposure at work than are experiments employing any other route. Inhalation toxicity studies in experimental animals have been conducted for many years and have long been employed in the setting of atmospheric exposure limits (Smyth, 1959; Gage, 1970; Carney, 1979). But there are still uncertainties to be resolved in the extrapolation of data from inhalation studies in animals to man. The received dose of a chemical absorbed by the body is not just a simple function of the prevailing concentration and duration of exposure. Other factors come into play (MacFarland, 1976; Sharratt, 1976).

There are differences between species in minute volume, respiratory pattern, the percentage of inhaled contamination absorbed or deposited on the lung surfaces, the rate of absorption of the chemical from the surface of the lungs, its distribution about the body, metabolic rate and elimination by excretion, exhalation or other means. These factors influence the relationship between exposure and dose received at the critical site. Additionally, when considering aerosols, penetration into the respiratory system of a species and deposition therein is influenced by the size of particles, their shape, whether rounded or fibrous and their density. A special technique sometimes used to help elucidate these aspects is to employ a system of tubes, valves and respirator for each animal so as to separate inhaled and exhaled air for measurement and analysis so as to establish the retained dose by difference. There may be further differences between species not only in the site of deposition but also, in the case of aerosols, in their subsequent clearance from the lungs.

The test protocols employed in industrial toxicology are designed to reveal potential long term hazards. Additionally they may incorporate sensitive, non-invasive tests of pulmonary function, liver function and kidney function. Tests for the detection of cardiovascular, teratogenic or adverse reproductive effects may be included (Carney, 1979). They do not exclude long term mortality studies as low intensity exposure may be life shortening (Dayan and Brimblecombe, 1978). These various types of lifetime test provide more direct information for use in setting occupational exposure limits than do acute toxicity tests.

Toxicology is multi-faceted. It is important that all the information on the toxicity of a substance be assimilated. A given piece of information is not judged in isolation, but is considered alongside the whole body of information. It is desirable that such an appraisal of the toxicological information be conducted in consultation with an expert, or at least with others of independent mind. There are four major aspects; the animal species employed in the experiments, the characteristics of dose-response curves, the duration of the animal exposures and the biological significance of the effects observed (Calabrese, 1983; ECETOC, 1984).

Which species

Animal species differ widely in their genetic make-up, their anatomy and physiology. They differ widely in their capacity to react to hazardous substances, through differences in absorption, distribution, metabolism and elimination. Of the many species available rats and mice have been used most commonly in the past mostly because of their ease of handling, relative cheapness and the accumulated knowledge about their value as human models; their similarities and their differences. Unusual animal species are of limited value as a human model in the present instance (Gibson and Starr, 1988). Nevertheless other species than rats or mice may be preferred in certain cases. For example, other species may be more sensitive indicators of a particular effect. Thus the cat may be preferred for nervous effects, the guinea pig for skin sensitisation, the rabbit for skin and eye irritants.

The critical organ for the behaviour of a substance in experimental animals may be different from that in humans. For example, a substance producing tumours at a specific site in rodents may also produce tumours in humans, but not at the same site. Such inconsistencies also occur when comparing experimental data obtained in different animal species. Different targets or end points are usually attributable to differences between species in metabolism or toxico-kinetic parameters of the substance under study. The rate processes may differ in magnitude or kind as between species; absorption, distribution, metabolism and elimination. For example, the biological half time for carbon monoxide uptake and release in humans is about five hours whereas it is only about one hour in rabbits due to differences in absorption/elimination rates (Saltzman and Fox, 1986). As an illustration of differences in kind, it has been found that the metabolism of beta-napthylamine in humans is similar in dogs but different in rodents. It was studies on dogs which showed that this substance could cause bladder cancer (Heuper, Wiley and Wolfe, 1938). To take another example, animal models do not as yet offer a satisfactory predictive toxicological screen for occupational asthma (Trizio, Basketter, Botham, Graepel, Lambre, Magda, Pal, Riley, Ronneberger, Van Sittert and Bontinck, 1988).

The coherent relationship among anatomical and physiological characteristics of different species provides the basis for extrapolation of toxico-kinetic data to humans from laboratory animals (Adolph, 1949; Dedrick, 1973; Lutz, Dedrick and Zaharko, 1980). Comparative toxico-kinetics is then needed to establish the relationships between exposure and the body burden of substance as between the man and animals (Fiserova-Bergerova and Hughes, 1983; Ramsey and Andersen, 1984; Paustenbach, Clewell, Gargas and Andersen, 1988).

If an animal species shows peculiar susceptibility to the chemical under study and its metabolism is not dissimilar from that of humans it is reasonably safe to assume that humans are as susceptible as that animal species (Zapp, 1977; Brown and Fabro, 1983). More specifically, the US Food and Drug Administration has stated its opinion that where no human experience existed it would be prudent to assume that man is ten times as sensitive as the most sensitive test animals (Lehmann and Fitzhugh, 1954; Dedrick, 1973).

Dose-response relationship

When extrapolating from the results obtained by animal experiments with hazardous substances to predict the effects of employee exposure, the slope of a dose-response curve is a major consideration in assigning safety factors. A dose-response curve for a single animal refers to the spread between the level of dose producing minimal or questionable change and the dose producing frank effect. The less the spread the steeper is the slope of the dose-response curve and the smaller the safety factor applied to the level producing minimal response in order to estimate the 'no-effect' level in the individual animal. A dose-response curve for a group of animals usually refers to response of the group measured in terms of the percentage of the group exhibiting a given change, plotted against group exposure. The steeper the slope of this type of dose-response curve the less is the variation in tolerance between different animals in the group when exposed to the particular hazardous substance. The distribution of tolerances is commonly lognormal.

In natural processes many random variables exhibit a lognormal distribution. In all cases where the reaction of the body to a given stimulus is proportional to both the intensity of the stimulus and the size of the body the form of the distribution is lognormal. In practice, this applies closely to toxicological and other similar biological studies, where logarithmic transformation of the dose variable is nowadays a matter of routine. The most widely used model assumes that the log-tolerances have a normal distribution and this is perfectly satisfactory so long as it is descriptive of the observations. The geometric standard deviation expresses the extent of the biological variation between the tolerance of different animals to the administered substance and governs the slope of a lognormal dose-response curve. The slope can also be expressed in terms of the dose corresponding to a stated percent response divided by the dose to produce a 50% response. In Table 7.1 useful examples are given of typical ratios from lognormal dose-response curves. The geometric standard deviation of the distributions is in the final column.

Typical values of geometric standard deviation of tolerance distribution of rodents range over 1.5 to 3.0 according the substance in question. Experimental animals are bred and selected for uniformity. Employees by their very nature are also to some extent a selected population but would be expected to vary more than experimental animals. An increase in the geometric standard deviation by 50% might be a fair basis for extrapolation from a rodent dose-response curve to humans.

TABLE 7.1. *Ratio DP:D50 where DP is the dose for P percentage response, lognormal dose-response curves*

D10/D50	D1/D50	D0.1/D50	D0.01/D50	Geometric standard deviation
0.68	0.50	0.40	0.33	1.35
0.61	0.40	0.30	0.24	1.48
0.51	0.30	0.20	0.14	1.68
0.41	0.20	0.12	0.079	2.00
0.28	0.10	0.047	0.026	2.69
0.24	0.075	0.032	0.014	3.04
0.19	0.050	0.019	<0.010	3.62
0.13	0.025	<0.01	<0.010	4.88

Note: Geometric standard deviation of tolerance distribution of (a) laboratory rodents = 1.5–3.0, (b) employees = 2.3–4.5.

Duration of experiments

The quantity of substance administered to an experimental animal is the dose. Dose rate is the dose delivered per unit of time. It may refer to a single means of administration or several, but especially ingestion, inhalation and dermal absorption. A given dose may have markedly different consequences according to whether it is administered in a single shot, several spread over a period of time, by a series of successive bursts or at a steady rate over a period of time, with or without breaks. Common examples of different primary effects from short term exposure as compared with those from long term exposure include acute narcosis from chlorinated hydrocarbon solvents as distinct from chronic liver damage from long term exposure; acute respiratory irritation from a single exposure to a high concentration of toluene-2,4-diisocyanate as distinct from chronic allergic sensitisation of the respiratory tract from repeated exposure; narcosis from short term exposure to high concentration of benzene as distinct from leukaemia from long term exposure to low concentration.

The ideal schedule for exposure to or administration of the chemical to assess chronic toxicity would cover the normal lifetime of the animal which, for example, is approximately two years for the rat and at least 5 years for dogs. Lifetime studies are particularly important for the detection of carcinogenic effects because of the characteristic long latent period for the development of cancer even though potent carcinogens often express their effects within a shorter period (Gold, Slone, Backman, Magaw, Da Costa and Ames 1986; Gold, Backman, Hooper and Peto 1987). However, since such experiments do take a long time to perform they are expensive. Nevertheless they are fully justified when there is a suspicion that the chemical may cause an adverse effect after prolonged exposure which is not revealed by a short term study (Rall, 1979).

Between lifetime and acute exposures lie sub-chronic and sub-acute exposures. In these latter experiments exposure continues for up to 90 days and 28 days respectively. Aside from carcinogenic effects the vast majority of adverse effects appear within 90 days in rodents (McNamara, 1976). Naturally, a longer experimental period allows greater confidence for predicting the effects of chronic exposure on humans.

Irritation and narcosis are common effects from atmospheric exposure to hazardous substances which are elicited by short term exposure. Unfortunately, symptoms as described by humans cannot be elicited from animals who cannot tell the observer how they feel. However, there is a growing interest in recent years in the possibility of using experimental animals for testing for sensory irritation and narcosis by means of measurements of respiratory function (Kane, Barrow and Alarie, 1979; Nielsen and Alarie, 1982).

Modification of the respiratory pattern or body movements in animals exposed to volatile materials have been reported by numerous investigators over many years, starting in the nineteenth century (Kratschner, 1870; Allen, 1928). For estimating the relative toxicity of volatile hydrocarbons a method of measuring respiratory rate and depth of respiration may be employed in which the animals are placed with their heads in an exposure chamber and their bodies in individual plethysmographs. When high concentrations of irritants are inhaled, that is, of the order 10 to 50 times the employee exposure limit the principle change that is observed is a decrease in respiratory rate (Swann, Kwon, Hogan and Snellings, 1974). A pressure transducer is connected to a counter and recorder (Alarie, 1966). The decreased respiratory rate in relation to the inhaled concentration of respiratory irritants is apparently correlated with the exposure limit for employees. The atmospheric concentration necessary to depress the respiratory rate of mice by 50% (RD_{50}) is approximately proportional to the ACGIH Threshold Limit Value of respiratory irritants (Alarie, 1981a, 1981b):

$$\text{ACGIH TLV} = 0.03 \times RD_{50} \quad \text{(approx.)} \tag{1}$$

It has been suggested that such tests might be used to estimate exposure limits prior to employee exposure (Alarie, 1981a; DeCeaurriz, Micillino, Bonnet and Guenier, 1981; Nielsen and Alarie, 1982). A compilation of the results of respiratory rate reduction tests on 40 irritant gases and vapours has been prepared by Alarie (1984). The correlation displayed is variable. The ACGIH TLV of two of the forty irritants was one tenth or less of that given by equation (1) and for two others it was more than twenty times so that the test should not be relied upon to predict the ACGIH TLV closer than an order of magnitude. Nevertheless it could be argued that at least some of the variability is accounted for by ACGIH error.

Short term, *in vitro* experiments are sometimes undertaken to investigate the mechanism of toxic action of hazardous substances. Unfortunately, the results do not give a very reliable indication of the potency of the chemical in an intact biological system, nevertheless *in vitro* biological tests do help in predicting whether or not a long term effect such as carcinogenesis might occur (Purchase, Longstaff, Ashby, Styles, Anderson, Lefevre and Westwood, 1976; Frohberg, 1981).

Biological significance of effects

Account has somehow to be taken of the biological significance of the myriad of reported minor effects. Other things being equal more weight is obviously given to heritable or irreversible changes such as mutagenic or carcinogenic effects than those showing reversible changes such as hair loss, reduction in body weight or transient changes in enzyme activity.

Data quality

Toxicological information from a wide range of experiments may be available on a given chemical. The relative value of each item of information in the setting of atmospheric exposure limits differs. The information is more valuable for this purpose if, for example, the substance was administered to mammals, by inhalation and the study was continued for their lifetime. An experienced assessor would normally collect, review and weigh the results of a multiplicity of animal studies relevant to the prediction of human response. Even so, the information is seldom, if ever complete and safety factors will need to be introduced to produce an atmospheric exposure limit for employees that could be used with confidence (Smyth, 1959; Zielhuis and van der Kreek, 1979a; Dourson and Stara, 1983; Hallenbeck and Cunningham, 1986).

Questions which should be asked about the quality of animal data on hazardous substances include:

- What was the animal species tested?
- What was the route of exposure, as compared to employees?
- Were animals in a series of tests assigned to the test groups at random?
- Is the organ system affected comparable with humans?
- Were the doses confirmed by analysis?
- Were the doses quantitatively appropriate?
- Was exposure long enough for disease to develop?
- How many animals survived and how many were tested?
- How many animals and which sites were examined pathologically?
- Were examinations performed 'blind'? Was the dose-response significant?

(Task Force of Past Presidents of the Society of Toxicology, 1982; Martonik, 1983; Gad and Weil, 1986).

Extrapolation procedures, no matter how sophisticated, cannot make an unreliable data base any the less so. The application of extrapolation to the results of experiments which have been badly designed, poorly conducted, with inadequate pathological appraisals, unsatisfactory animal care or badly kept records will result in unreliable exposure limits. Advice should be taken from experienced toxicologists as to which experimental data are appropriate for use in extrapolation from experimental animals to employees at work. The toxicologist should be familiar with the objectives of the work, and experienced in extrapolation. Given the complexity of this issue and the relative paucity of empirical data from which sound generalisations can be constructed, emphasis must be placed on flexibility, judgement and a clear articulation of the assumptions and limitations in any risk assessment that is developed.

Stepwise extrapolation factors

Having regard to comparative toxico-kinetics as between rodents and man and the relationships between experimental animal data and exposure limits in Western countries it is suggested that the following factors might be employed alone or jointly to fill gaps in the information (Zeilhuis and Van der Kreek, 1979a, 1979b; ; US Environmental Protection Agency, 1985; Hallenbeck and Cunningham, 1986; Johnson, 1988; Rubery, Barlow and Steadman, 1990; Illing, 1991). This classification is not meant to be applied rigidly or mechanically. It is necessary to make an overall, balanced judgement of the totality of the available evidence. The factors are meant to supplement, not to replace the judgement of an experienced toxicologist.

(a) Given:

The value of the lowest atmospheric concentration that causes an adverse effect of least severity in the most susceptible rodents after exposure lasting 8-hours every day of the adult rodent's lifetime. This level of concentration is markedly dependent on the number of animals tested, 0-out-of-5 and 0-out-of-1,000 unaffected clearly having different interpretations.

Factor:

In order to find a suitable exposure limit for employees the value is divided by 10 to allow for principal differences between species.

(b) Given:

The value of the median dose or inhalation exposure which produces an adverse effect in rodents.

Factor:

In order to make due allowance for the same effect but for possible differences between species alone the value is divided by 10.

(c) Given:

That an adverse effect has been observed in rodents two weeks after an experimental atmospheric exposure or other means of administration, and that the administration itself took one day or less to complete.

Factor:

In order to make due allowance just for a difference between a day and a month in the period of administration to produce the same effect, the exposure or dose rate is divided by 10. Similarly, in order to allow just for a difference between a month and a lifetime in the period of administration to produce the same effect, the exposure or dose rate is divided by 10. To allow for both, the exposure or dose rate is divided by 100.

(d) Given:

That the observed adverse effect on experimental animals is death.

Factor:

In order to make due allowance just for the difference in outcome between mortality and morbidity the exposure or dose is divided by 10.

(e) Given:

The value of the concentration of a gas or vapour which gives rise to a 50% reduction in respiratory rate of mice.

Factor:

The concentration is divided by 300 in order to obtain a value to which employees may be exposed and be reasonably confident that for the same substance the ACGIH TLV set to protect against irritation will be higher.

(f) When making calculations for extrapolation from rodents to humans the following representative values are used, when needed:

Species	*Weight (kg)*	*Respiratory Ventilation (m^3 per 8 hr)*
Human	70	10
Rat	0.4	0.07
Mouse	0.03	0.015

Direct extrapolation from mortality data

To understand the extrapolation procedure from animal mortality data on short term atmospheric exposure or oral administration of a hazardous substance it has to be held in mind that lethal atmospheric concentration or lethal oral dose is being used essentially as a comparative index of toxicity. It is an initial test to place the substance approximately in the wide spectrum of toxicity. Furthermore, it refers to toxicity resulting from only a brief exposure or short course of administration of the substance. The nature of the response to a high dose administered in a short space of time can be quite different from the response to long term exposure. Acute and chronic actions of a hazardous substance are often exerted on different systems.

An index of toxicity can be employed to derive a secondary exposure limit for employees provided that the animal mortality data and exposure limit data are of uniform quality, that the two parameters are correlated and there are sufficient data points on enough substances of a given type to establish the degree of correlation with confidence. The number of different substances for which there is both a published exposure limit and a published toxicity index is limited.

We elect to start with the knowledge that in many parts of the world the exposure limits of the American Conference of Governmental Industrial Hygienists (ACGIH) are generally accepted as representing the latest professional thinking on hazardous substances at work. Other lists could, if preferred, be used as a base line. Some may feel more comfortable with

ACGIH values when employing a fixed multiple of TLVs such as 0.5, 0.25 or 0.1 × TLV as a base line (Roach and Rappaport, 1990). Given one or more indices of acute toxicity of a substance predictive factors can be reasoned out for deriving an exposure limit with which the degree of protection would be comparable with published limits, yet have an appropriate, calculated safety margin.

The two principle indices of acute toxicity in animals in practice available for this work are the LC50 and LD50. The LC50 is the median lethal concentration, that is, the atmospheric concentration, exposure to which, for a specified length of time, would cause the death of 50% of the entire population of specified animals. The LD50 is the median lethal dose, that is, the dose which would cause the death of 50% of the entire population of specified animals. Other lethal dose percentiles such as LD1, LD10, LD30, LD99 and so on are quoted occasionally. Formally, the administration or exposure should have been completed within 24 hours and the animals subsequently observed for two weeks. In order to determine the LC50 or LD50 a series of tests is made employing different concentration or dose levels. Often, 3 to 5 levels have been used, involving 10 to 20 animals in all (Miller and Tainter, 1944; Weil, 1983). Under carefully controlled conditions, with the same kind of animals maintained in the same environment and fed the same diet, the LC50 and LD50 of hazardous substances calculated from such results is moderately reproducible. Typically an LC50 or LD50 has a standard error of about 25%. The LCLo is a less precise index than LC50 but is available for some substances whose LC50 has not been determined. It is the lowest reported concentration which has caused lethal effects in some or all of the animals under study.

Such indices as the acute inhalation LC50 and acute oral LD50 for experimental animals have significant even though limited predictive power for determining exposure limits appropriate for employees (Ljublina, 1973; Sanotzki and Siderov, 1973; Krichagin, 1977). They are also correlated with one another. Obviously, the indices relate more directly to mortality than morbidity or other such expressions of toxicity. Moreover, short term mortality measurements may still be gross underestimates of the mortality that would occur with prolonged exposure. Few occupational exposure limits are designed only to provide protection against possible mortality. They are mostly aimed against any of a wide variety of the very earliest adverse effects.

An acute inhalation LC50 and/or oral LD50 in rats and/or mice is published for most of the substances included in the list of ACGIH Threshold Limit Value-Time Weighted Average exposure limits (TLV-TWA). Additionally, the LCLo in rats or mice is published for many of the gases and vapours on that list. There are too few aerosol LCLos available to clearly establish their correlation with ACGIH exposure limits as a whole.

In the analysis of the data which follows, gases and vapours were grouped together and aerosols were treated as a separate group as their ACGIH TLV-TWAs were, as a whole, markedly lower. A further breakdown into smaller groups by chemical or physical properties was judged to be of increasingly doubtful value.

More of the gases and vapours on the ACGIH list had been allotted high exposure limits than had aerosols on the list. The precise reasons for this are not immediately obvious. The acute LC50s and LD50s in rats and in mice are of much the same order. In addition to overt toxicological effects it is believed by limit-setting committees that the simple nuisance of high dust concentrations would not be easily accepted by employees. The notion of adverse health effect is stretched to include simple nuisance. Those setting in-house exposure limits should have regard to the apparent reluctance of limit-setting authorities to adopt high concentrations of aerosols for their exposure limits.

The information analysed was limited to substances on the ACGIH 1986–87 TLV list and the toxicity indices for rats and mice from the then current NIOSH Registry of Toxic Effects of Chemical Substances (RTECS) (National Institute for Occupational Safety and Health, 1985). The ratio (TLV-TWA): (acute toxicity index) was calculated for all substances for which there was an ACGIH TLV-TWA and an RTECS LC50, LCLo or LD50. To find the value of these ratios concentrations were expressed in weight/volume units. Concentrations originally in volume/volume units were converted by employing the equation:

$$1 \text{ ppm} = \frac{\text{molecular weight}}{24.45} \text{ mg m}^{-3}. \qquad (2)$$

Inhalation toxicity indices were expressed in terms of the mass concentration of substance in air, grams per cubic metre, g m^{-3}, using the conversion equation (2) where necessary. Oral toxicity indices were expressed in grams per kilogram body weight. The variability of the ratio (TLV-TWA):(acute toxicity index) of a substance reflects the limited predictive power of these indices of toxicity. Factors are given in Table 7.2 for the percentiles of the ratio (ACGIH TLV-TWA):(RTECS acute toxicity index).

Now, suppose the ACGIH TLV committee had set a TLV-TWA for a certain substance. Given one of the parameters in Table 7.1, column 2, it is possible, from columns 3 and 4, to work out quite simply the probability that the committee set the limit above or below a value x.

TABLE 7.2. *Distribution parameters of the population of ACGIH TLV-TWAs for substances with a given rodent toxicity index*

Physical form of substance	Index of acute toxicity in rodents (RTECS)	TLV-TWA / Toxicity index: Median	Decile	GSD
Gas or vapour	Inhalation LC50 (g m^{-3})	4	0.4	6.0
	Inhalation LCLo (g m^{-3})	5	0.5	6.0
	Oral LD50 of liquid (g kg^{-1})	20	1.0	10.3
Aerosol	Inhalation LC50 (g m^{-3})	1	0.4	2.0
	Oral LD50 of solid (g kg^{-1})	4	0.6	4.4

Note: RTECS = NIOSH Registry of Toxic Effects of Chemical Substances. Oral LD50 is given in terms of g per kg body weight (g kg^{-1}). TLV-TWA = Threshold Limit Value. Time Weighted Average in mg m^{-3}. GSD = Geometric Standard Deviation.

Problem 7.1.

Carbon tetrachloride has an acute oral LD50 in rats of 1.77 g per kg body weight. Devise an appropriate exposure limit, given *only* this information.

Solution

From column 3 of Table 7.2

$$\frac{\text{Median TLV}}{\text{LD50}} = 20$$

Substituting LD50 = 1.77, Median TLV-TWA = 35.4 mg m^{-3}. Therefore the ACGIH TLV-TWA of the vapour of a liquid which has an LD50 of 1.77 g per kg body weight is equally likely to be above or below 35.4 mg m^{-3}. This would be the 'best' estimate of the limit if all that was known about the substance was the acute oral LD50 in rats. In this instance reference to the published list shows that the TLV-TWA of carbon tetrachloride was 30 mg m^{-3}.

The various other and more important uses of Table 7.2 are, perhaps, best illustrated by following through the first line in the body of the table. This indicates that given, for example, the LC50 of a vapour in rats, the 'expected' TLV-TWA (in mg m^{-3}) would be about four times the vapour LC50 expressed in g m^{-3}. The term 'expected' is used here in the sense that 50% of vapour TLV-TWAs are less than 4 × LC50 and 50% are more than 4 × LC50. The ACGIH TLV-TWA of a substance is thought of as a primary standard whereas the exposure limit derived from the acute inhalation LC50 is thought of as a secondary standard, based on the broad correlation between the two. Also, from the fourth column, 10% of vapour TLV-TWAs are less than 0.4 × LC50. The distributions fit lognormal criteria. It follows that about 10% of vapour TLV-TWAs are over 40 × LC50. Other factors corresponding to different percentiles can be worked out, if desired, from these data.

The ratio (TLV-TWA):(acute toxicity index) shows less variability when the acute toxicity index is LC50 or LCLo than when it is LD50, as indicated by the geometric standard deviations. Accordingly, where both are available a good, combined estimate of the TLV-TWA is the weighted mean in which the estimate from the LC50 or LCLo is given a weight four times the estimate from the LD50. The uses of Table 7.2 are further illustrated by the following examples:

Problem 7.2

Suppose vapour of liquid X, molecular weight 100, has an acute LC50 by inhalation of 2,930 ppm. There is no published exposure limit for this vapour. Calculate a suitable exposure limit.

Solution

From equation (1), the LC50, when expressed in mg m^{-3} is

$$\frac{2930 \times 100}{24.45} = 12{,}000 \text{ mg m}^{-3}.$$

That is, 12 g m^{-3}.

From Table 7.2, column 3, the median of the population of TLV-TWAs of substances with this toxicity was 4 × 12 = 48 mg m^{-3}. That is, the TLV-TWAs of substances with an LC50 of 12 g m^{-3} were just about equally common above and below the level of 48 mg m^{-3}. This is the best estimate of the expected TLV-TWA, but the estimate has a wide confidence interval by any criterion.

From Table 7.2 column 4, which gives the decile of the distribution, the probability is 0.9 that the TLV-TWA of such a vapour would have been greater than $0.4 \times 12 = 4.8$ mg m^{-3} and, by the same token, the probability is 0.1 that the TLV-TWA of such a vapour would have been less than 4.8 mg m^{-3}.

From the molecular weight of the vapour, employing equation (2) it may be calculated that these concentrations are respectively 12 ppm and 1.2 ppm. Taking into consideration the need to be conservative with exposure limits, the lower figure is advisable, rounded down to 1 ppm. The increased assurance of adequate protection is gained at some cost of increased possible error, namely the possibility of declaring exposure unacceptable when the health risk is in fact insignificant. Just how conservative one should be is a value judgement. Too conservative could lead to manufacture of the substance being abandoned or at least restricted and the loss of a possibly valuable chemical; insufficiently conservative could lead to undue health risks.

In conclusion, the in-house secondary exposure limit suggested for the vapour of liquid X is 1 ppm, 8-hour time weighted average. This value would be adopted unless there are special factors regarding technical feasibility or costs incurred in the circumstances of use which make it more reasonable to adopt a lower limit.

Problem 7.3

Suppose a new substance Y, a solid, has an acute oral LD50 in rats of 80 mg per kg body weight. Recommend a suitable exposure limit for employees engaged in its manufacture.

Solution

The LD50 of the substance is 0.08 g per kg body weight. From Table 7.2, column 3 the TLV-TWAs of such substances were found equally often above and below the value $4 \times 0.08 = 0.32$ mg m^{-3}.

From column 4 it also follows that had there been an ACGIH TLV-TWA for airborne dust of this substance there is a chance of 1 in 10 that it would have been set less than $0.6 \times 0.08 = 0.05$ mg m^{-3} and a chance of 9 in 10 that it would have been set higher than 0.05 mg m^{-3}. The in-house secondary exposure limit suggested is thus 0.05 mg m^{-3}, 8-hour time weighted average, subject only to the proviso that there are no special factors in the circumstances of use which make it reasonable to adopt a lower limit. Vigilance should be maintained with regard to the health of exposed employees to confirm the new exposure limit.

Problem 7.4

Liquid Z has an acute oral LD50 in rodents of 100 g per kg body weight. Its molecular weight is 200. In pilot trial manufacture the maximum employee exposure to the vapour was to a concentration of 5 ppm, 8-hour time weighted average. Would this be an acceptable exposure limit?

Solution

From Table 7.2, column 3 the expected ACGIH TLV-TWA of the vapour would be $20 \times 100 = 2{,}000$ mg m^{-3}. Converting the maximum exposure in the pilot trial from a concentration in terms of ppm to one in terms of mg m^{-3}; the concentration was, by equation (1)

$$\frac{5 \times 200}{24.45} = 41 \text{ mg m}^{-3}$$

From Table 7.2, column 5 the geometric standard deviation of the distribution of TLV-TWAs for vapours of liquids with a given LD50 is 10.3. The distribution is lognormal, so that with the aid of tables of the Normal distribution it is readily calculated that the probability that the TLV TWA would have been 41 mg m^{-3} or more is 0.95. Accordingly there is ample justification for employing an exposure limit of 5 ppm, 8-hour time weighted average. This value may be used with confidence

Suitable exposure limits may be calculated in this fashion for any substance whose LD50, LC50 or LCLo in rodents is known. There are more than 25,000 such substances listed in the NIOSH Registry of Toxic Effects of Chemical Substances, which is updated annually (National Institute for Occupational Safety and Health). Thus extrapolation from indices of acute toxicity to exposure limits, based on correlation with published limits, can make considerable inroads into the 70,000 or 100,000 long list of chemicals awaiting occupational hygiene control (EINECS, 1990).

Dermal toxicity

Experiments on animals are also conducted using the dermal route, which is especially pertinent to establishing skin absorption and skin toxicity in humans. Skin exposure ranks first in the production of mild occupational disease. The hazardous substance may react with the skin surface and cause primary irritation there, it may penetrate the skin and cause sensitisation to repeated exposure or may penetrate the skin, become distributed about the body and cause systemic poisoning. Dermal absorption occurs readily with skin exposure to nitroglycerine, aniline, parathion and benzidine, for example. Substances hazardous by skin contact may be identified by their dermal toxicity. A dermal LD50 on rabbits is performed by removing a portion of the fur and retaining the dose in contact with the skin beneath an impervious plastic film or watch glass for 24 hours and observing the animals for 14 days (Smyth, Carpenter, Weil, Pozzani and Striegel, 1962; Dutkiewicz and Tyras, 1978).

Atmospheric exposure limits for substances hazardous by skin contact should carry a notation to this effect. As guidance for deciding whether a substance is hazardous in this regard, skin protection should be provided for work with solids and liquids whose dermal LD50 in rabbits is less than 2g per kg body weight and the work is such that skin contamination is liable to occur.

References

Adolph E F. Science **109** 579 (1949).
Alarie Y. Arch. Environ. Health **13** 433 (1966).
Alarie Y. Food Cosmet. Toxicol. **19** 623 (1981a).
Alarie Y. Toxicological evaluation of airborne chemical irritants and allergens using respiratory reflex reactions. In: Proceedings of the inhalation toxicology and technology symposium (Edited by B K J Leong), p. 207. Ann Arbor Science, Ann Arbor, Mich, USA (1981b).
Alarie Y. Establishing threshold limit values for airborne sensory irritants from an animal model and the mechanisms of action of sensory irritants. In: Occupational and industrial hygiene: Concepts and methods (Edited by N A Esmen and M A Mehlman), p. 153. Princeton Scientific Publishers Inc, Princeton, New Jersey 08540 USA (1984).
Allen W F. Amer. J. Physiol. **87** 319 (1928).
Bliss C I. Science **79** 38 (1934).
Bliss C I. Ann. Appl. Biol. **22** 134 (1935).
Boyd E M. Toxic. Appl. Pharmac. **1** 229 (1959).
Brown N A and Fabro S. Clin. Obstet. Gynecol. **26** 467 (1983).

Carney I F. Ann. Occup. Hyg. **22** 163 (1979).
DeCeaurriz J C, J C Micillino, P Bonnet and J P Guenier. Toxicol. Lett. **9** 137 (1981).
Dedrick R L. J. Pharmacokin. Biopharm. **1** 111 (1973).
Dedrick R L. J. Pharmacokin. Biopharm. **1** 435 (1973).
Dourson M L and J F Stara. Reg. Tox. Pharmacol. **3** 224 (1983).
Dreser H. Arch. Ges. Physiol. **76** 306 (1899).
Dutkiewicz T and H Tyras. Brit. J. Ind. Med. **35** 43 (1978).
ECETOC. Considerations regarding the extrapolation of biological data in deriving occupational exposure limits. Technical Report No. 10. European Chemical Industry Ecology Toxicology Centre, Brussels (1984).
EINECS. European inventory of existing commercial chemical substances. Commission of the European Communities, EEC, Brussels (1990).
Feron V J, P J Van Bladeren and R J J Hermus. Fd. Chem. Toxic. 28 (1990).
Fiserova-Bergerova V and H C Hughes. Species differences in bioavailability of inhaled vapors and gases. In: Modeling of inhalation exposure to vapors: uptake, distribution and elimination, Vol. II (Edited by V Fiserova-Bergerova), pp. 97–106. CRC Press, Boca Raton, Florida, USA (1983).
Frohberg H. Krebserzeugende arbeitsstoffe: Identfizierung und bewertung. In: Deutsche forschungsgemeinschaft: wissenschaftliche grundlagen zum schutz vor gesundheitsschaden durch chemikalien am arbeitsplatz, pp. 49–63. Boppard, Boldt, Germany (1981).
Gaddum J H. Reports on biological standards III—methods of biological assay depending on a quantal response. Spec. Rep. Ser. Med. Res. Coun., No. 183. London (1933).
Gage J C. Brit. J. Ind. Med. **27** 1 (1970).
Gibson J E and T B Starr. Environ. Health Perspect. 77 99 (1988).
Gold L S, G M Backman, N K Hooper and R Peto. Environ. Health Perspect. **76** 211 (1987).
Gold L S, T H Slone, G M Backman, R Magaw, M Da Costa and B N Ames. Environ. Health Perspect, **67** 161 (1986).
Hueper W C, F H Wiley and H D Wolfe. J. Ind. Hyg. **20** 16 (1938).
Illing H P A. Ann. Occup. Hyg. 35 569 (1991).
Johnson E M. Regul. Toxic. Pharmac. 8 22 (1988).
Kane L E, C S Barrow and Y Alarie. Am. Ind. Hyg. Assoc. J. **40** 207 (1979).
Kratschner F. Sitzber Akad. Wiss. **62** 147 (1870).
Krichagin V. The principles of threshold estimation for setting maximum allowable concentrations (MAC) or threshold limit values (TLV). In: Standards setting (Edited by P Grandjean). Arbejdsmiljofondet, Copenhagen (1977).
Lehmann A J and O G Fitzhugh. Assoc. Food Drug Off. US Quarterly Bulletin **18** 33 (1954).
Ljublina E J. Gig. truda i Prof. Zabol. **1** 31 (1973).
Ljublina E I and V A Filov. Chemical structure, physical and chemical properties and biological activity. In: Methods used in the USSR for establishing biologically safe levels of toxic substances. World Health Organisation, Geneva, Switzerland (1975).
Lutz R J, R L Dedrick and D S Zaharko. Pharmacol. Ther. **11** 559 (1980).
MacFarland H N. Respiratory toxicology. In: Essays in toxicology (Edited by W J Hayes), pp. 121–154. Academic Press, New York (1976).
McNamara B P. Concepts in health evaluation of commercial and industrial chemicals. In: M A Mehlman (Editor). New concepts in safety evaluation, Vol. I, pp. 61–140. Halsted Pressbook, New York (1976).
Martonik J F. Ann. Am. Conf. Gov. Ind. Hyg. 5 105 (1983).
Miller L C and N L Tainter. Proc. Soc. Exper. Biol. & Med. 57 261 (1944).
National Institute of Occupational Safety and Health. Registry of toxic effects of chemical substances. US Government Printing Office, Washington DC, 20402 USA (up-to-date).
Nielsen G D and Y Alarie. Toxicol. Appl. Pharmacol. **65** 459 (1982).
Paustenbach D J, H J Clewell, M L Gargas and M E Andersen. Toxicol. Appl. Pharmacol. **96** 191 (1988).
Purchase I F H, E Longstaff, J Ashby, J A Styles, D Anderson, P A Lefevre and F R Westwood. Nature **264** 624 (1976).

Rall D P. The role of laboratory animal studies in estimating carcinogenic risks for man. In: Carcinogenic risk strategies for intervention, p. 179. IARC Scientific Publications, No 25, Lyon, France (1979).
Ramsey J C and M E Andersen. Toxicol. Appl. Pharmacol. **73** 159 (1984).
Roach S A and S M Rappaport. Am. J. Ind. Med. (1990)—to be published.
Rubery E D, S M Barlow and J H Steadman. Food Addit. Contam. 7 287 (1990).
Saltzman B E and S H Fox. Environ. Sci. Technol. **20** 916 (1986).
Sanotzki J V and K Siderov. Gig. truda i Prof. Zabol. **9** 11 (1973).
Sharratt M. Uncertainties associated with the evaluation of the health hazard of environmental chemicals from toxicological data. In: The evaluation of toxicological data for the protection of public health. Proc. Int. Colloq. Commission of the European Community. Luxembourg (1976).
Smyth H F. Am. Ind. Hyg. Assoc. J. **20** 341 (1959).
Smyth H F, C P Carpenter, C S Weil, U C Pozzani and J A Striegel. Am. Ind. Hyg. Ass. J. **23** 95 (1962).
Swann H E, B K Kwon, G K Hogan and W M Snellings. Am. Ind. Hyg. Ass. J. **35** 511 (1974).
Task Force of Past Presidents of the Society of Toxicology. Fundam. Appl. Toxicol. **2** 101 (1982).
Trizio D, D A Basketter, P A Botham, P H Graepel, C Lambre, S J Magda, T M Pal, A J Riley, H Ronneberger, N J Van Sittert and W J Bontinck. Fundam. Chem. Toxic. **26** 527 (1988).
US Environmental Protection Agency. Federal Register, National Primary Drinking Water Regulations; Volatile Synthetic Organic Chemicals; Final Rule and Proposed Rule (50:46830–46901, 1985). National Primary Drinking Water Regulations, Synthetic Organic Chemicals, Inorganic Chemicals and Microorganisms; Proposed Rule. (50:46936–47025, 1985).
Weil C S. Drug and Chem. Toxicol. **6** 596 (1983).
Zapp J A. J. Tox. Environ. Health. 2 1425 (1977).
Zaugolinicov S and M Kotchanov. Gig. truda i Prof. Zabol. **1** 28 (1974).
Zielhuis R L and F van der Kreek. Int. Arch. Occup. Environ. Health. **42** 191 (1979a).
Zielhuis R L and F van der Kreek. Int. Arch. Occup. Environ. Health. **42** 203 (1979b).

Bibliography

Boyd E M. Predictive toxicometrics. Scientechnica (Publishers) Ltd., Bristol, UK (1972).
Calabrese E J. Principles of animal extrapolation. Wiley, New York (1983).
Dayan A D and R W Brimblecombe (Editors). Carcinogenicity testing. MPT Press, (1978).
Gad S C and C S Weil. Statistics and experimental design for toxicologists. Telford Press, Caldwell, NJ, USA (1986).
Gerarde H W. Toxicology and biochemistry of aromatic hydrocarbons. Elsevier Publishing Co., New York (1960).
Hallenbeck W H and K M Cunningham. Quantitative risk assessment for environmental and occupational health. Lewis Publishers Inc, Chelsea, Michigan, USA (1986).

CHAPTER 8

Everything is a Mixture

Most published exposure limits refer specifically to single substances. However, no material described as a single substance is in practice 100% pure. Impurities are variable; possibly less than 1% of the bulk material, but sometimes 10% or even more. Furthermore the impurities may pose the greater hazard to health of those exposed. It is imperative, therefore, that some consideration be given to the full composition of a material and of the consequent exposure of employees to a mixture of hazardous substances. Many commercial products are mixtures. They may take the form of isomers, homologues or chemically related compounds resulting from a reaction mixture which may be offered for sale without separation, as for example, petrol and pesticide formulations. Another common type is the proprietary mixture of substances used in fixed proportion to meet a specific purpose, as in paints, disinfectants or fertilisers.

In assessing the health risks which may be associated with exposure to mixtures of two or more substances consideration should be given to the applicability or otherwise of single substance atmospheric exposure limits. Advice appended to different published lists of limits is very limited and poorly documented. Where given it differs in important respects from one authority to another. The function, Concentration/Exposure Limit, may be calculated for each ingredient in the mixture. One formulation is that the sum of such functions corresponding to each of the ingredients should be less than unity.

The American Conference of Governmental Industrial Hygienists (ACGIH) give a brief discussion of basic considerations involved in developing TLVs for mixtures of different substances in its booklet on Threshold Limit Values and Biological Exposure Indices (American Conference of Governmental Industrial Hygienists, 1990). The additive formula is recommended by the ACGIH for calculating TLVs of mixtures of substances which are believed to act upon the same organ system. However, when the effects of each substance on the body are believed to be independent of one another, provided no individual function, Concentration/Exposure Limit, exceeds unity the TLV of the mixture is not exceeded. The rationale for TLVs of mixtures of different substances is, unfortunately, not elaborated in the ACGIH Documentation of its TLVs (1986). Historically, the ACGIH rationale appears to have developed from the ideas presented by Elkins (1962), who thought that the exposure limit of a mixture of independently acting substances should lie somewhere between that given by the additive formula and the highest value of the function, Concentration/Exposure Limit.

The US Occupational Health and Safety Administration has advocated the additive formula for every mixture of substances whose limits are listed (Federal Register 23502, 1975). The US Environmental Protection Agency has recommended dose additivity for systemic toxicants which induce similar effects and effect additivity for others. Dose additivity reduces to the additive formula (Federal Register 34014, 1986). The USSR requirements have specified the simple additive formula just for mixtures of substances with similar actions (Bukowskij, Zokov, Kuzukova, Sanockih and Sidorov, 1977). The UK Health and Safety Executive has advised that this formula is applicable where the constituents of a mixture are 'additive' and where the exposure limits are based on the same health effects. It is ambivalent about mixtures of substances having different health effects. Where no synergistic or additive effects are considered likely the UK Health and Safety Executive believes that it is sufficient to ensure compliance with each of the individual exposure limits but adds that to treat all non-synergistic systems as though they were additive would be 'the more prudent course' (Health and Safety Executive, 1989).

The FDR Senate Commission for the testing of Industrial Materials has expressed dislike of simple formulae, believing mixtures should be evaluated on a case by case basis (1979). MAK-values apply only to single substances. A little guidance on exposure to mixtures of different substances has been published by the World Health Organisation (WHO). The WHO guidance suggests that mixtures whose components produce similar systemic effects could be assessed by the additive formula but that mixtures whose components possess a potentiating type of action should be assessed individually (World Health Organisation, 1981).

It is noteworthy that in industrial processes it is quite often found that exposure to one substance is dominant to such an extent that the additional hazard posed by other substances is probably insignificant in comparison. Nevertheless this is far from being the rule and it would be a mistake to make such an assumption without proper validation. Furthermore an individual employee may be exposed to air contamination from several processes concurrently or sequentially. Few, if any, work on the same process for the whole of their working lives. The question thus cannot be ignored indefinitely as to how the exposure limits for a mixture of different substances are combined when exposure to the different substances occurs concurrently or sequentially.

Different combinations even of just two substances could require a complete family of exposure limits to cover the full range of possible mixtures in air. Exposure limits are needed for the many situations when exposure is to several substances in different proportions, each with its own atmospheric exposure limit. Animal experiments might be instructive although the impracticability of experimentally determining the toxicity of every conceivable mixture of hazardous substances makes it imperative that theoretical approaches be developed for the prediction of combined toxicity (Ball, 1959). Human experience of mixtures could be sought but in companies handling many different chemicals there are so many mixed exposures that the number of people exposed to only one particular combination is too few to be of epidemiological value on its own.

Some mixtures are of substances whose effects on the body are additive in the same organ. Other mixtures are of substances which have completely different and independent actions on different parts of the body. The question of how to derive exposure limits for mixtures of substances having independent effects seems to cause the most lively debate. Mixtures of substances which have synergistic or possibly antagonistic effects have long been thought to pose more problems (Dautrebande, 1939; Amdur, 1954; Veldstra, 1956; Levine, 1973; Goldstein, Aronow and Kalman, 1974; NRC, 1980).

It may be questioned at the outset whether exposure limits can properly be included in mathematical equations as to some extent they represent social judgements about the perceived importance of different health effects. On the other hand an exposure limit does indicate a certain concentration and duration of exposure which will cause a certain minimum response in a group of individuals. With this latter consideration in mind, for the purposes of the present discussion the exposure limit is regarded as a toxicological parameter, a derived constant comparable with the LD50, for example. This opens up a number of questions. How can exposure limits for mixtures be sensibly derived from exposure limits of single substances? How can one decide which substances interact; which have additive effects, which have independent effects and which are synergistic? What should be done about sequential mixtures? In this chapter an endeavour is made to refine intuitive experience by use of appropriate mathematical models, with a view to recommending a risk assessment procedure that is appropriate in the light of current knowledge about the health risks from exposure to mixtures.

Composition of bulk material as compared with air contamination

It is important to analyse the air as soon as possible to determine the composition of its contamination and not to rely on analysis of the bulk material. For example, it is fairly obvious that the more volatile liquids in a liquid mixture tend to evaporate first. The rate of evaporation of each component is directly proportional to the mole fraction of the component in the liquid, directly proportional to the vapour pressure of the pure component and inversely proportional to the cube root of its molecular weight (Gmeling and Weidlich, 1989). The vapours from drying paint change progressively. The pigments and many other solids that are added in making a commercial product do not contribute a significant inhalation risk to painters unless the paint is sprayed. On the other hand in manufacture all components are potential airborne hazards and, at the end of the life of paint the solvent vapours are negligible and the health hazard is entirely from airborne solids when rubbing down. Spray and mist droplets are also likely to differ markedly in composition from the original liquid mixture due to differential evaporation. Consequently the composition of air contamination by vapours will differ from the bulk composition unless, exceptionally, the operation is one in which the whole liquid rapidly evaporates to dryness.

Understanding the relationship between the composition of bulk solids and their airborne counterparts also presents problems. Finer particles and the least dense materials are generally more readily airborne. The chemical analysis of solid raw materials such as crushed rock or bulk powders is not a very satisfactory basis for deciding the exact composition of air contamination from operations and processes involving these materials. The composition often varies from one size to another; so the finer fraction may be quite different in its chemical or mineralogical make-up from the coarser fraction and from the parent material. The analysis of settled dust on beams and ledges is also of questionable value since coarse and heavy dust settles preferentially. Welding, brazing, soldering, arc burning and like processes are prolific sources of air contamination but the analysis and particle size distribution of the contamination is difficult to predict from the analysis of the electrodes and metal parts involved (Gray, Hewitt and Hicks, 1980; Tandon, Crisp, Ellis and Baker, 1983; Thorne and Hewitt, 1985; Hewitt and Hirst, 1991).

In the light of the size-selective characteristics of the respiratory system which govern initial dust deposition, and the differences in behaviour of the clearance mechanisms which operate to remove deposited particles from the upper respiratory tract and from the lungs, it is evident that the variation in the composition of aerosols greatly complicates the relationship between the rate of respiratory intake of dust and the effective rate of uptake of the particles at the critical site within or beyond the lungs. Air should be sampled, the contamination collected and analysed directly.

Mixtures with additive effects

It is said that when two or more hazardous substances are present, which act upon the same organ system, their combined effect, rather than that of either alone, should be given primary consideration (American Conference of Governmental Industrial Hygienists, 1988). It might be added that the kind of effect on the organ by the two substances should also be the same. Otherwise it will be "like adding apples and pears". An example of a group of substances which attack the same organ and are additive are tetrachloroethane, ethylene dichloride and chlorinated naphthalenes, all of which are hepatoxic. Since the mechanisms of action for most compounds are not well understood, the justification of the assumption of additive effect will often be limited to similarities in pharmaco-kinetic and toxicological characteristics. Equally, good examples of effects which are obviously of a different kind are irritation, narcosis and lung fibrosis.

Irritant gases, vapours and dusts differ widely in chemical and physical properties. However what irritants have in common is that they all induce the same effect in tissues with which they come in direct contact, namely inflammation. The differences in the nature of the symptoms which result from the action of different irritants are mainly due to differences in the localities of the body on which the irritants act. It is the physical properties, solubility of gases and vapours, and particle size of dust, which govern where they are absorbed and deposited. Coarse, irritant airborne dust and the most soluble irritant gases such as ammonia, hydrogen chloride and hydrogen bromide act primarily upon the eyes, nose and upper respiratory tract. Finer dusts and less soluble gases such as sulphur dioxide penetrate into the respiratory tract and extend their action to the bronchi or bronchioles. The finest dust particles and least soluble gases such as nitrogen dioxide penetrate to the alveoli and act there.

Different irritants which act at a given locality display different degrees of potency for producing inflammation. This may be expressed in units of the reciprocal of the exposure limit, provided that limit is based upon prevention of irritation. In broad terms, the higher the exposure limit, the less potent the irritant.

The concentrations of the different irritants are first standardised with respect to their single substance exposure limits. First, for each single substance the value of the function, Concentration/Occupational Exposure Limit, is calculated. For the purposes of this discussion the function, Concentration/Occupational Exposure Limit, may be thought of as representing the quantity of adverse effect on the employees caused by a single substance, in this instance a quantity of inflammation. This is not unlike the concept of Hatch (1968) who suggested that basic doses of such substances might be expressed in terms of equivalent physiological 'units of irritation'. The joint effect on each individual employee exerted by two or more irritants acting at the same locality in the body is indicated by the simple sum of all the units of irritation, that is, the sum of all the single substance functions, Concentration/Occupational Exposure Limit. There is an implicit assumption here that the units of irritation attributable to a substance are not altered by the presence or absence of other substances. Most students of the subject feel comfortable making such an assumption but it must be recognised for what it is. Actual experimental results would be needed to prove beyond doubt the validity of the assumption in particular cases. The additivity model assumes that the probability of a response to such agents depends only on the sum of the amounts of each present, when the amounts are expressed in standardised units, as represented by the function; Concentration/Occupational Exposure Limit.

Therefore for employees exposed to a mixture of n irritant substances concurrently, which act at the same locality in the body, whose individual occupational exposure limits are OEL_1, OEL_2, OEL_3, ..., OEL_n and whose concentrations in air are C_1, C_2, C_3 ..., C_n respectively, and the effects in the body are simply additive, the function, C/OEL, of the mixture is given by:

$$\frac{C}{\mathrm{OEL}} = \frac{C_1}{\mathrm{OEL}_1} + \frac{C_2}{\mathrm{OEL}_2} + \frac{C_3}{\mathrm{OEL}_3} + \cdots \frac{C_n}{\mathrm{OEL}_n}, \tag{1}$$

where C is the total concentration of contamination and OEL is its exposure limit.

To avoid an excessive risk of irritation from all the substances together the criterion is:

$$\frac{C}{\text{OEL}} < 1$$

Furthermore, where C_1, C_2, C_3 . . . , C_n are in consistent units:

$$C = C_1 + C_2 + \cdots C_n$$

Common air sampling instruments do not, in general, respond equally to different hazardous substances so that there is only limited practical use for a combined exposure limit in terms of the sum of all the contaminating substances. However, there are exceptions, as, for example, dusts whose exposure limit is expressed in terms of total mass. Furthermore, there are control procedures which may operate equally on all contamination, as, for example, dilution ventilation in engineering control or restrictions on the period of exposure in personnel control. An explicit combined exposure limit for additive substances is thus sometimes helpful.

Dividing both sides of equation (1) by C we have,

$$\frac{1}{\text{OEL}} = \frac{f_1}{\text{OEL}_1} + \frac{f_2}{\text{OEL}_2} + \frac{f_3}{\text{OEL}_3} + \cdots \frac{f_n}{\text{OEL}_n}, \qquad (2)$$

where $f_1, f_2, f_3, \ldots f_n$ are the fractional concentrations of the components of the mixture, and $f_i = C_i/C$. In applying this formula consistent units are essential; where f_i is the fractional concentration by volume OEL_i must be expressed in terms of volume concentration units, parts per million by volume; where f_i is the fractional concentration by mass OEL_i must be expressed in terms of mass concentration units, mg m^{-3}.

The OEL of the mixture as a whole is thus

$$\frac{1}{\dfrac{f_1}{\text{OEL}_1} + \dfrac{f_2}{\text{OEL}_2} + \dfrac{f_3}{\text{OEL}_3} + \cdots \dfrac{f_n}{\text{OEL}_n}} \qquad (3)$$

Problem 8.1

Employees are exposed to air contamination coming from two separate processes, one of which uses isoamyl alcohol and the other cyclohexanol. The vapour of both of these solvents is irritating to the eyes, nose and throat. The atmospheric concentration of the isoamyl alcohol vapour to which the employees are exposed is 70 ppm (OEL 100 ppm) and of the cyclohexanol vapour it is 25 ppm (OEL 50 ppm). The irritative effects of these solvents are believed to be additive. Is the occupational exposure limit for such a mixture exceeded? Calculate the exposure limit.

Solution

Equation (1) yields

$$\frac{70}{100} + \frac{25}{50} = 1.2$$

This exceeds 1.0, showing that the exposure limit of the mixture is exceeded. The exposure limit of the mixture of contamination, given by formula (3) is

$$\text{OEL} = \frac{1}{\dfrac{0.74}{100} + \dfrac{0.26}{50}} = 79 \text{ ppm}$$

Narcosis is produced by a wide variety of gases and vapours. The precise mechanism by which they exert their actions probably differs from one to another, however all exert their action upon the nervous tissues. Consequently, when the different narcotic substances are inhaled simultaneously such effects are additive and the additive equation (1) is appropriate. Examples of narcotic substances are toluene, trichloroethylene, perchloroethylene, naphtha and methyl chloroform.

Problem 8.2

A solvent mixture of 70% perchloroethylene and 30% methyl chloroform is used for cleaning parts on an assembly line. At 20C the vapour pressure of perchloroethylene is 19 mm mercury and of methyl chloroform is 100 mm mercury. Employees in the vicinity are exposed to a mixture of the vapours originating from the solvent tank and drying parts. The atmospheric concentration to which employees are exposed is 35 ppm perchloroethylene (OEL 50 ppm) and 150 ppm methyl chloroform (OEL 350 ppm). Both vapours give rise to narcosis at atmospheric levels in excess of the single substance exposure limits. The effects on the body of the two vapours are believed to be additive. Is the occupational exposure limit for such a mixture exceeded? Calculate a suitable exposure limit.

Solution

The additive equation (1), yields

$$\frac{35}{50} + \frac{150}{350} = 1.13$$

Neither of the single substance exposure limits is exceeded but the combination of the two produces an atmosphere which is unacceptable. The exposure limit of the mixture is given by formula (3). On substituting the values of the fractional concentrations, namely 35/185 for perchloroethylene, OEL 50 ppm, and 150/185 for methyl chloroform, OEL 350 ppm, formula (3) yields

$$\text{OEL} = \frac{1}{0.0038 + 0.0023} = 164 \text{ ppm.}$$

In the case of lung fibrosis the value of the function, Concentration/Occupational Exposure Limit, may be thought of as representing a quantity of fibrous tissue produced by exposure to that substance for one work shift. Equation (1) then represents the sum quantity of fibrous tissue.

Problem 8.3

A sample of respirable dust collected from the atmosphere in the vicinity of a mica grinding operation was found to contain 5% quartz. Calculate an exposure limit for the mixture, given that the exposure limit of mica is 3 mg m^{-3}, respirable dust, and of quartz is 0.1 mg m^{-3}, respirable dust.

Solution

The effect of the two substances is additive. Assuming that the respirable dust contains 95% mica and 5% quartz, equation (3) applies. A suitable OEL of the mixture is

$$\frac{1}{\frac{0.95}{3}+\frac{0.05}{0.1}} = 1.2 \text{ mg m}^{-3}.$$

Mixtures with independent effects

An appropriate treatment of mixtures of substances with independent effects may be developed by focussing attention on the joint risks to the individuals in a common group rather than on the combined toxicological effect on one individual. In toxicological theory and practice the mortality and the morbidity from independently acting substances are both additive parameters. This is not an issue of co-toxicity but one of probability. The theory regarding mortality is examined first.

It is fairly easy to show that the mortality from small doses of each constituent is additive in a mixture whose constituents act independently. Suppose, for example, that the doses of the two substances in a mixture are separately capable of producing mortality P_1 and P_2 respectively. Since they act independently a proportion P_2 of the test animals which survive one substance will nevertheless succumb to the other and the expected total mortality is expressed by

$$P = P_1 + P_2(1-P_1), \tag{4}$$

which can be rewritten and extended to three or more constituents as:

$$P = 1-(1-P_1)(1-P_2)(1-P_3) \cdots (1-P_m) \tag{5}$$

(Ball, 1959). Provided that $P_1, P_2, P_3, \ldots P_m$ are small quantities this reduces to the following approximation,

$$P = P_1 + P_2 + P_3 + \cdots P_m. \tag{6}$$

By exactly the same reasoning the morbidity from small doses of each constituent of a mixture is also additive.

The theory is supported in practice. In experimental toxicology on non-human animals it is well established that the toxicity of a mixture is usually the sum of the toxicity of the ingredients provided the presence of one ingredient of a mixture neither augments nor diminishes the absorption, elimination or organ effects of another (Pozzani, Weil and Carpenter, 1959; Smyth, Weil, West and Carpenter, 1969, 1970; Boyd, 1972; Murphy, 1980).

These principles might be extrapolated to produce the principle of additive risks for groups of humans exposed to mixtures of substances whose effects are independent of one another. The same conclusion can be reached by risk analysis applied to exposure limits.

The ultimate objective is to harmonise exposure limits of atmospheric mixtures of several substances acting independently with those for the same substances present alone. Examples of such atmospheric mixtures of two independently acting substances would be mixtures of carbon monoxide with sulphur dioxide, sulphuric acid mist with airborne lead oxide dust, toluene with benzene, and carbon tetrachloride with manganese dioxide in air. A risk analysis basis is supported by consideration of the following propositions:

Proposition 1 **Risks of adverse health effects are additional one to another, whether or not the same organ or same effect is involved.**

This proposition, which stems from a holistic view of health, is sometimes questioned along the following lines.

'How can one add together in the same equation risks from such diverse effects as irritation, systemic effects and cancer? They have nothing in common.'

It may help to resolve doubts by reflecting on the obvious, namely, that sickness, disability and death remain just so, whatever dysfunction is the cause.

Suppose there were three factories, A, B and C, each with 1,000 employees. Suppose at factory A the concentration of substance *X* in the atmosphere is everywhere at its exposure limit, where it produces kidney complaints in, say, 1 in 100 employees (that is, 10 employees). Suppose, next, in factory B the concentration of substance *Y* in the atmosphere is everywhere at its exposure limit, where it produces mild throat irritation in, say, 1 in 10 employees (that is, 100 employees). Suppose, finally, in factory C substances *X* and *Y* are both present in the atmosphere at their exposure limits and they do not interact in any way. There will be 9 employees there with kidney complaints alone, 99 with throat irritation alone and, by chance alone, 1 employee with both. The situation at factory C clearly involves more risk than at either factory A or B. (Note that if the effects are only partially independent in the sense that the individuals sensitive to kidney complaints are precisely the same individuals sensitive to throat irritation then at factory C there will be 10 employees with both throat irritation and kidney complaints and 90 with throat irritation alone).

Proposition 2 **As long as a substance is present in the atmosphere there is a definite risk or at least a possibility of it that some employees may be adversely affected by it, no matter how low the concentration.**

Biological variation is a fact of life. Human variation in tolerance to hazardous substances has no definite upper or lower limit so far as is known. As a practical matter variation in true exposure either side of the measured value also enhances the apparent variation in tolerance.

Proposition 3 **The risk of adverse effects can be reduced by reducing the concentration of contamination in inspired air.**

Other things being equal the lower the concentration of contamination in inspired air the lower will be the received dose. All adverse effects from hazardous substances are dose related.

Proposition 4 **Exposure limits are arrived at by explicit or implicit regard to the acceptability or otherwise of risks of adverse effects.**

The acceptability of a risk is dependent on the nature of the effect. The more severe an effect the less is the acceptability of a risk of it. It is also a matter of common observation that the greater the ease with which a risk is reducible the less acceptable is the risk. Other things being equal, avoidable risks are avoided. An increasing awareness of the risks decreases their acceptability.

In conclusion, whenever the substances in a mixture have independent effects on the body the residual risks to health below the single-substance exposure limits are additive one to another. Notwithstanding these remarks the question remains as to the magnitude of those residual risks. Suppose the risks at the exposure limits of substances 1, 2, 3, . . . m are P_{OEL_1}, P_{OEL_2}, P_{OEL_3}, . . . P_{OEL_m}, respectively. In terms of the theory, in order to proceed it remains to establish appropriate values for the functions; P_1/P_{OEL_1}, P_2/P_{OEL_2}, P_3/P_{OEL_3}, . . . P_m/P_{OEL_m} for each mixture.

Now, the prevalence of adverse effects amongst employees exposed at levels equal to the exposure limits are not known very precisely. The order of magnitude of the prevalence is higher than some would expect (Roach and Rappaport, 1990). Nevertheless adverse health effects generally would be expected to affect less than 1 in 10 employees at levels of concentration at or below the atmospheric exposure limit. Exposure to hazardous substances which produce the more severe effects would be expected to affect less than 1 in 100 employees exposed at or below the exposure limit.

It is safe to assume that any amount of exposure may carry some probability of harm to those exposed. At reduced exposure levels the risks would be proportionately reduced if hazardous substances exhibit a straight-line exposure-response curve through zero. Otherwise the risks will be disproportionately reduced. The exact shape of exposure-response curves at these low levels of exposure is inevitably somewhat speculative as published data is scanty.

Where tolerance is expressed as the minimum exposure which will produce an adverse effect in an individual, however defined, the tolerance distribution as between one individual and another would be expected to be approximately lognormal in the central part of the distribution. This is not to deny that some other form of distribution may fit the limited available data just as well but that in the light of present knowledge if a different distribution is truly operative not enough is known about it to permit an inference to a significantly different risk from that inferred from the lognormal distribution. For technical reasons only a limited downward extrapolation is needed for the present purposes; from the response at the exposure limit down to one-tenth the exposure limit at most.

The parameters of a distribution serve to define the precise position and slope of the cumulative distribution curve. Where the parameters of the lognormal tolerance distribution for each substance in a mixture are known, the risks from a mixture can be calculated fairly readily by standard procedures (Bliss, 1939; Finney, 1952). But the parameters of the tolerance distribution are seldom known with any precision. The most that is usually known is the incidence of adverse effects at a certain exposure; that is, a single point on the exposure response curve. More often, the situation is even more tenuous, and all that may be known is that at the exposure limit 'nearly all' people are believed to be protected.

To overcome this dilemma consideration is given to the tolerance distribution of the least adverse effect that is known about, whether that adverse effect be irritation of the throat, headache, reduced kidney function, a chest radiograph classified as Category II pneumoconiosis, a single tumour, death or anything else. The people who are experiencing the adverse effect after a given exposure must have first become affected at some prior point in time. The proportion affected is somewhere on the lower part of the cumulative tolerance distribution, a sigmoid curve.

The exposure limit of a given substance corresponds to a certain proportion of people being affected through that exposure. Were one to assume proportionate reduction of risk when concentration is reduced below the exposure limit, that is, risk proportional to $(C/\mathrm{OEL})^s$, where $s = 1$, the risk calculations would certainly err on the high side, and very much so at the lowest concentrations. This ultra-conservative procedure has been proposed for extrapolating downwards for low-dose risk assessment of environmental carcinogens where it is known as the Mantel-Bryan procedure (Mantel and Bryan, 1961; Mantel, Bohidar, Brown, Ciminera and Tukey, 1971; Gaylor and Kodell, 1980; US Office of Science and Technology Policy, 1985; US Environmental Protection Agency, 1986). However, insofar as lognormal tolerance distributions approximate to the truth in the range below the exposure limit but above one-tenth the limit a closer assumption would seem to be to assume a higher power law than unity for risks from exposure to concentrations below the exposure limit. The reason for this is that a characteristic of the lognormal exposure-response curve is that as exposure decreases to comparatively low values, zero response is approached with ever increasing rapidity, until eventually the level of response is decreasing more rapidly than any finite power of exposure. Lognormal exposure response curves are plotted in Figure 8.1 using log-log scales where they show the convex property in the range of risk levels and tolerance distribution likely to be met in the range 0.1–$1.0 \times$ Occupational Exposure Limit for hazardous substances.

By adopting a square law or even cube law for some substances, that is, risk extrapolation proportional to $(C/\mathrm{OEL})^2$ or even $(C/\mathrm{OEL})^3$ rather than C/OEL the estimate of risk contribution at the lowest concentrations would still tend to err on the high side and the greater the extrapolation the greater the safety factor. For practical purposes of assessing exposure to mixtures at work what is wanted is such a power law, or something similar, which does not involve too much intricate mathematics each time it is used and which still overestimates risks so that the exposure limit calculated for the mixture errs on the safe side, if anything. The tolerance distribution for different adverse effects is likely to be somewhere between geometric deviation 1.5 and 3.5. The likely geometric deviation for most substances would result in at least a ten-fold reduction in risk for a three-fold reduction in exposure below the exposure limit as may be

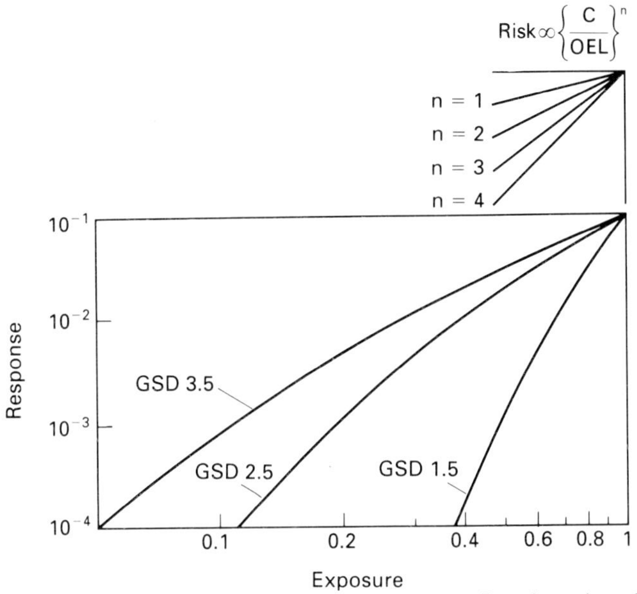

FIG. 8.1. Lognormal exposure vs response curves. Log-log plot of arcs from 10^{-1} to 10^{-4} risk level. Examples illustrated are of geometric standard deviation (GSD) 1.5, 2.5 and 3.5. In the top right hand corner, for comparison, are lines with slopes corresponding to a simple power law.

calculated by calculating the reduction in risk from a range of likely start points (Table 8.1). As a general rule use of the square law would be a safe course. Exceptionally, a cube law could be used provided there was special evidence to justify it.

The function, (Concentration/Occupational Exposure Limit)^s, may be thought of as representing the quantity of risk of adverse effects in the population of employees exposed below the single substance exposure limit. When exposure is equal to the occupational exposure limit the risk is one which is the maximum that would be accepted in the work environment. When calculating the exposure limit for a mixture, letting $s = 2$ would give reasonable assurance that the exposure limit would err, if anything, on the safe side. The joint risk to the population of employees exerted by being exposed to two or more substances acting independently on the body is indicated by the sum of all the individual functions, (Concentration/Occupational Exposure Limit)^2.

TABLE 8.1. *Ratio $DP_1 : DP_2$ where DP_n is the dose for P_n percentage response, lognormal dose-response curves, $D50/D1$ = 2, 10 and 20*

D50/D1	D1/D10	D0.1/D1	D0.01/D0.1	Geometric standard deviation
2	0.74	0.80	0.83	1.40
10	0.36	0.47	0.55	2.70
20	0.26	0.40	0.45	3.66

Therefore for employees exposed concurrently to m independently acting substances whose occupational exposure limits are OEL_1, OEL_2, OEL_3, ..., OEL_m and whose concentrations in air are C_1, C_2, C_3 ..., C_m respectively, each alone producing some acceptably small risk of adverse effects from that substance, the combined risk from all components of the mixture is represented by:

$$\left\{\frac{C}{\mathrm{OEL}}\right\}^s = \left\{\frac{C_1}{\mathrm{OEL}_1}\right\}^s + \left\{\frac{C_2}{\mathrm{OEL}_2}\right\}^s + \cdots \left\{\frac{C_m}{\mathrm{OEL}_m}\right\}^s, \qquad (7)$$

where C is total concentration of contamination in air and OEL is its exposure limit.

$$C = C_1 + C_2 + \cdots C_m$$

To avoid an excessive risk of adverse effects from all the substances together the criterion is:

$$\left(\frac{C}{\mathrm{EOL}}\right)^s < 1$$

An explicit combined exposure limit for mixtures of substances which have independent effects is sometimes helpful for ventilation design purposes.

Dividing both sides of equation (7) by C^s we have,

$$\frac{1}{(\mathrm{OEL})^s} = \left\{\frac{f_1}{\mathrm{OEL}_1}\right\}^s + \left\{\frac{f_2}{\mathrm{OEL}_2}\right\}^s + \cdots \left\{\frac{f_m}{\mathrm{OEL}_m}\right\}^s, \qquad (8)$$

where $f_1, f_2, f_3, \ldots f_m$ are the fractional concentrations of the components of the mixture. In applying this formula consistent units must be used; where f_i is the fractional concentration by volume OEL_i must be expressed in terms of volume concentration units, parts per million by volume; where f_i is the fractional concentration by mass OEL_i must be expressed in terms of mass concentration units, mg m^{-3}.

The OEL of the mixture as a whole is thus

$$\frac{1}{\left[\left\{\frac{f_1}{\mathrm{OEL}_1}\right\}^s + \left\{\frac{f_2}{\mathrm{OEL}_2}\right\}^s + \cdots \left\{\frac{f_m}{\mathrm{OEL}_m}\right\}^s\right]^{1/s}} \qquad (9)$$

When $s = 1$ expression (9) reduces to the OEL of mixtures of substances with additive effects, formula (3). At the other extreme, as $s \rightarrow \infty$ expression (9) reduces to the statement that the OEL is exceeded when the concentration of one or other of the ingredients exceeds its single substance limit. This is precisely the ACGIH TLV of mixtures of substances when the effects of the different substances are not additive, but are believed to be independent. The OEL of mixtures given by expression (9) when $s = 2, 3, 4, \ldots$ is intermediate between these two extremes.

The interplay of the single substance exposure limits and the power, s, is illustrated in three dimensions (f_1, $1/s$, C) by a surface shown in Figure 8.2 for mixtures of two substances; both of whose single substance OEL is 200 ppm. At the back of the block, $s =$, along arc AB the single substance exposure limit of substance Y is equalled and along arc BC the single substance exposure limit

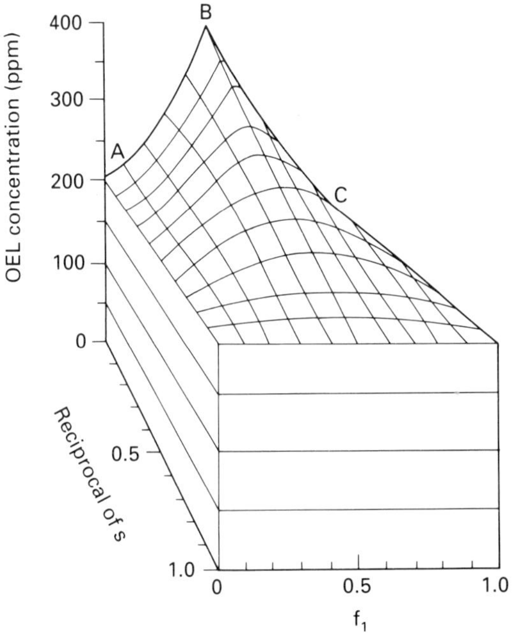

FIG. 8.2. Occupational exposure limit of mixtures of two substances, X, Y, whose single substance exposure limits are both 200 ppm. $OEL_1 = OEL_2 = 200$. f_1 = fractional concentration of substance X.

$$\text{OEL} = \frac{1}{\left[\left\{\frac{f_1}{\text{OEL}_1}\right\}^s + \left\{\frac{f_2}{\text{OEL}_2}\right\}^s\right]^{1/s}}$$

of substance X is equalled. At the front of the block, $s = 1$, the substances are additive and the OEL of the mixture is 200 ppm. At points elsewhere on the surface the concentrations of the ingredients are each below the corresponding single substance exposure limits. The surface shown in Figure 8.3 is for mixtures of two substances; one whose single substance OEL is 200 ppm and the other whose single substance OEL is 50 ppm.

Problem 8.4

Air contamination in a workshop consists of a mixture making up 0.1 mg m^{-3} of lead arsenate in air (OEL 0.15 mg m^{-3}) and 0.8 mg m^{-3} of sulphuric acid mist in air (OEL 1.0 mg m^{-3}).

Lead arsenate causes systemic reactions whereas sulphuric acid is a primary irritant and causes tooth erosion. These are independent effects but the residual risks below the respective exposure limits are additional one to another. Is the exposure limit of the mixture exceeded?

Solution

Equation (7) applies. Prudence dictates $s = 2$ rather than $s = 3$. Substituting the given values in the equation,

$$\left\{\frac{C}{\text{OEL}}\right\}^2 = \left\{\frac{0.10}{0.15}\right\}^2 + \left\{\frac{0.8}{1.0}\right\}^2 = 1.08$$

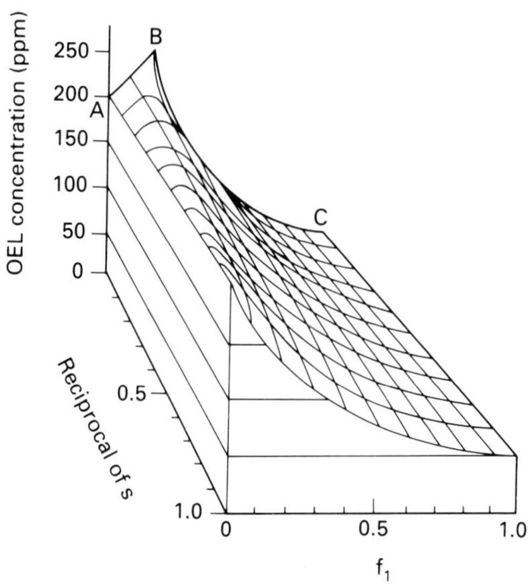

FIG. 8.3. Occupational exposure limit of mixtures of two substances, X, Y, whose single substance exposure limits are 50 ppm and 200 ppm, respectively. $OEL_1 = 50$, $OEL_2 = 200$. f_1 = fractional concentration of substance X.

$$\mathrm{OEL} = \frac{1}{\left[\left\{\frac{f_1}{\mathrm{OEL}_1}\right\}^s + \left\{\frac{f_2}{\mathrm{OEL}_2}\right\}^s\right]^{1/s}}$$

The exposure limit of the mixture is just exceeded. The exposure limit of the mixture, if needed explicitly, may be found directly by substituting the example values in expression (9):

$$\frac{1}{\left[\left\{\frac{0.11}{0.15}\right\}^2 + \left\{\frac{0.89}{1.0}\right\}^2\right]^{1/2}} = 0.87 \text{ mg m}^{-3}.$$

A suitable exposure limit of the mixture in air would therefore be 0.87 mg m^{-3}.

Mixtures with additive and independent effects

When mixtures of hazardous substances contaminate air, n of which act additively and m of which act independently, their joint effect can be assessed. It is first noted that the joint effect of n additive substances is independent of the effects of the m other substances. Thus, the joint effect of the n substances which act additively, together with the effects of the m other substances which act independently, is tested by employing the following representation:

$$\left\{\frac{C}{\mathrm{OEL}}\right\}^s = \left\{\frac{C_1}{\mathrm{OEL}_1} + \frac{C_2}{\mathrm{OEL}_2} + \cdots \frac{C_n}{\mathrm{OEL}_n}\right\}^s + \left\{\frac{C_{n+1}}{\mathrm{OEL}_{n+1}}\right\}^s + \left\{\frac{C_{n+2}}{\mathrm{OEL}_{n+2}}\right\}^s + \cdots \left\{\frac{C_{n+m}}{\mathrm{OEL}_{n+m}}\right\}^s, \quad (10)$$

where C is the total concentration of contamination in air and OEL is its exposure limit.

$$C = C_1 + C_2 + \cdots C_n + C_{n+1} + C_{n+2} + \cdots C_{n+m}.$$

To avoid an excessive risk of adverse effects from all the substances together the criterion is:

$$\frac{C}{\text{OEL}} < 1$$

Problem 8.5

As in problem 8.4 air contamination consists of a mixture making up 0.1 mg m^{-3} of lead arsenate in air (OEL 0.15 mg m^{-3}) and 0.8 mg m^{-3} of sulphuric acid mist in air (OEL 1.0 mg m^{-3}), but sulphur dioxide is also present at a concentration of 1 ppm sulphur dioxide in air (OEL 2 ppm). Is the occupational exposure limit for such a mixture exceeded? Devise a suitable limit.

Solution

Both sulphuric acid and sulphur dioxide are respiratory irritants and act additively. Lead arsenate acts independently of the other two. The example values are substituted in the equation

$$\left\{\frac{C}{\text{OEL}}\right\}^2 = \left\{\frac{0.8}{1.0} + \frac{1}{2}\right\}^2 + \left\{\frac{0.10}{0.15}\right\}^2 = 2.134$$

Thus,

$$\frac{C}{\text{OEL}} = 1.46.$$

The occupational exposure limit for such a mixture is therefore exceeded.

The exposure limit of the mixture may be found from this solution by first expressing the concentration of the constituents of the mixture in consistent units (mg m^{-3}) and adding them together; thus,

$$C = 0.8 + 2.5 + 0.1 = 3.4 \text{ mg m}^{-3},$$

from which OEL $= 3.4/(2.134)^{1/2} = 2.33$ mg m^{-3}.

There is little question that a square law ($s = 2$) would overestimate health risks from exposure to independently acting substances contaminating air in concentrations below their exposure limits. Where the combined exposure limit is so low that it gives rise to serious doubt about its practicability, feasibility or cost-effectiveness some careful consideration may be justified of the square law assumption at low concentrations. The knee of lognormal exposure-response curves, for example, is especially sharp for steep curves, that is for populations whose tolerance distribution has a relatively small geometric deviation. Thus, a cube law might be adopted for those substances for which there is good reason to believe the tolerance distribution of each of the substances is approximately lognormal with a geometric deviation less than 2.5.

Sequential mixtures

Each work shift the employees may be exposed regularly to a set of different hazardous materials in sequence by the nature of the process on which they are working or because their job requires them to move from place to place. The sequentially mixed exposure they thereby encounter can be assessed by logical extensions to the previous considerations regarding concurrent exposures.

Additive effects

To determine the time weighted average of a sequence of exposures to different substances whose effects are additive the concentrations of the different substances are first standardised with respect to their single substance time weighted average exposure limits. To do this, each individual function, Concentration/OEL, is calculated. The next step, the treatment of exposure duration, is more subtle. Additive substances with a long biological half-time accumulate in the body in proportion to the period of exposure. Thus, for these, the quantity of effect is proportional to duration of exposure each shift as well as to concentration during exposure, that is, to the product of the two.

With regard to substances which have a short biological half time the argument is different. The quantity of effect is proportional to its intensity and its duration. In a variable environment, as in industry, the expected period of time an employee is exposed above a given level of concentration is directly proportional to the duration of the exposure. So that the risk of exceeding a given body burden of a short biological half time substance and the duration of the adverse effects from it (narcosis, irritation, and so on) are in direct proportion to duration of exposure.

The joint effect on an employee exerted by two or more additive substances acting at the same locality in the body is therefore indicated by the time weighted average of all the individual functions, Concentration/OEL.

Employees exposed to a sequence of n different substances whose effects are *additive* and they are exposed to these substances for periods of time t_1, t_2, t_3, $\ldots$, t_n respectively, and the individual time weighted average occupational exposure limits are OEL_1, OEL_2, OEL_3, $\ldots$, OEL_n and the concentrations in air are C_1, C_2, C_3 $\ldots$, C_n, respectively, then the function, C/OEL, of the sequential mixture is given by:

$$\frac{C}{\text{OEL}} = \frac{\dfrac{t_1 C_1}{\text{OEL}_1} + \dfrac{t_2 C_2}{\text{OEL}_2} + \cdots \dfrac{t_n C_n}{\text{OEL}_n}}{8} \qquad (11)$$

where C is the time weighted average concentration of contamination in air and OEL is its exposure limit.

$$C = \frac{t_1 C_1 + t_2 C_2 + \cdots t_n C_n}{8}$$

In order to avoid an excessive risk of adverse effects from all the substances together the criterion is:

$$\frac{C}{\text{OEL}} < 1.$$

The OEL of the mixture as a whole is thus

$$\text{OEL} = \frac{1}{\dfrac{f_1}{\text{OEL}_1} + \dfrac{f_2}{\text{OEL}_2} + \cdots \dfrac{f_n}{\text{OEL}_n}}, \tag{12}$$

where $f_1, f_2, \ldots f_n$ are the fractional concentrations of the n constituents as might, in theory at least, be determined on a sample obtained by sampling air at a constant rate throughout the work-shift:

$$f_1 = \frac{t_1 C_1}{tC_1 + t_2 C_2 + \cdots t_n C_n}, \quad f_2 = \frac{t_2 C_2}{t_1 C_1 + t_2 C_2 + \cdots t_n C_n},$$

etc.

Problem 8.6

Employees are exposed to air contamination coming successively from two separate processes, one of which employs isoamyl alcohol and runs for 2 hours each work-shift, the other employed cyclohexanol and runs for 6 hours each work-shift. The vapour of both of these solvents is irritating to the eyes, nose and throat. The atmospheric concentration of the isoamyl alcohol vapour to which the employees are exposed is 70 ppm (TWA OEL 100 ppm) and of the cyclohexanol vapour it is 25 ppm (TWA OEL 50 ppm). Is the occupational exposure limit of this sequential mixture exceeded? Devise a suitable exposure limit.

Solution

The irritative effects of these solvents are additive and the additive equation (11) yields

$$\left[\frac{(2)(70)}{100} + \frac{(6)(25)}{50}\right]\left[\frac{1}{8}\right] = 0.55.$$

This does not exceed 1.0, showing that the exposure limit of the mixture is not exceeded.

$$f_1 = \frac{140}{140 + 150} = 0.48, \quad f_2 = \frac{150}{140 + 150} = 0.52.$$

The exposure limit of the mixture of contamination, given by formula (12) is therefore:

$$\text{OEL} = \frac{1}{\dfrac{0.48}{100} + \dfrac{0.52}{50}} = 66 \text{ ppm}$$

Independent effects

Single substance occupational exposure limits usually refer to the time weighted average concentration over 8 hours. The function, (Concentration/Occupational Exposure Limit)^s, may be thought of as representing the quantity of risk of adverse effects in the population of employees resulting from 8 hours of exposure. The risk to health would be less from a shorter period of exposure in much the same way as it would be less from a lower concentration. The combined health risk to the population of employees exerted by two or more substances acting independently on the body and sequentially is therefore indicated by

$$\sum \left\{ \frac{t_i}{8} \frac{C_i}{\mathrm{OEL}_i} \right\}^s,$$

where t_i is the period of exposure to substance i, in units of hours, C_i is the atmospheric concentration of the substance and OEL_i is its exposure limit, 8-hour time weighted average.

Therefore, for employees exposed sequentially to m independently acting substances for periods of time $t_1, t_2, t_3, \ldots, t_m$, whose occupational exposure limits are $\mathrm{OEL}_1, \mathrm{OEL}_2, \mathrm{OEL}_3, \ldots, \mathrm{OEL}_m$ and whose concentrations in air are $C_1, C_2, C_3 \ldots, C_m$ respectively, each alone producing some acceptably small risk, the criterion, C/OEL, for the sequential mixture is given by:

$$\left\{ \frac{C}{\mathrm{OEL}} \right\}^s = \frac{\left\{ \frac{t_1 C_1}{\mathrm{OEL}_1} \right\}^s + \left\{ \frac{t_2 C_2}{\mathrm{OEL}_2} \right\}^s + \cdots \left\{ \frac{t_m C_m}{\mathrm{OEL}_m} \right\}^s}{8^s}, \tag{13}$$

where C is the time weighted average concentration of atmospheric contamination and OEL is its exposure limit.

$$C = \frac{t_1 C_1 + t_2 C_2 + \cdots t_m C_m}{8}$$

To avoid an excessive risk of adverse effects from all the substances together the criterion is:

$$\frac{C}{\mathrm{OEL}} < 1$$

The OEL of the mixture as a whole is thus

$$\frac{8}{\left[\left\{ \frac{t_1 f_1}{\mathrm{OEL}_1} \right\}^s + \left\{ \frac{t_2 f_2}{\mathrm{OEL}_2} \right\}^s + \cdots \left\{ \frac{t_m f_m}{\mathrm{OEL}_m} \right\}^s \right]^{1/s}} \tag{14}$$

where $f_1, f_2, \ldots f_m$ are the fractional concentrations of the m constituents as might, in theory at least, be determined on a sample obtained by sampling air at a constant rate throughout the work-shift.

$$f_1 = \frac{t_1 C_1}{t_1 C_1 + t_2 C_2 + \cdots t_m C_m}, \quad f_2 = \frac{t_2 C_2}{t_1 C_1 + t_2 C_2 + \cdots t_m C_m},$$

etc.

Problem 8.7

Employees are exposed to air contamination coming successively from two separate processes, one of which employed lead arsenate and runs for 2 hours each work-shift, the other involves exposure to sulphuric acid mist and runs for 6 hours each work-shift. The atmospheric concentration of the lead arsenate to which the employees are exposed for 2 hours is 0.2 mg m^{-3} (OEL 0.15 mg m^{-3}) and of the sulphuric acid mist to which the employees are exposed for 6 hours is 0.8 mg m^{-3} (OEL 1.0 mg m^{-3}).

Lead arsenate causes systemic reactions whereas sulphuric acid is a primary irritant and causes tooth erosion. These are independent effects but the residual risks below the respective exposure limits are additional one to another. Is the 8-hour time weighted average exposure limit exceeded?

Solution

The level of lead arsenate concentration is above its 8-hour exposure limit but exposure is for only 2 hours. Health risk from each independently acting substance varies as the square of the period of exposure to that substance. Equation (13) applies. Substituting the given values in the equation,

$$\left\{\frac{C}{\text{OEL}}\right\}^2 = \frac{\left\{\frac{2 \times 0.2}{0.15}\right\}^2 + \left\{\frac{6 \times 0.8}{1.0}\right\}^2}{64} = 0.47$$

The exposure limit of the sequential mixture is not exceeded. The exposure limit of the mixture, if needed, may be found directly by substituting the example values in expression (14).

$$f_1 = \frac{2 \times 0.2}{0.4 + 4.8} = 0.08, \quad f_2 = \frac{6 \times 0.8}{0.4 + 4.8} = 0.92.$$

$$\text{OEL} = \frac{8}{\left[\left\{\frac{2 \times 0.08}{0.15}\right\}^2 + \left\{\frac{6 \times /.92}{1.0}\right\}^2\right]^{1/2}} = 0.67 \text{ mg m}^{-3}$$

In personnel control advantage may sometimes be taken of the operation of a non-linear, square or possibly cube law relating health risk to period of exposure. Favourable circumstances occur when the exposure period can be split into two or more separate periods in such away that exposure is to a different, independently acting substance in each period.

Problem 8.8

Employees are in two different job-exposure groups. One group is exposed to lead arsenate and the other to sulphuric acid mist. The atmospheric concentration of the lead arsenate to which the employees in one group are exposed is 0.2 mg m^{-3} (OEL 0.15 mg m^{-3}) and of the sulphuric acid mist to which the employees in the other group of employees are exposed is 1.5 mg m^{-3} (OEL 1.0 mg m^{-3}). Both groups are exposed above the respective exposure limits and the two substances act independently. It is not feasible to reduce the concentration to which either group is exposed and the provision of respiratory protection is contemplated. What should be done?

Solution

The solution to this problem may lay in personnel control. Thus consideration should be given to the possibility of splitting the tasks so that each group is exposed to just 4 hours to each substance. The criterion, C/OEL, for the sequential mixture is given by equation (13). Substituting the values presently under consideration:

$$\left\{\frac{C}{\text{OEL}}\right\}^2 = \frac{\left\{\frac{4 \times 0.2}{0.15}\right\}^2 + \left\{\frac{4 \times 1.5}{1.0}\right\}^2}{64} = 1.0$$

Accordingly, it is concluded that by such an arrangement the exposure limit of the sequential mixture is not exceeded. Provided there are no insurmountable administrative problems this solution is recommended.

Additive and independent effects

Sequential mixtures of substances, n of which act additively and m of which act independently can be combined by observing first that the n ratios are additive one to another and second that the combined effect of the n substances is independent of the effects of the m other substances. Thus, the combined effect of the n additive substances together with the m other, independently acting substances is tested by employing

$$\left\{\frac{C}{\mathrm{OEL}}\right\}^s = \left[\left\{\frac{t_1C_1}{\mathrm{OEL}_1} + \frac{t_2C_2}{\mathrm{OEL}_2} + \cdots \frac{t_nC_n}{\mathrm{OEL}_n}\right\}^s + \left\{\frac{t_{n+1}C_{n+1}}{\mathrm{OEL}_{n+1}}\right\}^s + \left\{\frac{t_{n+2}C_{n+2}}{\mathrm{OEL}_{n+2}}\right\}^s + \cdots \left\{\frac{t_{n+m}C_{n+m}}{\mathrm{OEL}_{n+m}}\right\}^s\right] \times \left\{\frac{1}{8}\right\}^s \quad (15)$$

where C is the time weighted average concentration of contamination and OEL is its exposure limit.

$$C = \frac{[t_1C_1 + t_2C_2 + \cdots t_nC_n + t_{n+1}C_{n+1} + t_{n+2}C_{n+2} + \cdots t_{n+m}C_{n+m}]}{t_1 + t_2 + \cdots t_n + t_{n+1} + t_{n+2} + \cdots t_{n+m}}$$

Problem 8.9

As in problem 8.7 air contamination consists of a sequential mixture of 0.2 mg m^{-3} of lead arsenate in air (OEL 0.15 mg m^{-3}) and 0.8 mg m^{-3} of sulphuric acid mist in air (OEL 1.0 mg m^{-3}). The period of exposure to lead arsenate is 2 hours and this is followed by a period of exposure to sulphuric acid of another 2 hours. Finally there is a period of exposure of 4 hours to air contamination at a level of 1 ppm sulphur dioxide in air (OEL 2 ppm).

Both sulphuric acid and sulphur dioxide are respiratory irritants and act additively. Lead arsenate acts independently of the other two. Is the exposure limit of this combination of additive and independently acting substances exceeded? Calculate a suitable exposure limit.

Solution

$n = 2$, $m = 1$, $s = 2$. The example values are substituted into equation (15):

$$\left\{\frac{C}{\mathrm{OEL}}\right\}^2 = \left[\left[\frac{(2 \times 0.8)}{1.0} + \frac{(4 \times 1)}{2}\right]^2 + \left\{\frac{2 \times 0.20}{0.15}\right\}^2\right] \times \frac{1}{64}$$
$$= 0.31.$$

Thus $C/\mathrm{OEL} = 0.56$, and the exposure limit is not exceeded.

The time weighted average concentration of the sequential mixture may be found by first expressing the concentration of the constituents in consistent units; thus,

$$C = \frac{(2 \times 0.8) + (4 \times 2.5) + (2 \times 0.2)}{8} = 1.5 \text{ mg m}^{-3},$$

from which, $\mathrm{OEL} = (1.5)^{1/2}/(0.31) = 2.2$ mg m^{-3}.

Trivial constituents

In published lists of exposure limits it is not uncommon to express the limit in terms of the constituent which is presumed to be solely or primarily responsible for the associated health hazard, as for example chromium compounds and quartz. In the present context the issue is raised when it is realised that there is no upper limit to the number of terms in the equations for testing the exposure limit of a mixture of substances with additive or independent effects [equations (9) and (15)]. There are dozens of components in common paints and even in the vehicle there may be ten different solvents (Hansen, 1982). In some instances the mixtures are highly complex, consisting of scores of compounds that are generated simultaneously as by-products from a single source or process, as in coke oven emissions and diesel exhaust. It is theoretically possible to have hundreds of small terms which sum to much more than 1.0. But the concentration of a substance in air becomes progressively more difficult to detect and measure accurately the lower it is. A procedure is needed for deciding which constituents may safely be ignored. These will be ones which are intrinsically relatively innocuous or in such small amounts as to be of negligible importance.

The usual sensitivity required of industrial hygiene measurements is down to $0.1 \times \text{OEL}$. Implicit in the acceptance of this rule is the belief that risks of adverse effects at these levels are negligible. This is not denied here. Substances present in air at a concentration below some small fraction of their exposure limits may reasonably be ignored. Accordingly it is suggested that components whose value of $(C/\text{OEL})^s$ is less than 0.1 may safely be ignored and treated as zero for present purposes. This would effectively impose an absolute upper limit of 10 to the number of significant substances below their respective exposure limits and rarely would there be more than three or four.

Problem 8.10

As in problem 8.2, a mixture of atmospheric contamination making 35 ppm perchloroethylene in air (OEL 50 ppm) and 150 ppm methyl chloroform in air (OEL 150 ppm) but there is also present a third constituent, 3 ppm trichloroethylene in air (OEL 50 ppm). All three vapours are narcotic and therefore have additive effects. Is the exposure limit of this combination exceeded? Compare the result with the solution to problem 8.2.

Solution

The additive equation (1) yields

$$\frac{35}{50} + \frac{150}{350} + \frac{3}{50} = 1.19.$$

The standardised concentration of the mixture is 19% in excess of unity. Clearly, the exposure limit is exceeded and corrective action is required. Had the third term been ignored equation (1) would have yielded 1.13, 13% in excess of unity, as in problem 8.2 and the recommendation would have been much the same. Formula (3) yields

$$\text{OEL} = \frac{1}{0.0037 + 0.0023 + 0.0004} = 156 \text{ ppm},$$

as compared with 164 ppm in problem 8.2.

Complex mixtures

There are practical limitations to the formulae based upon exposure limits of single substances. Exposure limits may not be known for all the constituents of a complex mixture or the mixture may be very variable. Naturally occurring ores, rocks and vegetable dusts are complex mixtures. One way of dealing with these is to express the limit in terms of the total quantity of dust, as for example coal dust, or cotton dust. This proves satisfactory provided the potency of the material does not vary widely. Another technique is to express the limit in terms of the component which presents the greatest hazard. An example is coal tar pitch volatiles, whose published exposure limit is commonly expressed in terms of just the benzene soluble fraction.

There are situations where the potency of the material varies wildly. One example is mineral oil mist, whose potency is often governed by the composition and quantity of a variety of additives designed to confer beneficial metal working properties. Practical ratings are being developed for groups of these oily mixtures (Deutsche Forschungsgemeinschaft, 1985). Welding fume is another material which has varying composition and many constituents, some of them very hazardous (Spee and Zwennis, 1987; Wal, 1990). It is not technically feasible to keep track of all the possible constituents of welding fume. Control of health hazard in these situations relies upon 'worst case' control; using engineering and personnel control measures which are designed for a worst case situation (American Industrial Hygiene Association, 1984; Stott, Champion, Wallis, Lodge and Sims, 1982). Monitoring of atmospheric concentration is of secondary value. Health risk surveillance is concentrated upon the engineering and personnel control measures.

Another example of very variable composition is in chemical laboratories where any of several hundred or more chemicals may be handled. Again, this is a case where continual measurement of all possible substances is not technically feasible nor for that matter very helpful. The preferred approach is to rely essentially on laboratory management and good ventilation to secure adequate control over the worst case condition. Operations involving the release of especially hazardous materials or large quantities of moderately hazardous substances are conducted in dedicated, high performance laboratory hoods. The most hazardous operations are conducted in glove boxes.

Synergistic and antagonistic substances

The formulae for substances with additive or independent effects should not be used where two or more of the substances are known to exert synergistic or antagonistic effects. Toxicological interaction between different hazardous substances in industry is probably rare, although there are several examples of synergism in the literature (La Belle, Long and Christofano, 1955; Calabrese, 1978). Where synergism or antagonism is suspected each mixture should be considered on its own merits (Du Bois, 1972; US Environmental Protection Agency, 1986).

An interesting example is the potentiation of the toxicity of a gas or vapour with the concurrent inhalation of an aerosol. Absorption of the gas on the particles may result in changes in the rate of absorption in the lungs and in chemical behaviour after inhalation (Amdur, 1957, 1959, 1961). Sulphuric acid layered on the surface of solid particles is carried into the lungs where it deposits, affects respiratory function directly and increases the sensitivity of the lungs to other insults (Lippmann, 1986; Lippmann, Gearhart and Schlesinger, 1987; Amdur, 1989).

Acidic aerosols greatly increase lung sensitivity to ozone (Osebold, Gershwin and Zee, 1980; Warren and Last, 1987). Mineral oil mist enhances the upper respiratory tract irritant properties of gases and vapours (Dautrebande,

Shaver and Capps, 1951). Quartz dust is suspected of having a synergistic action when inhaled with coal dust (Seaton, Dick, Dodgson and Jacobsen, 1981).

Another good example of synergism is the relationship between cigarette smoking, combined with inhalation of asbestos fibres giving rise to a greatly increased occurrence of lung cancer in humans (Saracci, 1977). In a study by Hammond, Selikoff and Seidman (1979) the relative risk of lung cancer attributable to smoking was 11, while the relative risk associated with asbestos exposure was 5. The relative risk of lung cancer from both smoking and asbestos exposure was 53, indicating a substantial synergistic effect.

A type of synergism is sometimes found in the interaction between chemicals in suspension in air forming a more hazardous compound. An example of this is formaldehyde and hydrochloric acid, which interact to form the potent carcinogen bis-chloromethyl ether.

Alcohol has been regarded as a potentiating or a predisposing substance in relation to diseases from such diverse materials as carbon tetrachloride, calcium cyanide and carbon monoxide (Pecora, 1959). Synergism is also possible between chemical substances and viable organisms. Graham and co-workers have reported that ozone and nitrogen dioxide increased the mortality in mice challenged with bacterial infections (1987).

Examples of antagonism are rare but there is sufficient animal evidence to show that it is not just a theoretical concept (Kay, 1960). It has been shown in guinea pigs that an aerosol of naphthenic medicinal oil or paraffinic mineral oil can protect against the increase in pulmonary flow resistance which would otherwise result from inhaling sulphur dioxide (Costa and Amdur, 1979). The anti-irritant properties of mineral oils are widely accepted in the cosmetic industry (Hoekstra and Phillips, 1963).

References

Amdur M O. Public Health Repts. **69** 503 (1954).
Amdur M O. Am. Ind. Hyg. Assoc. J. **18** 149 (1957).
Amdur M O. Int. J. Air Pollution **1** 170 (1959).
Amdur M O. In: Inhaled particles and vapours (Edited by C N Davies). Pergamon Press, Oxford (1961).
Amdur M O. Appl. Ind. Hyg. **4** 189 (1989).
American Conference of Governmental Industrial Hygienists. Documentation of the Threshold Limit Values and Biological Exposure Indices. ACGIH, 6500 Glenway Ave., Bldg. D-7, Cincinnati, Ohio 45211, USA (1986).
American Conference of Governmental Industrial Hygienists. Threshold Limit Values and Biological Exposure Indices for 1990–91. ACGIH, 6500 Glenway Ave., Bldg. D-7, Cincinnati, OH 45211, USA (1990).
American Industrial Hygiene Association. Welding health and safety resource manual. American Industrial Hygiene Association, 475 Wolf Ledges Parkway, Akron, OH 44311–1087, USA (1984).
Ball W L. Am. Ind. Hyg. Assoc. J. **20** 357 (1959).
Barr J T. The calculation and use of carcinogenic potency: A review. Regulatory Toxicology and Pharmacology. 5, 432 (1985).
Bliss C I. Ann. Appl. Biol. **26** 585 (1939).

Bukowskij M L, V I Zokov, T V Kuzukova, I V Sanockih and K K Sidorov. Maximum admissible concentrations of harmful substances in ambient objects. All-Union Scientific Research Institute of Labour Hygiene and Occupational Diseases of the USSR Academy of Medical Sciences, Sverodenetak, USSR (1977).

Costa D L and M O Amdur. Am. Ind. Hyg. Assoc. J. **40** 681 (1979).

Dautrebande L and R Capps. Arch. Int. Pharmacodyn. Ther. **82** 505 (1950)

Dautrebande L, J Shaver and R Capps. Arch. Int. Pharmacodyn. Ther. **85** 17 (1951)

Deutsche Forschungsgemeinschaft. Maximum concentrations at the workplace and biological tolerance values for working materials 1985. Report No. XXI of the Commission for the Investigation of Health Hazards of Chemical Compounds in the Work Area. VCH Verlagsgesellschaft, Postfach 1260/1280, D-6940 Weinheim, FRG; VCH Publishers, Deerfield Beach, FL, USA (1985).

Du Bois K P. In: Multiple factors in the causation of environmental disease (Edited by D H K Lee and P Kolin). Academic Press, New York (1972).

Elkins H B. Am. Ind. Hyg. Assoc. J. **23** 132 (1962).

Gaylor D W and R L Kodell. J. Environ. Pathol. Toxicol. 4 305, (1980).

Gmeling J and C S Weidlich. Staub Reinhalt—Luft **49** 227, 295 (1989).

Graham J A, D E Gardner, E J Blommer, D E House, M G Menache and F J Miller. J. Toxicol. Environ. Health **21** 113 (1987).

Gray C N, P J Hewitt and R Hicks. The prediction of fume compositions in stainless steel metal inert gas welding. In: Weld pool chemistry and metallurgy. The Welding Institute, London (1980).

Hammond E C, I V Selikoff and H Seidman. Ann. NY Acad. Sci. **330** 473 (1979).

Hansen C M. Solvent technology in product development. In: Advances in environmental toxicology, Vol II, Occupational health hazards of solvents (Edited by A. Englund, K. Ringen and M. A. Mehlman). Princeton Scientific Pubs, Inc., Princeton (1982).

Hatch T F. Arch. Environ. Health **16** 214 (1968).

Health & Safety Executive. Occupational exposure limits 1989, Guidance Note EH 40/89. H M Stationery Office, UK (1989).

Hewitt P J and A A Hirst. Ann. Occup. Hyg. **35** 223 (1991).

Hoekstra W G and P H Phillips. J. Invest. Dermatol. **40** 79 (1963).

Kay K. Occ. Health Review **12** 2 (1960).

La Belle C W, J E Long and E E Christofano. AMA Arch. Ind. Hyg. & Occ. Med. **11** 297 (1955).

Lippmann M. Respiratory tract deposition and clearance of aerosols. In: Aerosols (Edited by S D Lee, T Schneider, L D Grant and P Verkerk). Lewis Publishers, Chelsea, MI, USA (1986).

Lippmann M, J M Gearhart and R B Schlesinger. Appl. Ind. Hyg. **2** 188 (1987).

Mantel N, N R Bohidar, D C Brown J L Ciminera and J W Tukey. Cancer Res. **35** 759 (1971).

Mantel N and W R Bryan. J. Nat. Cancer Inst. **27** 455 (1961).

Murphy S D. Assessment of the potential for toxic interactions among environmental pollutants. In: C I Galli, S D Murphy and R Paoletti (Editors). The principles and methods in modern toxicology. Elsevier/North Holland Biomedical Press, Amsterdam (1980).

Osebold J W, L J Gershwin and Y C Zee. J. Environ. Path. Toxicol. **3** 221 (1980).

Pecora L J. Am. Ind. Hyg. Assoc. J. **20** 235 (1959).

Pozzani U C, C S Weil and C P Carpenter. Am. Ind. Hyg. Assoc. J. **20** 364 (1959).

Roach S A and S M Rappaport. Am. J. Ind. Med. **17** 727 (1990).

Saracci R. Int. J. Cancer **20** 325 (1977).

Seaton A, J A Dick, J Dodgson and M Jacobsen. Lancet **ii** 1272 (1981).

Smyth H F, C S Weil, J S West and C P Carpenter. Toxicol. Appl. Pharmacol. **14** 340 (1969).

Smyth H F, C S Weil, J S West and C P Carpenter. Toxicol. Appl. Pharmacol. **17** 496 (1970).

Spee T and W C M Zwennis. Scand. J. Work Environ. Health **13** 52 (1987).

Stott M D, A Champion, R Wallis, P Lodge and B J Sims. Ann. Occup. Hyg. **25** 279 (1982).

Tandon R K, P T Crisp, J Ellis and R S Baker. Aust. Welding J. 27–30 (Autumn) (1983).

Thorne B D and P J Hewitt. Ann. Occup. Hyg. **39** 181 (1985).

US Environmental Protection Agency. Guidelines for carcinogen risk assessment. Federal Register **51** 33992 (1986).

US Environmental Protection Agency. Guidelines for the health risk assessment of chemical mixtures. Federal Register **51** 34014 (1986).
US Office of Science and Technology Policy. Chemical carcinogens: review of the science and its associated principles. Federal Register **50** 10372 (1985).
Veldstra H. Pharmacol. Rev. **8** 339 (1956).
Wal J F, van der. Ann. Occup. Hyg. **34** 45 (1990).
Warren D L and J A Last. Toxicol. Appl. Pharmacol. **88** 203 (1987).
World Health Organisation. Health effects of combined exposures in the work environment, Technical Report Series No. 662. WHO, Geneva, Switzerland (1981).

Bibliography

Boyd E M. Predictive toxicometrics. Bristol Scientechnica (Publishers) Ltd, Bristol, UK (1972).
Calabrese E J. Methodological approaches to deriving environmental and occupational health standards. Wiley, New York (1978).
Dautrebande L. Bases experimentales de la protection contre les gas de combat (Edited by J Ducolot). Geneblaux, Belgium (1939).
Finney D J. Probit analysis: A statistical treatment of the sigmoid response curve. Cambridge University Press, London (1952).
Goldstein A, L Aronow and S M Kalman. Principles of drug action: the basis of pharmacology, 2nd ed. John Wiley and Sons, Inc., New York (1974).
Levine R E. Pharmacology: drug actions and reactions. Little, Brown and Company, Boston, MA, USA (1973).
NRC (National Research Council). Principles of toxicological interactions associated with multiple chemical exposures. Academy Press, Washington DC, USA (1980).

CHAPTER 9

Nobody Works Eight Hours

Few employees are exposed to hazardous substances at work for precisely 8 hours each day and for 5 days each and every week as specified in most published occupational exposure limits. Many who work a standard 5-day week of 8-hour shifts are exposed intermittently to hazardous substances, for only a fraction of their time at work, perhaps as little as an hour or less per week. A similar pattern is followed by those who experience sporadic exposure to short term high atmospheric concentration episodes on a background of relatively low exposure (Figure 9.1). The health risks from such an isolated, short term exposure or isolated peak may be gauged in terms of its influence over the body burden as compared with exposure for 8 hours each work shift and 5 days each week.

Atmospheric exposure limits are implicitly based upon a 'normal' continuous process with its associated more or less random concentration variability about the grand mean, having a geometric deviation of about 1.5–2.5 within

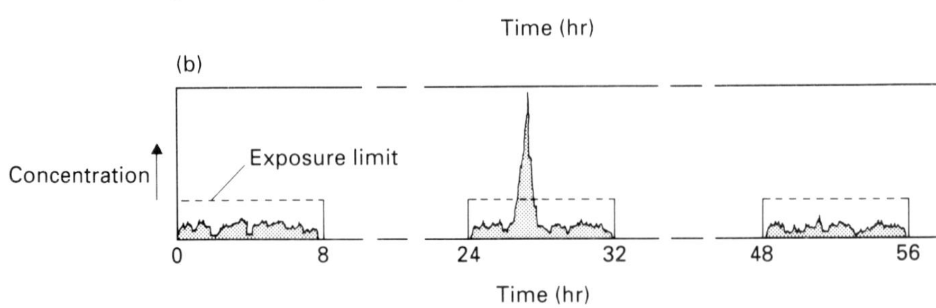

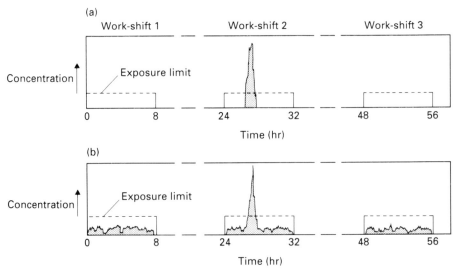

FIG. 9.1. Trace of concentration of hazardous substance in air, against time. Short term exposure. Isolated episode. (a) Exposure normally insignificant. (b) Exposure normally low.

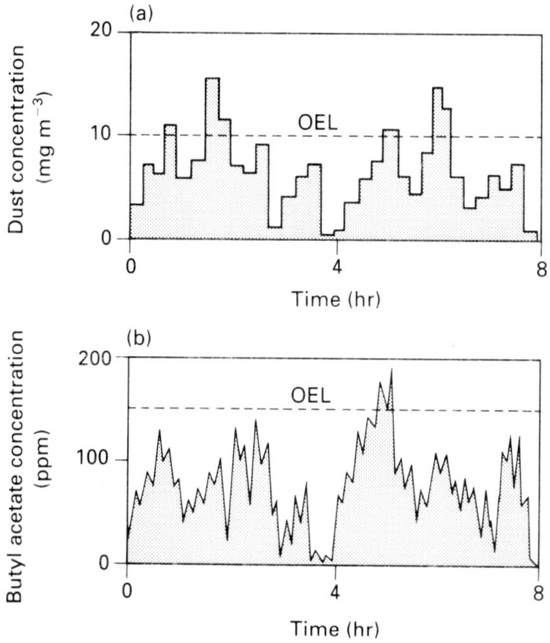

FIG. 9.2. Random variation of atmospheric concentration over time. (a) Sequential 15-minute time weighted averages of airborne limestone dust concentration during one work shift at a grinding plant. (b) Continuous read-out meter readings of atmospheric concentration taken at intervals of six minutes in a work room at an artificial leather factory where butyl acetate was in use. (------) 8-hour time weighted average exposure limit.

each work shift. But there are two major types of non-standard, within shift, temporal exposure pattern which will be considered in this chapter; one type is characterised by unusually high but still random variation of concentration and the other is characterised by marked systematic variation. Procedures are needed for handling both of these types.

Random variation of atmospheric concentration of hazardous substances in industry is illustrated in Figure 9.2. Sequential data of this kind, resulting from air contamination emitted from a continuous process, features a fluctuating concentration with no regular pattern discernible except, perhaps, for a dip during the mid-shift break. The data are often strongly autocorrelated. The first example, Figure 9.2a, is concentration of dust present in air sampled in the vicinity of a limestone rock grinding plant. The results were in the form of 15-minute time weighted averages. The second example, Figure 9.2b, shows readings taken from a continuous read-out meter of the concentration of butyl acetate present in air next to a work bench in an artificial leather factory. An unusual degree of variability, as in this example, could possibly result in increased health risk from exposure peaks (Ulfvarson, 1987; Zielhuis, Noordam, Roelfzema and Wibowo, 1988).

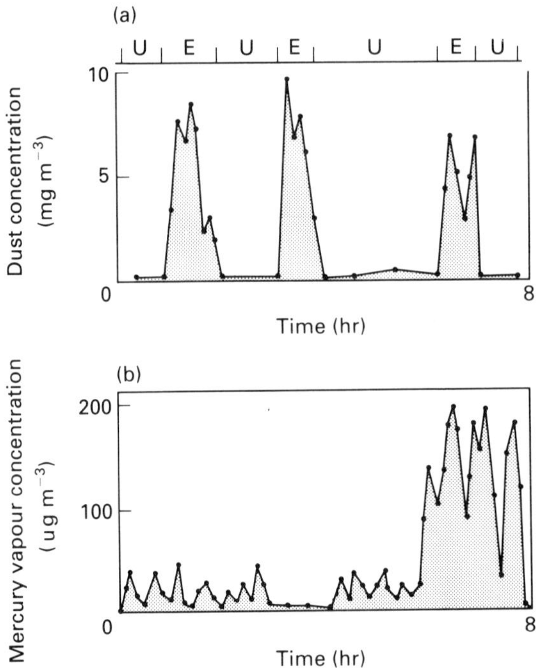

FIG. 9.3. Concentration of hazardous substance in air, against time. Systematic variation during a work shift. (a) Example of a simple work cycle; U = unloading bags, E = emptying bags. (b) Example of a step change. Screens were started 2 hours before the end of a work shift at a fertiliser plant.

Systematic temporal variation of atmospheric concentration is illustrated in Figure 9.3. This is exposure with a distinct pattern, usually following a cycle of operations or tasks. The type includes cyclic exposures of various periodicity up to 8 hours. The example shown in Figure 9.3a is from a study of employees alternately unloading and emptying bags of powder. The periods spent emptying bags are clearly marked by the high concentration values in air samples taken during that operation. Brief inhalation of high concentration of air contamination every work-shift with no other exposure during the work-shift is an extreme example of this general type, as is brief inhalation of a high concentration during prolonged exposure to low concentrations each shift. Step changes during each work-shift or other kinds of systematic variation could also result in a health risk that is absent in a standard exposure pattern. The example shown in Figure 9.3b is of readings of a mercury vapour meter through which air was drawn from the breathing zone of an operative whose job included work with inorganic mercury in a different and poorly ventilated part of a work room for the last 2 hours of each work shift.

In studies of exposure over several days or weeks it is found that periods of exposure to hazardous substances may be somewhat less than the full work shift for employees working on continuous processes. In practice small differences from the standard 8 hours duration of exposure are customarily ignored by those assessing health risks or accommodated by making proportionate reductions or increases in the exposure limits and biological exposure indices allowed. But some employees work a considerable amount of overtime, some work at week-ends and others are on shift systems which involve long periods at work and a long 'week-end'. An example is work on offshore oil installations in the North Sea, where the work schedule is usually 12 hours a day for 14 consecutive days. Between tours of duty there is an off-duty period, which in the British and Danish sectors is 2 or 3 weeks, and in the Norwegian sector either 3 weeks or 3 and 4 weeks alternately (Eide, 1990). A formula or procedure is needed to be reasoned out from general toxico-kinetic principles with which an exposure limit for a 5-day week of 8 hours each day on a continuous process can be adapted to suit a pattern of work periods and breaks which is markedly different from standard (Mason and Dershin, 1976; Lowe and Chambers, 1983; Veng-Pendersen, 1984; Paustenbach, 1985).

Lifetime maximum body burden

Exposure to hazardous substances at work adversely affects the health of an employee through excessive quantities of the substance being inhaled, deposited and retained in the lungs, or absorbed through the surface of the respiratory tract, or absorbed through the skin or eyes. Hazardous substances may also produce ill effects by contaminating food and hands, then being ingested and absorbed through the surface of the alimentary tract. In any case, by whatever route it enters, the consequent body burden of the substance or its metabolites in the critical organ would be a more direct indication of whether an employee is likely to be adversely affected by it than would the concentration in the air of the work room.

It is axiomatic that if there is a maximum work room concentration, to which exposure for 8 hours each day for 5 days each week for a working lifetime would be without adverse effect on health, there is a corresponding lifetime maximum body burden of a substance or metabolite which would be without adverse effect (Roach, 1966). The body burden is related to the biological exposure index of the substance.

Exposure limits represent conditions for a working lifetime under which it is believed significant risk of adverse health effects will be prevented. Corresponding to each exposure limit there will be a maximum lifetime burden of the substance or metabolite in the critical organ. It seems reasonable to assume as a point of departure that equal lifetime maximum burdens in the critical organ will be associated with equal risk of adverse effects. This assumption is a decisive one. Once made, the values of concentration in exposure limits for non-standard exposure periods may be modified in such a way as to ensure compliance with the modified limit would prevent this maximum lifetime body burden being exceeded. Persons exposed should be at no greater level of risk. This does not imply an acceptance or rejection of any specific hypothesis about the process by which the body burden of a particular substance produces adverse health effects. It simply provides what is believed to be a procedure for assessing relative risks from exposure to hazardous substances, superior to those procedures in which the toxico-kinetics is ignored.

The minimum information required is an estimate of the biological half time in the critical organ so that single compartment toxico-kinetics may be applied. Any uncertainty in this estimate may be accommodated by employing the greatest or least likely value and by accepting the safety factors thereby incorporated in the testing schedules.

Brief exposure—for much shorter than 8 hours

Given an exposure limit for exposure to some substance for 8 hours per day and 5 days per week, it may be asked what is a suitable exposure limit for an operation involving an isolated brief exposure. By brief exposure is meant exposure for an isolated period, well short of 8 hours, the exposure being negligible at other times. This is notably different from a short term excursion above (or below) an average value of continuous exposure, which will be considered shortly.

Short period inhalation exposure limits may also be required in the design of safe operations (Bridges, 1985). For example:

(a) The concentration to which an employee is exposed by a foreseeable accident can be constrained to meet the short period exposure limit by maintaining controls on the quantity of hazardous material stored or held and its proximity to employees.

(b) By advance calculations of concentrations which can occur following specific accidents, operating rules can be formulated to regulate the time someone may work in an emergency before reducing their effectiveness by donning respiratory protection and before retreating to safety (American Industrial Hygiene Association Toxicology Committee, 1964; Smyth, 1966).

It is shown in Chapter 4 that where the body is modelled by a single, well-mixed compartment the accumulation and decay of the body burden over time may be examined algebraically. The rate of input of hazardous substance is directly proportional to its concentration in inspired air and the rate of output is directly proportional to the concentration of substance in the compartment. The concentration of hazardous substance in the compartment after a time t from the start of the exposure event is given by

$$x = kC[1 - \exp(-at/V)]/a\,, \qquad (1)$$

where C = concentration in inspired air (mg m^{-3})
x = concentration in compartment (mg m^{-3})
k = minute volume × percent retention (m^3 min^{-1})
t = time from start (min)
a = rate of removal from compartment (m^3 min^{-1})
V = volume of compartment (m^3).

Where C_L is the maximum concentration in inspired air which could be tolerated indefinitely, the corresponding maximum concentration in the compartment is kC_L/a mg m^{-3}. At higher concentrations in inspired air the time it takes to reach this maximum will be shorter. This is the maximum tolerable duration. It is dependent on the mean residence time, $V/a = 1.44T$. The biological half time is T minutes.

We have

$$\frac{kC_L}{a} = \frac{kC[1 - \exp(-at/V)]}{a},$$

that is

$$\frac{C}{C_L} = \frac{1}{1 - \exp(-at/V)},$$

$$= \frac{1}{1 - \exp(-0.693t/T)}. \tag{2}$$

Explicitly, the maximum tolerable duration of exposure to atmospheric concentration C is given by

$$t = \frac{V}{a} \ln \frac{C}{C - C_L}, \tag{3}$$

where C = concentration in inspired air (mg m^{-3})
C_L = maximum concentration in inspired air that could be tolerated indefinitely (mg m^{-3})
t = time from start of exposure (min)
a = rate of removal from compartment (m^3 min^{-1})
V = volume of compartment (m^3).

Given the half time of a hazardous substance in the critical organ, for brief exposures the maximum tolerable concentration in inspired air, C_{max}, can be estimated. Thus, employing the single compartment approximation,

$$C_{max} = \frac{C_L}{1 - \exp(-0.693t/T)}, \tag{4}$$

where C_{max} = maximum concentration in inspired air that could be tolerated for t minutes (mg m^{-3}),
C_L = maximum concentration in inspired air that could be tolerated indefinitely (mg m^{-3}),
T = biological half time (min).

Corresponding values of the ratio C_{max}/C_L and the ratio t/T from equation (2) are in Figure 9.4. It will be observed from Figure 9.4 that in the range where the brief period of exposure is $T/2$ or less the maximum concentration in inspired air which can be tolerated is given approximately by the relationship

$$C_{max} = 1.4C_L T/t \quad \text{(approx.)} \tag{5}$$

that is, the maximum tolerable concentration in inspired air is inversely proportional to the period of exposure to it. On the other hand, where the

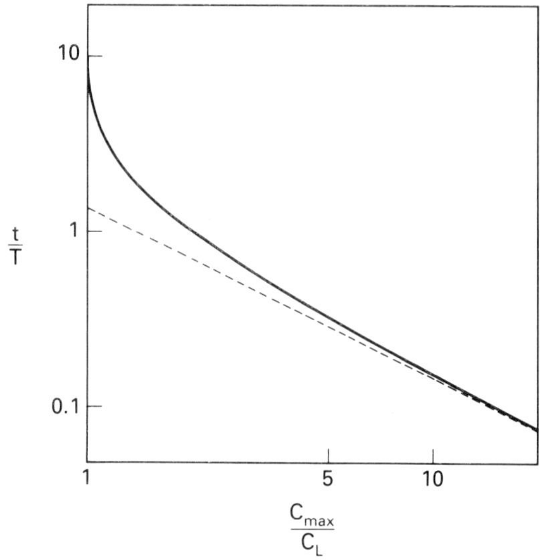

FIG. 9.4. Maximum concentration tolerable for a brief exposure. Single compartment toxico-kinetics. t = duration of brief exposure (hr), T = biological half time (hr), C_{max} = maximum concentration tolerable for time t (mg m^{-3}), C_L = maximum concentration tolerable indefinitely (mg m^{-3})

period of exposure is $5T$ or greater, the maximum tolerable concentration in inspired air is given approximately by

$$C_{max} = C_L \quad \text{(approx.)} \tag{6}$$

For the interesting intermediate or more exact values the exact equation (4) should be employed. Where the maximum concentration that can be tolerated for 8 hours per day, 5 days per week, 50 weeks per year is known it may be calculated, as will be shown shortly, that this is not sensibly different from C_L for substances whose biological half time is less than 3 hours.

Problem 9.1

Exposure is to a substance whose biological half time is 2 hours. It is known that the concentration that can be tolerated indefinitely is 100 ppm, that is, for 24 hours per day and for 7 days per week. Recommend the maximum concentration that could be tolerated for 30 minutes.

Solution

For a substance with a biological half time of 30 minutes,

$$\frac{a}{V} = \frac{0.693}{T} = 0.347 \text{ min}^{-1}.$$

Entering this in equation (4) with $t = 0.5$, and $C_L = 100$,

$$C_{max} = 628 \text{ ppm}.$$

Such a value, over six times C_L, should be used with caution. The reliability of this estimate depends upon the correctness of the assumption of single compartment toxico-kinetics over the range of inspired air concentration from 100 ppm to 630 ppm. Furthermore, other organs or parts of the body may have adverse effects which, while not present at 100 ppm for 8 hours, may be present at 628 ppm for 30 minutes. An example of such a substance would be a secondary irritant, secondary narcotic or a substance of any other kind having a very short half time in the critical organ for that effect. For such reasons, as a precautionary measure it is recommended that C_{max} should not be allowed to exceed 5 times the concentration which can be tolerated indefinitely.

The dose, D, of air contamination received by the body in time t is kCt mg, so that, rearranging equation (1) in terms of received dose, the concentration of hazardous substance in the compartment,

$$x = \frac{D[1 - \exp(-at/V)]}{at}. \tag{7}$$

It follows from this equation that for any value of at less than $0.2V$, that is when the duration of exposure t, is less than $0.3T$, the body burden would be in excess of 90% of the dose taken, as may be confirmed by substituting back into equation (7). This result may be employed to advantage when considering the possibility of adverse effects caused by a single brief exposure. The increment in body burden from a brief exposure to a high concentration cannot exceed the dose no matter how high the atmospheric concentration.

In these terms within the limitations that the body burden behaves as a single well mixed compartment, that t was no greater than $0.3T$ and the total dose over the period was no more than its tolerable maximum, the body burden could not at any time have been more than 10% above its tolerable maximum, no matter how variable the concentration had been during the period t. Consequently, by limiting the time weighted average concentration over the sampling period the body burden would be prevented from exceeding its tolerable maximum by any more than 10%. Provided only that t was no more than $0.3T$, however great the magnitude of any peaks and troughs of shorter duration they could have given rise to only minor medical consequences and even these would have been avoided if the time weighted average concentration was below 90% of its tolerable maximum. This means that although the sampling period may have completely spanned the exposure the time weighted average result may still warrant giving assurance that there have not been undue risks to health. The key requirement is that the sampling period did not exceed $0.3T$. In short, the minimum sampling period of significant medical consequence is about $0.3T$.

From the basis of a critical body burden it can thus be shown that during an isolated exposure lasting less than one third the biological half time any fluctuations about the time weighted average concentration are of quite negligible importance. This also puts a lower limit to the useful averaging period for atmospheric exposure measurements. At the other extreme, if the averaging period of exposure measurements is made too long, let us say more than 10 times the biological half time, important fluctuations will almost certainly be missed. Advantage may be taken of this conclusion when monitoring and assessing risks from organic solvents, for example, as their half time in the critical organ, the brain is of the order of minutes (Fernandez, Droz, Humbert and Caperos, 1977; Andersen, 1981; Ward, Travis, Hetrick, Andersen and Gargas, 1988; Olsen and Seedorf, 1990). This is not to deny that for purposes of identifying a source of excessive exposure there may be merit in employing short period or even grab sampling techniques associated with simultaneous time studies of the tasks being carried out.

Random excursions from the running mean

By the term random excursions is meant chance variations in exposure either side of the longer term mean in continuous processes. The limits to such excursions given in published lists of exposure limits are built partly on the notion that with some chemicals a given time weighted average concentration including major peaks during a shift is more likely to injure than is the same average concentration during substantially constant exposure, that is, in the absence of peaks (Ninth Annual Congress on Industrial Health, 1950).

There are several quite different approaches in published lists of exposure limits. Excursion limits may be based upon the geometric deviation of short term averages found in processes under good control, or governed by the biological properties of the substances or may be a simple multiplier for all substances irrespective of industrial realities and biological properties.

The American Conference of Governmental Industrial Hygienists have published three exposure limits which operate simultaneously; 8-hour TLV-TWAs, 'Excursion Limits' which are three times the TLV-TWA concentration for a total of 30 minutes each work shift and a provision that at no time should the concentration exceed five times the TLV-TWA. These multiple limits are based on the moot concept that if a process displays a greater geometric standard deviation than 3.13 it is not under 'good' control (American Conference of Governmental Industrial Hygienists, 1990). The German list of exposure limits has included a variety of limits to 'Short-Term Excursions' ranging from twice to ten times the 8-hour time weighted average, based on the biological half time of the substance (Breuer and Henschler, 1975; Henschler, 1979, 1984; Bolt and Drope, 1983; Deutsche Forschungsgemeinschaft, 1985). Netherlands OELs have included health-based STELs, somewhat similar to the German system but considerably more flexible (Zielhuis, Noordam, Roelfzema and Wibowo, 1988). The UK Health and Safety Executive has recommended a general practical upper limit to the 10-minute time weighted average, being three times the 8-hour time weighted average (Health and Safety Executive, 1991). The situation regarding excursions to published limits has been in a state of flux in each of these lists and the current position should be established when the appropriate excursion limit is needed to design appropriate compliance testing procedures.

Where the purpose of the sampling is to produce the best risk assessment the aim might be to obtain results which reflect the second-by-second fluctuations in health risk and a single compartment toxico-kinetic approach might be considered. In order to make sampling results of exposure to different substances equally sensitive to excursions of importance to health it would be necessary for the averaging period for sampling different substances to range from several months for some substances to a few seconds for others. On the other hand even if such a second-by-second measurement were feasible it could not easily be matched with a second-by-second response in the engineering controls. The speed with which adverse conditions can be rectified is strictly limited by technological and cost constraints. Similarly, a reference period of several months has practical disadvantages. With these considerations in mind, in-house exposure limits in the form of 8-hour time weighted averages are recommended in this book for the purposes of risk management rather than risk assessment, that is, for what are essentially practical reasons of control. The following considerations are also pertinent:

- Normal variability is assumed in the setting of 8-hour time weighted averages, but as the geometric standard deviation increases so does the arithmetic mean. Time weighted averages of samples from lognormal distributions are highly correlated with the frequency and magnitude of short term peaks. A constraint on the arithmetic mean of a lognormal distribution automatically provides a strong constraint on peak values.

- Human evidence that random variability *per se*, on a time scale of 10–15 minutes, causes adverse health effects is very sparse. Few data from experimental toxicology have been published on this aspect of exposure limits.

- The most convenient unit period for personal sampling is one work shift, equal to the unit length of time employees are at work. In continuous processes additional, supplementary exposure limits for shorter sampling periods make for progressively greater testing costs.The means of engineering control generally available in industry are not very responsive to unexpected changes on a time scale shorter than one shift.

- The variation in concentration between one job and another or between one place and another is greater in importance than the variation between one time and another. Priority in the allocation of atmospheric sampling resources should follow the greatest sources of variation.

For the aforementioned reasons it is doubtful whether limits to short term, purely random excursions are warranted.

Systematic or cyclic variations in exposure

By 'systematic' variations is meant exposure with a distinct pattern, usually following a cycle of operations. This type includes cyclic exposures of various periodicity. The work cycle could include brief inhalation of high concentration of air contamination with no other exposure during the day or brief inhalation of a high concentration during prolonged exposure to low concentrations. Crane driving is a job which may give rise to episodes of this kind (Figure 9.1b). However, these patterns usually occur when the job of an employee includes visiting a high concentration locality once or twice during a work shift, possibly for maintenance or inspection purposes. Examples of maintenance duties of this kind might be cleaning down a dusty process, filter cleaning or a degreasing job. Inspection duties may be any of a great variety of semi-automatic processes which are segregated in rooms set aside for the purpose and require occasional inspection and adjustment. The question repeatedly comes up as to what exposure limits would be reasonable to impose, should respiratory protective equipment be supplied for use during such episodes and if so, what degree of protection should be provided.

Another example is illustrated by the following scene. Suppose there were two rooms in Factory A, one in which the atmospheric concentration was twice the 8-hour time weighted average exposure limit and another in which the concentration was one half the 8-hour time weighted average exposure limit. Should employees be told to divide their time each shift between 4 hours in one room and 4 hours in the other?

The answer to this problem is first to recall that practical engineering control measures should always take priority over personnel control, as a matter of policy, so as to reduce atmospheric concentration rather than reduce duration of exposure. The policy should be implemented, therefore, by reducing the concentration in the second room in Factory A to below the value in the exposure limit. Only when this is not reasonably practical should personnel control measures be considered. It is at this stage that the various means of health protection be considered in relation to associated health risks, if any.

Suppose the employees in another factory, Factory B, spent most of their time undertaking tasks in which the atmospheric concentration was one half the exposure limit and there was one task, lasting 30 minutes, when the concentration was typically six times the exposure limit. Should this be condoned? Again, the answer to the question lies first in technical practical considerations, rather than in considerations of health risks. The weapons in the armoury of the engineering control specialists are equally effective for 30 minutes as for 8 hours, sometimes more so. Thus endeavours should be made to reduce the concentration in the 30-minute task to below that in the 8-hour exposure limit. The consequent reduction in 8-hour time weighted average is a bonus.

It is therefore recommended that discontinuous exposure is thus handled by having the same time weighted average concentration for the limit for each task with a distinct exposure wherever it is reasonably practical to do so. Only when it is not reasonably practical to control the concentration on undertaking specific tasks should consideration be given to treating the exposure pattern as out of the ordinary, that is, as an extraordinary work schedule, below.

Extraordinary work schedules

Work schedules differing markedly from 5 days each week and 8 hours each day are increasingly common. In parallel with these trends there is a growing need for exposure limits adapted to suit altered, unusual or extraordinary work schedules. This problem was introduced by the author in Britain and by Brief and Scala in USA (Roach, 1966; Brief and Scala, 1975, 1986).

Suppose, first, an employee is exposed in a work cycle which differs from the standard, whether through regularly working more hours per week or through working a complicated long-shift system. The employee's lifetime maximum body burden would probably become greater than would have occurred with the standard work cycle and the exposure limit should be adjusted downwards proportionately (Hickey and Reist, 1977; Roach, 1978). The amount of the necessary adjustment may be worked out by first considering the body burden at the end of some particular shift in the work cycle.

Sometimes the work schedule involves exposure for a brief period only, shorter than a work shift, being zero at other times. This might happen, for example, when an employee's job, normally without significant exposure, entails the employee going into another room or building so that exposure is for a short period only. Because the period of exposure is limited to a small fraction of the work shift in this instance a somewhat higher atmospheric concentration than the 8-hour exposure limit could perhaps be inhaled without undue harm. The duration of exposure that could be tolerated for this limited period may be estimated from the biological half time of the substance.

An employee's normal attendance on a standard 7-day work cycle might give rise every week to repeated exposure to a constant concentration of air contamination for 5 work shifts of 8 hours each day, Monday to Friday, followed by 2 days without exposure, Saturday and Sunday. In these circumstances the body burden of the substance or metabolite in the critical organ would be at its maximum each week when the employee finishes work on Friday, while the minimum would be attained immediately before the employee starts work each Monday. However, suppose, more realistically, the employee is exposed in a 7-day work cycle but the pattern of exposure differs somewhat from the standard 5 shifts of exactly 8 hours each shift. Or suppose the work cycle is changed more drastically to a 3-week work cycle or even longer. The question arises of whether the maximum body burden under this regimen would then be higher or lower and by how much.

The concentration of hazardous substance in the body, x_e, at the end of a shift is due in part to exposure in work shifts which began less than one work cycle previously and in part to exposure in work shifts which began more than one work cycle previously. The mathematics is easier to follow if these two parts are considered separately.

On exposure to concentration C mg m^{-3} for t hours the concentration of hazardous substance in a single compartment model of the body is $(kC/a)[1-\exp(-at/V)]$, and when exposure has subsequently ceased for time t' the concentration, x, falls exponentially:

$$\frac{kC[1-\exp(-at/V)]\exp(-at'/V)}{a} \tag{8}$$

Suppose every work shift is m hours in duration. The concentration of hazardous substance in the body remaining from exposure during that particular work shift in the work cycle which ended n hours previously is

$$\frac{kC[1-\exp(-ma/V)]\exp(-na/V)}{a}, \tag{9}$$

so that the part of the body burden concentration remaining from all work shifts which began less than one work cycle previously is

$$\frac{kC[1-\exp(-ma/V)]}{a}\cdot\Sigma\exp(-na/V), \tag{10}$$

including just those work shifts which began less than one work cycle previously.

With regard to the remaining part of the body burden, the property is employed that where the work cycle is l hours duration x_e repeats itself at intervals of l hours. Thus the part of the body burden concentration due to exposure in work shifts which began more than one work cycle previously is

$$x_e\exp(-la/V). \tag{11}$$

Summing the two parts,

$$\frac{kC[1-\exp(-ma/V)]}{a[1-\exp(la/V)]}\cdot\Sigma\exp(-na/V). \tag{12}$$

In a standard work cycle exposure is idealised as occurring at the same time on five successive days, Monday to Friday, followed by 2 days without exposure. Furthermore, exposure each work shift is assumed to last for a continuous period of 8 hours, the concentration remaining at the same level throughout the 8 hours. Under such conditions and after many work cycles have been completed the body burden fluctuations would settle into a standard pattern each week in which the body burden is at its highest at the end of the Friday work shift. For the present purposes this is taken as the point in time marking the end of the cycle. Working from this point, backwards in time, $m = 8$, $n = 24$, 48, 72 and 96, $l = 168$. Inserting these values in equation (12) it is found that the body burden concentration at the end of a standard work cycle reduces to

$$\frac{kC[1-\exp(-8a/V)][1-\exp(-120a/V)]}{a[1-\exp(-168a/V)][1-\exp(-24a/V)]}. \tag{13}$$

The ratio of the body burden at the end of a standard work cycle to the body burden at the end of a particular shift in some other work cycle will be denoted R. Combining (12) and (13),

$$R=\frac{[1-\exp(-8a/V)][1-\exp(-120a/V)][1-\exp(-la/V)]}{[1-\exp(-168a/V)][1-\exp(-24a/V)]\times[1-\exp(-ma/V)]\cdot\Sigma\exp(-na/V)}. \tag{14}$$

In this formula, the work shifts included in the final term of the denominator are those in one complete work cycle. The work cycle is the one prior to the shift end in question and

l = total number of hours for a complete work cycle

m = number of hours duration of each work shift

n = interval between the end of some previous shift in the cycle and the end of the given shift (hours)

e = the exponent of natural logarithms, 2.718

$$\frac{a}{V} = \frac{\ln 2}{T} = \frac{0.693}{T}.$$

The minimum value of R occurs at the end of that shift in the cycle for which Σn is a minimum. With this in mind, the systematic way to proceed is first to calculate Σn for each shift in the cycle to find which gives the least value. R is then calculated from equation (14). The value of R thus obtained is the factor by which the standard exposure limit for exposure 8 hours per day, 5 days per week would have to be multiplied to obtain an exposure limit which would give the same maximum body burden for the particular work schedule in question.

Problem 9.2

Exposure occurs on a 348-hour work cycle of 7 × 12-hour work shifts separated by the following intervals:

Shift	Hours at work	Hours off work	n
A	12	12	252
B	12	60	228
C	12	12	156
D	12	12	132
E	12	72	108
F	12	12	24
G	12	84	0
			900

Set up a procedure for calculating R.

Solution

The first step is to find the work shift in which the maximum body burden occurs. Taking the sequence as presented in the above table, the number of hours, n, from the end of each shift to the end of shift G are calculated in reverse order, starting from the last shift. Thus,

Shift G, $n = 0$,
Shift F, $n = 12 + 12 = 24$,
Shift E, $n = 12 + 72 + 24 = 108$,
Shift D, $n = 12 + 12 + 108 = 132$,
. . . and so on.

Adding the values in the last column gives Σn when shift A is taken as the start of the work cycle. The calculation of Σn is then repeated for the cases when shift B, C, D, E, F or G are taken as the start of the work cycle. This shows:

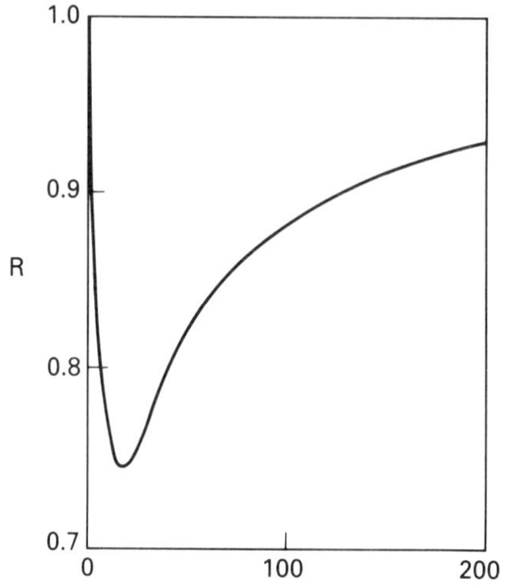

FIG. 9.5. Body burden ratio R, for the standard work cycle as compared with the work cycle in Problem 9.2.

$$R = \frac{\text{Body burden under the standard work cycle}}{\text{Body burden under the new work cycle}}$$

	Shift at start of work cycle						
	A	B	C	D	E	F	G
Σn	900	1224	1044	1200	1020	840	1080

The least value of Σn occurs when shift F is taken as the start of the work cycle from which it follows that the maximum body burden occurs at the end of shift E. Accordingly values of n are calculated for this shift. These are 0, 24, 48, 120, 144, 240, 264 hours. Equation (14) is therefore entered with the following values:

$$l = 348,$$
$$m = 12,$$

and

$$\begin{aligned}\exp(-na/V) = 1 &+ \exp(-24a/V) + \exp(-48a/V) \\ &+ \exp(-120a/V) + \exp(-144a/V) \\ &+ \exp(-240a/V) + \exp(-264a/V).\end{aligned}$$

The relationships between R and T thus calculated are shown graphically in Figure 9.5 where it can be seen that for such a work cycle the appropriate value of the multiplying factor R would be about 0.75 for substances whose biological half time is 10 to 20 hours. On the other hand R is between 0.9 and 1.0 for substances whose biological half time is less than 2 hours or is in excess of 100 hours.

As illustrated by the above example, substances which have a very short biological half time ($\ll$5 minutes) will have the occupational exposure limit based on such effects altered very little when the work schedule is changed in this way. Examples of such substances are those associated with irritation of the upper respiratory tract or odour. Substances which have a very long biological half time ($\gg$1,000 hours), such as mineral dusts, will have the exposure limit based on such properties modified in proportion to the average hours worked per week. With such substances, the duration of any practical work cycle is short in comparison with their biological half time, and the exposure limits will not be altered provided the average hours worked per week remain at 40 (Hickey and Reist, 1977, 1979; Roach, 1978; Hickey, 1980).

When evaluating complex work schedules in which work shifts or periods of exposure are of unequal duration equation (14) may still be employed by first dividing the work shifts into segments of equal duration, such as segments each of 2 hours, 1 hour or 0.5 hour duration. Each segment is then treated as a separate work shift for the purpose of entering equation (14).

Problem 9.3

The regular work cycle is 5 shifts of 8 hours worked per week but the job is one in which the period of exposure to hazardous substances is for a fraction of the work shift according to the following cycle each week:

Monday	8.00 hrs to 10.00 hrs
Tuesday	8.00 hrs to 10.00 hrs
Wednesday	6.00 hrs to 10.00 hrs
Thursday	6.00 hrs to 14.00 hrs
Friday	8.00 hrs to 10.00 hrs

Set up a work sheet for n, the number of hours between the end of previous exposure periods and the end of the present exposure period. Calculate the factor R by which the standard exposure limit would have to be multiplied to give the same maximum body burden for the new exposure cycle.

Solution

Inspection of the exposure cycle shows that the work shift in the cycle in which the maximum body burden occurs will be Thursday each week. This may be verified, if desired, by splitting the work shifts of exposure into unit exposure periods of 2 hours each (Wednesday, 6.00–8.00, 8.00–10.00; Thursday, 6.00–8.00, 8.00–10.00, 10.00–12.00, 12.00–14.00) and determining Σn for each of the nine unit exposure periods as in problem 9.2 to see which yields the least value. A table for calculating n is set up for a cycle ending on Thursday work shift end as follows:

Shift	Hours at work	Hours off work	n
Friday	2	70	148
Monday	2	22	76
Tuesday	2	20	52
Wednesday	2	0	30
	2	20	28
Thursday	2	0	6
	2	0	4
	2	0	2
	2	18	0

Equation (14) is entered with

$$l = 168,$$
$$m = 2,$$

and

$$\begin{aligned}\exp(-na/V) = 1 &+ \exp(-2a/V) + \exp(-4a/V)\\ &+ \exp(-6a/V) + \exp(-28a/V)\\ &+ \exp(-30a/V) + \exp(-52a/V)\\ &+ \exp(-76a/V) + \exp(-148a/V).\end{aligned}$$

The relationship between R and T thus calculated for this example work schedule is shown graphically in Figure 9.6. It can be seen that for such a work cycle the appropriate value of R

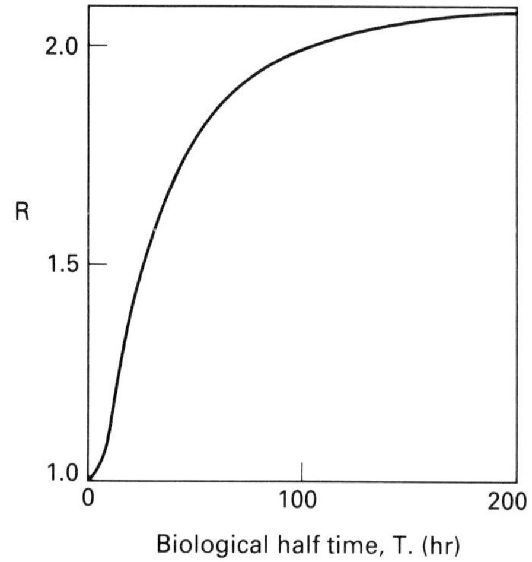

FIG. 9.6. Body burden ratio R, for the standard work cycle as compared with the work cycle in Problem 9.3.

$$R = \frac{\text{Body burden under the standard work cycle}}{\text{Body burden under the new work cycle}}$$

would be about 1.0 for substances whose biological half time is less than 5 hours. R then increases, tending to 2.22 for very long biological half times as the change in period of exposure each week is as 40:18 = 2.22.

Equation (14) is quite general and may be applied to evaluate work cycles of any duration and complexity. One-week (7-day) work cycles are most common, of course, and a wide variety of useful examples of the appropriate value of R is given in Table 9.1 for 1-week cycles.

Where the work cycle is 1 week and s shifts are worked each week, equation (14) simplifies to

$$R = \frac{[1 - \exp(-8a/V)][1 - \exp(-120a/V)]}{[1 - \exp(-ma/V)][1 - \exp(-24sa/V)]}, \tag{15}$$

where R = the ratio of the body burden at the end of a standard work cycle to the body burden at the end of another weekly work cycle

s = number of shifts worked per week

m = number of hours of exposure during each work shift

$$\frac{a}{V} = \frac{\ln 2}{T} = \frac{0.693}{T}$$

7-day work cycle.

TABLE 9.1. *Examples of factor (R) by which to multiply the exposure limit for the standard work cycle to find the exposure limit for a different work cycle which would yield the same maximum body burden. Equation (15), one-week (7-day) work cycles*

$$\text{Formula: } R = \frac{[1-\exp(-8a/V)][1-\exp(-120a/V)]}{[1-\exp(-ma/V)][1-\exp(-24sa/V)]}$$

where s = number of shifts worked per week
m = number of hours of exposure during each work shift
$\frac{a}{V} = \frac{\ln 2}{T} = \frac{0.693}{T}$
7-day work cycle.

No. of shifts per week	Exposure each shift	Hours each week	R when half-time is 1 h	10 h	100 h	1000 h
7	24	168	1.0	0.52	0.31	0.30
7	8	56	1.0	1.0	0.82	0.73
7	6	42	1.0	1.3	1.1	0.97
6	8	48	1.0	1.0	0.89	0.84
6	7	42	1.0	1.1	1.1	0.96
5	10	50	1.0	0.85	0.81	0.80
5	8	40	1.0	1.0	1.0	1.0
5	1	5	2.0	(6.4)	(7.8)	(8)
5	0.5	2.5	3.4	(13)	(16)	(16)
5	0.1	0.5	(15)	(61)	(78)	(80)
1	8	8	1.0	1.2	4.0	(6.0)
1	1	1	2.0	(7.8)	(31)	(48)
1	0.5	0.5	3.4	(15)	(62)	(95)
1	0.1	0.1	(15)	(76)	(310)	(480)

Note: A value in brackets ( ) should be used with caution. It may be employed provided there is good evidence that single compartment toxico-kinetics closely describes the behaviour of the substance in question and if it is supported by practical experience of the health status of employees under the unusual work cycle. Otherwise employ the value $R = 5$ as a precautionary measure.

Where there are five shifts worked each week but the duration of the work shift differs from 8 hours per shift, due to regular overtime, for example, equation (14) simplifies even further to

$$R = \frac{[1-\exp(-8a/V)]}{[1-\exp(-ma/V)]} \tag{16}$$

where R = the ratio of the body burden at the end of a standard work cycle to the body burden at the end of another 5 work shifts per week work cycle
m = number of hours of exposure during each work shift
$\frac{a}{V} = \frac{\ln 2}{T} = \frac{0.693}{T}$

7-day work cycle, 5 shifts worked per week.

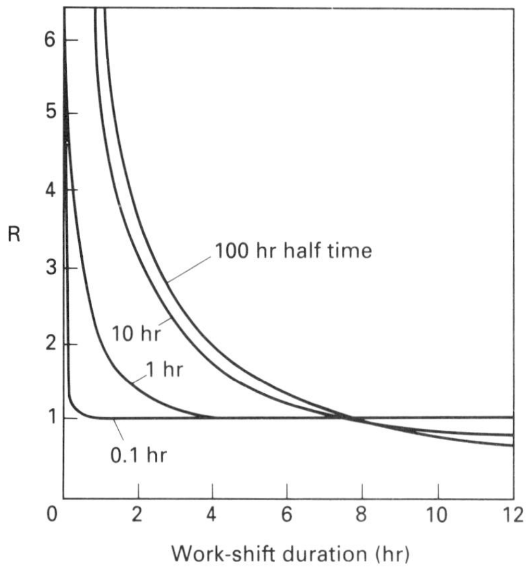

FIG. 9.7. Values of the ratio R for 5 work shifts per week work cycles and different biological half times.

$$R = \frac{\text{Body burden under the standard work cycle}}{\text{Body burden under the new work cycle}}$$

Values of R from equation (16) are shown in Figure 9.7 for different work shift duration and half time. The family of curves pass through the common point, work shift duration = 8 hours, $R_{min} = 1$.

Problem 9.4

Exposure is to a mineral dust, exposure limit 0.3 mg m^{-3} for exposure 5 days each week and 8 hours each day. Calculate an appropriate exposure limit for exposure 6 days each week and 8 hours each shift. The biological half time of the dust is in excess of 1,000 hours.

Solution

Where the biological half time of a substance is in excess of 1,000 hours the appropriate multiplying factor is closely proportionate to the inverse of the hours worked per week. The exposure limit multiplying factor is 40/48 = 0.83. The appropriate exposure limit is, therefore,

$$0.3 \times 0.83 = 0.25 \text{ mg m}^{-3}.$$

Problem 9.5

Exposure is to a substance whose biological half time in the kidney is 10 hours. Its exposure limit for exposure 5 days each week for 8 hours each day is 150 ppm to protect against kidney damage. The substance is known to cause mild throat irritation in some people at 500 ppm. An employee has a job through which the period of exposure is just 1 hour each week. Recommend an appropriate exposure limit.

Solution

Reference to Table 9.1 shows that $R = 7.8$ for a substance which has a biological half time of 10 hours. Therefore the concentration giving the same maximum body burden in the kidney on exposure to the work schedule in question would be 150 × 7.8 = 1,320 ppm. Some caution should be exercised before employing such a high multiplying factor in practice. Many hazardous substances have multiple effects. In this instance a concentration of 1,320 ppm may give rise to little risk of kidney damage but would certainly cause excessive throat irritation. An exposure limit of 500 ppm should be used for the job in question.

The multiplying factor, R, may be higher than 1.0, especially when exposure per week is less than 40 hours and the substance has a long biological half time. Some caution should be exercised before employing high multiplying factors in practice. Many hazardous substances have different effects on different parts of the body and different exposure limits would apply for each distinct effect. This is acknowledged to some extent in the list of limits published by the American Conference of Governmental Industrial Hygienists in which some substances have limits for both 8-hour and 15-minute time weighted averages. The 15-minute time weighted average or 'short term exposure limit' is not a separate independent exposure limit, rather it supplements and is additional to the 8-hour time weighted average limit where there are recognised acute effects from a substance whose toxic effects are primarily of a chronic nature (American Conference of Governmental Industrial Hygienists, 1988). An adjustment to an in-house limit made to suit one effect could allow another effect to appear which was hitherto suppressed, such as irritation or narcosis.

There are two aspects of the rationale presented here which merit careful consideration. First, the rationale is based upon the simplifying assumption of exponential accumulation and decay in a single compartment. There may be clear evidence relating to particular hazardous substances that more than one compartment is involved in the passage of the substance through the body and that at least two exponentials can be detected in washout curves (Andersen, MacNaughton, Clewell and Paustenbach, 1987). In this case the biological half time in the critical organ, if known, would be the one to employ in the formulae presented for calculating a suitable multiplying factor. Otherwise, if the critical organ is uncertain, a safe course would be to employ whichever biological half time yields the lowest value of R_{min}. It is theoretically possible to develop formulae similar to equation (14) for any network of compartments but they are seldom necessary and are beyond the scope of the present review (Andersen, Clewell, Gargas, Smith and Reitz, 1987).

Second, and in some ways a more important factor than the single compartment simplification, the exposure limit modifications have been designed to limit accumulation of substances in the body, assuming this will thereby limit the consequential adverse effects caused by those accumulated substances. This linkage may, in reality, not be a simple one. The full biological effects of a quantity of foreign substance in an organ of the body may be realised almost instantaneously, or may be limited, fixed and cumulative. Otherwise, in the course of weeks, months or years, the effects may progress even in the absence of further exposure or may regress. The significant effects may be subtle and not become evident until many years after exposure. The effect may be indirect and, for example, cause the employee to be more liable to adverse effects from other substances or infection from viable organisms in the environment. The effect of excessive exposure to a substance may have any or all of these characteristics and others besides. For these reasons, although the modifications to exposure limits calculated above will, if implemented, successfully limit the maximum lifetime body burden to no more than it would be in the standard work schedule, it is conceivable that adverse effects could arise from the somewhat different manner in which the body burden has fluctuated in reaching this maximum. When applying these adjustments to practical exposure limits and the corresponding biological exposure indices, as an additional precaution, appropriate medical surveillance to detect any adverse effects would be advisable if the work schedule is such that R_{min} is less than 0.5 or greater than 2.0.

References

American Conference of Governmental Industrial Hygienists. Threshold Limit Values and Biological Exposure Indices for 1990–1991. ACGIH, P O Box 1937, Cincinnati, Ohio 45201 USA (1990).

American Industrial Hygiene Association Toxicology Committee. Am. Ind. Hyg. Assoc. J. **25** 578 (1964).

Andersen M E. Neurobehav. Toxicol. Teratol. **3** 383 (1981).

Andersen M E, H J Clewell III, M L Gargas, F A Smith and R H Reitz. Toxicol. Appl. Pharmacol. **87** 185 (1987).

Andersen M E, M G MacNaughton, H J Clewell and D J Paustenbach. Am. Ind. Hyg. Assoc. J. **48** 335 (1987)

Bolt H and E Drope. Toxikologischarbeitsmedizinische berundung von MAK-Werten. Verlag Chemie, Weinheim, Germany (1983).

Bridges J W. The assessment of toxic hazards. In: The assessment and control of major hazards. The Institution of Chemical Engineers, EFCE Publication No. 42. Pergamon Press (1985).

Brief R S and R A Scala. Am. Ind. Hyg. Assoc. J. **36** 467 (1975).

Brief R S and R A Scala. Am. Ind. Hyg. Assoc. J. **47** 199 (1975).

Breuer W and D Henschler. Arb. Med. Soz. Med. Prav. Med. **10** 165 (1975).

Deutsche Forschungsgemeinschaft. Maximum concentrations at the workplace and biological tolerance values for working materials 1985. Report No. XXI of the Commission for the Investigation of Health Hazards of Chemical Compounds in the Work Area. VCH Verlagsgesellschaft, Postfach 1260/1280, D-6940 Weinheim, FRG; VCH Publishers, Deerfield Beach, FL, USA (1985).

Eide I. Ann. Occup. Hyg. **34** 13 (1990).

Fernandez J G, P O Droz, B E Humbert and J R Caperos. Brot. J. Ind. Med. **34** 43 (1977).

Health & Safety Executive. Occupational exposure limits 1991, Guidance Note EH 40/91. H M Stationery Office, London (1991).

Henschler D. Arb. Med. Soz. Med. Prav. Med. **14** 191 (1979).

Hickey J L S. Am. Ind. Hyg. Assoc. J. **41** 261 (1980).

Hickey J L S and P C Reist. Am. Ind. Hyg. Assoc. J. **38** 613 (1977).

Hickey J L S and P C Reist. Am. Ind. Hyg. Assoc. J. **40** 727 (1979).

Lowe L M and D B Chambers. Pollut. Eng. 36 (Nov. 1983).

Mason J W and H Dershin. J. Occup. Med. **18** 603 (1976).

Ninth Annual Congress on Industrial Health. Report of the Panel on Environmental Hygiene. AMA Arch. Ind. Hyg. & Occup. Med. **1** 601 (1950).

Olsen E and L Seedorf. Ann. Occup. Hyg. **34** 379 (1990).

Paustenbach D J. Occupational exposure limits, pharmacokinetics and unusual work shifts. In: Patty's Industrial hygiene and toxicology, Vol III pp. 111—277 (Edited by Cralley and Cralley). John Wiley & Sons, New York (1985).

Roach S A. Am. Ind. Hyg. Assoc. J. **27** 1 (1966).

Roach S A. Am. Ind. Hyg. Assoc. J. **39** 345 (1978).
Smyth H F. Arch. Environ. Health **12** 488 (1966).
Ulfvarson U. Scand. J. Work Environ. Health **13** 389 (1987).
Veng-Pedersen P. J. Pharm. Sci. **230** 101 (1984).
Ward R C, C C Travis, D M Hetrick, M E Andersen and M L Gargas. Toxicol. Appl. Pharmacol. **93** 108 (1988).
Zielhuis R L, P C Noordam, H Roelfzema and A A E Wibowo. Int. Arch. Occup. Environ. Health **61** 207 (1988).

PART III

ASSESSMENT OF HEALTH RISKS

Introduction

Formal assessments of health risks from exposure to hazardous substances are increasingly being required under law (EEC, 1988). Assessments of risks from hazardous substances in industry are required by the national law of many countries. In the UK the Control of Substances Hazardous to Health Regulations 1988 and Approved Code of Practice require an assessment to be made of the health risks created by work with any substance which may be hazardous to the health of employees (Health and Safety Commission, 1988). The assessment is further required to include the steps needed to be taken to control exposure so as to avoid these risks. An integral part of such assessments would be an appraisal of the extent of employee exposure. The exposure evaluation identifies populations exposed to the agent, describes their composition and size, and presents the types, magnitudes, frequencies and durations of exposure to hazardous substances. The results of the exposure evaluation are combined with appropriate exposure limits to make a quantitative assessment of the health risks. West German Regulations of Hazardous Substances, made under the German Chemicals Law, require assessment of the risks connected with the handling and use of all hazardous substances and the determination of the necessary precautions to avoid them (Verordnang uber gefahrliche Stoffe, 1986). An example from Canada is the Ontario Health and Safety Act 1978 under which Regulations regarding 'designated' substances require that an assessment be made of the likelihood of worker exposure and that a control programme be developed and be put into effect (Ontario Ministry of Labour, 1986). In the USA the Occupational Safety and Health Act requires every employer to provide employment free from recognised hazards (1970). Prudence dictates that this condition be confirmed by a systematic appraisal of exposure to hazardous substances at work.

Principles of risk assessment from exposure to hazardous substances at work are put forward in this Part. Identifying the employees who are exposed to them and assessing the health risks in a quantitative manner is considered first, in Chapter 10.

A central task is the appraisal of atmospheric exposure. Objective measurements are needed in environments where employee exposure approaches the relevant exposure limits. New methods of sampling and analysis may have to be developed for new substances and existing substances lacking established methods. The methods which are available for atmospheric sampling and the specialised instruments used are in Chapter 11. Where and when to take representative air samples is an art. The sampling strategy is designed to cope with the highly variable environment and the movement of employees within it. Various procedures employed to measure employee exposure are presented in Chapter 12.

Health risk surveillance is in Chapter 13. Just as engineering and personnel controls are the first line of defence against occupational disease so the first element of a health risk surveillance programme is surveillance of the controls. Supplementary to this is implementation of the principle that adverse health effects may be largely prevented by holding employee exposure below the exposure limits. This is the practice of occupational hygiene. It includes regular measurement of exposure to spot trends. Because of the variability of exposure as between employees and from time to time this necessarily entails a multiplicity of measurements. Air sampling may usefully be supplemented with measurement of biological determinants for hazardous substances or their metabolites whose biological exposure indices have been established.

Finally, this leads to consideration of the possible merits of regular medical examinations. It is theoretically possible to keep control over health risks by giving the employees medical examinations and removing from further exposure any who show early signs of disease from excessive exposure. In principle it would be necessary at least to couple this with an investigation into the cause and rectification of the faults found. Furthermore, permanent damage may be done before adverse health effects are apparent. Medical examinations are very much a last resort when other controls prove inadequate, and best kept in reserve to detect substances which have unexpected effects on health and individuals who are especially sensitive to exposure to particular substances.

The control of exposure at work has as its aim to help the control of risks to the health of employees which might otherwise occur. The control of health risks from hazardous substances involves contributions from a wide variety of disciplines; occupational hygiene, industrial medicine, engineering safety, toxicology, chemistry, personnel management and so on. The need for an integrated approach to health protection is increasingly recognised in industry. The multi-disciplinary approach and teamwork are widely advocated. The systems approach to the study of hazard control has much to commend it. The monitoring of exposure to hazardous substances may be thought of as just one of many mutually dependent elements or sub-systems of the control system. Other elements or sub-systems include the design, procurement, installation, operation and maintenance of the engineering and personnel controls; their management, the education and training of the workforce in its operation, and so on. The organisation of assessment schemes within companies is considered in Chapter 14.

References

EEC. Council Directive 80/1107/EEC, of 27 November 1980 on the protection of workers from the risks related to exposure to chemical, physical and biological agents at work (as amended by Directive 88/642/EC of 24.12.88, the 'exposure limit directive'). Official Journal of the European Communities No L 356, 24. 12. 1988, p. 74 (1988).

Health and Safety Commission. Control of Substances Hazardous to Health Regulations and Approved Code of Practice. ISBN 0 11 885468 2. H M Stationery Office, London (1988).

Ontario Ministry of Labour. Designated substances in the workplace: A general guide to the Regulations. Occupational Health and Safety Division, Publications Services Section, 5th Floor, 880 Bay St., Toronto M7A 1N8 (1986).

US Public Law 91–596, 91st Congress (84 Stat. 1590) "Occupational Safety and Health Act of 1970"

Verordnung uber gefahrliche Stoffe (Gefahrstoffverordnung-GefStoff V), Von 26 August (1986).

CHAPTER 10

Assessment Procedures

A risk assessment is the complete process by which the chance of harm or injury is estimated. In the present context the probability of occurrence of adverse health effects resulting directly from or as a consequence of exposure at work is estimated. The assessment typically includes four components: hazard identification, dose-response assessment, exposure assessment and risk characterisation (National Research Council, 1983). The assessment may be aided by medical examinations of employees.

The health risks from atmospheric exposure are most readily evaluated by making a comparison of exposure measurements against established exposure limits. With the advent of modern occupational hygiene technology this is now feasible. Exposure limits are recommended by national authorities for several hundred substances. Normally, a published limit of this kind would be adopted for a substance so listed. In-house exposure limits are set for the remainder. Examination of the plant for likely sources of exposure is undertaken in comparison with a survey of the geographical distribution of employees about the plant and their time spent exposed.

The standard procedure may include four distinct phases:

1. Identification of materials which, by their quantity and nature may pose a significant health risk from exposure at work.
2. Selection or setting of exposure limits at levels which, provided they are respected, would prevent significant risks to health.
3. Undertaking of an examination of existing plant and new plant designs, wherever there is believed to be a possibility of significant atmospheric exposure to substances of interest. The places hazardous materials are used, worked on, handled or stored and the places employees could be exposed are identified.
4. Study of the demography of exposure. Appraisal of employee exposure, its distribution between different employees and its variation in time in relation to exposure limits.

There is a certain amount of information about safety and health aspects available from manufacturers and suppliers of hazardous substances as required for example by the UK Health and Safety at Work etc Act 1974 or in the form of Material Safety Data Sheets as required by the US Hazard Communication Standard (Occupational Safety and Health Administration, 1983). Such advice may be all that is needed to supplement an employer's own information held in technical and personnel departments and enable him to make an informed judgement about whether special precautions need to be taken. Where there is uncertainty the larger companies employ the services of an occupational hygienist.

Occupational hygienists make professional judgements of the state of hygiene in work places. They make appraisals of employee exposure to chemical and physical agents sentiently, supplemented where necessary by actual measurements. This is at the heart of occupational hygiene. However, companies with full time, professional hygienists on their staff are few in number and in the smaller company the employment of a specialist in this field may not be justified. For these to undertake occupational hygiene the work has to be spread amongst other specialists (Health and Safety Executive, 1988).

Identification of sources

The majority of sources of exposure to hazardous substances are fairly obvious, although in complex situations it is sometimes hard to be sure. The first step is to consider the processes and plant item by item, following the hazardous material in its path through the area and identifying on a plan the points where contamination is probably being emitted. The sources could be ranked on a scale of 1 to 10 according to the apparent rate of emission.

The precise contribution to personal exposure made by each of several sources is sometimes uncertain. The direction of air flow in the vicinity may be measured with a hot wire anemometer or similar instrument and this may enable some sources to be ruled out. Commercially available smoke tubes or pellets are sometimes helpful to visualise turbulent flows. Sometimes it is possible to turn off a particular source temporarily, or to arrange that a particular operation be discontinued. Measurements of atmospheric concentration with and without the source in these circumstances enable the contribution from specific sources to be determined by difference.

A systematic examination of the plant would naturally be aided by a plan or drawing showing the position of the principal features to be scrutinised. In any event a drawing of some sort will be helpful, if not essential to describe the arrangement of processes, location of machines, employees and hazardous substances. To this end a simple architectural layout plan should be prepared for each work area.

Transparent overlays could be marked up with employee density, walkways, pipe-ways, conveyors, ventilation arrangements and other factors relevant to employee exposure. Another overlay should identify, list and be accompanied by a schedule showing where and when employees make key operations on the plant like sampling, adding chemicals and drawing/blowing down. A similar process should be undertaken in respect of all hazardous substances present in significant quantities (Jackson, 1979; Egan, 1979).

A valuable reference with respect to atmospheric contamination is a dossier compiled on the mechanical and natural ventilation arrangements. It should include:

- Location, size, configuration and volume flow rate from air supply openings, including provisions for air conditioning and re-circulating air, where present.
- Location, size, configuration and volume flow rate into extract openings. Hood static pressures of exhaust hoods. Static pressure upstream and downstream of air cleaners. Static pressure upstream and downstream of fans.
- Normal values of fan power and/or fan amps and location of meters.
- Location, size and configuration of louvres, windows, doors, ventilation shafts and gravity air stacks, wherever reliance is placed on natural ventilation.

Outline diagrams of the ventilation arrangements should be appended including pertinent elevations through work areas.

The check list in Table 10.1 of common plant and processes found in a wide range of industries and which are liable to emit air contamination may be useful. To this list should be added special plant and processes which are found in the particular industry under study, as, for example, coke ovens, a foundry

TABLE 10.1. *Plant and operations liable to liberate hazardous substances into work-place air*

	Type of materials handled	
	Solids	*Liquids*
Plant	Crusher	Valve
	Screens	Flange
	Transfer point	Vent
	Shaker conveyor	Storage tank
	Mixer	Mixer
	Saw	Plating tank
	Grinding wheel	Drain
	Sander	Vat
	Bagging machine	Spill tray
	Shake-out	Degreasing tank
Operations	Abrasive blasting	Spraying
	Welding	Boiling
	Soldering	Anodising
	Emptying bags	Pickling
	Crushing	Filtering
	Grinding	Paint dipping
	Sanding	Drying
	Weighing	Pouring
	Filling bins	Sampling
	Filling cartons	Filling cans
	Clearing up spills	Mopping up spills

shake-out, carding engines in the cotton and flax industry, kilns in the ceramic industry and so on. The list in Table 10.1 is divided into items of plant and processes. It is not meant to be exhaustive but such a list could be employed, for example, to give a crude priority ranking for the investigations according to the number of checked items in a given area.

Occupational exposure limits

The judgement as to whether or not atmospheric exposure is sufficient to be hazardous is usually made by reference to occupational exposure limits for the substances in question. Indeed, in many situations it is difficult to make valid and justifiable conclusions about exposure and risks in any other way. There are published occupational exposure limits recommended by national authorities only for 1,000 or so chemical substances important in commerce. An integral part of the assessment process for most substances is thus the development of suitable exposure limits. It is first necessary to find out enough about the materials being handled to assess which, if any, of its constituent substances could be hazardous in the conditions of use. Suppliers literature such as a material safety data sheet is an invaluable guide for this purpose. The substances identified there may have established exposure limits which have been published. There may otherwise be a recommended exposure limit in suppliers literature or one may be obtainable from the supplier on request. Assurance should be sought that it is suitable for the user's needs. Even so, there are many thousands of other substances without recognised exposure limits whose usage may not be great but where used constitute a potential health risk. To assess the possible health risks from air contamination by these substances in-house occupational exposure limits have to be devised and implemented.

Work place inspections for health risks

A good inspection of processes can be made on a walk-through survey. It should follow a planned route based on knowledge of the process, plant and operations. The route is designed to take in the the major sources of emission, specific operations liable to cause emissions, and jobs likely to involve heavy exposure. The immediate object of scrutinising the operations and work activities is to make an appraisal of the normal atmospheric exposure of the employees, having regard to the following aspects:

—nature of the substances present and in production
—machines in operation and equipment in the vicinity
—location of employees in relation to potential sources of exposure
—working methods
—rate of production and use of hazardous substances.

The inspections should take due account of the interaction between these different aspects and the cumulative effect of exposure to a wide variety of hazardous substances.

Making use of the odour of gases and vapours is generally to be discouraged but it can be a useful indicator in certain circumstances. If the odour of a substance can be detected it may or may not be at a concentration in air which is harmful. The odour of some substances cannot be detected until its atmospheric concentration reaches a level well in excess of the occupational exposure limit. But many substances can be detected by smell at a level well below the exposure limit (Leonardos, Kendall and Barnard, 1969). The absence of an odour is only useful for gases and vapours which have an odour threshold at or below one tenth the exposure limit. Its presence is only useful for those which have an odour threshold approaching or just above the limit. Thus, of 214 gases and vapours whose odour thresholds have been tabulated 113 are in the former category and 52 in the latter (Amoore and Hautala, 1983). The absence of employee complaints is a poor guide to odours. Employees often become so used to an odour they no longer detect it. This tolerance may be termed olfactory fatigue, desensitisation or accommodation. Whatever the term used the phenomenon is well known (Hunter, 1974). When interpreting the presence or absence of odour it should be borne in mind that generally the nose will become fatigued quickly to any odour although remaining responsive to fresh, different odours.

Closely related to odour detection is irritation of the eyes and upper respiratory tract. Over 1 in 3 exposure limits of the American Conference of Governmental Industrial Hygienists for example, are based on this criterion (Stokinger, 1970). Any irritation that is experienced during the walk-through survey is good grounds for assigning the group of employees in the area an exposure in excess of the appropriate exposure limit.

Visual examination of dusty processes is aided by a dust lamp. A lamp with a parallel beam can make clouds of fine dust readily visible by shielding the direct beam but viewing the light at an angle of 5–10 degrees to the line of the beam. Many gases and vapours can be studied with portable, direct reading instruments. These are invaluable when there is no detectable odour at levels of concentration of hazardous substances present in air below the exposure limit.

The second, but no less important objective of carrying out an inspection is to put together an inventory of emissions. The aim of this is to identify leaks, spills, releases, emissions and discharges and quantify their rate as far as possible. A record should be made of their location and of the processes involved. Special vigilance should be maintained during dismantling and rebuilding of plant and processes; a regular feature in the chemical industry. These operations are liable to sporadic, unforeseen emissions of air contamination under poor engineering control.

Regular work regimes should be studied. Features like valves, glands, flanges, joints, access doors, inspection ports and seals around vessels should be examined as should conveyors, screens and transfer points because these are particularly prone to leaks. Heaps of material spilt on the floor beneath conveyors should be noted and tabulated with a visual estimate of the weight in each pile. The assessor should be on the look-out for process changes, breakdowns, staff increases and shortages, besides technical factors such as fumes from hot processes, including discharges from furnaces, ovens and vehicles. Work involving entry into tanks, chambers, pits, vessels, pipes and other confined spaces should be considered for a possible accumulation of toxic air contamination as well as flammable vapours and, of course, oxygen deficiency. Work for which respiratory protective equipment is provided should be examined to confirm that masks, respirators and other safety equipment are worn when required and that the protection factor afforded by the equipment provided is adequate in relation to the air contamination present.

On completing the work place inspection the recommendations will fall into four classes corresponding to exposure classes of employees in those work places.

Class 1 [$<0.1 \times$ exposure limit]. Where it is judged that the exposure of a given job-exposure group is such that no action is warranted except, perhaps, for deciding the necessary re-assessment interval the job-exposure group is in Class 1.

Class 2 [0.1–1.0 $\times$ exposure limit]. Where it is judged that exposure is not excessive but it is advisable to institute surveillance of air contamination for a job-exposure group, exposure is in Class 2.

Class 3 [$>1.0 \times$ exposure limit]. Where it is judged that exposure is excessive and will continue to be so there is little to be gained by further measuring. Priority goes to instituting remedial measures forthwith. A job-exposure group in such an environment is in Class 3.

Class 4 [uncertain exposure]. The work places of those groups in Class 4, the uncertain class, should be given priority for a ventilation analysis as this may remove the uncertainty and allow confident classification.Where exposure classification of a job-exposure group is uncertain measurement of employee exposure should be put in hand.

Representative employees from similar exposure groups

For the purposes of simplifying assessment of the health risk to employees there is distinct advantage in establishing groups of persons with similar exposure. In many cases, the risks to each individual can be reliably determined by considering groups with similar working characteristics and concentrating on a few employees who are representative of the limited range of exposures typified by the group, from highest to lowest.

Employees sharing the same work area may have the same exposure even though they are doing different work and have different job titles. Personnel records in a company are created for administrative purposes. They are closely related to pay rates and are rarely satisfactory for delineating employees with distinctive exposure. There is considerable inter- and intra-company variation in job definitions and job coding (Gamble and Spirtas 1976; Marsh, 1987).

Within a defined area such as a building, floor, floor area, hall or room there might, of course, be different plants or processes where employees doing the same work have a very different exposure. Sometimes particular individuals on a plant find themselves regularly allocated special tasks giving rise to them also having a distinctive exposure.

Job-exposure groups

Rarely do employees all do exactly the same work and those that do rarely have the same exposure. The atmospheric exposure of a given employee comes from up-wind and may be poorly correlated with the work the employee is doing or the hazardous substances he/she handles. Employees who work within a specified area on a large site may or may not have the same exposure to hazardous substances and those who have the same exposure may or may not have the same job title (Gamble and Spirtas, 1976). The concept is illustrated in Figure 10.1. From the point of view of engineering control the classification of employees according to work area, indicated by the top circle has distinct advantages since most engineering controls are applied to buildings and machines fixed in place. From the point of view of personnel control the management of the individuals would be best aided by classification according to job title, the bottom left circle. Those sharing the same job title and work area are in the overlapping portion. From the point of view of exposure to hazardous substances the classification would allocate employees differently again, the bottom right hand circle. The employees sharing the same work area, job title and exposure are represented by the shaded portion where the three circles overlap. Such classifications have important advantages for organising the assessment of health risk and control of employee exposure in larger companies. Classification of employees by exposure may under some sampling schemes be imprecise and at best semi-quantitative (Woitowitz, Schacke and Woitowitz, 1970; Corn and Esmen, 1979; Esmen, 1984; Boleij, Heederik and Kromhout, 1987; Kromhout, Oostendorp, Heederik and Boleij, 1987). Such imprecise groupings are not unacceptable, but care must be taken to sample right across each such group to be sure to cover the full range of actual exposure (Rappaport, 1991).

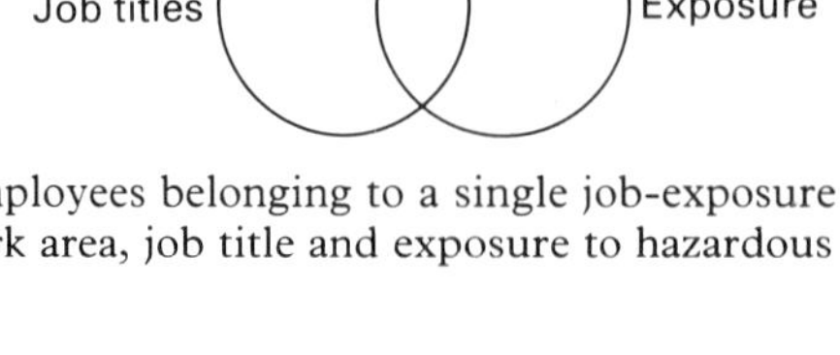

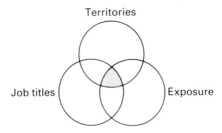

FIG. 10.1. Employees belonging to a single job-exposure group share the same work area, job title and exposure to hazardous substances.

A 'job-exposure group' includes all those undertaking work of a given type, within a given work area, associated with exposure to similar hazardous substances and to a similar level of exposure. For practical purposes the most ready means of defining the type of work of employees is by job title. The group exposure boundaries should preferably be defined uniformly, in terms of a specific range of exposures. The wider the range the fewer the number of different job-exposure groups and this will greatly simplify the task of risk assessment and control. But if the range is made too large the grouping will lose much of its purpose. Considerable simplification of the organisation of the monitoring of atmospheric exposure is obtained by locating the group exposure boundaries at a multiple of the relevant exposure limits. In the present system, advocated here, the boundaries of the groups are set fairly wide apart but defined specifically in these terms.

A list of current employees in a common job-exposure group has to be specially compiled for each work area. An important feature of this procedure is nesting of jobs within the physical boundaries of individual plants. This has advantages for exposure control purposes. Engineering control procedures in different plants rely on different items of equipment and the organisation of personnel control is generally by plant. Thus, taking each work area and using best judgement a list would be compiled of job titles of those employees expected to have a similar exposure to hazardous substances, not necessarily identical. No two individuals have identical exposure. Ancillary personnel such as cleaners, maintenance and repair crew, research and development staff, crane drivers, storekeepers, and so on should be included as appropriate.

This author recommends that employees are initially assumed to have atmospheric exposure in the band 0.1–1.0 × exposure limit and are assigned a different job-exposure group when the employee's atmospheric exposure is firmly expected and/or actually found to be more than the exposure limit or less than one tenth the exposure limit. All employees exposed to less than 0.1 × exposure limit could be thought of as falling in one exposure class. These are persons whose risk to health from atmospheric exposure is insignificant. Employees who have higher exposure than this could be conveniently divided into small job-exposure groups according to their work area. Narrower or even wider exposure bands could be employed in particular circumstances. Employees on every shift round the clock should be studied. Employees working on different shifts may or may not have the same exposure.

Aside from job-exposure groups for a simple assessment, some of the issues to be decided when drawing up groups in different ways for special investigations are illustrated by considering a partly fictional example, below. This was put together in such a way as to combine data which have been found in a variety of actual situations.

Problem 10.1

There are 150 employees working in one large room, where they are exposed to substance X. Each employee has one of five different job titles. The results of an exposure reconnaissance showed that the number of employees in each exposure class was as follows:

	No. of employees in exposure class			
Job title	Class 1 (<0.1 × OEL)	Class 2 (0.1–1 × OEL)	Class 3 (>1 × OEL)	Total
A	2	54	4	60
B	10	20	0	30
C	1	25	4	30
D	18	2	0	20
E	1	1	8	10
Total	32	102	16	150

One conclusion is that employees with job titles A, C and E have a particularly wide range of exposures, covering at least tenfold. Reasons for this should be sought. Most employees with job title D have insignificant exposure, Class 1, and most employees with job title E have excessive exposure, Class 3.

The issues to be decided are

(a) A sampling strategy for deciding on improvements to engineering control.

The whereabouts of those with excessive exposure, Class 3, especially those with job title E gives the best lead. The principle causes of excessive exposure may be obvious. Excessive exposure may be clearly associated with specific tasks, plant, processes, operations or locations. Remaining doubts may be dispelled by following up with area sampling in relation to suspected sources of contamination. Improved engineering control has top priority to remove the risk to health in such a job.

(b) Which employees should be considered for personnel control measures.

Employees identified as being exposed above the exposure limit, Class 3 exposure band, should, in such a case, immediately be provided appropriate respiratory protection for use when they are so exposed. All employees having the same job title, E, should be included. The respirator use should not be discontinued until adequate engineering controls have been installed, tested and found satisfactory.

(c) The best grouping for continuing surveillance of employee exposure.

Those employees in exposure Class 2 are eligible for continuing surveillance. Of these, the ones having job titles A and C could be grouped together for administrative convenience.

Those employees having job title D should first be investigated in relation to the engineering control provisions, as there is a distinct possibility with such a distribution of exposures that the employees in Class 2 exposure band are ones who work in a particular region.

Those employees with job title E should be considered for continuing surveillance after installation and commissioning of new controls.

Itinerant employees

Itinerant employees whose work takes them to different work areas and buildings, possibly covering the whole site, should be treated as such. It should be appreciated that their exposure can be onerous to assess as it may include exposure to very different substances and levels of atmospheric concentration.

The demography of exposure

The presence of a significant risk to health from exposure to hazardous substances used on a work activity or process will obviously depend not only on how hazardous the substance is but also on exactly where the employees work and the magnitude of their exposure there. The jobs of employees in the chemical industry often entail them continually moving from one place to another about a works or process. This is especially true of the numerous employees engaged in repair and maintenance, process operators, fitters, electricians, plumbers, welders and the like. Another common example is the supervisor who leaves his office to inspect the work at regular intervals.

There will not be significant exposure everywhere inside the factory fence. Offices, kitchens, general stores and the like will often be rapidly ruled out. Nevertheless the assessor should not dismiss jobs without thought. For example, solvent stores are sometimes the worst places for vapour accumulation and need special attention. Experience repeatedly reveals health risks resulting from exposure in the most unlikely places. Indeed, a totally risk free environment is unattainable.

The nature and degree of the risks to health have to be assessed. This is a function both of exposure and toxicity. The nature and degree of the different kinds of toxicity possessed by a substance need to be taken into account. Moreover special precautions are necessary if the substance displays some degree of mutagenicity, carcinogenicity or reproductive toxicity. All substances can be harmful if taken in excessive quantities, even food and drink. Assessment of risks to health from hazardous substances thus acknowledges how employees are distributed about a works, how long an individual stays in each place, ventilation of the workshops, respiratory protection, protective clothing worn, the way materials are handled, the quantity to which the employees can thereby become exposed and the hazardous nature of the substances involved.

When making an assessment it is helpful to assign employee exposure levels into one of four broad classes defined in relation to the relevant exposure limits:

Class 1. Likely to be less than 0.1 × exposure limit.

Class 2. Likely to be between 0.1 × limit and 1.0 × limit.

Class 3. Likely to be more than 1.0 × limit.

Class 4. Uncertain and in need of more data.

Grounds for assignment to Class 2 may be found as follows. Employee exposure will normally be assigned this class only on grounds of earlier measurements by 'personal' sampling in that particular work area. 'Area' sampling often yields results lower than those from personal sampling so that such results should be treated with some caution. The full tenfold range of atmospheric contamination concentration encompassed by exposure Class 2 in the above table may also need explanation. About 1 in 15 ACGIH TLV-TWAs, for example, are changed every year. They are nearly always lowered substantially, often by 50% and sometimes by as much as 90%. This is characteristic of exposure limits from whatever authority. It would therefore be prudent to have exposures monitored over a broad range such as is recommended for Class 2 exposures.

To summarise the findings the approximate number of employees in each class should be estimated for each room, floor, building or other logical work area. The duration as well as intensity of exposure should be determined as should the shift system. Where exposures are to mixtures of two or more hazardous substances and the exposures are simultaneously in excess of 0.1 × limit it is advisable to take account of their combined effects. This is discussed in Chapter 8. Where the duration of exposure differs from 8 hours in the day or 40 hours in a week because of shorter than normal work-shifts or overtime working, a pro rata adjustment could be made to the limits up or down. Where the biological half time of a substance is known more exact adjustments may be made for hours of exposure to various substances as discussed in detail in Chapter 9.

A helpful guide when deciding the appropriate class can be exposure data about similar substances and processes elsewhere. Care must be exercised with this approach as the data may be limited in scope and represent a very specific scenario. The assessor must judge whether the data fully represent the substances, processes and state of exposure control under scrutiny. There might be some important characteristic where there is a poor match (Craig Mattheissen, 1986).

Insignificant exposure—Class 1

Work being undertaken by a single individual or job-exposure group could be assessed as being conducted in such a place and for such a duration that the concentration of substances in inhaled air and/or its duration is insignificant. Illustrations which might satisfy this requirement are:

(a) During work operations the points of a process at which hazardous substances may be given off are totally enclosed or otherwise barred from access by the job-exposure group.

(b) Work operations with hazardous substances are undertaken in a glove box or equivalent, kept under negative pressure of at least 3 mm water gauge.

(c) The points of a process at which hazardous substances may be given off are partially enclosed and the enclosure is exhausted at all times. The extracted air volume flow rate is maintained at such a level that the maximum credible 10-minute time weighted average concentration in air inhaled by any employee in the job-exposure group is less than half the 8-hour time weighted average exposure limit of the substances.

    Note that the maximum 10-minute time weighted average concentration in the vicinity of an employee is usually 3–5 times the 8-hour time weighted average but may be 10 times this for an individual employee close to a source of air contamination. Air inhaled by an employee in a highly variable environment may be best measured by sampling through a probe whose entrance is located close to the cheek.

(d) The duration of exposure to hazardous substances is less than 2 minutes each work-shift, during which time the maximum credible concentration of hazardous substances in inhaled air is less than 10 times the 8-hour time weighted average occupational exposure limit. Such possibilities occur, for example, when exposure is limited to a small room which is visited rarely or the employees are mostly in a clean control room or office which they rarely leave.

In making assessments of insignificant exposure particular care should be taken with regard to employees engaged in maintenance. Their work is very different from that of production workers and the nature of their work may easily lead to loss of process containment.

Note the use of the word 'credible' in the phrase 'maximum credible concentration' in examples (c) and (d) above. This term is used deliberately. The theoretical maximum possible concentration that could be physically sustained, given the necessary extraordinary circumstances may be 10^4-10^6 ppm gas or vapour and 10^3-10^5 mg m^{-3} solids. In reality the possibility of this happening is so remote as to be virtually impossible. Rather, a commonsense judgement about the credibility of a set of circumstances is advocated. A worst case situation is envisaged whilst remaining both reasonable and realistic about the chances of its occurrence.

Medicals

Industry is in a rapid state of change with the introduction of new technology. The chemical industry introduces new chemical products daily. It is especially important with hazardous substances to keep a constant watch for any deterioration in the control over health risks and for the emergence of hitherto unsuspected adverse health effects. Medical examinations provide the means for this (International Labour Organisation 1988). There are major considerations which caution against reliance on the exactitude of atmospheric exposure limits, however seemingly well founded.

- Some individuals are intrinsically susceptible to particular hazardous substances.
- Some individuals are by their nature, disposed to choose work in the highest concentrations, to work with a high energy expenditure and breathe at high minute volume.
- There is a largely unknown possible synergism between different hazardous substances taken into the body at the same time or sequentially.
- There is a largely unknown contributory dose from skin absorption, from ingestion and from exposure away from work.

If harmful concentrations are prevalent, occasional reports of sickness or complaints of minor ill effects will be received. Adverse effects diagnosed by medical examinations of employees provide a means for obtaining direct evidence of possible health risks from exposure to hazardous substances. Account needs to be taken of the effects from non-occupational causes and from previous employments. Linkage of the prevalence of specific adverse effects and their severity with occupational exposure in present employment enables a quantitative assessment of health risks at work.

Some medical examinations at work are statutory or form part of the terms of employment although most are purely voluntary. Pre-planned medical examinations at work can be the best source of data on occupational health hazards. A 'passive' medical service whereby the medical officer sits in the surgery waiting for the cases to come in of their own volition is a poor substitute.

Medical examinations conducted to detect adverse effects of exposure on health have little in common with those for job placement, general health checks or medical retirement. Also, therapy plays little part in such examinations, patients with abnormalities being referred to their general practitioner. The pre-employment examinations, commonly used for confirming fitness for work, do not provide the required type of evidence, although when coupled with subsequent examinations for specific effects may help in a subsidiary fashion to confirm deleterious effects from hazardous substances (Taylor and Raffle, 1973). It should also be cautioned that sickness absence records are often considered as possible indices of occupational illness but they are notoriously unreliable for such purposes. Sickness absence is strongly correlated with sex, age, status and a host of other factors which are not related to exposure to hazardous substances (Bellaby, 1989). Notwithstanding these remarks, medical examinations of employees conducted after they return from an extended spell of sickness absence, say 1 month or more, are very valuable for finding adverse health effects from the working conditions.

References

Amoore J E and E Hautala. J. Appl. Toxicol. **3** No. 6. (1983).

Bellaby P. Ann. Occup. Hyg. **33** 423 (1989).

Corn M and N A Esmen. Am. Ind. Hyg. Assoc. J. **40** 47 (1979).

Craig Mattheissen R. Chemical Engineering Progress. **82** No. 4, p. 30 (1986).

Esmen N. On estimation of occupational health risks. In: Occupational and industrial hygiene concepts and methods (Edited by N A Esmen and M A Mehlman). Princeton Scientific Publishers, Princeton, New Jersey, USA (1984).

Gamble J and R J Spirtas. Occup. Med. **18**, 399 (1976).

Health & Safety Executive. COSHH Assessments—A step-by-step guide to assessment and the skills needed for it. H M Stationery Office, London (1988).

International Labour Organisation. The impact of new technology on safety and health protection in the chemical industries. Chemical Industries Committee, Tenth Session. International Labour Office, Geneva, Switzerland (1988).

Kromhout H, Y Oostendorp, D Heederik and J Boleij. Am. J. Ind. Med. **12** 551 (1987).

Leonardos G, D Kendall and N Barnard. J. Air Pollut. Contr. Ass. **19** 91 (1969).

Marsh G M. Am. Ind. Hyg. Assoc. J. **48** (5) 414 (1987).

Rappaport S M. Ann. Occup. Hyg. **35** 61 (1991).

Stokinger H. Criteria and procedures for assessing the toxic responses to industrial chemicals. In: Permissible levels of toxic substances in the working environment. International Labour Office, Geneva, Switzerland (1970).

Taylor P J and P A B Raffle. Preliminary, periodic and other routine medical examinations. In: Occupational health practice (Edited by R S F Schilling), pp. 59–72. Butterworths, London (1973).

Woitowitz H J, G Schacke and R Woitowitz. Staub-Reinhalt. d. Luft **30** 15 (1970).

Bibliography

Boleij J, D Heederik and H Kromhout. Karakterisering van blootstelling aan chemische stoffen in de werkomgeving. PurdocWageningen, The Netherlands (1987).

Egan B. Safety inspections. The New Commercial Publishing Company Limited, 4 St. John's Terrace, London W10 (1979).

Hunter D. The diseases of occupations. Little, Brown and Co., Boston, USA (1974).

Jackson J. Health and safety—The law. The New Commercial Publishing Company Limited, 4 St. John's Terrace, London W10 (1979).

National Research Council. Risk assessment in the federal government: managing the process. National Academy Press, Washington DC. (1983).

CHAPTER 11

Apparatus for Measuring Atmospheric Exposure

When designing measurements of exposure a prime consideration is the nature of the available sampling apparatus and analytical procedures. Results are liable to vary according to the exact combination of methods employed. This factor is critical with aerosols but is also important with gases and vapours. A thorough discussion of atmospheric sampling instruments and their calibration has been published by the American Conference of Governmental Industrial Hygienists (1978, 1988a).

The methods employed for atmospheric sampling and analysis divide broadly into those which yield the concentration of a hazardous substance at a moment in time and those which yield a time weighted average over a specified period. Methods for obtaining an instantaneous measurement of atmospheric concentration in work rooms are of two types; one consists of direct reading instruments, which give a reading on the instrument dial in the work area and the other is grab sampling, in which the sample of air is taken back to the laboratory for analysis. Time weighted average measurements are generally accomplished with an instrument which is running continuously for a measured period of time and accumulating contamination collected by it. Sometimes it is done by storing data from a continuous read-out meter for subsequent statistical analysis.

The easiest way to measure personal exposure to air contamination is normally to employ 'personal' sampling. Such a 'roving' sampling station is made possible using instruments in which a pocket-sized pump draws air through a tube leading to a small sampling head on the employee's lapel. A key issue with these methods is whether there is available an adequate analytical method at reasonable cost for the small quantity of contamination collected. If not, and where high flow rates become essential for analytical reasons, larger, motor driven, hand held instruments may be employed with their sampling probes held by the sampling technician in the breathing zone of the employee. Analytical methods for the small quantities collected improve progressively, over the years. The techniques of physical chemistry, including ultraviolet, infrared and visible spectrometry, gas and liquid chromatography, mass spectrometry and ion chromatography have all been engaged to increase the accuracy and specificity of laboratory analysis of air contamination samples (Grob, 1977; Bodek and Menzies, 1981; Baker, 1982; Gurka, Billets, Brasch and Riggle, 1985; Swarin and Lipari, 1983; Schomberg, 1987).

Since sampling results are so dependent upon the precise methods employed, the testing of compliance with exposure limits is closely tied to the sampling apparatus and analytical procedures used in their development. When testing compliance with exposure limits is the sole purpose of the measurements, as may be the case with enforcing agencies, there is little choice for them but to employ standard procedure (Health and Safety Executive, 1986). The effect of any deviation from standard procedure, however technically desirable, has to be established by the most careful calibration.

On-the-spot methods

The instruments most highly prized by occupational hygienists are direct reading instruments. They are especially useful for finding the sources of emission of hazardous substances, for demonstrating local transient variations in the concentration of hazardous substances in air and for walk-through inspections of control measures.

There is a class of direct reading instruments designed with laboratory use in mind. These are too heavy to move, except, perhaps, with the aid of two men and a horse. Others are sufficiently rugged to move with a trolley. Some of these, supplied with a handle on top, could be described as 'luggable'. There are still others, weighing less than 10 kg, properly described as portable. It is the truly portable instruments which have the greatest appeal to occupational hygienists.

These instruments have the advantage that they can be used like a metal detector or geiger counter to search for 'hot spots' and locate the sources of air contamination precisely. They usually have a dial calibrated in terms of atmospheric concentration of a specific substance. A useful feature is a pre-set level at which a visual or audible alarm is triggered. Such instruments can also be rigged up in a suitable place in a work-room to warn employees of hazardous situations. Other direct reading instruments also produce a strip chart recording of the results or a tape printout which can be examined at leisure. Some have a data logger which can be connected to a desk computer for ease of data analysis. Caution should be exercised in interpreting the readings literally as many direct reading instruments are non-specific.

Some instruments are more specific than others but in any event strict attention to possible interferences and calibration against known concentrations of the substance in question is needed before great reliance can be placed on the exact numerical results. At the same time, a non-specific instrument has versatility in the sense that it can be employed for several and perhaps many different hazardous substances in air. The portable instruments tend to be expensive.

Closely related to direct reading instruments are those which integrate over a period of time and can yield what is in effect a time weighted average over a few minutes. Such devices also enable measurements to be made on the spot and where a short term time weighted average is the aim the time taken is no great disadvantage.

Direct reading instruments for gases and vapours

Gases and vapours are assessed by adapting an astonishing variety of physical/chemical phenomena to operate direct reading instruments. These include notably infrared absorption, ultraviolet absorption, flame and photo-ionisation, chemiluminescence and combustibility. Gas or liquid chromatography may be employed in sophisticated systems to separate interferences and thereby improve the specificity of the methods (Grob, 1977).

Photometry at an appropriate wave length is useful for many gases and vapours. There are two common gases with significant absorption in the visible region, but they are the exception. Nitrogen dioxide is a reddish brown gas and chlorine is yellowish green. On the other hand many gaseous hazardous substances have significant absorption bands in the infrared or ultraviolet region, and this phenomenon is applied in several direct reading techniques. Good selectivity may be achieved by employing special light sources, filters or detectors with similar spectra to the gas or vapour under study. Typical direct reading instruments include infrared devices for measuring a wide variety of gases and vapours in air, such as carbon monoxide, ammonia, halothane and chloroform amongst many others (Figure 11.1). Ultraviolet devices include ones suitable for mercury vapour, ozone and other gaseous constituents.

When a volatile organic compound is introduced into a hydrogen flame burning in air a great number of ions are created in proportion to the concentration of the compound in air. Collection of the ions under the influence of an electric field and measuring the current can provide a means to detect the concentration of almost all organic compounds which enter the burner, save for carbon monoxide and carbon dioxide, which produce too few ions in this manner. This is the basis of flame ionisation direct reading instruments for total hydrocarbons (Figure 11.2). Compounds measured include methyl ethyl ketone, octane, acetone, toluene, ethyl alcohol, ethyl acetate and many others. Photo-ionisation instruments work in a somewhat similar way but they employ an ultraviolet lamp as an ioniser rather than a flame and this method is suitable for a limited number of gases and most organic vapours (Figure 11.3).

Some substances introduced into a hydrogen flame exhibit strong chemiluminescence which can be picked up by a photomultiplier. Compounds with sulphur in the molecule are a good example. They can be measured in concentrations ranging down to parts per billion in air in this manner. Examples of hazardous substances measured with chemiluminescence direct reading instruments operating in this manner are sulphur dioxide, hydrogen sulphide and carbon disulphide. At room temperature the reaction of nitrous oxide with ozone and of ethylene with ozone also exhibits chemiluminescence.

Air containing halogen compounds passed through an AC spark gives off radiation in the ultraviolet and blue region. This principle is employed for measuring the halogenated hydrocarbons with direct reading instruments,

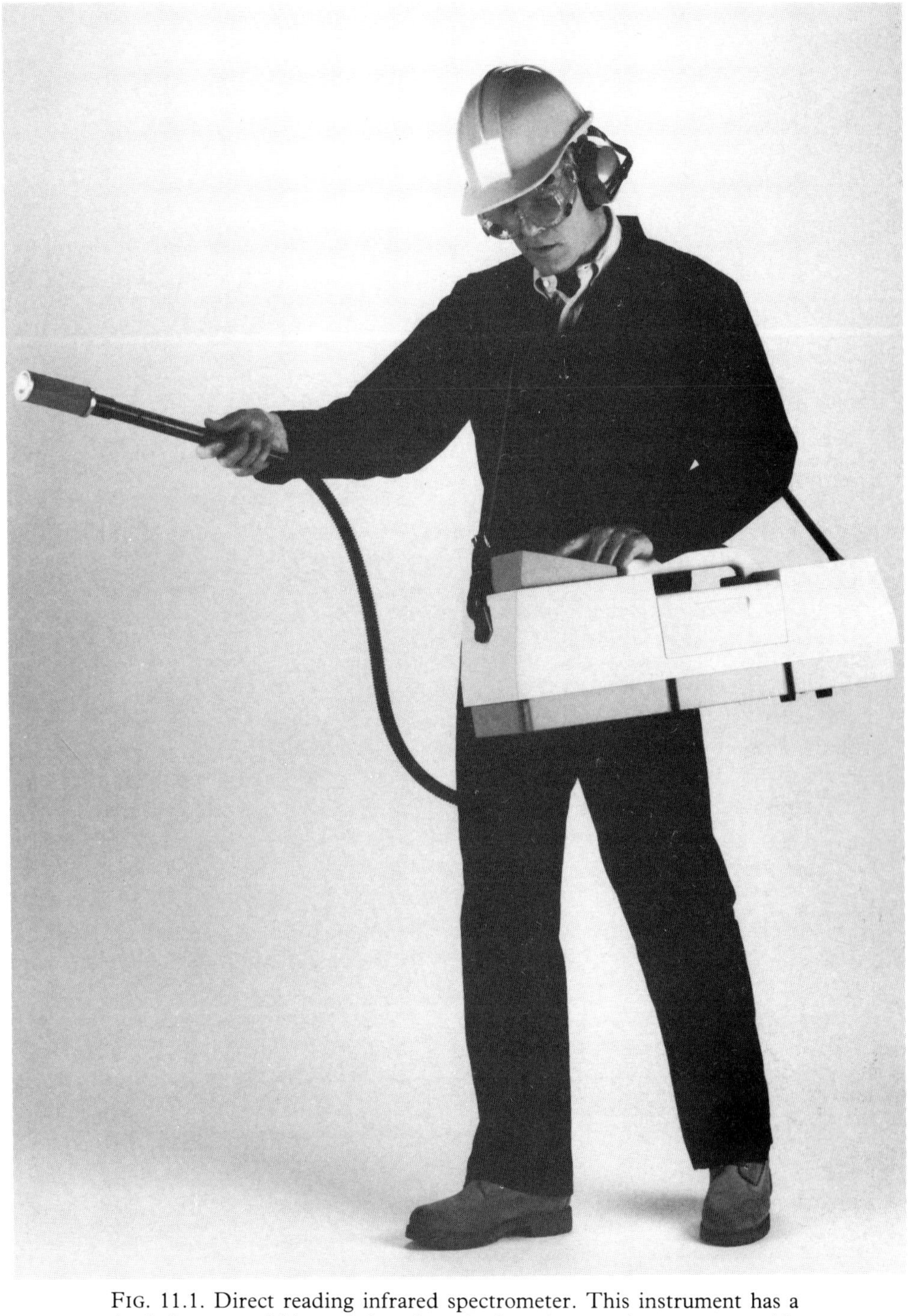

FIG. 11.1. Direct reading infrared spectrometer. This instrument has a single beam, variable path length and a broad spectral range to provide a combination of high sensitivity with good specificity. (By courtesy of Quantitec Limited)

FIG. 11.2. Organic vapour analyser being used to monitor gaseous air contamination in a burned out building. A non-specific direct reading instrument based upon flame ionisation of the vapour in air drawn through the instrument. It is sensitive to levels of the order of 1 ppm and up to 1,000 ppm on logarithmic scales. (By courtesy of Quantitec Limited)

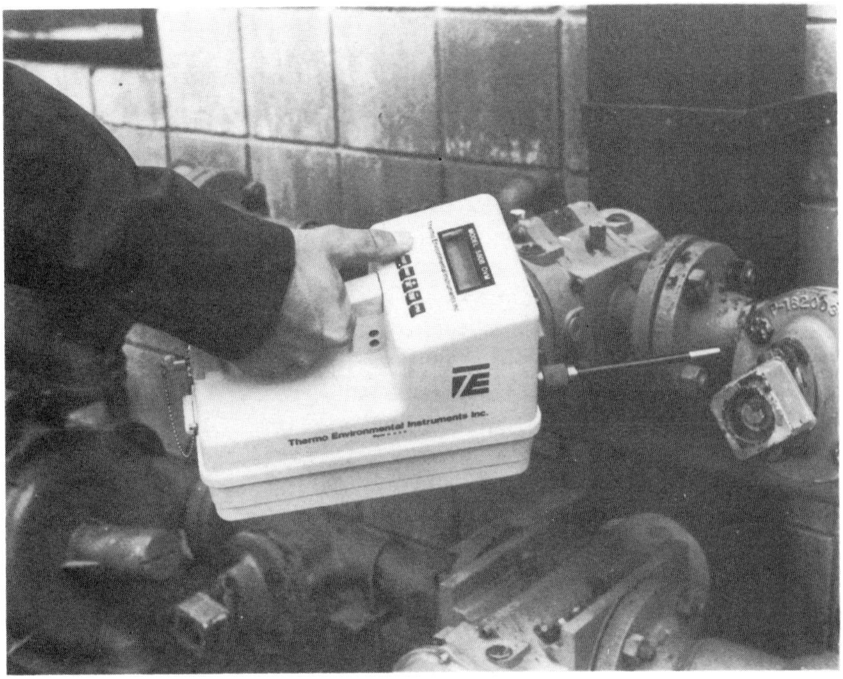

FIG. 11.3. Direct reading non-specific photoionisation organic vapour meter sensitive to 0.1 ppm benzene. The instrument incorporates a data logger with location, time and date, compatible with IBM-PC for data analysis. (By courtesy of Thermo Electron Limited)

including trichloroethylene, perchloroethylene, carbon tetrachloride and a host of other gases and vapours with a halogen in the molecule.

Sensitive combustible gas indicators are commercially available. These devices rely on catalytic combustion. Usually one arm of a Wheatstone bridge is of platinum. The combustible gas oxidises around the platinum wire or bead, heats it differentially and its increased resistance reflects the concentration of combustible gas or vapour in air.

Direct reading instruments for aerosols

A class of on-the-spot dust sampling instruments rely upon deposition by impaction from an air jet against a surface. The Konimeter was among the first instruments of this kind used for environmental evaluation in South African mines (Michaelis, 1890; Kotze, 1919). By means of a spring-operated piston pump, in this instrument a known volume of air, typically 2.5, 5.0 or 10 cm^3, is drawn through a fine, circular jet located close to a greased glass slide. In modern Konimeters, the deposited particles are counted under a built in microscope and others have a size selecting pre-impactor so as to separate off the coarse particles (Quillium and Kruss, 1972). An optical size distribution may also be performed on the spot with these instruments by comparing the image of the particles with graduated circles on an eyepiece graticule (Silverman, Billings and First, 1971).

New direct reading instruments for measuring aerosols commonly rely upon their light scattering properties. A typical instrument may sense light scattered through only a few degrees off the forward direction from a fairly broad beam of light. This is the so-called 'Tyndall' effect, named after J. Tyndall who employed it with great effect in the last century in investigations of contamination by airborne viable organisms (Tyndall, 1882). The finest particles scatter more light close to the forward direction than do coarser particles. Coarse particles are seen in a beam of light mainly because of their reflected light (Tolman, Gerke, Brooks, Herman, Mulliken and Smyth, 1919).

In modern instruments for sensing fine aerosols light is collected over a surface positioned to receive that portion of a parallel beam scattered away from the forward direction but scattered over a small angle to it. This combination yields an integrated measurement of all particles in a relatively large illuminated volume (Lilienfeld and Stern, 1982; Armbruster, Breuer, Gebhart and Meulinger, 1984; Vinson and Williams, 1987). A dark stop is positioned to obscure light in the forward direction. The scattered portion of arc which is sensed is so chosen as to make the instruments particularly sensitive to fine, 'respirable' particles (Figure 11.4). Even so, the reading these instruments yield does not match that obtained with gravimetric samplers exactly so that their calibration against mass concentration of respirable dust varies with the overall size distribution and the composition of the dust (Chung and Vaughan, 1989; Roebuck, Vaughan and Chung, 1990).

Detector tubes for gases and vapours

Detector tubes, sometimes called 'indicator' tubes, contain granules of adsorbent in the tube which change colour when contaminated air is drawn through the tube. They are used for gases and vapours, not aerosols. The first patent for detector tubes was taken out in 1919 (Lamb and Hoover, 1919). The earliest ones were made for carbon monoxide, to replace the use of a canary to detect 'white damp' in coal mines (Lamb, Bray and Frazer, 1920).

Different types of detector tube are nowadays available for hundreds of different gases and vapours (Leichnitz, 1982). They mostly rely on a hand operated piston pump, bellows pump or squeeze bulb of 100 cc capacity to draw the air through the tube. The total volume of air drawn through a tube is determined by the number of strokes employed. The tubes are about 15 cm long and 5–10 mm diameter. The indicating layer itself generally fills 2–10 cm of the tube length. In use the sealed ends of the tubes are broken open and the tubes, when fitted on the pump, can be held with the inlet in the breathing zone when sampling (Figure 11.5). It usually takes about 2 minutes to take a sample, which is subsequently evaluated by comparing the stain length or, sometimes, intensity of colour against a scale printed on the tube or a chart provided for the purpose. Because of spatial variation in concentration the 'breathing' zone is

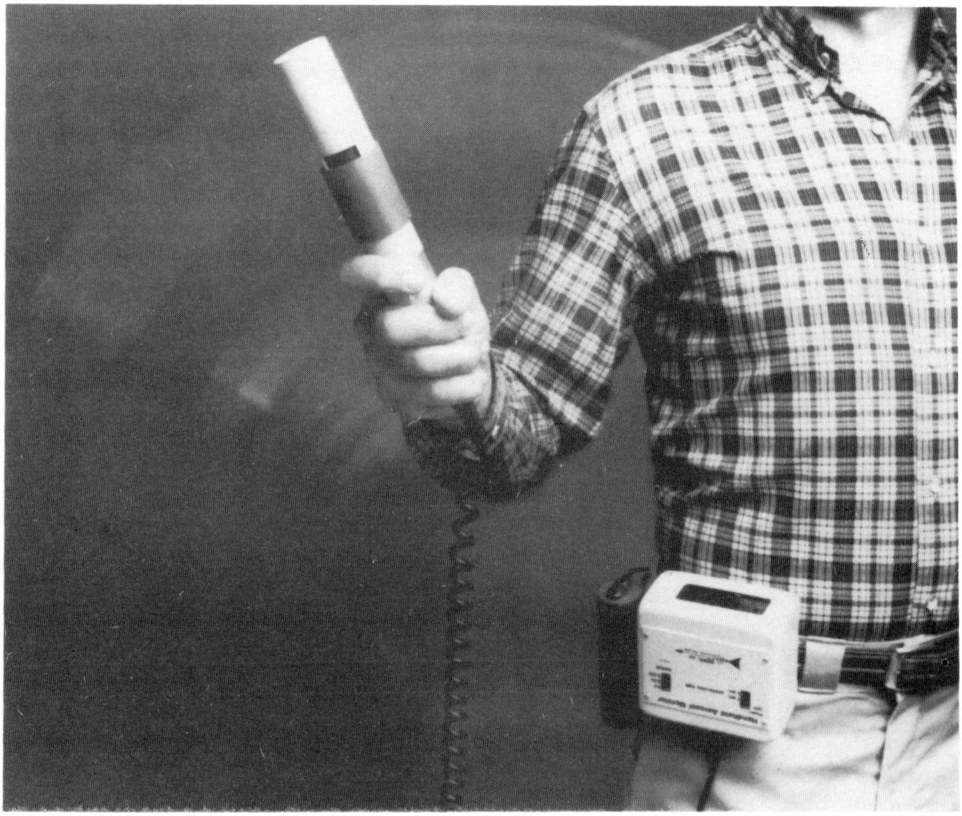

FIG. 11.4. Hand-held aerosol monitor for measuring the mass concentration of respirable dust. The instrument relies upon near forward light scattering from light emitting diodes. (By courtesy of Environmental Monitoring Systems Ltd)

defined as being the region within 50 cm of the nose and mouth, upstream. This may be difficult to sample for two minutes without interfering with the work. A procedure sometimes employed to overcome this particular problem is to have a sample bag on a shoulder harness worn by the employee, which is filled by means of a belt mounted pump drawing air from an inlet clipped to the lapel. Subsequently the average concentration in the collected sample is measured by sampling from it with a standard indicator tube.

Being relatively simple and low first cost, detector tubes are deservedly popular. However, the method becomes progressively more expensive when many samples are taken as may be necessary in a variable environment. Long duration (8 hours) indicator tubes are available for a few common gases and vapours, using a lightweight battery-powered air pump drawing between 20 and 200 ml min^{-1}. Detector tubes are simple, rapid and convenient but they have limitations and inherent potential errors. It is advisable to use tubes and pumps which meet National standards as far as possible (US Public Health

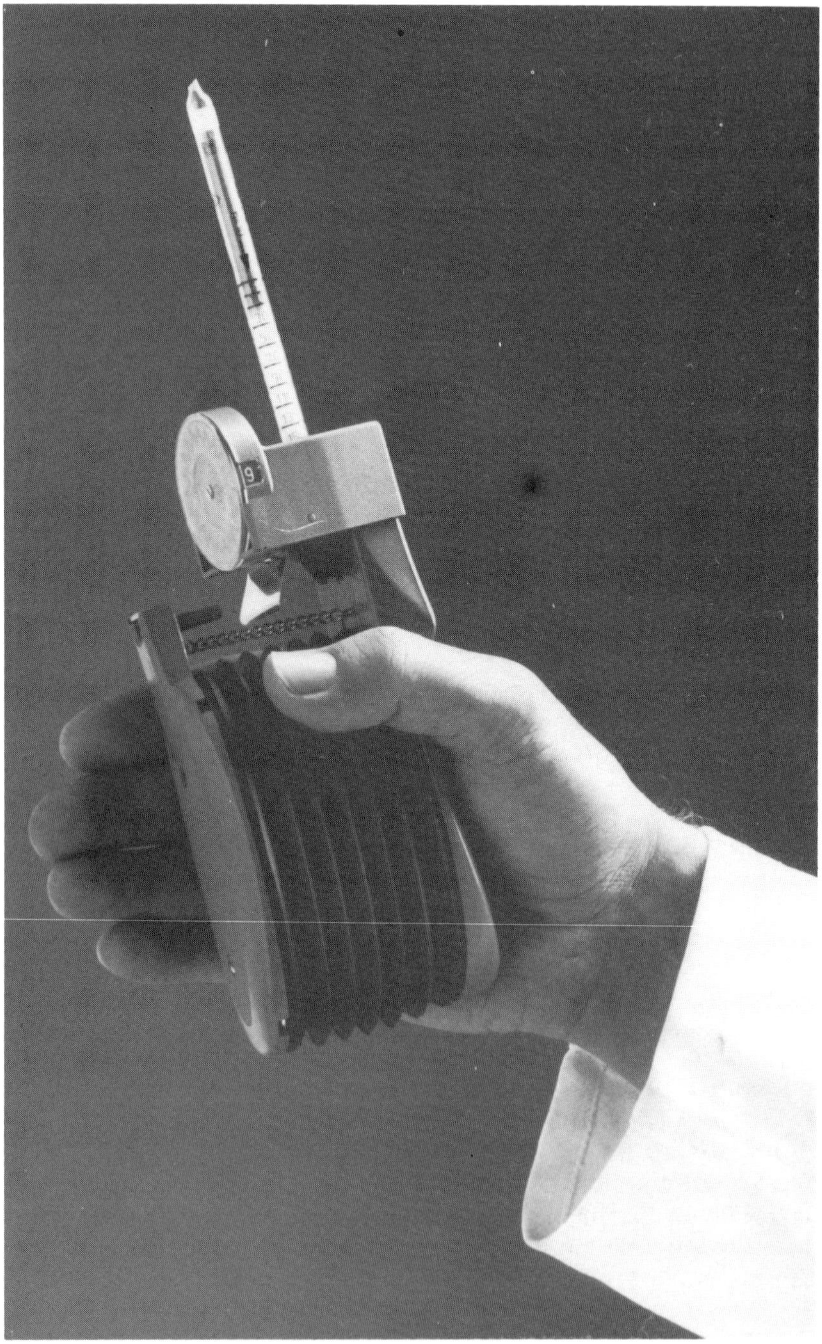

FIG. 11.5. Detector tube sampling for gases and vapours. Short period sampler read on the spot. Adsorbent granules in the tube change colour on drawing a known volume of contaminated air through the tube. (By courtesy of Draeger Limited)

(1) Leak tests.

(2) Tests against standard atmospheres to confirm the applicability of the bags for the gas or vapour in question.

(3) Evaluation of storage decay curves for levels of concentration in the range 0.1–1.0 × exposure limit.

Aerosols

When using a standard Konimeter the spot of dust deposited on the glass slide by impaction from the jet is examined in the laboratory. When the samples are taken back to the laboratory the slides may be heated to 600C to remove the grease and other organic matter and can then be treated with water, alcohol, hydrochloric acid and so on to remove other unwanted substances, as desired. The labour of counting particles by eye may be largely eliminated by employing an automatic particle scanner (Flint, 1965; Chmara, 1969; Verma, Sebestyen and Muir, 1987). The Konimeter is not only of historical interest. Konimeters are still in use in South Africa, Canada and to a limited extent in Germany.

An improvement on the standard Konimeter was developed in Britain in which the sampled air passes first through a chamber lined with moist blotting paper to humidify the air. On emerging from the jet moisture condenses on the particles, so obviating the use of grease to trap the particles. The jet was also shaped in the form of a slit, which reduced overlapping amongst the deposited particles (Owens, 1922).

Continuous sampling methods

The amount of contamination by a particular substance in work-room air is of the order which an analytical chemist would normally regard as being in trace quantities. Work-room air analysis is a specialised field and the number of hazardous substances for which accurate, inexpensive and reliable methods have been devised is still strictly limited. The easiest way to measure personal exposure to air contamination is normally to employ 'personal' sampling. Such a 'roving' sampling station is made possible using instruments in which a pocket-sized pump draws air through a tube leading to a small sampling head on the employee's lapel. The air contamination is absorbed on solid absorbent, or by the liquid in a fritted bubbler, or, if an aerosol, is filtered out on one of several types of filter. For gases and vapours there is also an expanding range of the even simpler 'diffusive' samplers which do not rely on a mechanical pump but depend for their action on diffusion from the atmosphere on to the absorbent in the sampling head fixed to the lapel. Nevertheless, the flow rate obtainable from a small battery-powered pump is limited and more so from a diffusion tube or badge.

For others a continuously running, high volume flow rate sampling train may have to be employed to collect enough material for quantitative analysis in the laboratory. Such a sampling train may be bought as such, may be custom built, or made from a commercially available device modified for the purpose.

Air sampling trains, whether of high or low flow rate have the critical elements in a certain order:

—air inlet orifice, followed by
—aerosol pre-selector, when required, followed by
—aerosol separator or gas/vapour absorber, followed by
—air flow meter, followed by
—flow rate control valve, followed by
—suction pump.

The flow meter is downstream of the aerosol separator to prevent contamination depositing in or reacting with the flow meter parts. The flow rate control is more reliable upstream of the pump as common air sampling pumps are not leak tight.

Other important elements in the train are the motor and power supply for the pump and, in the case of an electrostatic precipitator, an additional power supply for the aerosol separator. In work-places where there is a risk of explosion, or the materials are combustible, only electrical equipment which is certified as safe for use in such conditions may be used.

The design of a sampling train for aerosols is more critical than one for gases and vapours as the deposition of particles depends on their size, shape, electrification and other disturbing factors (Shaw, 1978; Liu, Pui and Fissan, 1984). Aerosol sampling is divided into two main categories, namely those with and those without a pre-selector to separate particles which are larger than about 5 microns diameter. In general, exposure limits for dusts which cause pneumoconioses from particles originally deposited in alveoli are based on mass sampling with a pre-selector. A comprehensive review of the theory and practice is given by Vincent (1989). An important exception is asbestos dust, which has exposure limits in terms of the number of fibres in unit volume of air.

Air inlet

The size and shape of the air inlet orifice does not affect the efficiency of a gas or vapour sampling train but when sampling aerosols in a side draught unless the air flow into the inlet is isokinetic there will be errors in sampling large particles. On the other hand it is arguable that when work-room air is inhaled the flow into the nose and mouth is far from being isokinetic and when working in a high air velocity the tendency is to turn one's head away from the dust. A type of inlet now coming into use for aerosols is incorporated in 'inhalable' aerosol samplers, which have an entry shaped to reject some of the largest particles, as only a proportion of them are normally inhaled (Figure 11.8) (International Standards Organisation, 1983; American Conference of Governmental Industrial Hygienists, 1984; Mark and Vincent, 1986; Vincent and Mark, 1987). It is also argued that exposure limits for aerosols would be more closely related to health risk if they were expressed in terms of inhalable aerosol samplers (Vincent and Armbruster, 1981).

The tube leading from the air inlet orifice of a sampler should be as short as possible to keep wall losses in the tube to a minimum. This is especially

FIG. 11.8. Personal sample pump with 'seven hole' filter head for inhalable dust sampling. (By courtesy of SKC Ltd)

important when sampling for aerosols. Nevertheless it is sometimes necessary to have a probe tube of some length connected to the air inlet as in the following cases:

(1) When concentration of contamination is highly localized, near the face of an employee;

(2) When automatic, multi-point sampling is employed, requiring a system of fixed tubes running from a central sampling instrument to various key points in a work room;

(3) When investigating the source of air contamination. Having the entry positioned close to the floor, bench or on someone's lapel might result in an inconclusive outcome.

Wall losses of gases and vapours may be kept small by using appropriate tubing material and by allowing sufficient time for the surface to become saturated before taking readings. Wall losses of aerosols may be excessive in probe tubes which are longer than about 1 m or have sharp bends or in which the air velocity is less than 1 m s^{-1} In such cases dust in the tube may be dissolved or washed off and added to the sample. If the aerosol separator is small enough and light, it may be connected to the air flow meter by a suitable length of tubing inserted between it and the air flow meter. This avoids the problem of wall losses in a probe tube upstream of the separator.

Aerosol pre-selector

In 1943 the British Medical Research Council suggested that the mass concentration of dust be limited to particles smaller than 5 microns diameter when sampling in relation to coal workers pneumoconiosis (Medical Research Council, 1943). In USA Hatch and Hemeon extended this idea to all dusts causing pneumoconiosis (1948). The design requirements for instruments to select the fraction of inhaled dust most likely to be deposited in the pulmonary air spaces were described by Davies (1949). The stage was set for the development and use of respirable dust samplers.

Airborne coarse dust larger than 5–10 microns aerodynamic diameter may be separated from the finer particles by employing a pre-selector of any of a variety of designs. The two principal types of such a pre-selector in use are the cyclone and the horizontal elutriator. Cyclones 10 mm to 50 mm diameter are employed when measuring the atmospheric concentration of 'respirable' dust, that is, dust smaller than about 5 microns aerodynamic diameter, which could penetrate to alveolar regions of the lungs. The characteristics of a size selector should preferably match that specified in the current relevant exposure limit and, conversely, exposure limits should be expressed in terms of available instrumentation (Lippmann and Harris, 1962; Aerosol Technology Committee of the American Industrial Hygiene Association, 1970). If the match is a poor one a selector which oversamples could still be employed, recognising that it has a built in safety factor. The size collection efficiency for Respirable

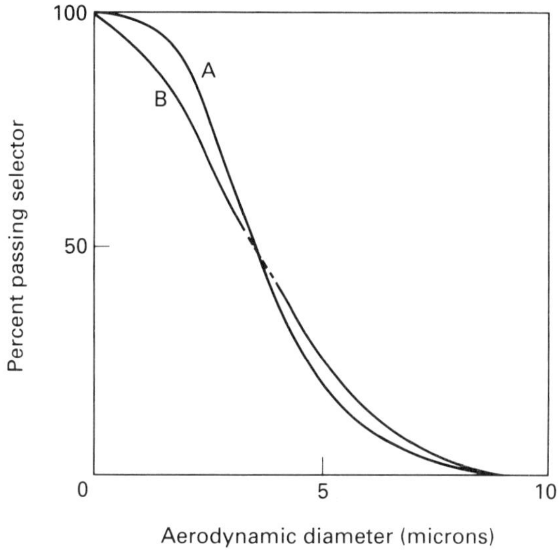

FIG. 11.9. Aerosol size selector characteristics. Curve A—ACGIH Respirable Particulate Mass TLVs. Curve B—Typical cyclone selector for practical sampling.

Particulate Mass in the published, 1988 list of ACGIH Threshold Limit Values for mineral dusts, for example, is described by a cumulative lognormal function with a median aerodynamic diameter of 3.5 microns and with a geometric standard deviation of 1.5 (± 0.1) (American Conference of Governmental Industrial Hygienists, 1988b). It is plotted in Figure 11.9 (American Conference of Governmental Industrial Hygienists, 1984).

The air enters a cyclone tangentially at the side of a cylindrical or inverted cone shaped body, swirls around inside and leaves along the axis from a tube at the top connected to a filter holder. The coarse dust is thrown to the side of the cyclone by centrifugal force and collects in the base. In the Conicycle the coarse dust is thrown out of the entrance (Wolff and Roach, 1961; Carver, Nagelschmidt, Roach, Rossiter and Wolff, 1962). The air velocities in a cyclone are very high, giving centrifugal accelerations in excess of one hundred times gravitational acceleration and the flow is highly turbulent. All cyclones have a characteristic, somewhat similarly shaped, size-efficiency performance curve, differing in slope according to design. However, the detailed pattern of air flow through a cyclone depends so much on the exact design adopted that the performance of particular designs must be checked experimentally. The size-performance curve of a typical cyclone designed to allow 50% of particles of 3.5 microns equivalent diameter to pass through is shown in Figure 11.9 for comparison.

The standard definition of 'respirable' dust for the purpose of designing sampling instruments was originally based by the British Medical Research Council upon the graded separation characteristics of an elutriator matched against alveolar deposition (Davies, 1952). There are two classes of elutriator employed in sampling air for aerosols: horizontal elutriators and vertical elutriators.

A single element of a horizontal elutriator used in this work is a thin, horizontal, rectangular duct. Commonly, the horizontal elutriator package consists of several such elements stacked one above the other, connected in parallel to a common exit (Walton, 1954; Wright, 1954). The theory is fairly easy to understand and performance closely matches theory.

Suppose a volume flow rate, Q, of air is passing uniformly along a horizontal duct of length L, width W and height H. The time, T, it takes for air to pass through the duct is LWH/Q. Among the particles entering the duct those which, in time, T, would fall freely under gravity a distance greater than H would all deposit on the floor of the duct. Thus, the minimum terminal velocity, V_c, for 100% capture by the elutriators is

$$V_c = H/T = Q/LW \quad (1)$$

Also, among those particles with terminal velocity V, [V less than V_c] a proportion would be captured by deposition on the floor of the duct. The percentage captured in this way is $100[V/V_c]\%$. The proportion, P, not captured by the elutriator would be $100[1 - V/V_c]\%$ of those with terminal velocity under gravity less than V_c.

$$P = 100[1 - V/V_c]\% = 100(1 - VLW/Q)\% \quad (2)$$

Thus, the proportion passing through the elutriator, P, is independent of the height of the duct, provided the floor area and flow rate are constant. It may also be shown that under ideal streamline flow conditions errors in the above formulae through assuming uniform flow cancel out. Flow is maintained streamline by making H small.

A vertical elutriator is a single vertical tube, either parallel sided or in the shape of an inverted, truncated cone. The air with particles in suspension is drawn upwards through the tube.

Suppose a flow rate, Q, of air is passing uniformly up a parallel sided, vertical duct, cross-sectional area A. The upward air velocity is Q/A, equal to V_c. None of the particles which have a terminal velocity in air exceeding V_c would be carried up the tube by the air. Those particles with a terminal velocity, V, lying between zero and V_c. would be carried up at a velocity $V_c - V$. The percentage, P, passing through the tube,

$$P = 100[1 - V/V_c]\% = 100(1 - VA/Q)\% \quad (3)$$

P is independent of the length of the tube, provided the cross-sectional area and flow rate are constant. It is particularly important to maintain streamline flow in vertical elutriators and to do this their cross-sectional area must be small. However, since the pass, P, is dependent on cross-sectional area, the volume flow rate must be correspondingly small and this type is therefore only used for sampling at very low volume flow rates.

The conical form of vertical elutriator, with a small entry at the tip at the bottom, is used to promote smooth flow through the elutriator. A parallel portion is usually arranged at the top, where the cross-sectional area is largest and the final performance therefore determined.

Perfect streamline flow is not realised in practice, even in a narrow horizontal elutriator. The effects of a disturbance to streamline flow on the collection characteristics of a horizontal elutriator may be understood by considering deposition under conditions of perfect mixing. Inside a horizontal elutriator element of width W, length L and height H at a point distance l from the entrance the average velocity of the air stream,

$$\frac{dl}{dt} = \frac{Q}{WH}.$$

The horizontal base area of a small element length dl is Wdl. The number of particles of terminal velocity V falling on this area in time dt is $CVWdl$ particles per minute, where C is the concentration of particles in the air over the element. The volume of air above the element is $WHdl$, so that the rate of change of concentration with time,

$$\frac{dc}{dt} = -\frac{CVWdl}{WHdl} = -\frac{CV}{H} \tag{4}$$

From which,

$$C = C_o \exp[-Vt/H] \tag{5}$$

where C_o is the concentration at t_o. when the air enters the elutriator. The time it takes for the air to pass through the elutriator is LWH/Q, so that the concentration at the exit is,

$$C_o \exp[-VLW/Q] \tag{6}$$

Thus the percentage, P, of particles in suspension in the air entering the elutriator which pass through it without depositing,

$$P = 100 \exp[-VLW/Q]. \tag{7}$$

Elutriators are commonly designed on the assumption of uniform flow and so as to capture 50% of particles of 5 microns aerodynamic diameter. The terminal velocity of particles of aerodynamic diameter, D, settling in air under gravity is proportional to D^2. Therefore, the expected performance of such an elutriator under streamline flow [equation (2)] is

$$P = 100[1 - 0.02D^2], \tag{8}$$

and its performance under perfect mixing conditions [equation (7)] is

$$P = 100 \exp[-0.02D^2]. \tag{9}$$

These two equations are shown in Figure 11.10.

The performance of a real horizontal elutriator will lie somewhere between these two curves. By careful design streamline flow is often nearly achieved and the performance comes close to theoretical (Hamilton, 1967). Such elutriators have an inherent fail-safe feature in that if, due to distortion of the plates or poor design streamline flow is seriously disturbed, capture of coarse particles is reduced and the estimated sample concentration errs on the high side. On the

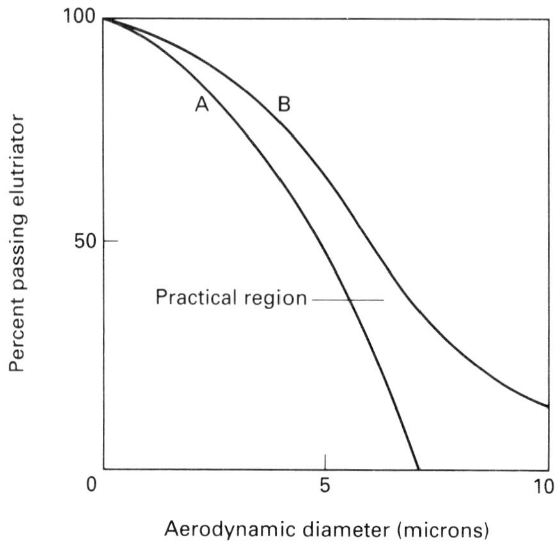

FIG. 11.10. Size selection with horizontal elutriators. Curve A—Perfect streamline flow conditions. Curve B—Perfect mixing conditions.

other hand, flow rate control through an elutriator is particularly important. If the flow rate is low, coarse particles are separated to a greater extent, giving an additional error on the low side. The reverse occurs when the flow rate is high. This does not occur with cyclones which allow more of the coarse particles to pass when the flow rate is low and less when the flow rate is high. Orientation of the horizontal elutriator is important, and it should be kept within 5–10 degrees of the horizontal. Because of this, cyclones rather than horizontal elutriators are used for personal sampling of respirable dust.

The air flow resistance of an elutriator is lower than that of a cyclone for the same flow rate. Cyclones are less affected by orientation and more suited for fastening to an employee's clothing and thus enable personal exposure measurements to be undertaken. In both types of pre-selector the resistance is largely independent of dust loading and small in comparison with the resistance of customary filters downstream.

Gas/vapour absorber

A gas or vapour absorber should be preceded by a filter to remove any aerosol which might interfere in the absorption or analysis. Historically, liquid absorbent has been found to be satisfactory when the contamination is chemically reactive with the liquid or is readily soluble in it. Simple bubblers, fritted glass bubblers and glass bead columns provide progressively greater contact between the gas and liquid phases and are increasingly efficient liquid absorbers.

Absorption of gases and vapours by chemical reaction is limited only by the quantity of reagent solution and the speed of reaction. If sufficient excess of reagent is maintained and the reaction is rapid, efficiency of collection is close to 100%. Examples of systems employing chemical reaction are; acid gases in alkalis, ammonia in sulphuric acid, hydrogen sulphide in cadmium sulphate solution and toluene-2,4-diisocyanate in Marcali solution.

The efficiency of collection of non-reactive gases and vapours by solution is dependent upon three principal factors: the volume of air sampled, the volume of absorbing liquid and the vapour pressure of the gas or vapour in the absorbing liquid (Gage, 1960). Examples of systems employing solution in this way are: butanol in water, esters in alcohol and organic chlorides in butanol.

The maximum volume of air which may be drawn through a bubbler whilst maintaining efficient collection is limited. The minimum flow rate is about 50–100 cm^3 min^{-1} according to the gas or vapour to be collected with the result that personal monitoring is not always feasible.When it is desired to collect insoluble or non-reactive vapours, adsorption on solids is frequently the method of choice. So much so that adsorption in small glass or metal tubes is now standard practice for many organic vapours. Activated charcoal and silica gel are common adsorbents and can provide high collection efficiency. Activated charcoal is a good adsorbent for most organic vapours and is generally preferred to silica gel as it is less liable to interference from water vapour (Fraust, 1975). The charcoal is subsequently desorbed, employing thermal desorption, or a suitable solvent such as carbon disulphide, for subsequent analysis by gas chromatography. Silica gel is used for substances which are difficult to remove from activated charcoal, such as amines, aniline, anisidine, toluidine and xylidine (Wood and Anderson, 1975). Thermal desorption of silica gel may be employed or a solvent chosen to suit the particular gas or vapour. Activated alumina, molecular sieve or porous polymer adsorbents such as Tenax and the Chromosorb and Poropak series are also increasingly used for certain applications as they may be conveniently desorbed thermally. The adsorbent granules are commercially available packed in sealed glass tubes ready for use containing 50 mg to 1000 mg according to size. Most such tubes have two sections in series: one for sample collection and the second serving as a check that the first has removed all the contamination.

A sample is collected by breaking off the end tips and fitted inside a closed holder which may be clipped to an employee's lapel and connected to a sample pump drawing 20–200 cc min^{-1} (Figure 11.11). After sampling is completed the tube is sealed with a push-on cap and returned to the laboratory for analysis, typically employing thermal desorption coupled with assessment by gas chromatography.

The collection efficiency of activated charcoal and silica gel tubes is dependent primarily upon the air volume flow rate employed, 50 ml per minute to 1 litre per minute, and the duration of sampling, 5 minutes to 8 hours. Standard procedures for preparing tubes, sampling, desorption and subsequent analysis have been worked out for most common gases and vapours (Kupel and White, 1971; NIOSH, 1985). These procedures should be controlled within narrow limits. Batches of prepared sampling tubes should be themselves sample tested against standard atmospheres before use.

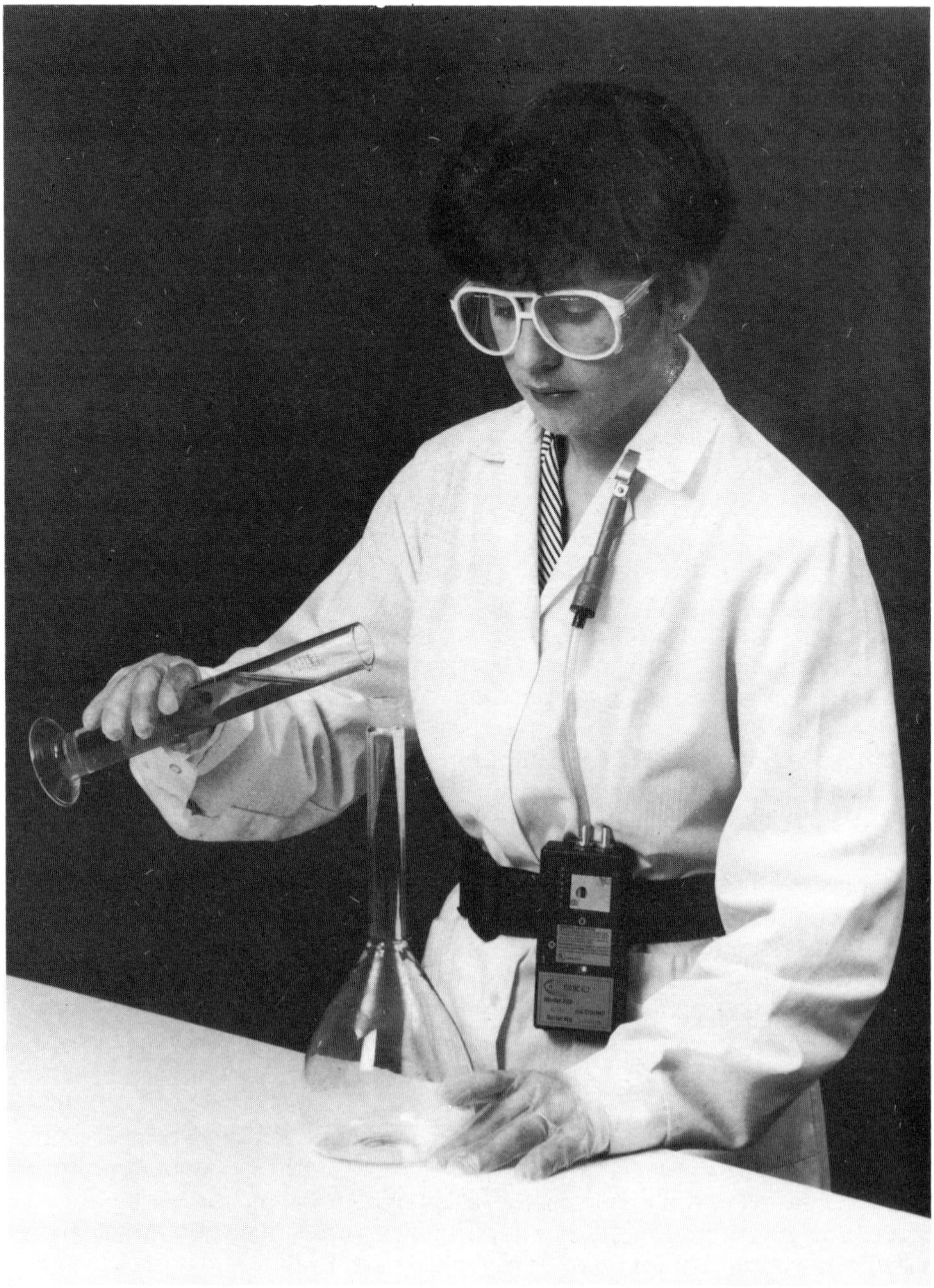

FIG. 11.11. Adsorbent tube clipped to the lapel for vapours. Air is drawn through the tube with a personal sample pump. The contents are analysed in the laboratory. (By courtesy of SKC Ltd)

FIG. 11.12. Diffusive sampler—badge type. Adsorbent in a membrane capsule is placed in a reusable badge housing and is removed and analysed after being worn clipped to the lapel for a work shift. (By courtesy of SKC Ltd)

Diffusive samplers

For gases and vapours there is an expanding range of the simple 'diffusive' or 'passive' samplers which do not rely on a mechanical pump to draw air through an absorber but depend for their action on molecular diffusion from the atmosphere on to the absorbent or adsorbent. This promising technique, originally only semi-quantitative, has been used in occupational hygiene monitoring for over 50 years but has been slow to develop into a reliable field method (Barrer, 1939). Atmospheric sampling by this means is especially suitable for personal sampling of employees at work and modern methods employ a sampling head attached to the lapel or collar (Berlin, Brown and Saunders, 1987). In one type of diffusive sampler the air contamination diffuses across a permeable membrane at the front of the badge housing, for collection on a solid or liquid absorbent in the cavity behind (Figure 11.12). Liquid collection medium is especially suitable for inorganic compounds. In another type of sampler the gas or vapour is allowed to diffuse through an air space at the front of a charcoal, silica gel or other solid adsorbent sampling tube (Palmes and Gunnison, 1973; Reiszner and West, 1973; Ellwood, Groves and Pengelly, 1990). There are diffusive samplers now available for some 150 different gases and vapours. The effective rate of sampling is proportional to the molecular diffusivity of the gas or vapour in air, proportional to the open face area of the sampling head and inversely proportional to the diffusive path length or thickness. The basic theory is fairly simple.

By Fick's law of diffusion, the net mass flow rate of diffusing substance in the steady state is given by,

$$M = -DA.\frac{dc}{dl} \qquad (10)$$

where M = net mass flow rate of diffusing substance,
D = molecular diffusivity of the substance in air,
A = cross sectional area,
c = concentration of substance,
l = distance measured in the direction of diffusive substance flow.

The concentration at the surface of unsaturated adsorbent is near zero. From which, on integrating equation (10), if $c \rightarrow C$ as $l \rightarrow 0$ and provided there is sufficient adsorbent to prevent back diffusion $c \rightarrow 0$ as $l \rightarrow L$, it follows that

$$C = \frac{LM}{DA} \qquad (11)$$

where C = concentration at the open end of the tube,
and L = length of diffusion pathway.

After sampling for a measured length of time the mass flow of diffusing substance is found by analysis of the adsorbent. Molecular diffusivity of the substance in air may be tabulated in the literature. In any event the effective diffusivity operating in a given configuration of diffusive sampler is determined experimentally by calibration against known atmospheric concentrations (Lugg, 1968). End affects at the beginning and end of sampling can be kept within reasonable bounds by sampling for at least 30 minutes (Hearl and Manning, 1980, Underhill, 1983; Bartley, Doemeny and Taylor, 1983).

The principles of detector tubes have been combined with diffusive sampling in a commercial tube which can be 'read' in the same way as a detector tube (May, 1989).

Diffusive gas migration is not as certain as air flow induced with a mechanical pump. During periods of low concentration the gas or vapour previously collected is slowly eliminated from solid adsorbents in a manner analogous to compartmental decay, discussed in Chapter 4 (Bartley, Doemeny and Taylor, 1983; Coutant, Lewis and Mulik, 1985; Bartley, Woebkenberg and Posner, 1988). Liquid absorbents relying upon chemical reaction with the gas or vapour being measured are not prone to this effect (Hallberg and Rudling, 1989). Some diffusive samplers are unduly influenced by moderately high ambient air velocity, which reduces the thickness of the surface boundary layer at the entrance to the sampler and hence reduces the length of the diffusive pathway (Palmes, Gunnison, Di Mattio and Tomczyk, 1976; Palmes, 1979; Hollingdale-Smith, 1979; Brown, Charlton and Saunders, 1981; Coleman, 1983; Samimi and Falbo, 1983). Others are 'starved' in stagnant flow environments and will read low (Jones, Billings and Lilas, 1981). Reports of adverse environmental conditions based upon results from diffusive samplers should, if possible, be confirmed by other methods. Diffusive samplers are relatively inexpensive, light, robust and easy to use. The method still has great potential and should find increasing use as the reliability of commercially available types steadily improves (Harper and Purnell, 1987).

Aerosol separator

This is the most important element in an aerosol sampling train. The efficiency of separation must be high and reliable but the pressure drop across the separator must be low in order to keep to a minimum the size of the necessary suction pump and motor. The separator may consist of a single element such as a filter, impinger, thermal precipitator or electrostatic precipitator and there may be two or more elements in series, so as to separate the different sizes of particles (Courbon, Wrobel and Fabries, 1988).

Filters

Many instruments used in assessing the mass concentration of aerosols incorporate a filter. The most commonly used filters are made of fibrous filter mats of any of a wide variety of organic or inorganic fibres. Membrane filters are also used in industrial hygiene. These membranes are generally made of organic materials, although a membrane filter made of sintered silver metal is obtainable, has a high collection efficiency and is used in special applications for solvent extraction.

Common filter paper consists of an irregular mesh of fibres which are about 20 microns in diameter or less. Air passing through the filter changes direction around the fibres and particles in suspension impinge there. With particles greater than 0.5 micron diameter deposition efficiency generally increases markedly with the velocity of the air stream and with the density and diameter of the particles. A face velocity in excess of 0.1 m sec^{-1} is satisfactory for most purposes. On the other hand, when sampling at low flow rates, as in personal air samplers, the face velocity may be below 10 cm sec^{-1} and the efficiency of cellulose filter paper falls away for fine particles. The largest particles, greater than about 30 microns, deposit to a significant extent by direct interception with the filter fibres and by sieving action. Industrial airborne dust contains a wide range of particle sizes and the overall mass efficiency of filters employed in high volume flow rate air sampling is in excess of 95% at the commencement of sampling and the efficiency increases with the accumulation of solid particles on the filter surfaces.

Below 0.5 micron particle diameter there is significant deposition brought about by diffusion of the particles caused, in its turn, by Brownian motion. This phenomenon increasingly dominates over deposition by impaction for the very smallest particle sizes and increases further as the diameter of the particles decreases. Consequently, there is a size at which the combined efficiency by impaction and diffusion is a minimum. This size is usually less than 0.2 micron diameter. The weight of particles below 0.5 and certainly of those below 0.2 micron aerodynamic diameter is usually less than 2% of the airborne dust of hygienic significance so that in practice the amount of deposition by diffusion may be ignored as a rule. Aerosols which consist primarily of freshly formed metal fumes could be an exception, although they readily agglomerate.

Cellulose fibre filter papers are relatively inexpensive, are readily obtainable, have a high tensile strength and are low in ash content. They are used for high volume flow rate sampling, especially for static sampling. Their main disadvantage is their hygroscopicity, for which allowance must be made in weighing procedures. Whatman No. 41 is the most widely used cellulose filter as it combines good efficiency with low flow resistance.

Filters made of glass, asbestos, ceramic, carbon, quartz or polystyrene fibres less than 20 microns in diameter have a higher efficiency than cellulose filters and are favoured for atmospheric sampling at lower flow rates. However, their principal advantage over cellulose filters is the relative ease of determining the blank weight. This is the main reason why glass fibre papers, in particular, are very popular for gravimetric determinations.

The resistance of a filter increases with thickness and compression of the filter mat and with the dust loading. A good fibre filter for aerosol sampling is made of thickly matted fine fibres and is small in mass per unit face area.

A membrane filter is a micro-porous plastic film made by precipitation of a resin under controlled conditions. The polymers used are cellulose esters, polyvinyl chloride or acrylonitrile. Those made from cellulose esters are widely used in colorimetry, atomic absorption, emission spectroscopy, fluorescence, infrared and X-ray diffraction of aerosols. Those made from polyvinyl chloride are favoured for gravimetric determinations because of their low hygroscopicity.

Membrane filters have remarkable properties. Most particles are deposited on the surface so that they may be used to collect aerosols for examination under the microscope. The objective lens of a microscope has a limited depth of focus. Membrane filters may thus be used either to determine mass concentration or number concentration. Their main disadvantage is their high resistance to air flow as compared with equivalent fibre filters. A 1.0 micron pore diameter membrane filter has a resistance of 3 to 5 times that of a cellulose fibre filter at the same face velocity and a 0.2 micron pore size membrane filter 10 to 500 times.

Membrane filters are flabby and are accordingly supported on a metal gauze or other grid. Special types have an integral nylon support to give added strength. In manufacture the pore size is controlled within close limits and membrane filters are obtainable with mean pore sizes of from 0.01 micron to 10 microns in diameter. They are usually 140 microns thick and have an efficiency close to 100% for particles larger than the mean pore size. The ash content of these filters is negligible, which is a considerable advantage in certain applications. When immersion oil is placed on a membrane filter the membrane becomes transparent and the particles may be examined under the microscope by transmitted light. The refractive index of organic membrane filters is 1.4–1.6, according to type. Nevertheless some particles smaller than the nominal pore size are also trapped within the body of the filter (Megaw and Wiffen, 1964; Lindeken, Petrock, Phillips and Taylor, 1964; Spurny and Pich, 1964). Complications may also arise when examining dust of minerals whose refractive index is close to that of the filter material. They are obtainable with a grid stamped on, which facilitates counting over a known area. They may also be obtained black, which is advantageous for certain white opaque dusts. Common types are soluble in cyclohexane or other organic solvents, which facilitates separation of the particles when required.

Membrane filters are used in the determination of mass concentration by personal sampling provided the available analytical methods are sufficiently sensitive as to make it feasible to employ a low air sampling flow rate. The relatively low flow rate is necessary because of their high air flow resistance. The ability to dissolve a filter in a variety of solvents is advantageous for subsequent analysis. A common membrane filter used for sampling in relation to exposure limits, expressed in terms of mass concentration of all particle sizes, has a nominal pore size of 5 microns. These have a high efficiency for collecting particles smaller than 5 microns as well as larger particles (Lockhart, Patterson and Anderson, 1964; Liu and Lee, 1976).

Polycarbonate (Nuclepore) filters are made from polycarbonate sheet treated in a nuclear reactor. They are transparent, even without immersion oil, although the collected particles tend to penetrate further into the filter mass than they do in membrane filters. They are non-hygroscopic and lighter than membrane filters so that they are also good for gravimetric determinations.

In use, filters are supported in a sampling head or cassette. The joint between filter and cassette should be tight but not abrade the filter material. Plastic membrane filters and cassettes tend to build up an electric charge and significant quantities of aerosol may be found on the plastic (Mark, 1974; Blackford, Harris and Revell, 1985; Demange, Gendre, Herve-Bazin, Carton and Peltier, 1990). Anti-electrostatic treatment can reduce the deposits on the cassette walls. A solvent rinse of the inside surfaces should be employed as a precautionary measure (Wood and Birkett, 1979; Costello, Eller and Hull, 1983; Cohen, Harley and Lippmann, 1984). Metal cassettes do not build up an electric charge but are still not entirely free of dust deposited on the walls. One solution is to weigh the cassette with the filter still in place (Vaughan, Milligan and Ogden, 1989; Mark, 1990).

Impingers

The principle of operation of impingers is a jet impactor in which the jet is submerged in liquid, usually water or alcohol. Reagent may be added. In USA the standard mineral dust sampling instrument for some 50 years was the

Greenberg-Smith impinger, operating at a flow rate of 1 cfm [28.3 l min^{-1}], and its miniature counterpart, the midget impinger, operating at 0.1 cfm [2.83 l min^{-1}] (Greenberg and Smith, 1922; Littlefield and Schrenk, 1937). Much of the original exposure data for mineral exposure limits was obtained with the use of these devices. The samples were returned to the laboratory where a portion of the liquid was placed in a cell and the particles visible in a defined area were counted under a light-field microscope (Roach, Baier, Ayer and Harris, 1967). The result was expressed as millions of particles per cubic foot of sampled air. The efficiency of collection combined with the limited resolution of the microscope counting system was such that few particles smaller than 0.8 micron in diameter were counted. Counting particles viewed under a microscope is extremely tedious and is prone to large and variable errors. The impinger has been largely replaced for mineral dusts by the development of respirable dust, gravimetric sampling instruments, coupled with exposure limits expressed in the same terms. The midget impinger and its one-tenth size counterpart, the micro impinger, are used when it is advantageous to employ liquid absorption or reaction, as for example when measuring atmospheric concentration of ammonia, phenol, nitric acid or sulphuric acid mist.

Thermal precipitation

Just as the impinger was the standard instrument for mineral dusts in USA so the thermal precipitator became the standard in Britain for many years. The principle of operation of the two instruments is, however, quite different. An airborne particle in a thermal gradient becomes subject to a differential pressure from molecular bombardment either side, such that the particle moves down the thermal gradient and deposits on cold surfaces. The finest particles tend to be deposited first (Fuchs, 1964). The dark deposit on the cold wall above a radiator demonstrates deposition of indoor air pollution by thermal precipitation.

The presence of a dust-free zone around a hot rod was employed by Aitken to precipitate smoke quantitatively (Aitken, 1923). The principle was adapted by Green and Watson who designed a thermal precipitator for use in workshop atmospheres (Green and Watson, 1935). The design was based upon employing a heated wire stretched across a 0.5 mm vertical slit formed by two microscope cover glasses in a brass sampling head. Thermal precipitators are used primarily for collecting particles on a microscope slide for examination, counting and size analysis under the microscope, as the deposition is gentle but highly efficient for the finest particles. A small horizontal elutriator may be fitted to the inlet (Burdekin and Dawes, 1956). Counting and sizing particles viewed under the microscope can be very laborious when used routinely and the method is mostly restricted to research applications. Due to the temperature involved the method is not suitable for mists except those of liquids with a high boiling point. A high flow rate gravimetric thermal precipitator incorporating a vertical elutriator has been developed by Wright (1953).

Electrostatic precipitation

That high tension current has a coagulating action on smoke was first reported by Hohfeld in the first half of the 19th century (1824). This action, coupled with precipitation was applied to the collection of smelter fume by Lodge towards the end of the 19th century (Lodge, 1886). An airborne dust sampling instrument for collecting a weighable quantity of dust from workshop air by electrostatic precipitation was developed in USA by Barnes and Penney over 50 years ago (1938). Electrostatic precipitation systems have the advantage of negligible flow resistance, no clogging and precipitation of the dust on a metal tube whose weight is unaffected by humidity. A special power pack is needed to supply the high voltage and precautions have to be taken to guard against electric shock.

The wire and tube or 'tulip' system is used in the electrostatic precipitator sampling instruments. The tube is a light alloy cylinder about 6 in long and $1\frac{1}{2}$ in diameter, held horizontally. The tube is grounded. A stiff wire, supported at one end, is aligned along the centre of the tube and serves as the charged electrode. The tip of the wire is sharpened to a point. A high DC voltage, between 10 kV and 20 kV, is applied to the central electrode. The corona discharge from the tip charges the particles in suspension in air drawn into the tube. Under the action of the potential gradient between the wire and the tube the charged particles migrate to the inside surface of the tube.

The migration velocity of the charged particles greater than 1 micron in diameter increases in proportion to particle diameter. On the other hand, migration velocity is approximately independent of particle diameter for particles smaller than 1 micron diameter.

Electrostatic precipitators do not have a pre-selector ahead of the collecting tube. They are suitable for aerosols whose exposure limits are in terms of the mass concentration of all particle sizes, including nuisance dusts. Because of their high efficiency for separating particles smaller than 1 micron diameter, they are very often used when sampling for fumes. In use the ionising voltage is maintained sufficiently high to collect all the particles but not so high as to produce arcing between the central electrode and the collecting tube itself. A check that the aerosol has all been collected is to observe that the downstream end of the collecting tube is clear of dust. There is a practical limit of about 100 mg to the amount of dust that can be collected on each tube, as a thick layer of dust is easily dislodged and may be lost on handling. For very high dust concentrations, a coiled filter paper liner may be used to enable higher dust loads to be carried successfully.

Air flow meter

The air flow meter is commonly an air rotameter. This instrument is a variable area fluid flow meter consisting of a vertical, tapered tube in which a metal or plastic float is free to move vertically. The tube is normally of glass but may be acrylic plastic for field use instruments. The position of the float in the tube is governed by the equilibrium between the weight of the float and the velocity pressure of the fluid flowing through the space between the float and tube wall.

Where the flow rate is constant or automatically controlled, there may be only an on-off indicator that the device is functioning. In such cases the flow rate through the sampling train is measured in the laboratory and checked after sampling. The flow rate should be checked with a sample in place. Aerosol on a filter, for example, can sometimes affect the flow rate drastically.

Similar considerations apply in sampling trains with an integral flow meter. The value shown by a rotameter or dry-gas meter is partly dependent upon the air pressure at the entry to the meter and upon the magnitude and frequency of pulsations in the air flow. Furthermore the short rotameters used on field instruments generally have poor accuracy. Measurement of the free air flow rate, or calibration of the flow meter should be performed in the laboratory before and after sampling, with all the elements of the sampling train in circuit. A pulsation damper may be necessary.

Flow control

The flow rate control may be manually operated if in the form of a needle valve or a simple pinch valve. When filters are employed the flow rate control may need repeated adjustment while sampling, since aerosols clog the filter. This effect may be mitigated by using a sampling train with a high internal resistance. Automatic flow-rate control may be obtained with a critical orifice. Otherwise, electrical or pneumatic means may be utilized.

A critical orifice is a simple and popular means of achieving constant flow. The principle of the method is to draw the air through the orifice under 'critical' flow conditions and constant upstream pressure. The volume flow rate of a gas into an orifice increases with the pressure differential across it until a point is reached when the air is moving through the orifice at a velocity equal to the velocity of sound through the gas. The volume flow rate then stays the same for any further increase in pressure differential. The downstream to upstream pressure ratio below which the volume flow rate becomes constant is known as the critical pressure ratio. This is approximately 0.53 for air through a well-rounded orifice. At atmospheric pressure upstream the flow becomes critical as the downstream pressure is reduced below 400 mm mercury. The critical pressure drop, 400 mm mercury, reflects a resistance to flow. This may be reduced by making a gradual enlargement or evase on the discharge side of the orifice in the form of a 1-in-5 enlargement for 15 orifice diameters. This reduces the necessary overall pressure drop because of the pressure recovery in the expansion piece. In practice, an orifice having an overall pressure drop of 100 mm mercury at critical flow is made fairly easily.

A critical orifice used for sampling aerosols should be calibrated from time to time as it may become worn by the dust passing through it. If a critical orifice is not used the flow rate may be maintained constant with devices downstream of the absorber or separator, such as by having a flow regulating valve followed by a pressure regulating valve downstream. The pressure drop necessary to maintain control with this latter system is about 250 mm mercury. Even with absorbers which have a resistance independent of the loading of contamination it is advisable to have a pressure regulator as the performance of most pumps tends to wander with time.

Pump

The most versatile suction pump for a custom built sampling train is a motor driven rotary pump but other, less powerful kinds may be adequate for some purposes, including centrifugal fans, diaphragm pumps and piston pumps. However, the pump must produce sufficient air horse power and draw the necessary flow rate through the sampling train under the most adverse conditions of air flow resistance as when a filter has a high dust load. The weight of the necessary pump and motor increases roughly in proportion to the pressure drop and the flow rate through the system. When making up a sampling train, it is helpful to measure the flow resistance contributed by each element, as it is often found that a high resistance contributed by a secondary element such as tubing, elbows, connections or other fittings can easily be reduced.

The very lowest flow rates, less than 100 ml min^{-1}. are accommodated by water displacement apparatus. Next are light-weight, fully portable devices that are used for personal samplers and employ a sample volume flow rate up to 5 litres per minute. A battery powered diaphragm pump or piston pump can be used for this purpose. Mechanical damping may be required with these if a uniform flow rate is needed, as is the case when aerosol preselectors are fitted. The battery should be capable of driving the pump for 8 hours without recharge.

Mains powered pumps are necessary for still higher flow rates, although it has to be recognised that this is achieved at the expense of some portability. Centrifugal fans are suitable where the pressure drop through the sampling train is no more than 15 mm mercury. Higher pressure drops of up to 100 mm mercury can be sustained with small multi-stage centrifugal turbines. Rotary pumps utilised in this work to produce a fairly high pressure differential of 50–350 mm of mercury are generally the sliding vane type. Multi-lobe blowers or gear pumps are used for the very highest volume flow rates at greatest pressure differential.

Power supply

Nickel-cadmium rechargeable batteries are commonly used on the smaller sampling trains and are essential for the greatest portability. Disposable dry cells are not very suitable but they have been used occasionally. In flammable atmospheres, where there is a risk of ignition or explosion only instruments approved for use in these environments by the competent authority may be employed.

The power supply used with area sampling is typically line current. Sometimes electricity is not available on site or is banned for reasons of safety, so compressed air or water ejectors may be considered. Compressed air ejectors can provide flow rates up to 200 l min^{-1} against 400 mm mercury pressure drop, utilising compressed air at a pressure of 1.5 atmosphere. Low sampling flow rates, up to 10 litres per minute may be obtained with small ejectors working from bottled carbon dioxide or compressed air.

Instrument performance parameters

The combination of sampling instrument and analytical method has to be selected with care. Where the detailed description of the exposure limit allows some choice in sampling instruments or analytical procedures the particular combination employed should be one which is specific for the substance in question and sensitive down to one tenth of the exposure limit. After collection some substances are difficult to analyse with accuracy and affiliation to a quality assurance scheme may be advisable (World Health Organisation, 1981; Whitehead and Woodford, 1981; West, 1986; Health & Safety Executive, 1988). Deviation from approved procedures, however desirable, should only be contemplated after satisfactory calibration has been done against the standard.

The minimum performance that should be demanded of instruments is largely a matter of personal judgement. A set of desirable features of atmospheric sampling instruments for use in harsh conditions in a factory is given below. Before adopting a particular instrument satisfactory assurance should be obtained regarding each of the following aspects in the conditions of use:

Air flow

In sampling instruments which include a pump the volume of air sampled is given by the combination of flow rate with time-piece or pump volume with stroke counter. The combination should be accurate to within 10% of the true flow at room pressure. Pumps can be readily calibrated but care must be taken to check the flow rate or pump volume used both before and after sampling.

Weight

A pump and all its attachments which is to be carried by an employee during work should not weigh more than 1 kg. An instrument which is to be carried about by a sampling technician during a work shift should not weigh more than 10 kg. An instrument which is to be carried into a work place and set down in one place should not weigh more than 20 kg.

Ruggedness

A pump and all its attachments carried by an employee should withstand being dropped from a height of 20 cm on to a concrete floor.

Maintenance

The pump and other moving parts of an air sampling instrument should be capable of being maintained satisfactorily by the user for a total of at least 250 samples, each of one work shift duration.

Collection efficiency

The efficiency of collection for the air contamination involved should be >90%.

Sensitivity

The range of use should be at least 0.1–1.0 × exposure limit. Given the sensitivity of the analytical method, in terms of the least detectable mass of substance, the minimum volume of the sample of air must be

$$\frac{10 \times \text{analytical sensitivity}}{\text{atmospheric exposure limit}}. \tag{12}$$

When the analytical sensitivity is measured in milligram and the exposure limit in mg m^{-3} the required minimum volume of the sample of air is given by expression (11) in cubic metres. Thus, when grab sampling the expression gives the minimum volume of the sampling flask or plastic bag. When drawing air continuously with a pump and the sampling flow rate is fixed the required minimum duration of each sample can be calculated. Similarly, when the duration of each sample is fixed the required minimum flow rate can be calculated.

Problem 11.1

The analytical sensitivity for measurement of the weight of a certain dust on a given filter medium is 0.01 mg and the atmospheric exposure limit of the dust is 0.1 mg m^{-3}. The most easily available sampling instrument runs with a volume flow rate of 2 l min^{-1}. Calculate the minimum duration of each sampling. Calculate the minimum volume flow rate of a sampling instrument needed for a 2-hour sampling period.

Solution

From expression (12),

$$\begin{aligned} \text{minimum volume of sampled air} &= \frac{10 \times 0.01}{0.1} \\ &= 1 \text{ m}^3 \end{aligned}$$

The volume flow rate of the most easily available instrument is 0.002 m^3 min^{-1}. The minimum duration of sampling which should be employed with the available sampling instrument is thus

$$\begin{aligned} \frac{1}{0.002} &= 500 \text{ min} \\ &= 8.33 \text{ h}. \end{aligned}$$

Had the sampling period been limited to 2 hours an instrument should be sought with a minimum volume flow rate of

$$\frac{1{,}000}{2 \times 60} = 8.33\,l\ \mathrm{min}^{-1}$$

Specificity

Specificity is measured in terms of cross-sensitivity. Cross-sensitivity is the ratio of the measured value of an interfering substance to the measured value of the substance to be determined. Total interferences present in the work environment under study should contribute $<10\%$ of the value of the measurement. This condition might be relaxed if the interferences are constant and the method is calibrated with this constant interference in place.

Accuracy

In the range 0.1–1.0 × exposure limit the differences between the mean of an infinite set of measurements and the true value should not be more than 20% of the true value. Calibration procedures are discussed below.

Precision

The overall scatter of a set of measurements of the same concentration, lying in the range 0.1–1.0 × exposure limit should have a coefficient of variation of less than 15%.

Analytical methods

The methods employed for analysing samples of air contamination are ones normally regarded as the type necessary for trace quantities because of the underlying need to limit the air sampling apparatus to the very smallest, portable kind. The older methods play an important role. However, in the determination of common hazardous substances sensitive physical instruments have largely supplanted traditional chemical procedures. Many companies and consultants contract out any required analysis to an accredited laboratory which has the specialised equipment and skills to use it. The principal kinds of laboratory apparatus which have found extensive application in analysing air samples for hazardous substances at work include the following items:

- Gas chromatography is employed extensively for separating mixtures of closely related organic gases and vapours for subsequent quantitative determination. The separated components commonly pass directly to a flame ionisation detector or flame photometer, the output from which, the chromatogram, is recorded on a strip chart recorder or given by a digital read-out.
- Ion chromatography is employed for acids generally and some salts.

- The ultraviolet spectrophotometer is used for the aromatic hydrocarbons when interferences are known to be low.
- The polarograph is used for aldehydes, chromates and chlorinated hydrocarbons.
- The mass spectrometer is finding increasing use for analysing complex gaseous mixtures as, over the years, its cost and complexity has progressively been reduced.
- Gravimetric procedures are employed for many dusts where detailed chemical analysis is not needed. Semi-micro and micro balances are required, with a sensitivity down to 0.1 microgram.
- High power microscopes are employed routinely for assessment of samples of asbestos and other fibrous dusts. They are also used in special investigations of the size distribution of aerosols and optical properties of others.
- X-ray diffraction is the traditional method of choice for silica determinations although infrared spectrophotometry is an alternative favoured by many analysts for quartz in coal mine dust.
- The atomic absorption spectrophotometer is used in the determination of many metals and their compounds of interest to occupational hygienists. The intrinsic sensitivity of the technique is sufficient to enable determination of particulate material on air filters used in low flow rate personal samplers.

Why calibrate?

There is a rule, without any known exception, which could almost be elevated to the status of a scientific law. The rule is that whenever a series of air samples is taken in industry, the results are all over the place. When this happens, don't sack the chemist—at least not immediately. Analytical 'plus-or-minus' errors, a constant worry to chemists, are invariably swamped by the vast fluctuations in concentration of air contamination. Nevertheless, there may be a systematic error in the air sampling or analysis which goes undetected. A method which 'always' reads low or always high must be rooted out. The allure of a method which always reads zero has to be resisted with an iron will. Likewise, a method which always seems to yield 10 ppm, no matter what the concentration, is not unknown and must be exposed somehow. None of these characteristics is unknown, unfortunately. Indeed, from the outset a healthy scepticism should be adopted in regard to all 'makers' claims and all 'approved and tested' methods.

An incorrectly calibrated instrument will give measurements that are systematically above or below the true value. Regular calibration of air sampling volume flow rate indicators with a genuine sample in the train is essential in all air sampling routines. But this will not necessarily reveal all or the largest or even the most frequent systematic errors or bias. Poor efficiency of collection by the absorber or filter is a more frequent source of error. The determination of the amount of contamination collected may also present real difficulties.

The principles of primary calibration are simple but the apparatus employed is specialised and practical experience with the methods is essential to be able to calibrate with confidence. The calibration of instruments for measuring air contamination at work shows up common sources of inaccuracy. These are conveniently separated into three interrelated factors:

(1) Error in the volume of air sampled at room temperature and pressure.
(2) Less than 100% efficiency of collection of the contamination.
(3) Unreliable determination of the quantity of contamination collected.

In practice it is difficult to totally avoid all three sources of error and calibration against known atmospheric concentrations is needed. The errors may be variable in different concentration levels so that calibration over at least the 10-fold range, 0.1–1.0 × exposure limit, is necessary. Some errors are variable with the duration of sampling so that calibration over the full working range in this regard is also necessary. Furthermore, collected samples are prone to degrade with time, as, for example, samples of substances having a high vapour pressure, so that the calibration protocol should include investigation of the effect of the time interval between collection and analysis.

Many sampling machines are so designed that a check on the flow rate at the end of sampling or the calibration of a fitted flow meter is straightforward using an accurate rotameter or dry-gas meter as the basis. Rotameters are commonly used for day-to-day routine checking of sampling trains. Low flow rate sampling trains, below 50 cm^3 min^{-1}, may be accurately calibrated with a bubble flow meter. The device consists essentially of a graduated tube such as a laboratory burette held horizontally. The inside is first wetted with detergent solution and a bubble is formed at the open end. The pump or other device to be checked is connected to the other end. With the aid of a stop watch the time taken for the bubble to move between two marks is measured. The volume between the marks divided by the time taken is the air volume flow rate drawn by the pump. Flow calibration kits employing bubble flow meters are available for volume flow rates from 5 cc min^{-1} up to 5 l min^{-1} by employing large burettes.

Some instruments which employ a high volume flow rate but low pressure drop present serious difficulties when attempts are made to use the standard methods for checking the flow rate. This is the case when the pressure drop across the instrument in use is low compared with the pressure drop across standard flow meters. In these instances the 'null-point' method should be used by which during calibration both the inlet and outlet of the instrument are maintained at atmospheric pressure precisely (Roach, 1966). Other instruments employ flow pumps which produce noticeable pulsation in the flow, particularly at low flow rates, which may give rise to serious error in the flow rate indicated by rotameters. Bubble flow meters do not suffer from significant error from this cause and should be used for pulsating or irregular flow.

Where exposure limits are expressed in terms of specified instruments or methods it is imperative that calibration of different instruments or methods include comparison with the original specification since that specification may itself have built in errors. This is especially true for sampling aerosols.

Facilities for calibration against known concentrations are available in most well equipped laboratories. However this is not always the case, especially with aerosols. Furthermore many occupational hygienists work for long periods without recourse to laboratory facilities other than can be carried in a suitcase. Fortunately, aside from grab sampling and direct reading instruments which give readings of the momentary concentration there is a simple but searching internal calibration test which can be added to flow rate calibration of instruments, known as 'twicing'. In this linearity check two identical sampling instruments are set up side by side and samples are taken for twice as long through one as the other. The instrument used for the shorter period is switched on shortly after its companion so that the middle of the two sampling periods approximately coincide. A laboratory test chamber is not needed. The occupational hygienist may carry out the test unaided. Replicate sampling may advantageously be conducted in industrial conditions, where factors such as high or low temperature, humidity, air movement and vibration are commonplace (Siem and Dickeson, 1983).

There are many possible variations of this straightforward linearity check, the principle being simply to check that if twice the contamination is taken by an instrument twice is found. Unfortunately, too few trusted methods come through this test completely unscathed. Badge type diffusive samplers are commercially available for multiple sampling at different rates through two or more badges mounted on a single holder. Twicing is recommended for instruments and sampling trains for measuring the atmospheric concentration of all hazardous substances, whether they be gases, vapours or aerosols, particularly the latter. Special arrangements for primary calibration in the laboratory are warranted for instruments, sampling trains, methods and procedures which fail this linearity test by 10% or more.

Primary calibration—gases and vapours

In the case of instruments used for measuring gases and vapours a periodic overall calibration may be accomplished against standards produced in the occupational hygiene laboratory by dispersing a known mass or volume of the gas or vapour in a static chamber of known volume (Nelson, 1971; Health and Safety Executive, 1984; American Society for Testing and Materials, 1986). Direct reading instruments, grab sampling methods and indicator tubes requiring relatively small volumes of contaminated air are normally calibrated with the aid of static chambers. Large chambers are most convenient but the difficulty of making a large chamber leak tight cannot be overestimated. Surface adsorption may also be serious. Before-and-after trials to determine the decay rate through leaks and surface adsorption are essential. Flexible grab sampling bags can be used for small volumes, less than 100 litres, and have certain advantages as static chambers (Vinsjansen and Thrane, 1978; Melcher and Caldecourt, 1980). The standard is not diluted when air is withdrawn. Repeated filling and emptying of bags is necessary for the atmosphere to come into equilibrium with the container and much kneading of the bag is needed to ensure thorough mixing (American Society for Testing and Materials, 1980a). A hypodermic gas tight micro-syringe of appropriate size may be employed to introduce a known volume of pure gas or liquid. Two stages of dilution may be necessary. Otherwise, cylinders of known concentration in air are commercially available for some common gases and vapours and can be made to order for others. The moisture content of test gases under pressure is usually kept very low for reasons of stability. This is a distinct disadvantage with detector tubes which rely upon a moderate moisture content for the chemical reaction to proceed satisfactorily (Leichnitz, 1983).

Standard concentrations may also be produced dynamically where the apparatus to do this is available (Barratt, 1981). Gas or vapour is released continuously at a known rate into a metered stream of air so as to provide a continuous stream of contaminated air for sampling. Dynamic methods are rapidly adjustable and ample time may be taken to allow conditions to settle into equilibrium. Many ingenious methods have been devised for laboratory application, including a range of permeation devices to produce low flow rate reliably (American Society for Testing and Materials, 1980b). Some are commercially available (Drew and Lippmann, 1978; Barratt, 1981).

Rotameters may be employed to meter gases, whereas for vapours a means for vaporising liquid at a measured mass or volume flow rate is employed. One method for vapours is to employ a motorised syringe feeding the liquid on a wick or other vehicle for supplying contamination at a constant rate to a metered stream of air (Nelson and Griggs, 1968; British Standards Institution, 1970; Hunt, McNally and Smith, 1973; Brookes, 1981). Another is to use a 'diffusion' cell in which liquid evaporates from a central tube and the vapour mixes with a metered supply of air in a small mixing chamber (Fortuin, 1956; Altshuller and Cohen, 1966; McCollough and Worley, 1979; Ho, Guilbault and Rietz, 1980). The rate of evaporation of liquid is measured by loss in weight or volume over a period of time. The same methods may also be applied in an exhaust ventilation system whose flow rate has been accurately determined. A known volume flow rate of tracer gas is released at the hood entrance and the concentration is measured 8–10 duct diameters downstream.

Reasonable care needs to be taken over absorption on surfaces of the calibration apparatus or chamber and on the walls of the sampling probe, if any. In most instances the system comes into equilibrium within 2 or 3 minutes. Comprehensive calibration runs should include varying the temperature and humidity as gas and vapour measurements are liable to be affected by both these variables.

Primary calibration—aerosols

In the case of instruments for measuring aerosols primary calibration against a known atmospheric concentration of the aerosol is complex. With gases and vapours adsorption on the walls of the chamber, tubing, joints and so on can be kept small by appropriate choice of materials and by waiting for the system to come to equilibrium. However, surface deposition and agglomeration of aerosol in a test chamber or other apparatus is virtually impossible to prevent or predict with sufficient accuracy beforehand. It is virtually impossible to take a known weight of dust and hold it in dispersion in a known volume of air. With gases and vapours the air entry velocity into sampling apparatus is of little importance but with aerosols, particularly for coarse particles, the collection efficiency of an aerosol sampler varies markedly according to the orientation of the inlet in a draught (May, Pomeroy and Hibbs, 1976). In any event the performance of most instruments for measuring aerosols also varies markedly with the nature and size distribution of the aerosol of interest.

Comparison against a laboratory method employing isokinetic sampling is feasible in some laboratories. However, for most occupational hygienists the most practical form of calibration is comparison against an established instrument of known reliability and accuracy employed as a secondary standard. This may sometimes be done in a dusty workshop or, where feasible, in large diameter exhaust ventilation duct. Calibration in the field setting has the advantage of presenting the aerosol in the physical form, size distribution, state of electrification and so on which is desired. Such calibration should be performed in a range of concentrations, 0.1–1.0 × exposure limit.

References

Aerosol Technology Committee of the American Industrial Hygiene Association. Am. Ind. Hyg. Assoc. J. **31** 133 (1970).

Altshuller A P and L R Cohen. Anal. Chem. **38** 760 (1966).

American Conference of Governmental Industrial Hygienists. Threshold Limit Values and Biological Exposure Indices for 1988–1989. ACGIH, PO Box 1937, Cincinnati, Ohio 45201, USA (1988b).

American Society for Testing and Materials. Standard test method for C1 through C5 hydrocarbons in the atmosphere by gas chromatography. ASTM D2820–72 (Reapproved 1978). In: Annual book of ASTM standards, Part 26. American Society for Testing and Materials, Philadelphia, USA (1980a).

American Society for Testing and Materials. Standard practice for calibration techniques using permeation tubes. ASTM D3609–79. In: Annual book of ASTM standards, Part 26. American Society for Testing and Materials, Philadelphia, USA (1980b).

Armbruster L, H Breuer, J Gebhart and G Neulinger. Part. Charact. **1** 96 (1984).

Baker B B. Am. Ind. Hyg. Assoc. J. **43** 98 (1982).

Barnes E C and G W Penney. J. Ind. Hyg. & Toxicol. **20** 259 (1938).

Barratt R S. Analyst **106** 817 (1981).

Barrer R M. Trans. Faraday Soc. **35** 628 (1939).

Bartley D L, L J Doemeny and D G Taylor. Am. Ind. Hyg. Assoc. J. **44** 241 (1983).

Bartley D L, M L Woebkenberg and J C Posner. Ann. Occup. Hyg. **32** 333 (1988).

Blackford D B, G W Harris and G Revell. Ann. Occup. Hyg. **29** 169 (1985).

Bodek T and K T Menzies. Ion chromatographic analysis of formic acid in diesel exhaust and mine air. In: G Choudhary (Editor). Chemical hazards in the workplace—monitoring and control. American Chemical Society, Washington DC, USA (1981).

British Standards Institution. Gas detector tubes. BS 5343 (1976).

Brookes B I. Analyst **106** 403 (1981).

Brown R H, J Charlton and K J Saunders. Am. Ind. Hyg. Assoc. J. **42** 865 (1981).

Burdekin J T and J D Dawes. Brit. J. Ind. Med. **13** 196 (1956).

Carver J, G Nagelschmidt, S A Roach, C E Rossiter and H S Wolff. Min. Engr. No. 21 p. 601 (1962).

Chmara P. Can. Min. Metall. (CIM) November, 1171 (1969).

Chung K Y K and N P Vaughan. Ann. Occup. Hyg. **33** 591 (1989).
Cohen B S, N H Harley and M Lippmann. Am. Ind. Hyg. Ass. J. **45** 187 (1984).
Coleman S R. Am. Ind. Hyg. Assoc. J. **44** 929 (1983).
Costello R J, P M Eller and R D Hull. Am. Ind. Hyg. Ass. J. **44** 21 (1983).
Courbon P, R Wrobel and J-F Fabries. Ann. Occup. Hyg. **32** 129 (1988).
Coutant R W, R G Lewis and J Mulik. Anal. Chem. **57** 219 (1985).
David H A. Biometrika **43** 449 (1956).
Davies C N. Brit. J. Ind. Med. **6** 245 (1949).
Davies C N. Brit. J. Ind. Med. **9** 120 (1952).
Demange M, J C Gendre, B Herve-Bazin, B Carton and A Peltier. Ann. Occup. Hyg. **34** 399 (1990).
Dixon W J. Biometrics **9** 74 (1953)
Drew R T and M Lippmann. Calibration of air sampling instruments. In: Air sampling instruments for evaluation of atmospheric contaminants. pp. I1-I38. ACGIH, Cincinnati, Ohio, USA (1978).
Ellwood P A, J A Groves and M I Pengelly. Ann. Occup. Hyg. **34** 305 (1990).
European Council. On the manufacture of detector tubes to control the atmosphere in work places. Resolution AP (74) 4 (March, 1974).
Flint T R. Can. Min. J. October, 82 (1965).
Fortuin J M H. Anal. Chim. Acta **15** 327 (1956).
Fraust C L. Am. Ind. Hyg. Assoc. J. **36** 278 (1975).
Gage J C. Analyst **85** 196 (1960).
Green H L and H H Watson. Physical methods for the estimation of the dust hazard in industry, with special reference to the occupation of the stonemason. Medical Council of the Privy Council, Special Reprint 199, H M Stationery Office, London (1935).
Greenberg L and G W Smith. A new instrument for sampling aerial dust. US Bureau of Mines, R. I. 2392 (1922).
Grob R L. (Editor). Modern practice of gas chromatography. John Wiley and Sons, New York (1977).
Gurka D F, S Billets, J W Brasch and C J Riggle. Anal. Chem. **57** 1975 (1985).
Hallberg B-O and J Rudling. Ann. Occup. Hyg. **33** 61 (1989).
Hamilton R J. In: Inhaled particles and vapours II. Pergamon Press, Oxford, UK (1967).
Harper M and C J Purnell. Am. Ind. Hyg. Assoc. J. **48** 214 (1987).
Hatch T and W C L Hemeon. J. Ind. Hyg. Toxicol. **30** 172 (1948).
Health & Safety Executive. Methods for the determination of hazardous substances—Generation of test atmospheres of organic vapours by the syringe injection technique, MDHS Series, 3. H M Stationery Office, London (1984).
Health & Safety Executive. General methods for the gravimetric determination of respirable and total inhalable dust. MDHS Series, 14. H M Stationery Office, London (1986).
Health & Safety Executive. Workplace Analysis Scheme for Proficiency—WASP. H M Stationery Office, London(1988).
Hearl F J and M P Manning. Am. Ind. Hyg. Assoc. J. **41** 778 (1980).
Ho M H, G G Guilbault and B Rietz. Anal. Chem. **52** 1489 (1980).
Hohfeld M. Archiv. Naturlehre **2** 205 (1824).
Hollingdale-Smith P A. Ann. Occup. Hyg. **22** 86 (1979).
Hunt E C, W A McNally and A F Smith. Analyst **98** 585 (1973).
International Standards Organisation. Air quality—particle size fraction definitions for health-related sampling. Technical Report No. ISO/TR 7708. International Standards Organisation, Geneva, Switzerland (1983).
Jones L C, Q E Billings and C Lilas. Am. Ind. Hyg. Assoc. J. **42** 104 (1981).
Keenan R G. Direct reading instruments for determining concentrations of aerosols, gases and vapors. In: The industrial environment—its evaluation and control. p. 181.U.S. Department of Health, Education, and Welfare. Public Health Service, Center for Disease Control, National Institute for Occupational Safety and Health, US Government Printing Office, Washington DC, 20402, USA (1973).
Kim C S, D Triyillo and R McDonald. J. Aerosol Sci. **14** 633 (1983).

Kotze R H. (Chairman). Final report of the Miners' Pthisis Committee. GPSO, Pretoria, South Africa (1919).
Kupel R E and L D White. Am. Ind. Hyg. Assoc. J. **32** 456 (1971).
Lamb A B, W C Bray and J C W Frazer. Ind. Eng. Chem. **12** 213 (1920).
Lamb A B and C R Hoover. US-Patent 1 321 1062 (1919).
Leichnitz K. Chem. Int. **4** 12 (1982).
Lilienfeld P and R Stern. Personal dust monitor—Light scattering. US Bureau of Mines Contract Report, No. H0308132, US Dept. of the Interior, Pittsburgh, USA (1982).
Lindeken C L, F K Petrock, W A Phillips and R D Taylor. Health Physics, **10** 495 (1964).
Lippmann M and W B Harris. Health Physics, **8** 155 (1962).
Littlefield J B and H H Schrenk. Bureau of Mines midget impinger for dust sampling. US Bureau of Mines, R. I. 3360, Dept. of Interior (1937).
Liu B Y H and K W Lee. Environ. Sci. Technol. **10** 345 (1976).
Lockhart L B, R L Patterson and W L Anderson. Characteristics of air filter media used for monitoring airborne radioactivity. NRL Report No. 6054. US Naval Research Laboratory, Washington DC (1964).
Lodge O. J. Soc. Chem. Ind. **5** 572 (1886).
Lugg G A. Anal. Chem. **40** 1072 (1968).
Lynch J R. Measurement of worker exposure. In: Patty, Industrial Hygiene and Toxicology Vol 3A, pp. 569–615. John Wiley & Sons (1985).
Mark D. Ann. Occup. Hyg. **17** 35 (1974).
Mark D. Ann. Occup. Hyg. **34** 281 (1990).
Mark D and J H Vincent. Ann. Occup. Hyg. **30** 89 (1986).
May K R, N M Pomeroy and S Hibbs. J. Aerosol Sci. **7** 53 (1976).
May W J. Ann. Occup. Hyg. **33** 69 (1989).
McCollough P R and J W Worley. Anal. Chem. **51** 1120 (1979).
Medical Research Council. Chronic pulmonary disease of S. Wales coal miners. Special Report Series No. 244. H M Stationery Office, London (1943).
Megaw W J and R D Wiffen. Int. J. Air Wat. Poll. **1** 501 (1964).
Melcher R G and V J Caldecourt. Anal. Chem. **52** 875 (1980).
Michaelis H. Ztschr. Hyg. Infektionskr. **9** 389 (1890).
Nair K R. Biometrika **39** 189 (1952).
Nelson G O and K S Griggs. Rev. of Sci. Inst. **39** 927 (1968).
Olin J G and G J Sem. Atmospheric Env. **5** 653 (1971).
Owens J S. Proceedings Royal Soc. London, A. **101** 18 (1922).
Palmes E D. Ann. Occup. Hyg. **22** 85 (1979).
Palmes E D and A F Gunnison. Am. Ind. Hyg. Assoc. J. **34** 78 (1973).
Palmes E D, A F Gunnison, J Di Mattio and C Tomczyk. Am. Ind. Hyg. Assoc. J. **27** 570 (1976).
Quillium J H and J A L Kruss. J. Mine Vent. Soc. S. Afr. **25** 60 (1972).
Reiszner K D and P West. Environ. Sci. Technol. **8** 526 (1973).
Roach S A. Am. Ind. Hyg. Assoc. J. **27** 135 (1966).
Roach S A, E J Baier, H E Ayer and R L Harris. Am. Ind. Hyg. Assoc. J. **28** 543 (1967).
Roebuck B, N P Vaughan and K Y K Chung. Ann. Occup. Hyg. **34** 263 (1990).
Samimi B and L Falbo. Am. Ind. Hyg. Assoc. J. **44** 402 (1983).
Schomburg G. J. Liq. Chromatograph. Gas Chromatograph. **5** 304 (1987).
Siem H J and J A Dickeson. Am. Ind. Hyg. Assoc. J. **44** 562 (1983).
Spurny K and J Pich. Int. J. Air Wat. Poll. **8** 193 (1964).
Swarin S J and F Lipari. J. Liq. Chromatog. **6** 425 (1983).
Tolman R C, R H Gerke, A P Brooks, A G Herman, R S Mulliken and H D Smyth. J. Am. Chem. Soc. **41** 575 (1919).
Underhill D W. Am. Ind. Hyg. Assoc. J. **44** 237 (1983).
US Public Health Service. National Institute for Occupational Safety and Health: Certification of gas detector tube units. Federal Register **38** 11458–63 (May, 1973).
Vaughan N P, B D Milligan and T L Ogden. Ann. Occup. Hyg. **33** 331 (1989).
Verma D K, A Sebestyen and D C F Muir. Ann. Occup. Hyg. **32** 451 (1987).
Vincent J H and Armbruster. Ann. Occup. Hyg. **24** 245 (1981).

Vincent J H and D Mark. Am. Ind. Hyg. Assoc. J. **48** 454 (1987).
Vinsjansen A and K E Thrane. Analyst **103** 1195 (1978).
Vinson R P and K L Williams. Performance evaluation of two dust monitors. USBM IC 9162, US Bureau of Mines, Pittsburgh USA (1987).
Walton W H. J. of Phys. D: Applied Phys. Suppl. No. 3, 529 (1954).
West N G. Analyt. Proc. **23** 330 (1986).
Whitehead T P and F P Woodford. J. Clin. Pathol. 34 947 (1981).
Wolff H S and S A Roach. The conicycle selective sampling system. In: Inhaled particles and vapours (Edited by C N Davies), p. 460. Pergamon Press, London (1961).
Wood G O and R G Anderson. Am. Ind. Hyg. Assoc. J. **36** 538 (1975).
Wood J D and J L Birkett. Ann. Occup. Hyg. **22** 299 (1979).
World Health Organisation. External quality assessment of health laboratories. World Health Organisation, Copenhagen (1981).
Wright B M. Brit. J. Ind. Med. **11** 284 (1954).
Wright B M. Science **118** 195 (1953).

Bibliography

Aitken J. Collected scientific papers (Edited by C G Knott). Cambridge University Press, London (1923).
American Conference of Governmental Industrial Hygienists. Advances in air sampling. Lewis Publishers Inc., Michigan, USA (1988a).
American Conference of Governmental Industrial Hygienists. Air sampling instruments for evaluation of atmospheric contaminants. ACGIH, Cincinnati, Ohio, USA (1978).
American Conference of Governmental Industrial Hygienists. Particle size-selective sampling in the workplace. ACGIH, Cincinnati, Ohio, USA (1984).
American Society for Testing and Materials. Sampling and calibration for atmospheric measurements, STP-957. American Society for Testing and Materials, Philadelphia, USA (1986).
Berlin A, R Brown and K J Saunders (Editors). Diffusive sampling—an alternative approach to workplace air monitoring. Proceedings of an International Symposium in Luxembourg, 22–26 Sept., 1986. Royal Society of Chemistry, London (1987).
British Standards Institution. Methods for the preparation of gaseous mixtures. BS 4559 (1970).
Fuchs N A. The mechanics of aerosols. Pergamon Press, Oxford (1964).
van der Hulst H C. Light scattering by small particles. John Wiley and Sons, New York (1957).
Kerker M. The scattering of light and other electromagnetic radiation. Academic Press, New York (1969).
Leichnitz K. Detector tube measuring techniques. Ecomed-verlagsgesellschaft mbh, Justus-von-Leibig-Str. 1, Landsberg/Lech, Germany (1983).
Liu B Y H. (Editor). Fine particles. Academic Press, New York (1976).
Liu B Y H, D Y H Pui and H Fissan. Aerosols. Elsevier Science, Amsterdam (1984).
Nelson G O. Controlled test atmospheres: Principles and techniques. Ann Arbor Science Publishers Inc., Ann Arbor, Michigan, USA (1971).
NIOSH. Manual of analytical methods. National Institute for Occupational Safety and Health, Cincinnati, Ohio, USA (1985).
Shaw D T. (Editor). Fundamentals of aerosol science. John Wiley, New York (1978).
Silverman L, C E Billings and M W First. Particle size analysis in industrial hygiene. Academic Press, New York (1971).
Tyndall J. Floating matter of the air in relation to putrefaction and infection. Appleton-Century Co., New York (1882).
Vincent J H. Aerosol sampling science and practice. John Wiley & Sons (1989).

CHAPTER 12

Measuring Exposure at Work

The task of designing a programme of measurements to establish the pattern of exposure of an employee to hazardous substances is complicated by the mobility of many employees who move about, in and out of different parts of a work room or work area (Corn, 1985). Commonly there are large differences in exposure between one location and another. Most air contamination, for example, usually comes from a few concentrated sources which release widely meandering plumes (Jones, 1983). Furthermore, the tasks employees undertake often involve the liberation of air contamination in their vicinity.

A notorious characteristic of the air contamination in occupational environments is that at any one place the concentration of contamination is continually changing with respect to time. This means that air sampling must be spread over a sufficient period of time and over a sufficient number of locations to reveal the full picture. These features are generally more marked with aerosols than with gases and vapours, but they are always very striking.

A reconnaissance of atmospheric exposure is required early in the establishment of systematic control of hazardous substances, on the introduction of a new source of exposure or the curtailment of an old one, on the introduction of an atmospheric exposure limit or when the exposure limit is changed in some respect. This reconnaissance is of limited duration but broad based, lasting a few days or at most a month. Its principal purpose is to aid in setting priorities for continued surveillance. A more intense, focused type of survey might be termed a sampling 'campaign' designed in the light of previous results, whether from a brief reconnaissance or continued surveillance.

Limited spot sampling in the work environment may be sufficient for reconnaissance of groups working where atmospheric exposure is well below the exposure limit of the substance in question. Otherwise, exposure associated with a certain job-exposure group could be established with a comprehensive survey of personal exposures or by other, related means. In medium sized and large companies such a survey may take several weeks to complete satisfactorily.

Area sampling may be helpful in determining atmospheric exposure and should not be dismissed out of hand. The primary use of area sampling is to investigate individual sources of contamination. Furthermore, in circumstances where a high volume of air has to be drawn through the sampling device to collect enough contamination for accurate analysis area sampling may be the only feasible method because of the size, weight or complexity of the available sampling instruments.

Biological monitoring of blood, urine or exhaled air may be the most feasible means to make an assessment of exposure to hazardous substances in certain occupations where skin absorption and ingestion of hazardous substances are the most important routes of entry to the body. An obvious example of this is in spraying operations where the employees don masks simply to reduce their atmospheric exposure to the spray (Kay, Monkman, Windish, Doherty, Pare and Racicot, 1952). Dermal exposure to a hazardous substance may be investigated directly by employing patch tests. The patches are analysed for the substance itself or analysed for relatively innocuous tracers added to the hazardous substances at an earlier stage (Bonsall, 1985; Fenske, 1988).

What is the intensity and duration of exposure?

Developing even the simplest programme of atmospheric sampling or biological monitoring will involve a number of distinct tasks. These tasks will be examined in detail in this chapter. But the most important thing to realise at the outset is that the precise objective(s) of the programme should be identified and appraised before beginning to execute the programme, if possible, since the objectives govern the design of the programme. The discipline of 'thinking through' the programme should be maintained against all pressures to get the sampling started. A typical breakdown of activities in the development of a sampling programme of a fairly serious nature might be as follows:

(1) Deciding exactly what the programme is meant to achieve.

Analyse the objectives. Are they to measure the short term or long term exposure of employees and if so whom, when and why? Is the aim to test compliance and if so with what standard, rule or regulation? Are the objectives to measure emissions and if so from what machines, plant or processes.

(2) Deciding how the programme should achieve the objectives.

Design the programme.

(3) Preparing the programme.

Decide exactly who is going to do what and when.

(4) Checking that the programme works correctly.

Test it for faults.

(5) Writing instructions for those who are to conduct the programme.

Document the programme.

(6) Correcting, changing and improving the programme.

Periodic maintenance and amendment.

Personal and area atmospheric sampling

Until about 1960 the usual method of atmospheric sampling for hazardous substances in work places was by area sampling. The sampling instruments, which were often heavy and cumbersome, were placed somewhere in the vicinity of employees so as to reflect, as far as possible, the concentration of contamination in the inhaled air. The 1960s opened the era of the personal sampler for routine use (Sherwood and Greenhalgh, 1960). The standard personal sampler for air contamination by hazardous substances consists of a sampling head clipped to the lapel and a tube leading to a pocket-sized, battery-powered air pump, clipped to the belt. Passive personal samplers are sometimes employed. These do not use a mechanical pump and consist solely of a sampling head on the lapel into which air contamination penetrates by diffusion.

When personal samplers are used there is an implicit assumption that the atmospheric concentration of air contamination at the lapel is equal to that inhaled. In certain jobs this is not so and another technique which features a probe tube sampling closer to the mouth and nose must be used.

The air sampling strategy, precisely which individuals are studied, where samples are taken, and when, are all arranged so as to suit the perceived purpose of investigations. Most reconnaissance sampling and campaigns come under one of four main headings:

(a) *Employee exposure*

When the aim is to establish the exposure of all employees exposed significantly the unit period of exposure is the whole time at work from 'clock in' to 'clock out'. Personal sampling would be employed if reasonably practicable and on a sample representative of all the employees significantly exposed. Within this group the sampling may sometimes be concentrated on those individuals who are believed to have the highest exposure.

(b) *Compliance testing*

When the goal is to demonstrate compliance or otherwise with a regulation or rule the strategy employed is largely dictated by that regulation and by the methods employed by the enforcing authority. Testing compliance is essentially backward looking, to see whether or not the conditions complied at the time of the measurements. It is deliberately designed with scant regard to conditions long past or to future prospects (Roach, Baier, Ayer and Harris 1967; Coenen and Riediger, 1978; Rappaport and Selvin, 1987; Selvin, Rappaport, Spear, Schulman and Francis, 1987). By testing for compliance is often really meant testing for non-compliance. For this reason it is sometimes criticised as being essentially negative in character. But this may be precisely what is required. Viewed in that light it can be reduced to selecting the worst conditions after a period of work place observation, followed by very limited measurements (Roach, Baier, Ayer and Harris, 1967; Leidel, Busch and Lynch, 1977; Ulfvarson, 1983).

Unfortunately there is a basic flaw in the technique when by 'compliance' is meant that *no* measurement should exceed the exposure limit. In these terms making just one or two measurements in worst case conditions may be all that is needed to prove the opposite, that is, non-compliance (Tuggle, 1981; Rock, 1981,1982; Rappaport, 1984; Rappaport, Selvin, Spear and Keil, 1981). Also, the more measurements that are made the more likely the environment will fail the compliance test. The flaw lies in the definition by which non-compliance is the outcome when just one result is in excess of the limit. A more realistic definition of compliance would be one couched in terms of the probability of the limit not being exceeded (Tuggle, 1982; Rappaport, 1991).

(c) *Engineering control*

When the purpose is primarily to investigate engineering control measures the focus of attention is on the greatest source(s) of contamination and the ventilation provisions. A sampling campaign is needed. The sampling may be mostly area sampling, related to fixed machines and processes, rather than employees, so as to locate the control modifications that may be required (Hubiak, Fuller, VanderWerff and Ott, 1981). Direct reading instruments and short term samplers are especially useful to locate the time and source of processes giving rise to high employee exposure. The strategy is essentially forward looking; aiming to avoid future problems.

Some processes fluctuate so that air contamination is liberated in surges. Investigation of the circumstances causing the surges is necessary to indicate the steps necessary for their reduction. Local exhaust ventilation may be designed to control the maximum surge. A reconnaissance is required to determine the time, frequency and magnitude of all surges and grab sampling instruments can give more useful results than averaging instruments.

(d) *Personnel control*

When the purpose is to decide which jobs, tasks or employees need improved personnel control the focus of attention is on those employees who may benefit by that control. Personal sampling would be the rule in an exposure reconnaissance of specific personnel and would be employed during the conduct of the jobs and/or tasks in question.

The three main ways by which employee's exposure to air contamination is measured may be classified as:

- Personal sampling.
- Area sampling coupled with time studies.
- Area sampling alone.

A comprehensive reconnaissance would include measurements on or near at least 1 in 10 of the employees exposed to more than one tenth the exposure limit. At least 1 in 10 of these would be measured twice to ease analysis of the variation of exposure between different days and sampling positions.

An alternative to sampling in the breathing zone is to have a number of fixed positions or 'area' sampling stations, employing the highest volume flow rates. Area measurements, when coupled with time studies of the employees, enable an appropriate weighting to be given to the results from each sampling station. In the period 1935–1965, before personal sampling became popular, this procedure was standard practice (Bloomfield, 1933; Dallavalle, 1939; Drinker and Hatch, 1954; Breslin, 1966; Baretta, Stewart and Mutghler, 1969). Making time studies is time consuming and unpopular with employees but may be necessary if personal sampling is not technically feasible or practical. Area sampling techniques are not always capable of accurately measuring an individual's daily exposure experiences, especially should these involve exposure sources close to the breathing zone or unusual incidences such as chemical spills or exposures outside the monitored area.

The third procedure is suitable when the results from either of the two previous methods show, unusually, that the concentration does not vary significantly from one place to another in a work-room. In such a case a single, fixed point sampling station located in the middle of a work-room could suffice. Some thought would need to be given to the period of exposure. Cross questioning supervisory staff may enable the period of exposure to be established with sufficient accuracy.

A reconnaissance is essentially of a short term nature. Most manufacturing operations have a distinct weekly, monthly and annual cycle superimposed upon the process cycle. Sampling for the purpose of continuing health risk surveillance has to be spread out so that there is no part of the year which has not been covered. Continuing surveillance, which is discussed in detail later, is essentially of a long term nature.

Variation from time to time over a work-shift

The fluctuations in concentration of air contamination in work places are dependent primarily on the interaction between the varying time-rate of emission of contamination, the highly variable dispersion by air turbulence and the smoothing effects of general ventilation. The fluctuations are further smoothed to a limited extent by collecting samples with a pump operating for a significant length of time or through employing continuous recorders which have a significant time constant.

The following causes are significant:

- Process changes over time
- Individual work practices
- Large scale air turbulence
- Build-up at the start of operations and die-away at the end
- Accidents—spills, equipment/plant failure, shorts, operator error and so on

Whether it be on a person's lapel or at some fixed location the concentration at one sampling station varies continually with time. The most common method of sampling air is to draw the air through an absorber or filter at a constant flow rate for a measured period of time. The weight or volume of substance collected divided by the volume of air drawn through the filter is the time weighted average concentration over the period the sample was being taken. As a general rule, when measuring time weighted average concentrations the shorter the averaging period the greater the observed variability, that is the higher will be the observed maximum concentration and the lower will be the observed minimum concentration. When the atmospheric concentration of the material is constant or very nearly constant the evaluation of the environment is fairly straightforward. There can, however, be few industries in which this situation holds. The importance of a given atmospheric concentration of harmful material depends on its magnitude and its duration. Some knowledge of the variability of concentration with time is thus essential for a proper appraisal of the risk to health. Further, from the frequency and magnitude of past variations it may be possible to estimate the risk of a dangerous situation in the near future.

The ventilation of some factories and workshops is very dependent on the weather conditions. Even in places such as underground mines, where the ventilation current is rigidly controlled, the operations which give rise to air contamination are so numerous and variable that large fluctuations in concentration still occur.

It has been observed repeatedly in industry that at a given location the temporal distribution of atmospheric concentration of contamination is commonly lognormal in character, at least over the central portion of the distribution (Leidel, Busch and Lynch, 1977). Industrial hygiene data on atmospheric concentration of hazardous substances are generally reported to have geometric standard deviations ranging between 1.5 and 3.0 (Oldham, 1953; Juda and Budzinski, 1967; Hounam, 1965; Gale, 1965; Sherwood, 1966; Breslin, Ong, Glauberman, George and LeClare, 1967; Gale, 1967, Coenen, 1971; Jones and Brief, 1971; Langmead, 1971; Sherwood, 1971; King, Conchie, Hiett and Milligan, 1979; Kromhout, Ikink, De Haan, Kroppejan and Bos, 1988). Moreover the spatial distribution is also quasi-lognormal (Oldham and Roach, 1952). Possible explanations of this persistent property do not seem to have been discussed very much in the literature. The lognormal distribution would arise from the law of proportionate effect by which any change in a variate is a random proportion of the previous value of the variate (see Appendix 1). Nevertheless it still remains to explain convincingly why, in an industrial setting, the intensity of air contamination at different times should follow the law of proportionate effect very closely.

There are many factors which favour a lognormal distribution. For example, emission of contamination is intrinsically positive, and this is a necessary condition for a lognormal distribution. Furthermore, increments and decrements in emission rate are commonly through time rate processes such as material production rate, grinding rate, evaporation rate, leak rate and so on, which would give rise to proportionate rather than independent changes. Next, control of the emissions is commonly by ventilation; that is, either by dilution of room air with fresh air, causing proportionate decrease in concentration or by local exhaust causing removal of air contamination proportionate to its local concentration in the vicinity of the exit from the room. Over time the concentration distribution comes to a state of equilibrium between input and output. The conclusion from such considerations is that a lognormal characteristic would be robust, persistent and would show marked temporal auto-correlation, as indeed is the case.

There is movement of air into and out of a room as well as mixing within it. Insofar as the air at a particular location is the same air as was at a location upstream a short time previously it can be appreciated that concentration distribution over space is in some degree a reflection of the distribution over time and vice versa. Such considerations lead one to expect a skew, auto-correlated and quasi-lognormal spatial distribution of concentration.

Despite the favoured lognormal character of these distributions there are nevertheless other features of the work environment which would be expected to cause departures from lognormal behaviour at the extremes of the distribution. Thus individual workers exposed to airborne dust and odoriferous vapours would tend to avoid the nuisance of the very highest concentrations and seek the lowest. Planned intervention at moments of high concentration is also found in sophisticated control strategies. At its simplest, for example, a fan can be switched on and off by electrical or mechanical transducers sensitive to atmospheric concentration. The fan is switched on above a certain concentration and switched off below it. Another factor is the exposure during the period between arrival at work and the start of activity in the work place (getting ready) and a similar period at the end of the shift, before leaving the premises. It is necessary to reject all the zeroes before fitting a lognormal distribution. The precise moment exposure begins or ends is indeterminate.

Coping with extreme values lasting momentarily

Despite the general recognition of the lognormal distribution of temporal variation it is still quite common to find extremely high results have been rejected from a set before fitting the distribution. Occasionally they are even eliminated altogether as being 'obviously' false and not deserving serious consideration before any statistical analysis is attempted. This is clearly undesirable, for in some situations extreme values are the most informative ones in the set.

The following procedure is recommended. First logarithms are taken of all the data. The mean and standard deviation of the logged data is determined, which should be a set of at least 10 results, and preferably at least 20. If the highest value is less than 2.5 standard deviations above the mean it should not be rejected. A value lying between 2.5 and 3 standard deviations above the mean is grounds for alerting the investigator to consider whether any other aspect of the sampling and analysis has given cause for suspicion. If the highest value is more than 3 standard deviations above the mean it should be regarded as significantly too high and definitely warranting detailed investigation.

If an extreme value is found to be significantly too large its rejection is actually only justifiable when a check on the field work or laboratory analysis reveals a causal circumstance which accounts for its existence. Examples would be an error made in measurements or calculation, the unwitting inclusion of a maintenance man among a group of production workers being investigated, a sample being accidentally dropped on the floor, and so on. In such cases the extreme value may properly be excluded from further analysis of the set.

Otherwise, an important consideration to be borne in mind in judging extreme values is that the validity of such statistical tests is conditional on the population having a lognormal distribution. If this is not the case, the tests are practically worthless since they are very peculiarly dependent on the lognormal condition. Unless there is very good evidence that the result is false, no matter how high, it should not be rejected, but considered as evidence that the distribution underlying distribution is not lognormal.

Sampling from defined job-exposure groups

It is recommended that a total of at least 10 full shift or half shift samples are taken in each job-exposure group. It is not essential to measure the exposure of all persons listed in a job-exposure group but the more the better, and in large groups a cross section or representative sample of at least one in ten should be studied for at least one complete shift or half shift. The half shift option is acceptable provided there is good reason to believe each half shift of exposure is the same. For specific control purposes the first of the four ways described below is favoured here, although the others may be used very successfully for particular purposes (Petersen, Sanderson and Lecher, 1986). Whatever method is employed it is advantageous if some of the individual employees whose exposure is measured be each measured two or more times.

It is possible to employ any of at least four distinct ways of sampling amongst a group of employees:

(1) Choosing a set of employees representative of the whole spectrum of exposure through the group, from the highest exposure (worst case) to the lowest (best case). Experienced judgement is needed to ensure that the full range of exposures is covered. This method can give the most useful information from a limited survey.

(2) Choosing an employee believed to be representative of the normal or average exposure of all those in the group. This method is employed when the available sampling effort is minimal. Success depends critically upon the skill of the surveyor in choosing the employee who will get the average exposure. No information is gained about the range and variability of exposure between employees in the group.

(3) Choosing a set of employees at random from the job-exposure group and measuring their exposure in whatever task they are assigned. All shifts should be included; morning shift, afternoon shift, night shift, day shift, etc, and week-end working, where conducted. The selection of employees to be studied and on which day is made beforehand with the aid of random sampling numbers. There is often a facility for generating random numbers on hand calculators. This method is rarely employed routinely but

insofar as the sample is unbiased the mean and variance are also unbiased. The method tends to yield results having a relatively large standard error but is ideal for the less experienced surveyor.

Sampling is continued for the whole of the work shift, from clock-on to clock-off, irrespective of the tasks the employee undertakes. The random sample gives a fair cross section of the employees and the mean of the sample is an unbiased estimate of the mean for the whole job-exposure group. The results provide unbiased estimates of the mean concentration, the mean duration of exposure, the variations in exposure from employee to employee and thus from place to place. The results tell not only of the risk to employees but also the individual work places where the greatest exposure arises and thus pin-point the most important operations where preventive measures would be most effective in reducing the risk (Oldham and Roach, 1952; Roach, 1954).

(4) Stratified random sampling. This is a development of the previous method and is employed with advantage when there are clearly distinct exposure sub-groups within a given job-exposure group. Each sub-group or stratum is sampled at random. Experienced judgement is needed to recognise distinct sub-groups prior to the survey. This is a good compromise method when unbiased results are at a premium as when the findings are to be employed in epidemiological investigations.

Time at work

In order to assess the risk to health from studies of the employees' environment correctly it is necessary to consider the total period of their exposure to the hazardous substance. This is true of every substance, whatever its characteristic period of accumulation. For those materials which accumulate at a steady rate the amount accumulated by the body will be roughly proportional to the time spent at work each day, while for others, such as the irritant gases, the number of occasions for which the worker is exposed to concentrations above a specific level will again be roughly proportional to the time spent at work. The risk to health in both cases is higher for those workers who work the longest hours.

Individual employees vary amongst themselves in the time they spend at work each day. Some habitually work overtime while others get through their work as fast as possible. They also vary in the number of days they work each year. It is well known that some employees attend much more often than others. In following the attendance of the employees at one company over a year the author found that the number of shifts worked during the year varied from 150 shifts to 300 shifts.

The time spent at work is not wholly in active work and is surprisingly variable. Measurements limited to periods of obvious activity are likely to give a false picture of exposure. This is illustrated by a survey over one year of dust conditions at a colliery in South Wales during which it was possible to make simple time studies (Roach, 1959). The colliers attendance was scheduled as

TABLE 12.1. *Time occupied in working, travelling and resting on different days in a colliery*

Time occupied (hours)	Activity		
	Working	Travelling	Resting
	Number of days		
<0.5	—	7	5
0.5—	1	56	4
1.0—	2	68	25
1.5—	2	50	17
2.0—	4	12	12
2.5—	6	2	5
3.0—	11	1	3
3.5—	5	1	9
4.0—	9	—	3
4.5—	20	—	—
5.0—	21	—	—
5.5—	3	—	—
>6.0	7	—	—
Days studied	91	197	83
Mean time/day (h)	4.32	1.34	1.82

seven and a half hours underground each shift. However, colliers who become ill or have an accident and those who need to come out to repair or get a piece of equipment may leave the mine before the end of the shift. Some men also work overtime. There is thus a considerable variation in the duration of significant exposure, illustrated by the summary given in Table 12.1. The shortest time spent underground was 0.8 hours and the longest 9.5 hours. The average time spent underground was 7.52 hours.

Time studies were done on colliers chosen at random to measure the time spent travelling from the pit top to and from the place of work underground, and on half the days the time spent working and resting between spells of work was also measured. In other, surface occupations the time spent getting to the place of work is usually shorter but the time spent 'getting ready' at the beginning of the shift and 'tidying up' at the end of the shift may be considerably longer. The occupation of collier was clearly not sedentary by nature so that the time spent resting was obvious to the casual observer. However in a number of jobs underground the work is machine minding, particularly those connected with the transport of trucks and roof supports between the pit bottom and the coal-face. In these jobs the observer could not distinguish clearly between the working and the resting periods. This type of job was done on 19 days out of the 110 when detailed time studies were attempted.

The time spent in these different activities varied considerably as is illustrated by the results given in Table 12.1. The mean time spent working, during which the greatest dust exposure occurred, was 4.32 hours; the mean time spent travelling and resting were respectively 1.34 hours and 1.82 hours. On at least 10% of occasions the length of these periods was either less than half of the mean or more than one and a half times the mean.

Presentation of results

It is only possible to make precise statements about the conditions that occurred over specific periods of time at specific sampling stations. It is important to grasp the nature and limitations of these statements. There are five interdependent parameters which will describe the results. All five are needed for a full description:

The survey duration

This is the total duration of calendar time of the survey. It may take several weeks to complete the survey, one week or even less time. In any event the starting date of the survey and its duration should be recorded. There may be seasonal variation in the exposure to be taken into account. Also vigilance should be maintained for effects of an annual plant shut down on the exposure of the group as a whole and/or those who work during the shut down.

The reference period

This is the unit averaging period or reference period of interest. It is most commonly one shift. Results obtained within a given reference period are averaged or weighted so as to give a time weighted average concentration over the period. This may be done merely by having a sampling instrument drawing air at a constant flow rate over the reference period or it may be done by arithmetical calculation from several results obtained during the reference period.

The exposure measurements

The complete list of time weighted average concentrations, each over a stated reference period (one shift) is the basic data of the survey. This will usually be samples from several different people or sampling stations and several different days. The overall variability exhibited will generally be greater in surveys of many people and/or many days. A comprehensive survey would include measurements of at least 1 in 10 of the employees exposed to more than one tenth of the exposure limit. At least 1 in 10 of these would be measured twice to ease analysis of the variability between days and between employees.

The distribution of exposure results from one sampling station

Successive measurements at a fixed sampling station or in the breathing zone of an individual employee are summarised to show the frequency of occurrence of concentrations in appropriate ranges. A logarithmic scale of concentration is usual for this purpose as the distribution of concentration from time to time is quasi-lognormal in character. The results may also be summarised to show the frequency of occurrence of concentration above a stated value. A quick, convenient and deservedly popular method for examining the distribution is to plot the results on logarithmic-probability plotting paper (Coenen, 1966; Brief and Jones, 1976; Leichnitz, 1980). Marked deviations from a straight line are good grounds for close investigation of the cause as they are often the result of poor control. The exposure results may be further summarised by calculating their mean and standard deviation or geometric mean and geometric deviation.

Exposure measurements of each individual should be accumulated and studied to see if the exposures of that individual are repeatedly above or repeatedly below the defined band for the corresponding job-exposure group. If so the employee should be in another job-exposure group.

With fixed position, area sampling the results from each single sampling station may be averaged and the results tabulated according to the location. It is helpful to write these results on a layout plan overlay indicating where the sampling positions are located. The approximate position of concentration contours can then be drawn in.

The distribution of exposure results from one job-exposure group

Job-exposure groups are individuals with similar average exposure. The group boundaries need defining. They may be conveniently defined, for example, as individuals doing the same job and moreover whose exposure lies within a specific bandwidth. Because the group's boundaries are thereby more or less well defined in terms of exposure the subsequent distribution of exposures cannot be expected to be lognormal at the extremes.

The various atmospheric exposure results of a single employee may be averaged. The set of such averages obtained from one job-exposure group may be summarised to show the frequency of occurrence of employees according to their average personal exposure. Sophisticated statistical analysis of results is discussed by Leidel and Busch (1985). A shorter treatment is given by Roach (1986).

Biological exposure indices

There is a small but growing number of substances for which biological exposure indices comparable with exposure limits have been developed in terms of the concentration of a hazardous substance or metabolite in biological specimens, especially blood, urine and exhaled air. Measurement of such biological determinants is undertaken as a supplement to atmospheric exposure measurements wherever there is a risk of substantial skin absorption or ingestion of the substance. Organs are not available for analysis in life but specimens of blood, saliva, urine, faeces, hair, nails, sweat or exhaled air may be obtained from employees and analysed. Analysis of these specimens for the substance under study or a metabolite of it have all been tried and each may more or less well reflect the concentration in some critical organ. The concept of critical organ concentration is analogous to atmospheric exposure limit

TABLE 12.2. *Hazardous substances recommended for biological monitoring* (*Lauwerys*, 1983; *DFG*, 1986; *American Conference of Governmental Industrial Hygienists*, 1990)

Substance	Type of monitoring Blood	 Urine	 Exhaled air
Aniline	No	Yes	No
Arsenic	No	Yes	No
Benzene	Yes	Yes	Yes
Cadmium	Yes	Yes	No
Carbon disulphide	No	Yes	No
Carbon monoxide	Yes	No	Yes
Chromium	No	Yes	No
DDT	Yes	No	No
Dimethyl formamide	Yes	Yes	Yes
Ethyl benzene	Yes	Yes	Yes
Fluorides	Yes	Yes	No
Hexachlorobenzene	Yes	No	No
n-Hexane	No	Yes	Yes
Lead	Yes	Yes	No
Lindane	Yes	No	No
Mercury	Yes	Yes	No
Methyl chloroform	Yes	Yes	Yes
Methylene chloride	Yes	No	No
Methyl ethyl ketone	No	Yes	No
Nickel	Yes	Yes	No
Organophosphates	Yes	Yes	No
Parathion	Yes	Yes	No
Pentachlorophenol	No	Yes	No
Perchloroethylene	Yes	Yes	Yes
Phenol	No	Yes	No
Selenium	Yes	Yes	No
Styrene	Yes	Yes	Yes
Toluene	Yes	Yes	Yes
Trichloroethylene	Yes	Yes	Yes
Xylenes	Yes	Yes	Yes

(Task Group on Metal Accumulation, 1973). The most relevant and practical determinant in a particular instance could be the quantity of the substance found in any of these specimens, or a metabolite or some other biological change.

Established biological indices are discussed in Health & Safety Executive Guidance Note MS18 and in the introduction to the biological exposure indices of the American Conference of Governmental Industrial Hygienists (Health and Safety Executive, 1984; American Conference of Governmental Industrial Hygienists, 1990). Biological tolerance values are also promulgated in Germany (DFG, 1986). A short list of substances for which biological monitoring of different types is presently recommended is in Table 12.2. Indirect procedures for estimating the expected values of biological determinants from the concentration inhaled, based on known pharmaco-kinetic parameters, have been described by Leung and Paustenbach (1988).

A group of substances for which biological measurements are especially suitable comprises the organic solvents, since many of them, such as benzene, toluene, trichloroethylene and xylene are readily absorbed through the skin (Astrand, 1975). Exhaled air sampling is increasingly used for gases and vapours rather than blood or urine (Ho and Dillon, 1987). Exhaled air analysis is, however, attractive for only a limited range of substances. It is inherently unsuitable for a biological exposure index of exposure to aerosols, and gases or vapours which decompose on contact with body fluids (Fiserova-Bergerova, 1983a).

A good example of solids which are especially suitable for biological monitoring are inorganic lead compounds. Solids contaminate the hands and face and are liable to become ingested. Some of these compounds are also readily soluble, others moderately so and some are almost completely insoluble. There is a variable amount of lead ingested from skin contamination by soluble lead compounds. Fortunately lead in blood or urine is readily determined and moreover, because of its relatively long biological half time is inherently less variable than lead in air.

Blood samples

The amount of substance or metabolite in venous blood samples is a good index of body burden and is widely used. The technique does, however, have to be used with discretion as it is invasive in character and not every employee is willing to volunteer. It is favoured for substances which have a long biological half time such as metals as the precise moment the blood sample is taken is not critical with these. Biological exposure indices for volatile substances relate to venous blood and cannot be applied to capillary blood, which mainly represents arterial blood. Analysis of the whole blood is usual but some substances are concentrated in erythrocytes and for such reasons analysis of a specific blood constituent, plasma, serum or erythrocytes, may be advantageous.

Urine samples

The rate of excretion in urine may be estimated by collecting a sample of urine on emptying the bladder and analysing it for the substance or a metabolite of it. The rate of excretion of urine over an interval is estimated directly by measuring the volume collected at a voiding divided by the time interval since the previous voiding. In this way accumulation and decay curves can be investigated for appropriate substances. Regular measurements on employees may be designed upon the basis of such studies.

Urine samples are easier to obtain from employees than blood samples and their collection is usually free of legal and ethical complications, although to obtain them still causes a certain inconvenience to the employee. Urine analyses are also intrinsically more variable than blood analyses for a given body burden. They reflect what is being excreted from the body rather than stored in it. Analysis of urine from employees is generally performed on 'spot' specimens (Elkins, Pagnotto and Smith, 1974). Routine collection of 24-hour samples from employees is impractical. The concentration of the determinant in urine is partly dependent upon the rate of output of urine. A rough correction for this is made, where necessary, by measuring the creatinine content of the urine specimen, which is approximately inversely proportional to rate of urine output.

Exhaled air

All absorbed gases and volatile compounds which circulate in the blood stream are removed in part via the lungs. Where feasible, analysis of exhaled air may thus be an appropriate means of assessing the sum absorption by all routes; inhalation, skin absorption and ingestion. The breath concentration of such a substance is directly related to its blood concentration which in turn is a function of the amount absorbed and the elapsed time following exposure (Fiserova-Bergorova, 1983b). In the past, breath analysis has not been widely adopted, mainly due to technical difficulties in taking and storing breath samples. However, the technology is now available to use the technique more widely (Stewart, 1974; Berlin, Gage and Gullberg, 1980; Droz and Guillemin, 1986; Wilson, 1986; Fiserova-Bergerova, 1987).

Biological Exposure Indices based on analysis of exhaled air are being developed by the American Conference of Governmental and Industrial Hygienists (1990). Similar control values are enforced in Germany. Some of the advantages of breath analysis for assessing occupational absorption of gaseous compounds are, in summary:

- It is non-invasive.
- It is well accepted by employees.
- Using on-line procedures an almost immediate result is obtainable.
- It reflects solvent uptake by all routes.
- For several inert compounds, breath analysis is the only means of biological assessment other than analysing the substance itself in blood.

Methods of breath sampling and analysis can be divided into two main types. The first type, grab sampling, usually includes a glass tube with screw cap closures at either end. Absorption on water condensate from the breath can be avoided by maintaining the sampling tubes at 37C (Pasquini, 1978). Gas sampling bags have also been widely used (Gage, Lagesson and Tunek, 1977; Money and Gray, 1989). Such breath samples are analysed in the laboratory. With modern technology an on-line method using a gas sampling valve to collect a breath sample is preferred to grab sampling. Gas chromatography or mass spectrometry provide an almost immediate measurement of breath concentration of a solvent (Wilson, 1981). Indicator tubes have been popular in the past for carbon monoxide in exhaled air and, of course, ethyl alcohol, both in the past and to this day.

Three types of breath sample can be taken for analysis; whole breath, end-exhaled air and re-breathed air. A whole breath sample represents gas displaced from the dead space as well as from the alveoli. End-exhaled or end-tidal air samples are preferred since they are most reproducible and most accurately reflect the blood concentration of the compound. An end-exhaled air sample is one collected at the end of an exhalation. Exhaled air samples collected from employees with altered pulmonary function may be unsuitable for comparison with biological exposure indices.

Individuals included in biological samplings should be ones in occupational exposure groups with significant exposure. In terms of atmospheric exposure this would be those exposed to more than one tenth the corresponding exposure limit. The distribution of values of biological determinants over large population groups is usually skewed and sufficiently well described by a lognormal distribution (Zielhuis, 1979). However this is not always the case, especially in small sub-groups and the fit should always be examined. If the fit is poor or uncertain the main percentiles of the distribution should be reported, such as the levels exceeded in 5, 10, 50, 90, 95% of the population. The results may have to be corrected for background levels, and this applies particularly to urine analysis for metabolites, as, for example, phenol in urine from benzene exposure (Teisinger, 1980). Phenol is common in antiseptic medicines and household disinfectants, besides being a hazardous substance in its own right.

Dermal exposure

Published data relating adverse health effects in employees to the amount and duration of hazardous substances on the skin surface is scarce. This is partly because of the inherent technical difficulty of measuring skin exposure quantitatively. Also the occupations and jobs where skin exposure is the dominant source of intake to the body are relatively rare. Perhaps the most common job of this type is spraying; paint, glaze, pesticides and so on, in which there is an especially high atmospheric concentration and the sprayer has respiratory protection.

The means available to measure dermal exposure directly are comparatively crude. Washing and collecting the drainings is not very feasible for field use. There are two principal types of measurement on employees used in practice and reported in the literature. One type is patch testing in which gauze patches are attached to the clothing or skin prior to exposure and collected for analysis after exposure (Durham and Wolfe, 1962; Davis, 1980; World Health Organisation, 1986). The method is useful for solids, their solutions and low vapour pressure liquids. The results are peculiarly variable because in practice the contamination over various parts of the body is extremely uneven (Franklin, Fenske, Greenhalgh, Mathieu, Denley, Leffingwell and Spear, 1981).

The second type of measurement of dermal exposure is by the assessment of fluorescent tracers on contaminated individuals. The tracer is added to the process materials prior to them being handled. The fluorescent compound 4-methyl-7-diethylaminocoumarin (FWA) is commonly used (Fenske, Leffingwell and Spear, 1986; Fenske, Wong, Leffingwell and Spear, 1986; Fenske, 1988). The fluorescence is stimulated under long-wave ultraviolet light (320–400 nm) and viewed under standardised conditions with a video imaging system (Honeycutt, Zweig and Ragsdale, 1985). This method appears to be more reliable than patch testing (Fenske, Horstman and Bentley, 1987; Fenske, 1988).

Investigation of exposure sources

The results of personal sampling will generally indicate the region where the contamination level is at its highest. Area sampling may enable the fixed sources to be identified more precisely, to within 1 metre, since the air sampling instruments are stationary. The results may be indicated on a plan of the work rooms at the points where the sampling was undertaken. Concentration 'contours' may be drawn in to serve as a pictorial summary. If there is any doubt a further investigation without delay would be required to establish the precise source of contamination for places found to have a level of concentration in the air in excess of that in the occupational exposure limit. An investigation, with less urgency, would also be advisable for those places which are in excess of one tenth the level in the exposure limit.

Example 12.1

An example of personal monitoring correlated with position is illustrated by the plot in Figure 12.1 of data from measurements made by the author of airborne coal dust on an underground coal face where the coal was got by pneumatic picks operated by colliers spread along the face. The longwall system of getting coal was being used. The colliery had a simple layout of the coal faces and in this instance the face was 100 yards long with the intake air at one end and the return air from the face leaving at the other.

A coal face, although bounded on three sides by solid rock, namely, the roof, floor, and coal itself, is bounded on the fourth side by a series of stone packs at right angles to the face between

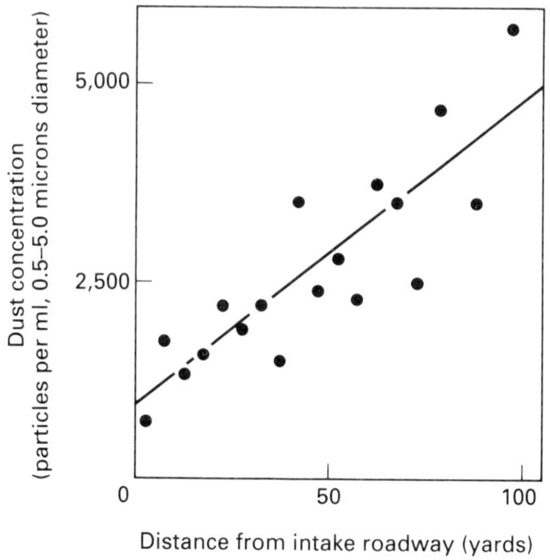

FIG. 12.1. The concentration of airborne dust during the coal-getting shift on the coal face at a colliery. Each point on the graph is the time weighted average concentration over the coal-getting period. In all, 467 samples were taken, spread over 149 shifts.

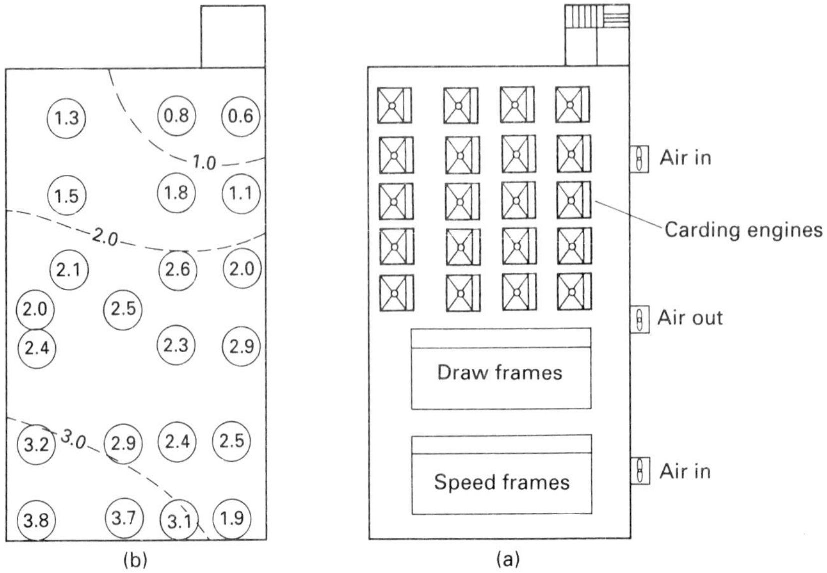

FIG. 12.2. Airborne dust concentration in the card room of a cotton mill. Results are given in terms of mg m^{-3} of total dust. (a) Plan layout of room. (b) Results from sampling at positions ringed; and interpolated concentration contours.

which the roof is allowed to dip and fall and there is some leakage of air through this loosely packed material which only partially mixes with the air passing the colliers (Roach, 1959). Nevertheless there is clearly a steady build-up of dust in the air along the face shown by the increase in average concentration from the intake to the return. The coal was shovelled from the face on to a belt conveyor along the face, from which it tipped to another conveyor in the intake roadway. The initial concentration of airborne dust at the intake end was from dust liberated at the transfer point.

Example 12.2

An illustration of simple concentration 'contours' was obtained from measurements made in a dusty cotton spinning mill (Wood and Roach, 1964). The carding engines, which 'card' the cotton and remove dirt (soil and dry leaf debris), were prolific sources of airborne dust and had recently been fitted with local exhaust hoods. They were arranged in a regular pattern in one half of the card room. Subsequent operations of drawing out and twisting the yarn were conducted in the other half of the same room. The basic layout plan is shown in Figure 12.2a. Three fans high up on one wall provided general ventilation to the room. The shift time-weighted average concentration of airborne dust was obtained by samplings made at fixed positions in a card room over a period of four weeks. The sampling positions and results are given on Figure 12.2b and to guide the eye concentration contours are drawn in by freehand interpolation.

In situations where the sources of air contamination are varied and on a relatively small scale direct read-out meters are needed to locate the sources readily. Where no direct read-out meter is available grab sampling bags may be used with advantage at the moments when the operations produce most air contamination. On the smallest scale, where sources are less than 2 cm diameter an instrument with an air sampling probe attached will be necessary to discern the smallest sources.

The occupational hygienist should always be on the alert for 'odd' results caused by the vagaries of the instruments, analytical procedures or, occasionally of employees tampering with instruments. There is no substitute for common sense guided by experience in sorting out these situations. Employees sometimes deviate from set procedures if another procedure is found to be faster, less troublesome or more feasible. At other times they may endeavour to ensure a high result by leaning over a tank or into a cloud of dust or even by pouring material into the sampler inlet. All hygienists build up a collection of stories to illustrate these occurrences.

Example 12.3

At a glue factory the employees in the mixing department showed high levels of trichloroacetic acid in their urine. This is a metabolite of trichloroethylene which is an important constituent of the glue. Visual inspection of the plant indicated that the process was well controlled. A minute by minute record of trichloroethylene concentration was sought to try to identify the operations causing the highest exposure.

For safety reasons the sampling instrument was set up in an adjacent office and a long probe tube lead to the work area. By chance the work area was not visible from the office. A very high peak of exposure was found at the end of every shift. It was subsequently observed that before leaving work the mixing department employees cleaned their shoes of glue with a can of trichloroethylene!

References

American Conference of Governmental Industrial Hygienists. Threshold Limit Values and Biological Exposure Indices for 1990–1991. ACGIH, Cincinnati, Ohio, USA (1990).

Astrand I. Scand. J. Work Environ. Health. **1** 199 (1975).

Baretta E D, R D Stewart and J E Mutchler. Am. Ind. Hyg. Assoc. J. **30** 537 (1969).

Berlin, M, J C Gage and B Gullberg. Scand. J. Work Environ. Health **6** 104 (1980).

Bonsall J L. Measurement of occupational exposure to pesticides. In: Occupational hazards of pesticide use (Edited by G J Turnbull). Taylor and Francis, London (1985).

Bloomfield J J. Preliminary surveys of the dusty environment. Pub. Health Rep. **48** 1343 (1933).

Breslin A J. In: Beryllium, its industrial hygiene aspects (Edited by H E Stokinger), pp. 245–321. Academic Press, New York (1966).

Breslin A J, L Ong, H Glauberman, A C George and P LeClare. Amer. Ind. Hyg. Assoc. J. **28** 56 (1967).

Brief R S and A R Jones. Am. Ind. Hyg. Assoc. J. 37 474 (1976).

Coenen W. Staub-Reinhalt.d.Luft **26** 39 (1966).

Coenen W. Staub-Reinhalt.d.Luft **31** 16 (1971).

Coenen W and G Riediger. Staub-Reinhalt. d. Luft 38 402 (1978).

Corn M. Scand. J. Work Environ. Health **11** 173 (1985).

Davis J E. Residue Rev. 75 33 (1980).

Dallavalle J M. The significance of dust counts. Pub. Health Rep. **54** 1095 (1939).

DFG. Deutsche Forschungsgemeinschaft. Maximale arbeitsplatzkonzentrationen und biologische arbeitsstofftolesanwerte. Verlag Chemie. Weinheim, Germany (1986).

Droz P O and M P Guillemin. J. Occup. Med. **28** 593 (1986).

Durham W F and H R Wolfe. Measurement of the exposure of workers to pesticides. Bull. WHO **26** 75 (1962).

Elkins H B, L D Pagnotto and H L Smith. Am. Ind. Hyg. Assoc. J. **35** 559 (1974).

Fenske R A. Am. Ind. Hyg. Assoc. J. **49** 438 (1988).

Fenske R A. Appl. Ind. Hyg. **3** 207 (1988).

Fenske R A, S W Horstman and R K Bentley. Appl. Ind. Hyg. **2** 143 (1987).

Fenske R A, J T Leffingwell and R C Spear. Am. Ind. Hyg. Assoc. J. **47** 764 (1986).

Fenske R A, S M Wong, J T Leffingwell and R C Spear. Am. Ind. Hyg. Assoc. J. **47** 771 (1986).
Fiserova-Bergerova V. Scand. J. Work Environ. Health **11** Suppl. 1, p. 7 (1983b).
Fiserova-Bergerova V. Appl. Ind. Hyg. 2 87 (1987).
Franklin C A, R A Fenske, R Greenhalgh, L Mathieu, H V Denley, J T Leffingwell and R C Spear. J. Toxicol. Environ. Health. 7 715 (1981).
Gage J C, V Lagesson and A Turek. Ann. Occup. Hyg. **20** 127 (1977).
Gale H J. The lognormal distribution and some examples of its application in the field of radiation protection. Atomic Energy Research Establishment Report AERE-R 4736, H M Stationery Office, London (1965).
Gale H J. Ann. Occup. Hyg. **10** 39 (1967).
Kay K, L Monkman, J P Windish, T Doherty, J Pare and C Racicot. AMA Arch. Ind. Hyg. Occup. Med. **6** 252 (1952).
Health & Safety Executive. Health surveillance by routine procedures. Guidance Note MS18. H M Stationery Office, London (1984).
Honeycutt R C, G Zweig and N N Ragsdale. Dermal exposure related to pesticide use. ACS Symposium Series 273, American Chemical Society, Washington DC, USA (1985).
Hounam R F. An application of the log-normal distribution to some air sampling results and recommendations on the interpretation of air sampling data. Atomic Energy Research Establishment Report AERE-M 1469, H M Stationery Office, London (1965).
Hubiak R J, F H Fuller, G N VanderWerff and M Ott. Occ. Health and Safety **50** 10 (1981).
Jones A R and R S Brief. Amer. Ind. Hyg. Assoc. J. 32 610 (1971).
Jones C D. J. Haz. Mat. 7 87 (1983).
Juda J and K Budzinski. Staub-Reinhalt.d.Luft **27** 12 (1967).
King E, A Conchie, D Hiett and B Milligan. Ann. Occup. Hyg. **22** 213 (1979).
Kromhout H, H Ikink, W De Haan, J Kroppejan and R Bos. The relevance of the cyclohexane soluble fraction of rubber dusts and fumes for epidemiological research in the rubber industry. In: Progress in occupational epidemiology (Edited by C Hogstedt and C Reuterwall), pp. 387–390. Elsevier Science, B.V. (Biomedical Division), Oxford, UK (1988).
Langmead W A. Air sampling as part of an integrated programme of monitoring of the worker and his environment. In: Inhaled particles III, Vol. 2 (Edited by W H Walton), pp. 983–995. Pergamon Press, Oxford, UK (1971).
Leichnitz K. Staub-Reinhalt.d.Luft **40** 241 (1980).
Leidel N A and K A Busch. Statistical design and data analysis requirements. In: Patty, Industrial hygiene and toxicology, Vol. 3A, Wiley and Sons, (1985).
Leung H-W and Paustenbach D J. Am. Ind. Hyg. Assoc. J. **49** 445 (1988).
Money C D and C N Gray. Ann. Occup. Hyg. **33** 257 (1989).
Oldham P D. Brit. J. Ind. Med. 10 227 (1953).
Oldham P D and S A Roach. Brit. J. Ind. Med. **9** 112 (1952).
Pasquini D A. Amer. Ind. Hyg. Assoc. J. **39** 55 (1978).
Peterson M R, W T Sanderson and S W Lecher. Am. Ind. Hyg. Assoc. J. **47** 655 (1986).
Rappaport S M. Am J. Ind. Med. **6** 291 (1984).
Rappaport S M. Ann. Occup. Hyg. **35** 61 (1991).
Rappaport S M and S Selvin. Am. Ind. Hyg. Assoc. J. **48** 374 (1987).
Rappaport S M, S Selvin, R C Spear and C Keil. An evaluation of statistical schemes for air sampling. In: Measurement and control of chemical hazards in the workplace (Edited by G Choudhary), pp. 431–455. American Chemical Society, Washington, DC, USA (1981).
Roach S A. Brit. J. Ind. Med. **11** 54 (1954).
Roach S A. Brit. J. Ind. Med. **16** 104 (1959).
Roach S A. Sampling for airborne toxic hazards. In: Handbook of occupational hygiene. Kluwer Publishing (1986).
Roach S A, E J Baier, H E Ayer and R L Harris. Am. Ind. Hyg. Ass. J. **28** 543 (1967).
Rock J C. The NIOSH action level: A closer look. In: Measurement and control of chemical hazards in the workplace (Edited by G Choudhary), pp. 472–489. American Chemical Society, Washington, DC, USA (1981).
Rock J C. Am. Ind. Hyg. Assoc. J. **43** 297 (1982).

Selvin S, S M Rappaport, R C Spear, J Schulman and M Francis. Am. Ind. Hyg. Assoc. J. **48** 89 (1987).

Sherwood R J and D M S Greenhalgh. Ann. Occup. Hyg. **2** 127 (1960).

Sherwood R J. Amer. Ind. Hyg. Assoc. J. **27** 98 (1966).

Sherwood R J. Amer. Ind. Hyg. Assoc. J. **32** 840 (1971).

Stewart R D. The use of breath analysis in clinical toxicology. In: Essays in toxicology (Edited by J Wayland and J Hayes), p. 121. Academic Press, New York (1974).

Task Group on Metal Accumulation. Environ. Physiol. Biochem. **3** 65 (1973).

Teisinger J. Benzene. In: Exposure tests in industrial toxicology. Avicenum, Prague, Czechoslovakia (1980).

Tuggle R M. Am. Ind. Hyg. Assoc. J. **42** 493 (1981).

Tuggle R M. Am. Ind. Hyg. Assoc. J. **43** 338 (1982).

Ulfvarson U. Int. Arch. Occup. Environ. Health **52** 285 (1983).

Wilson H K Scand J. Work Environ. Health. **12** 174 (1986).

Wilson H K and T W Ottley. Biomed. Mass. Spectrom. **8** 606 (1981).

Wood C H and S A Roach. Brit. J. Ind. Med. **21** 180 (1964).

World Health Organisation. Toxicol. Letters **33** 223 (1986).

Zeilhuis R L. General aspects of biological monitoring. In: The use of biological specimens for the assessment of human exposure to environmental pollutants (Edited by A Berlin, A H Wolff and Y Hasegawa), pp. 341–359. CEC Martinus Nijhoft Publishers, The Hague/Boston/London (1979).

Bibliography

Drinker P and T Hatch. Industrial dust. McGraw-Hill Book Company, Inc. New York (1954).

Fiserova-Bergerova V (Editor). Modelling of inhalation exposure to vapors: Uptake, distribution and elimination. Vols I and II. CRC Press, Boca Raton, Florida, USA (1983a).

Ho M H and H K Dillon (Editors). Biological monitoring of exposure to chemicals—Organic compounds. John Wiley & Sons (1987).

Lauwerys R R. Industrial chemical exposure: Guidelines for biological monitoring. Biomedical Publications, Davis, Calif. (1983).

Leidel N A, K A Busch and J R Lynch. Occupational exposure sampling strategy manual. National Institute for Occupational Safety and Health. US Department of Health, Education and Welfare (NIOSH). Publication 77–173 (1977).

CHAPTER 13

Health Risk Surveillance

Ideally exposure to hazardous substances could be so controlled that there is no significant risk to health. Employee exposure is related to the state of engineering control over it. Surveillance of the integrity of engineering control measures and to a limited extent of personnel exposure control measures marks the first line of defence. In favourable circumstances the comprehensive surveillance of engineering controls can make direct measurement of atmospheric exposure largely redundant. Surveillance of engineering control has priority. It is held by the author that monitoring of atmospheric exposure and biological indices should be regarded as a supplement to, not a replacement for engineering control surveillance.

At work places where there are hazardous substances and air contamination is below one tenth the atmospheric exposure limit a regular assessment of the engineering controls should be sufficient. Such an assessment would include the study of data from plant inspections for process leaks and checks of the ventilation system. A plant supervisor would be competent to do this. However, an integral part of this assessment is recommendation of the steps needed to rectify such faults as may be found. The assessor should be competent to do this directly or have access to others who are. An adequate hazard control system as a whole will include effective means to respond to such recommendations.

At higher exposures, that is 0.1–1.0 × exposure limit, some regular monitoring of the atmospheric exposure also becomes important to spot the development of adverse exposure conditions as early as possible. In addition, for those substances which have an established biological exposure index the appropriate biological monitoring may be introduced and in favourable circumstances may replace atmospheric monitoring [Fiserova-Bergorova (Thomas), 1990]. Biological monitoring establishes a link between, on the one hand, exposure to hazardous substances and on the other hand, the adverse effects on human health. Commensurate with the nature and severity of the health risk, appropriate medical surveillance may be advisable.

Inspection of engineering controls

An assessment of possible health health risks where hazardous substances are produced or handled should include assessment of the engineering control surveillance, that is, the regular plant inspections for process leaks and tests of the ventilation system.

Inspections of the plant and processes for emission of hazardous substances need not be very time consuming nor repeated very frequently. A weekly walk-through should suffice. Nevertheless, the walk-through should follow a deliberately planned and regular route. With dusts the inspection is usually visual; gases and vapours normally, but not always, necessitate an instrument to detect leaks. Portable flammable gas detectors are in common use for safety purposes. Regular inspection of plant for leaks and broken joints in order to pinpoint the sources of emissions is a feature of most chemical plants. A written record, however brief, should be kept of the results, in the same cover as a description of the route of the walk-through.

The results of weekly inspections should be reported within 24 hours to those responsible for maintaining the plant in good order. Serious leaks and dangerous work practices should, of course, be reported immediately. Testing and examination of local exhaust ventilation plant is in a Health and Safety Executive 'discussion document' on the subject (1985).

Where employee exposure is above one tenth the exposure limit, engineering control monitoring could be of two kinds; the weekly walk-through, as before and supplemented with a monthly systematic inspection and report of all items and work practices likely to cause excessive exposure to hazardous substances. The monthly inspections may also be semi-quantitative and might include, by way of illustration:

Valves

These are examined for signs of leaks around the valve stem, corrosion of the barrel and evidence on the floor. At plants with many valves, a regular count of leaky valves provides a running commentary of the level of maintenance.

Flanges and joints

Corrosion or faulty gaskets are signified by drips or pools on the floor. The location and number of leaks are noted.

Access doors, inspection ports and clean-out doors

Faulty latches or seals are identified as well as doors left open. Again, assessment of the number and severity of the leaks provides a semi-quantitative measurement of plant integrity.

Conveyors

Piles of loose material spilled from conveyors are noted and a visual estimate made of the weight of each pile. This provides a valuable record of steady improvement (or deterioration). A count is made of the number of covers left off conveyors.

Hot processes

This includes observing the presence or absence of fumes spilling from fixed furnaces and ovens, also from poorly tuned petrol-, diesel- or gas-driven vehicles. New or modified processes are examined with special care.

General cleanliness and housekeeping

Cleanliness and tidiness make for less hazardous methods of working. The cleanliness of floors, sills, ledges and other horizontal surfaces should be noted. Dry brushes or air hoses should, of course, not be used on hazardous substances. Tidiness of the work area and storage areas should be noted. Bad housekeeping is an encouragement to the spread of contamination.

Regular examination of ventilation systems

Fan blades wear and build up dust. Ducts corrode and develop leaks. Blocks occur by the gradual building up of dust on dampers, junctions and bends, or by the ingress of paper and other rubbish, insects and even birds. Clean-out openings may be left open and blast gates may be interfered with. Hoods and discharge grilles may be moved and others added. Constant surveillance to warn against gradual diminution of air flow is a critical element in system maintenance. Whether the performance changes slowly or without warning, some quantitative indication that a change has occurred is valuable. It is quite common to have a manometer fixed permanently across a filter, cyclone, washer or other air cleaner to show whether the pressure drop is normal. An additional manometer indicating the pressure upstream of the fan on an exhaust system, or downstream of the fan on a pressure system indicates the maintenance of pressure from the fan.

The effectiveness of the ventilation may be tested ultimately by comparing personal atmospheric exposure against the exposure limit. This supplements but does not replace direct tests of ventilation parameters. Monitoring of a ventilation system's performance may be carried out with pressure gauges. Reliance should not be placed upon merely noting whether the fan switch is on.

Relevant drawings, specifications, operating instructions and other papers created during the design, construction and initial testing of ventilation plant and equipment should be collected and recorded. There are two kinds of regular examination. The first kind is visual. In addition to the visual inspection specific objective tests of the ventilation could be undertaken 6-monthly (Fulwiler, 1973). The schedule should include the legal requirements and in any event the following items are recommended:

Daily — Check that all movable parts are in their correct position. Parts of the system not switched on, and the reason why, should be established. Check that all portable hoods are in use where needed and correctly positioned. Inspect flexible trunking for damage.

The off-to-on position of adjustable valves and blast gates should be checked. All removable covers, panels, clean-out doors and inspection ports should be examined to check whether they are in place. The locks and sealing of all access doors should be examined.

Weekly — Inspect the integrity of the system, including hoods, ducts, joints and supports. Look for pieces which have been damaged, broken, taken off, added or modified. Check valves and blast gates for free movement. Rusted and corroded joints and duct sections should be noted. The air cleaner and fan housing should be inspected for holes in the sheet metal work. Inspect the arrangements for disposal of dust, slurry and other collected material.

Monthly — In exhaust systems where the maintenance of volume flow rate through hoods is important for the preservation of safe working conditions, a diaphragm pressure gauge showing the hood static pressure is the preferred means for keeping a check on the volume flow rate. Any variation of pressure, whether it is an increase or decrease from the correct range may indicate something amiss in the system. A partially blocked hood is shown by an increase in hood static pressure and a block or leak downstream of the pressure point is shown by a decrease. Pressure gauges may actuate warning lights or an audible signal. In critical installations the hood static pressure point may be connected to a pneumatic or electrical transducer to switch off the process machinery when the hood static pressure drifts outside the normal working range. All such occurrences are recorded from the monthly inspections.

Read all the meters monthly and record the readings. The readings on all permanent pressure gauges should be noted and recorded. The fan pulley belts and fan blades should be examined for wear and/or corrosion. In the larger installations where electricity meters are a permanent feature fan power and/or fan amps should be recorded. Inspect the interior of the air cleaner.

6-Monthly — Check the air volume flow rate at the entrance to all hoods. Record the results.

The air volume flow rate through air registers, doors, windows and other openings should be measured with an anemometer. Record the results. It should be possible to make a rough air balance between air entering and leaving each room.

Annually — In addition to the above inspections a check should be made annually that the arrangements for maintenance of the fan, air cleaner and parts external to the building are satisfactory.

Supervision of personnel controls

Adequate supervision of personnel controls is an integral part of the controls themselves. It should include:

Protective clothing

Supervisory duties should include checking that the appropriate clothing is donned when necessary. Gloves, goggles, aprons, boiler suits and footwear all need regular cleaning daily or weekly according to circumstances.

Respiratory protection

Supervisory duties should include checking that the respiratory protection is donned when necessary. Disposable respirators should be replaced on demand or at least daily. Others should be cleaned, inspected and maintained daily, or perhaps weekly according to use.

Aside from disposable respirators the inspection, cleaning and maintenance should be done centrally by one or more persons given the responsibility. Respirators should be stored in a clean area, where the atmosphere is uncontaminated. Regular cleaning should be the rule and regular inspection of the interior and exterior of masks for cracked rubber is essential, as also is inspection of the valves, harness and air-purifying elements.

During regular use of a respirator the employee should get into the habit of testing the seal each time the respirator is donned. A convenient on-the-job test of the seal is accomplished by firstly covering the exhalation valve and breathing out to create a slight positive pressure. The pressure should keep the respirator inflated. Secondly by covering the cartridges with both hands and inhaling, a slight negative pressure is created. When this negative pressure is created the facepiece should stay collapsed while the employee holds his or her breath. If the seal fails either test the respirator should be re-adjusted until the fit is satisfactory.

Cleanliness

Supervisory duties should include checking that the washing and changing facilities are used and are in good order and checking that employees wash up and change into clean clothes before entering the canteen. Also check that the rules regarding eating and drinking are adhered to.

Exposure duration

Vigilance should be maintained by management that individuals or work groups do not develop into a habit of volunteering for excessive exposure duration, as by taking extended work shifts or overtime at week-ends in jobs which involve risk of exposure to heavily contaminated atmospheres.

Atmospheric exposure monitoring

As a rule, personal monitoring of the atmospheric exposure of an individual employee is very intermittent. Atmospheric sampling need not be carried out of all employee's exposure nor all the time. It is positive in nature, designed for the purpose of maintaining future conditions in a good state of control. Typically the total running time of all personal sampling instruments might add up to 1 in 25 or 1 in 50 of the shifts worked by employees exposed in the range 0.1–1.0 × exposure limit. A higher intensity of monitoring would be appropriate in plants where the job-exposure groups are small or the exposure levels close to the limit. A lower intensity would be appropriate where the exposure levels are very stable, as is sometimes the case in automatic continuous processes.

For effective personal monitoring over an extended period records should be kept up to date of all the employees in job-exposure groups whose exposure is more than one tenth the exposure limit. Personnel records kept for other purposes carry insufficient detail. The significance of keeping appropriate and accurate records of the whereabouts of employees should not be underestimated. Relying upon the memories of employees or supervisors is a poor substitute. Job-exposure groups should be carefully defined and coded at the outset. The list of employees in these groups will need to be brought up to date regularly. As a guide-line it is suggested that the list is revised initially every four months in 'snapshot' form; that is, by identifying those individuals in each relevant job-exposure group on, say, the first day of each four-month period.

Repeated measurements of the job-exposure groups at regular intervals are needed in industry since exposure typically varies so much from time to time according to changes in plant, processes, operating conditions and so on. Sampling programmes should extend over the full year so that seasonal patterns in output and plant maintenance are covered besides changes directly attributable to day-to-day weather conditions (Coenen, 1971). Statistical predictability of the future position is very limited. The sequence of concentration values is time dependent. The distribution of values of concentration to which employees become exposed from time to time is usually quasi-lognormal. But for predicting ahead of time the greater problem in the chemical industry is that the processes are not stationary in the statistical sense. Even in continuous manufacturing processes the atmospheric concentration level appears to vary in an analogous manner to a random walk (Coenen, 1976).

In manufacturing industries many factors which alter production also change the indoor environment, including employee exposure to hazardous substances. Examples of production alterations are output going up, or down, process improvements or deterioration taking place, operating procedures being modified, control measures failing, new products coming on stream, manning levels being altered, management policies changing and even individual employees changing jobs. In the chemical industry new chemicals and processes are continually being introduced. Such factors are in addition to changes in exposure controls whose effects on exposure to hazardous substances are more predictable. Equivalent changes occur in service industries.

Obviously, measurement of atmospheric exposure at regular intervals would be necessary to keep track of the effects. This is especially true for employees whose exposure to hazardous substances is in the range 0.1–1.0 × exposure limit. Statistically, the train of exposure levels is not expected to be a stationary process.

Technically, a random process which is strictly stationary is one which is stationary to all orders. This means that a process is stationary to order 1, 2, 3, ..., if its first, second, third, ..., order probability density function is independent of time.

A simple test for low order, non-stationary properties is to examine the difference between successive time weighted average measurements separated by an interval of time; minute, day, week, month, etc. In a stationary process the standard deviation or root mean square (RMS) of the difference in concentration between one measurement and the next would be independent of the length of the interval between the two measurements.

An analysis of 37 long series from continuous processes in the chemical industry is summarised in Figures 13.1 and 13.2. A representative set of six of these series was examined in detail. The results are plotted as six lines for reference periods of 1 minute, 10 minutes, 100 minutes, 1,000 minutes, 10,000 minutes and 100,000 minutes respectively. The lines for 1 minute, 10 minute and 100 minute reference periods in Figure 13.1 are each from the results obtained by personal monitoring of a single individual (two dusts, one vapour) and the others. The lines in Figure 13.2, are each from the results obtained at fixed position, multi-point sampling stations (three vapours). The intervals marked are those between the starting times in a train of successive measurements.

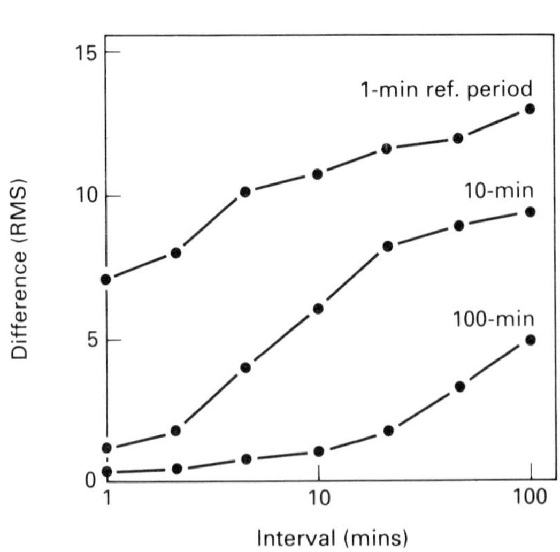

FIG. 13.1. Analysis of temporal variation of air contamination concentration. Root mean square (RMS) difference between pairs of measurements separated by an interval of time. Reference period 1 min to 100 min. Standardised mean concentration = 10.

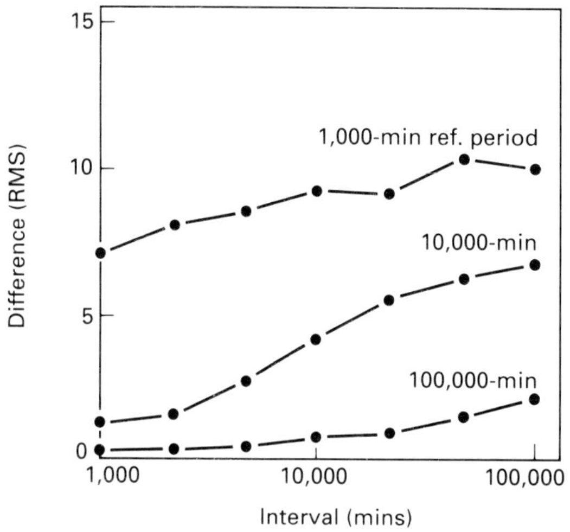

FIG. 13.2. Analysis of temporal variation of air contamination concentration. Root mean square (RMS) difference between pairs of measurements separated by an interval of time. Reference period 1,000 min to 100,000 min. Standardised mean concentration = 10.

Clearly none of the lines joining the points is even approximately parallel to the horizontal axis. These processes are not stationary to order 1. The longer the interval between successive measurements the higher the RMS difference, whether the reference period be 1 minute or 100,000 minutes. Furthermore, the author has analysed 31 other long series obtained by colleagues in controlled environments inside buildings in the chemical industry and none of these was stationary to order 1. This does not in any sense prove that stationary conditions cannot occur, but rather that limited experience in the chemical industry suggests they are unlikely to be common in exposure monitoring. Exactly similar results, although analysed in a different way, have been reported by Coenen (1976).

Such findings could conceivably be due to autocorrelation over both short and long periods of time (Rappaport and Spear, 1988; Francis, Selvin, Spear and Rappaport, 1989). Practical causes for the appearance of correlation between successive short-term values separated by a given interval are not hard to imagine. However, in so far as the lines are approximately straight the processes are approximately stationary to order 2, a condition unlikely to arise from autocorrelation alone. An approach in terms of separating a non-stationary component from an autoregressive element has been suggested by McCollister and Wilson (1975). The practical conclusion must be that whatever the cause, at least in these environments it is not possible to predict even a short period of time ahead what the concentration will be by the use of stationary statistics alone. This argues again that effective surveillance of atmospheric exposure is best carried out as far as possible by continual sampling at low frequency rather than by short, intensive campaigns.

Measurements should be spread more or less evenly throughout the year. In an endeavour to reduce costs it is tempting to suggest that surveillance could perhaps be by intensive sequential surveys lasting a few days, and repeated perhaps every 6 or 12 months with nothing in between. This may be adequate to check on trends taking place from one year to another, but the results are likely to be poorly related to those biological exposure indices or adverse health effects which are the result of exposure on a shorter time scale. To keep track of unexpected trends it is preferable to spread the monitoring effort evenly over time, albeit thinly. Air sampling strategies are discussed in Health and Safety Executive Guidance Note EH42 (Health and Safety Executive, 1989).

The most convenient and practical unit period for personal sampling is the work shift. Because of the cumulative nature of dusts causing pneumoconiosis it has long been argued that atmospheric exposure to these dusts each work shift should be assessed in terms of the product of time weighted average concentration and period of exposure (Oldham and Roach, 1952; Long, 1953; Roach, 1953; Wright, 1953). Sequential procedures were developed to test when a series of measurements of time weighted average over a work shift indicates an undue probability of excessive exposure to these dusts (Tomlinson, 1957; Galbas, 1979). As regards substances whose adverse effects are produced by peak concentrations during a work shift, in recent years it has been increasingly realised that advantage may also be taken of the fact that there is a high correlation between work shift arithmetic mean concentration and the frequency of peak concentrations. This is due to the skew, lognormal distribution of short term concentrations. Consequently, adherence to atmospheric exposure limits expressed in terms of time weighted average concentration provides good protection against unduly high short term peak concentrations as well as against excessive long term arithmetic averages (Rock, 1981, 1982; Rappaport, Selvin and Roach, 1988; Rappaport, 1991). The work shift is a practical standard unit sampling period.

It is therefore recommended that for a typical job-exposure group whose work shift exposure is in the range 0.1–1.0 × exposure limit an appraisal should be made each month of one or more employee's exposure over one shift. A different employee would be chosen each time. They may be studied in random order by employing random sampling numbers, although taking their names in alphabetical order is sufficiently random for most purposes. In statistical terms this is stratified random sampling; a variant of simple random sampling in which the population is first divided into sub-populations or strata. It may be appropriate to sample at a higher frequency from one stratum than from another. However, a sample mean for the whole population is readily determined, if desired, by weighting the mean from each stratum, or job-exposure group, by the number of employees.

The purpose of the surveillance is not so much concerned with providing a record of exposure as with providing a means to prevent excessive exposure in the near term future. The established techniques of quality control allied to negative feedback provide a good basis for risk control systems (Tebbens and

Spear, 1971; Roach, 1977). The following sequential rules illustrate the use of negative feedback employing air monitoring results.

1. When the highest of the last six in the train of monthly results is lower than the exposure limit but above one-tenth the limit the job-exposure group is in the 0.1–1.0 × exposure limit band.
2. When an exposure limit is first breached the plant manager is advised of the fact and warned that the state of control is deteriorating. The manager is warned that if the limit is also breached on the next test the operations will have to be classed as being no longer in the same state of control.
3. When two successive results are above the exposure limit the plant manager is again advised of the fact and immediate action is required to regain the state of control. Appropriate respiratory protection is provided to the employees, to be donned during operations where the concentration is liable to exceed the exposure limit.
4. When three successive results are above the limit this finding would be treated very seriously, with a warning that a plant shut-down may be required unless the situation is rectified immediately.
5. Finding four successive results above the limit, coupled with the previous warnings provides sufficient justification for serious consideration of immediate plant shut-down until the cause is traced and rectified.

These rules are buttressed less by statistical analysis than by trial and error in practical situations on continuous processes (Roach, 1977). It should be held in mind that statistical analysis of past results may aid interpretation but cannot replace professional judgement about future conditions and may even be undesirable in so far as it concentrates the mind on the past rather than the future. Consequently in industrial practice complex statistical calculations are seldom necessary and may be misleading. The purpose of health risk surveillance in general and regular measurements of air contamination in particular is, above all else, to provide a trigger for action commensurate with the foreseeable exposure conditions. The initiation of any particular action is truly an informed judgement taking into account the nature of the process, the amount of substance likely to be present, its rate of production, how and where it is used and any foreseeable changes in chemical processes or personnel.

A job-exposure group which is within the 0.1–1.0 × exposure limit band but yields results consistently below one-third the exposure limit could be sampled at a reduced frequency, namely once every two months. Furthermore if four successive results at one-month intervals or two at two-month intervals were below one tenth the exposure limit a change of classification to less than one tenth the exposure limit is indicated.

The general principles exemplified above could, in practice, be met in any number of different ways to suit local circumstances. Each design package is custom-built. Conformity with an official exposure limit may dictate a particular design. Equally, conformity with a company in-house exposure limit may itself constrain the design of the measurement regimen to a certain pattern.

The design chosen must be within the capabilities of the analytical function, the sampling instruments at hand and the hygiene personnel. Routine monitoring of personal exposure is limited to job-exposure groups whose exposures are in the range 0.1–1.0 × exposure limit. It is not normally required for exposure below one tenth of the exposure limit. For exposure over the exposure limit priority goes to rectifying adverse conditions.

Biological monitoring

To the extent that biological exposure determinants reflect the body burden they are inherently better correlated with health risks than are measurements of atmospheric exposure, which, furthermore, do not directly indicate how much absorption there is by skin contact or ingestion (Figure 13.3). There are up to 80–100 substances in published compilations of biological exposure indices, either of the substance itself or a metabolite of it (Alessio, Berlin, Roi and Boni, 1983; Lauwerys, 1975). The values of determinants do not, by themselves, indicate the origin of the substance being measured but they have considerable merit for health risk surveillance (Ho and Dillon, 1987). A biological exposure index is comparable with an atmospheric exposure limit, and in some ways is a superior limit. Air sampling results could be supplemented with, and in some circumstances replaced by measurement of biological determinants of those hazardous substances for which there are suitable biological exposure indices in blood, urine or exhaled air. Such an index would be one which is well correlated with exposure and is specific for the hazardous substance of interest.

A necessary precondition is that the relevant biological specimens are readily available and the analytical skills to assess them. Where biological monitoring is feasible it is superior to air measurements for indicating important aspects of exposure and its effects (Berlin, Wolff and Hasegawa, 1979; Aiito, Riihimaki and Vainio, 1984; King, 1990).

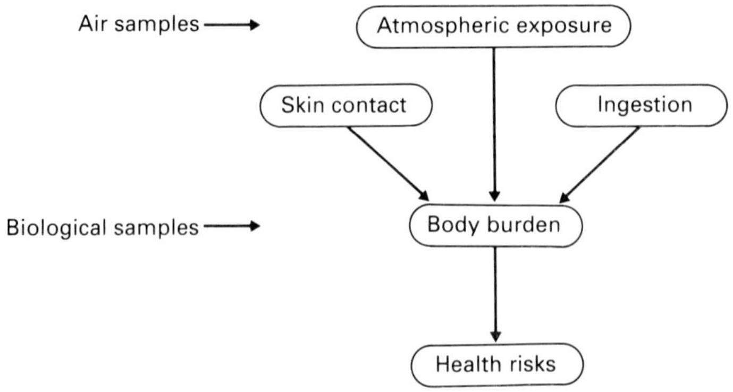

FIG. 13.3. Health risks are more directly related to biological exposure indices than to atmospheric exposure.

Examples are:

—Changes in an individual's body burden of substances. The health of the individual is more closely related to the body burden than to the atmospheric exposure.

—Exposure conditions at work where percutaneous absorption is significant. Biological exposure indices reflect body input by all routes. Exposure by skin contact is difficult to quantify directly and its consequences even more so.

—Exposure conditions at work where ingestion is significant. This is virtually impossible to quantify other than by biological exposure indices. Inorganic lead exposure is one where this can be a problem.

—Exposure when employees are wearing respiratory protective equipment. The value of respiratory protective equipment is often critically dependent on the fit of air purifying respirators in routine use. Regular air sampling behind respirators in use at work is not feasible.

—Individual bad working practices or excessive work load leading to especially high exposure or respiratory minute volume.

—Exposure conditions where exposure to the substance of interest outside of working hours is significant in comparison to exposure at work. The exposure outside of working hours may be by inhalation, skin contamination or ingestion. This includes air, food, milk, water, beverages, cosmetics, medicaments and so on. Ubiquitous substances such as carbon monoxide, lead, hexane and fluorides are examples.

—Changes in a variety of health parameters which become available for study once biological samples are obtained. For example, haemoglobin or cholesterol in blood samples.

The usefulness of a biological exposure index in urine, blood and exhaled air is based on there being some correlation between the level of substance or a metabolite of it and exposure levels or health effects. The correlation with exposure levels is of particular interest to the occupational hygienist. The correlations are variable as between individuals and from time to time so that before action to rectify environmental controls is justified, repeat biological sampling should be undertaken and individuals need to show repeatedly high values (Docter and Zielhuis, 1967). Alternatively, action should be triggered when the average level of the determinant for all employees in a job-exposure group exceeds the corresponding biological exposure limit.

Where biological monitoring is required by law the specified frequency and type of testing should be followed as should the requirements for remedial action in the event of there being unfavourable results. Otherwise the guidelines given earlier regarding frequency of testing and for triggering action on air sampling results may also be followed for biological sampling, with the proviso that adverse values should be confirmed within fourteen days. Because of the inherent variability of biological exposure determinants care needs to be taken to verify adverse results before taking precipitous action on isolated results (Gompertz, 1980). Similarly, the interpretation of the results is different according to the source of the published values. German Biological Tolerance Values are commonly two or three times higher than ACGIH Biological Exposure Indices for the same substance since the former are the highest values expected, whereas the latter are the 'most likely' values (Fiserova-Bergorova, 1990).

Timing of biological sampling

The precise time that samples are taken in relation to the work cycle needs careful consideration; whether during a work shift, at the end or beginning and when during the week. The time of sampling in relation to the periods of exposure is decisive for substances which have a short biological half time in various body compartments as the substance and its metabolites may be rapidly eliminated. The timing should be chosen which best reflects the body burden in the critical organ or critical organ for the substance in question (Droz, 1989; Sato, Nakajima, Fujiwara and Murayama, 1975). Substances which are eliminated most rapidly would normally be determined at the end of a convenient work shift. The interval between the cessation of exposure and taking the sample is increasingly critical for substances whose biological half time is a few minutes or less. When sampling urine the time that has elapsed since the previous voiding is also significant. A determination may also be made at the beginning of the same shift so that the employee serves as his or her own control.

A biological monitoring program for substances with a relatively short biological half time may fail in its objective unless a close relationship is maintained between the time the samples are taken and the time the exposure is expected to be at its highest. Substances which are not completely eliminated in the interval between work shifts would normally be determined at the beginning and the end of the work week. The beginning of the last work shift in the week may be preferred for substances with biphasic elimination as the longer half time component more closely reflects long term exposure. The timing is less critical with substances which have a very long half time. Substances which have a biological half time in excess of 50 hours may be determined at any convenient time.

When comparing results of biological monitoring against published biological exposure indices the recommended method and timing of the sampling should be strictly followed (American Conference of Governmental Industrial Hygienists, 1986, 1990; Commission for the Investigation of Health Hazards, 1989).

Validity of results

The combination of sampling routine and analytical method used for biological monitoring has to be selected with care. Where the detailed description of the biological determinant allows some choice in sampling or analytical procedures the particular combination employed should be one which is specific for the substance of interest and sensitive down to one tenth of the corresponding biological exposure index. Deviation from approved procedures, however desirable, should only be contemplated after satisfactory calibration has been done against the standard (Lauwerys, Buchet, Roels, Berlin and Smeets, 1975).

Before adopting a particular combination satisfactory assurance should be obtained regarding its specificity, sensitivity, accuracy and precision (Bullock, Smith and Whitehead, 1986). Each of these parameters can be quantified and should meet the following criteria in the conditions of use:

Specificity

Interferences present in the medium under study should contribute no more than 10% of the value of the measurement. This condition might be relaxed if the interferences are constant and the method is calibrated with this constant interference in place.

Sensitivity

The range of use should be at least 0.1–1.0 × biological exposure index. Given the sensitivity of the analytical method, in terms of the least detectable mass of substance, the minimum volume of the biological sample must be

$$\frac{10 \times \text{analytical sensitivity}}{\text{biological exposure index}}. \qquad (1)$$

When the analytical sensitivity is measured in milligrams and the biological exposure index in milligram per litre the required minimum volume of the sample is given by expression (1) in litres.

Accuracy

The differences between the mean of an infinite set of measurements and the true value should not be more than 20% of the true value. Particular care is needed with biological samples to prevent contamination of equipment, syringes and sample bottles.

Precision

The overall scatter of a set of measurements of the same concentration, lying in the range 0.1–1.0 × biological exposure index, should have a coefficient of variation of less than 15%.

Medical surveillance

In the case of a health screening survey the main objective is to detect the presence or absence of an adverse effect on health in the individual. Data on the individual employee is decisive. This contrasts with biological monitoring in which the distribution of data over a whole group of employees is decisive, not the individual result as such.

Disease may be caused or exacerbated by exposure to hazardous substances at work. The full effects need to be detected so that steps can be taken to protect, inform, advise, care for and compensate an individual with the condition and to revise as necessary the exposure limits in question. In a few processes specific medical surveillance including regular medical examination of all those exposed significantly to hazardous substances is a statutory requirement by national law. However, there are only 10–20 substances covered in this way and the medical surveillance required is not elaborate. The responsibility for deciding exactly what examinations to make largely rests with the employer.

At many works where large quantities of hazardous substances are used the medical service comprises a local physician who attends once per week and waits in the surgery for patients to call in, relying, perhaps, on a full-time industrial nurse in attendance during the rest of the week. A service of this kind may be perfectly adequate to ensure prompt diagnosis and treatment of those diseases which have symptoms which are obvious to the employee. The clinical observation of sick individuals who seek treatment or advice is a classic means for the detection of previously unrecognised health hazards (Schilling, Taylor and Jones, 1973). Such a passive service is however not sufficient by itself for *preventing* diseases due to hazardous substances at work. Medical surveillance for the purposes of prevention would normally be accomplished with the aid of specially designed medical examinations extended to cover all employees exposed significantly to hazardous substances in particular jobs irrespective of whether they are affected. They should be examined periodically incorporating procedures and tests designed for employees exposed to those specific substances (Figure 13.4).

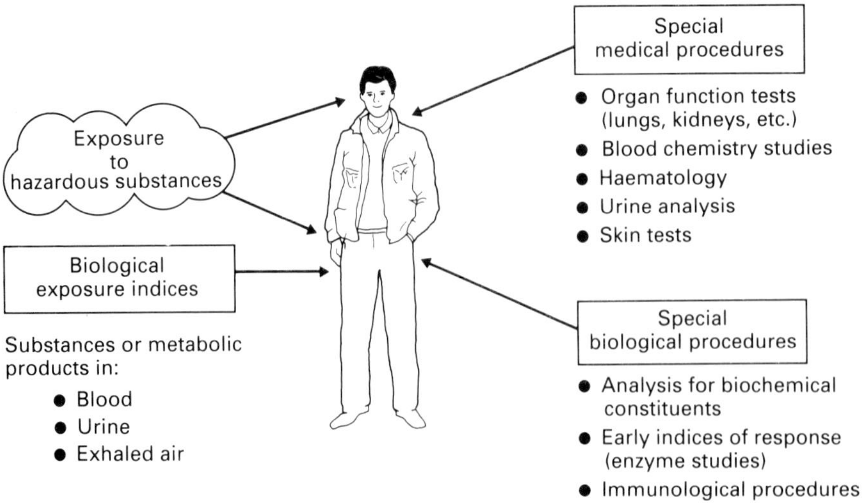

FIG. 13.4. Special procedures employed in medical surveillance at work.

What to look for

When the body is exposed to hazardous substances there is a variety of protective mechanisms that operate in the zone of homoeostasis. Further mechanisms come into play in the compensatory zone within which normal function is maintained without significant cost to health. It is the successful operation of these mechanisms whilst employee exposure is below the exposure limits that makes it arguable that it is possible to handle hazardous substances above zero exposure but without threat to health (Hatch, 1973).

There are available methods for detecting and measuring early changes and the degree of impairment, if any, caused by hazardous substances in terms of a variety of biochemical, physiological and psychological disturbances (WHO Study Group, 1975). Adverse effects which have been used in the past in setting exposure limits may be used as a guide. Some of the tests which may be employed are so sensitive that they can reveal responses within the zone where homoeostasis operates and these minor responses are indicators of continued good health rather than harbingers of injury. Others function on signs and symptoms at relatively gross levels at or immediately below clinical signs of impending illness and thus serve as unquestionable precursors of ill health. Measurement of increases in pulmonary flow resistance has proved to be a sensitive tool for assessing the response to irritant gases, vapours and aerosols. Pulmonary function tests are commonplace for assessing permanent changes caused by atmospheric exposure to hazardous substances. Renal clearance and other kidney function tests can serve to detect renal damage.

Over the years, as tests become more sophisticated through efforts to reveal earlier and earlier changes, they are increasingly valuable for tracking small changes in the average health status of a larger group of employees. On the other hand they tend to become less specific in revealing particular kinds of disturbance. Consequently they may lose significance as indicators of ill health in the individual except in combination with other tests. Animal studies should be used to identify early biochemical or physiological effects indicative of serious chronic disease in groups of employees believed to be exposed to hazardous substances.

Which employees to examine

Employees exposed at levels in excess of one tenth the exposure limit of hazardous substances should preferably be examined using tests specifically designed for employees exposed to the substances to which they are exposed. In deciding whom to examine it would be a mistake to imagine there is no risk in environments normally just below the atmospheric exposure limit. There are repeated reductions in 'authoritative' atmospheric exposure limits as new risks are found and old ones reassessed. There are many thousands of chemicals and millions of mixtures whose exposure limits are poorly understood and largely unknown. There is a strong case for independent medical examinations to seek out adverse working conditions (Rutstein, Mullan, Frazier, Halperin, Melius and Sestito, 1983). Take for example the American

Conference of Governmental Industrial Hygienists 1959 list of Threshold Limit Values. By 1974, that is over a period of 15 years, 23% of the values in that list had been lowered by a factor of 2 or more and by 1987, that is in a further 13 years, this was 39%. Furthermore, by 1987 7% of the values had been lowered by a factor of 10 or more. In all this time, that is a period of 28 years, only 2% of the values were increased.

When presenting the original ACGIH exposure limits Warren Cook wisely advised that maintenance of the limits should not be considered a substitute for medical monitoring (Cook, 1945). Exposure limits do not, unfortunately, guarantee a risk-free environment. This is understood by all who get to know the intimate workings of the exposure limit setting process.

Periodic medical examinations

Periodic medical examinations are especially pertinent when they are designed with specific toxicological hazards in mind. Whether all employees should be included or just those in jobs associated with high exposure is open to debate. They are undoubtedly best targeted on job-exposure groups whose atmospheric exposure has been in excess of $0.1 \times$ exposure limit.

The examinations would usually consist of

—a standard health questionnaire
—a structured interview to clarify occupational history
—tests to elicit adverse effects against which the exposure limit supposedly provides protection.

The precise form and design of special tests varies. It is usually possible to foresee the likely risk and choose the tests most likely to show it up. Examples of the most common occupational diseases have been given by the Health and Safety Executive (1983). Enlightenment as to the nature of the adverse effects and hence as to the appropriate tests may be gained by studying the references in the documentation of published exposure limits. Any of the senses and organs of the body may be damaged. The objective tests could include, for example; pulmonary function, diagnostic radiology, electro-cardiology, exercise tests or skin tests for allergy (Tyrer and Lee, 1979).

The ideal interval between successive examinations depends upon the time course of the development of the adverse effects in each case. Skin inspections may be advisable twice a week where the risk of dermatitis can develop quickly. Otherwise an appropriate interval may be six months, a year, two years or even longer. In job-exposure groups where significant adverse effects are not present or at least not detected there is little to be gained by further repetitious medical examinations in the short term. At the other extreme where the nature or prevalence of health risks which is uncovered gives cause for concern the introduction of engineering and/or personnel controls, as appropriate, takes priority over all other matters. Consideration should be given to making rigorous studies of cause and effect to enable the nature and scale of newly identified risks to be measured (Carter, 1988).

As a precautionary measure individual employees who are found to be particularly susceptible or sensitive to the agent in question may be redeployed or possibly advised to change jobs. Due regard should be paid to the ethical issues involved (Lappe, 1983). Personnel controls as well as engineering control are indicated. Those employees who become sensitised should be redeployed so as to avoid exposure altogether or their symptoms will recur. Traditional engineering controls are designed to prevent the acquisition of sensitivity by others. These include substitution by less hazardous materials, total enclosure of processes, local exhaust ventilation of sources of air contamination and, failing these, the wearing of respiratory protection.

Trained occupational health nurses and other medical auxiliaries can be taught to administer health questionnaires and occupational histories in appropriate detail for jobs involving significant exposure to hazardous substances. They can also undertake straightforward measurements of height, weight, blood pressure, respiratory function and the like, thus relieving the burden on the physician's time. They can also make superficial inspections of hands, skin, teeth, eyes and so on.

References

American Conference of Governmental Industrial Hygienists. Documentation of the threshold limit values and biological exposure indices (1986 with supplements up to date). ACGIH, Cincinnati, Ohio, USA (1986).

American Conference of Governmental Industrial Hygienists. Threshold limit values and biological exposure indices for 1990–1991. ACGIH, Cincinnati, Ohio, USA (1990).

Bullock D G, N J Smith and T P Whitehead. Clin. Chem. 32 1884 (1986).

Carter J T. Work related disease from exposure to substances hazardous to health. Toxic Substances Bulletin, Issue No. 9, Health and Safety Executive, Baynards House, London (1988).

Coenen W. Staub-Reinhalt.d.Luft **31** 16 (1971).

Coenen W. Application of long-term limit values based on short measuring periods. Commission of the European Communities. Doc. No. 964/76 (1976).

Coenen W. Staub-Reinhalt.d.Luft **36** 240 (1976).

Commission for the Investigation of Health Hazards of Chemical Compounds in the Work Area. Maximum concentrations at the workplace and biological tolerance values for working materials, Report No. XXV. VCH Publishers, Weinheim, Germany (1989).

Cook W A. Ind. Med. **11** 936 (1945).

Docter H J and R L Zielhuis. Ann. Occup. Hyg. **10** 317 (1967).

Droz P O. Appl. Ind. Hyg. 4 20 (1989).

Fiserova-Bergerova V. Ann Occup. Hyg. **34** 639 (1990).

Fiserova-Bergerova (Thomas) V. History and concept of BEIs. In: Biological monitoring of exposure to industrial chemicals (Edited by V Fiserova-Bergorova and M Ogata) pp. 19–23. Proceedings of the United States—Japan co-operative seminar on biological monitoring, Honolulu, 1989. American Conference of Governmental Industrial Hygienists, Cincinnati, Ohio, USA (1990).

Francis M, S Selvin, R C Spear and S M Rappaport. Am. Ind. Hyg. Assoc. J. **50** 37 (1989).

Fulwiler R D. Instruments and techniques used in evaluating the performance of air flow systems. In: The industrial environment—its evaluation and control. US Department of Health, Education and Welfare, National Institute for Occupational Safety and Health, US Government Printing Office, Washington, DC 20402, USA (1973).

Galbas H G. Staub Reinhalt.d.Luft **39** 463 (1979).

Gompertz D. Ann. Occup. Hyg. **23** 405 (1980).

Hatch T F. Arch. Environ. Health **27** 222 (1973).
Health & Safety Executive. Guidance Notes, Medical Series. H M Stationery Office, London (1983).
Health & Safety Executive. Monitoring strategies for toxic substances. Guidance Note EH42. H M Stationery Office, London (1989).
King E. Ann. Occup. Hyg. **34** 315 (1990).
Lappe M. J. Occup. Med. **25** 11 (1983).
Lauwerys R J, P Buchet, H Roels, A Berlin and J Smeets. Clin. Chem. **21** 555 (1975).
Long W M. Brit. J. Ind. Med. **10** 241 (1953).
McCollister G M and K R Wilson. Atmos. Environ. **9** 417 (1975).
Morton W E. Am. J. Ind. Med. **14** 721 (1988).
Oldham P D and S A Roach. Brit. J. Ind. Med. **9** 112 (1952).
Rappaport S M. Ann. Occup. Hyg. **35** 61 (1991).
Rappaport S M, S Selvin and S A Roach. Appl. Ind. Hyg. **3** 310 (1988).
Rappaport S M and R C Spear. Ann. Occup. Hyg. **32** 21 (1988).
Roach S A. Brit. J. Ind. Med. **10** 220 (1953).
Roach S A. Ann. Occup. Hyg. 20 65 (1977).
Rock J C. The NIOSH action level: a closer look. In: Chemical hazards in the workplace (Edited by G Choudhary), pp. 471–489. American Chemical Society, Washington DC, USA (1981).
Rock J C. Am. Ind. Hyg. Assoc. J. **43** 297 (1982).
Rutstein D D, R J Mullan, T M Frazier, W E Halperin, J M Melius and J P Sestito. Am. J. Public Health **73** 1054 (1983).
Sato A, T Nakajima, Y Fujiwara and N Murayama. Brit. J. Ind. Med. **32** 321 (1975).
Schilling R S F, P J Taylor and W T Jones. The functions of an occupational health service. In: Occupational health practice (Edited by R S F Schilling), p. 40. Butterworths, London (1973).
Tebbens B D and R C Spear. Am. Ind. Hyg. Assoc. J. 32 546 (1971).
Tomlinson R C. Appl. Stat. **6** 198 (1957).
Wright B M. Brit. J. Ind. Med. **10** 235 (1953).

Bibliography

Aiito A, V Riihimako and H Vainio (Editors). Biological monitoring and surveillance of workers. Hemisphere Publishing Corp., New York (1984).
Alessio L, A Berlin, R Roi and M Boni (Editors). Human biological monitoring of industrial chemicals. Commission of the European Communities, Industrial Health and Safety Series (EUR 8476 EN). Office for Official Publication of the European Communities, Luxembourg (1983).
Berlin A, A H Wolff and Y Hasegawa (Editors). The use of biological specimens for the assessment of human exposure to environmental pollutants. CEC Martinus Nijhoft Publishers, The Hague/Boston/London (1979).
Ho Mat H and H K Dillon (Editors). Biological monitoring of exposure to chemicals. John Wiley & Sons, New York (1987).
Tyrer F H and K Lee. A synopsis of occupational medicine. John Wright and Sons, Bristol, UK (1979).
WHO Study Group. Early detection of health impairment in occupational exposure to health hazards. World Health Organisation, Geneva, Switzerland (1975).

CHAPTER 14

Managing the Issues

Risk management is not new. The history of risk analysis and risk management can be traced back to 3000 BC (Cavello and Mumpower, 1985). An assessment of possible health risks from hazardous substances at work proceeds by first identifying the places where hazardous materials are handled. An essential part of the next stage of the assessment is the recognition of atmospheric exposure limits. Appropriate limits are published for some substances but must be set in-house for others in order to proceed with the assessment in a logical manner. Specialist advice may be needed; occupational hygiene, industrial medicine, toxicology, epidemiology, chemistry, design feasibility, local exhaust ventilation and so on (Western, 1986; Molyneux and Wilson, 1990). Measurement of atmospheric exposure against these limits is necessary unless that exposure is already known to be insignificant.

The elements in a company responsible for keeping control over health risks from hazardous substances involve several highly specialised professionals whose combined efforts in this direction constitute an integrated system (Bridge, 1979). An integrated approach to prevention, in which employers and employees cooperate fully, is called for by the International Labour Office Occupational Health Services Convention, 1985 (No. 161), and Recommendation (No. 171). This joint approach to the solution of problems posed by exposure to hazardous substances is an example of something which could properly be called a systems approach (Karnopp and Rosenberg, 1975).

Efficient control of exposure depends upon integrating exposure monitoring with other sub-systems in such a way as to make a smoothly working total system (Schroy, 1986). A systems approach to risk management highlights the interaction between sub-systems of this kind. The hazardous substances control system is constantly changing to meet external changes and internal technology developments. In common with other control systems the key to efficient control lies in effective negative feedback loops. The system is an open one, working within a 'super' system comprising the rest of the company, and as such features sophisticated communication and thorough networking. A central feature of the communication system for hazardous substances is the hazard data sheet.

A hazardous substances control system may be rudimentary for the smallest factory and highly specialised for a large multi-national company but the main elements are found in every system. It is instructive to work through an example system to show how the sub-systems link together.

A systems approach

The study of systems as such is of relatively recent origin. The concepts of management systems have slowly emerged in recent decades from the tenets of organismic biology. They assume a central importance in the thinking of many scientists nowadays and have played an important part in the evolution of organisation and management theory (von Bertalanffy 1950, 1962; Kast and Rosenzweig, 1970). In order that control measures shall become implemented they must fit into the organisational setting of a company. If this organisational setting is unsympathetic to the measures it may be that management has an organisational problem to solve first (Baram, 1984; Hosmer, 1987).

Historically, the major natural sciences; physics, chemistry and biology, slowly became subdivided into hundreds of individual subjects and areas of professional expertise. Knowledge is nowadays fragmented into small and seemingly more manageable compartments. Furthermore, in science and engineering the success of analytical techniques and in particular structural analysis has tended to concentrate attention on static and structural properties of systems. At least two groups find these trends limiting and an encumbrance; chemical engineers for one and occupational hygienists for another. Their interest is less in the chemical or man than in the interaction between them. Both groups want to build and run control mechanisms and devices which *work* in a changing, dynamic chemical industry in which all parts strongly interact with one another. Their interest is less in the molecular structure of a chemical or the anatomy of the human body than in the *behaviour* of the entire man-chemical system or the complete living organism. Theirs is essentially a holistic view.

Systems analysis draws attention to the interaction between sub-systems. It provides a basis for understanding and integrating knowledge from a wide variety of highly specialised fields, as is called for in the assessment, management and control of health risks from hazardous substances at work.

In a sense, the comprehensive systems of health hazard control outlined in this book are only illustrations. Others may be simpler or may be divided in a different way. Each sub-system may be divided into a number of different parts. The inter-connected sub-systems and parts could be defined, for example, in the following manner:

Sub-system	*Parts*
• Biological effects	—Portals of entry —Diseases —Effect thresholds —Toxico-kinetics

• In-house exposure limits	—Single substances —Mixtures —Unusual work schedules
• Assessment of health risks	—Data gathering —Appraisal of exposure —Work place inspection —Air sampling —Biological monitoring —Medical examinations
• Health risk surveillance	—Engineering controls —Personnel controls —Atmospheric exposure —Biological indices —Medical surveillance
• Information	—Layout plans —Inventory and throughput —Published exposure limits —Hazard data sheets —Labels
• Engineering control	—Substitution —Total enclosure —Segregation —Dilution ventilation —Local exhaust ventilation
• Personnel control	—Education and training —Protective clothing —Washing facilities —Exposure time —Respiratory protection

Managing change

Two properties of a system which are almost universally regarded as beneficial are stability and continuity. Each part of a sub-system has a tendency to change. Some changes are designed; made deliberately with the intention of improving the sub-system in some way, whilst others occur by drift or by chance. It takes time to institute a change and every change has interaction with other components and sub-systems. If each sub-system is allowed to change without regard to the system as a whole there may never be a moment when the complete system is stable. In order to allow healthy development and preserve system stability changes should be introduced in a planned manner with attention to timing and there should always be a period of overlap within the sub-system so that the effect of a change can be correlated (Christensen, 1987).

In industrial practice there are real limitations in the rapidity with which improvements to employee exposure to hazardous substances may be achieved in an orderly fashion. This procedure involves the hygienist from the health risk assessment sub-system in determining whether an engineering improvement is desired and in convincing those in the super-system with the powers to bring about an improvement, that such an improvement is needed. It is then necessary for a company to establish and agree on the precise way an improvement may be secured with minimum expenditure of resources by the engineering control sub-system. Time is needed to set priorities in relation to competing demands on time and resources. More time is consumed in making drawings and detailed engineering proposals. Orders for new equipment must then be placed and time allocated for making, supplying and installing the equipment and, in the personnel control sub-system, for training personnel in its operation and maintenance. The assessment sub-system then returns to the limelight to make sure that thorough before-and-after measurements are made and related to specific control improvements, thereby building a reservoir of data to guide future improvements.

In order to assist the implementation of company policy an assessment of risks to health from hazardous substances should include specific recommendations about the nature of any precautionary measures needed to prevent excessive employee exposure and, where reasonably practicable, to further reduce their exposure. Agreement on the form these recommendations might best take requires prior consultation and coordination with company management and technical services. It is advantageous if a member of senior management is allocated specific responsibility for seeing to it that suitable assessments of health risks are made and that adequate precautionary steps are taken as necessary, regarding the physical environment and personnel control. In any event it is important that:

- Where urgent action is believed to be required this is so stated.
- Short-term and long-term objectives and priorities are identified.
- Realistic target dates are determined for taking specific measures to achieve the desired conditions. It should be agreed who is to undertake what action and when, together with the audit arrangements to confirm their attainment.
- Schedules are established for exposure and ventilation surveillance and for medical surveillance.
- A programme is agreed for re-assessments commensurate with the precautionary steps taken.
- Company policy on control of hazardous substances is to continually seek ways of further reducing employee exposure and to seize the opportunity to do so wherever economically justified.

These procedures are such that it is reasonably practicable in industry to bring about a reduction in air contamination concentration over a whole works at an average rate of 25% per year, but it is unrealistic to suppose that improvements may be made at a much faster rate. Nevertheless, there are few situations so intractable that improvement at this rate is not feasible and there are isolated instances where dramatic improvements have been made almost overnight. Control charts are recommended to aid management of the working conditions regarding employee exposure. A long term reduction of atmospheric contamination by hazardous substances is both practicable and reasonable.

A property of an open system as opposed to a closed one is that the boundaries of the system between itself and other systems are flexible and penetrable both ways. This is healthy to the systems on either side of the boundary. A key function of the company organisation is to regulate the boundaries. The concept of interface between systems is useful in understanding boundary relationships.

Another beneficial property of an open system, such as a health hazard control system, is adaptability. The system must possess effective mechanisms which allow it to constantly respond to changing external and internal demands. Structures, committees, sub-committees, regular meetings and so on develop in a company to generate appropriate responses to external conditions (Katz and Kahn, 1966). However, committees have an inherent propensity to create sub-committees which by their very existence ensure the survival of the parent indefinitely and increase the boundaries of its operations. To balance this tendency such committee structures should include mechanisms which prevent unnecessary proliferation. Expressed more colourfully, they need built-in obsolescence or they take over everything.

Negative feedback

For any control system to work there has to be effective negative feedback loops (Hammer, 1972). This means that when the system deviates from the norm by more than preset amounts there is a mechanism which corrects the deviation. The principles of negative feedback can be seen operating in many physical activities. For example, in riding a bicycle the cyclist receives feedback in regard to direction and balance which cause the cyclist to take corrective action. The reaction must occur quickly, before the cyclist falls over, and must not be excessive, or the cyclist will fall over the other side. In the assessment and control of hazardous substances excessive exposure must be detected and prompt corrective action taken, commensurate with the deviation.

Negative feedback has two distinct parts. One part is the input of information which indicates that the system is deviating from a prescribed course and the other part is the mechanism by which the deviation is corrected. Both parts of a negative feedback loop are essential and both take a finite period of time to execute. Poor system control can often be identified as a weakness in one or other part of the negative feedback.

Care needs to be taken over selecting the size of deviation allowed in a system before a correction is instituted. Too small and the system operates in a continual state of turmoil, too large and the deviations may be too large to correct. Weakness in a negative feedback loop can sometimes be traced to an undue length of time being taken over the execution of one or other of its parts. Information about employee exposure that is late in coming may be ineffective if the conditions giving rise to that information no longer apply and have been largely forgotten. If action to correct a marked deviation in exposure levels is delayed too long the health of employees will suffer. This example illustrates the necessary systematic and graduated response to an increasingly unfavourable outlook which is the hallmark of negative feedback.

Problems with systems sometimes appear intractable because too little is known about the system as such; about what influences the behaviour of the system or about the design and operation of the system. Even the first step of drawing its boundaries can present formidable difficulties. Understanding the system for what it is, how the sub-systems influence the behaviour of the whole system, what its weak links are and how they can be corrected in a systematic fashion make more problems. But there is growing realisation of the advantages of a systems approach to handle the complexity of hazard control in a modern chemical industry and to integrate hazard control into overall corporate or site management policy. Ethical management is ultimately the most efficient (Rest and Patterson, 1986).

Communication

Interaction between the sub-systems of hazard control occurs through communications. These are integral parts of the system and are critical to its proper functioning. Faults in a system can very often be traced to inadequate or poor communications. An important function of those concerned with hazard control is the distribution of information in such terms that the people in need of the information will understand it. The system of hazard control does not work in complete isolation but is a part of a much larger, superior system, namely the company. The sub-systems rely upon specialists in their own field who must communicate freely with other specialists in related fields. The discipline of networking the communication is particularly valuable in hazardous substances control. It is an *open* system and in dynamic equilibrium with its surrounding systems within a company. It receives inputs from other systems in the form of information, resources, machines and materials. It transforms these inputs in various ways and exports outputs largely in the form of services to the systems with which it interacts. Communication between the hazard control system and the rest of the company is just as important as it is within the system.

The results are variously reported to one or more of the following:

- managers
- supervisors
- in-house committees
- individual employees
- trades union representatives
- enforcing authority inspectors

- designers
- production engineers
- maintenance engineers
- consultants

Communication with other systems should be two-way and is consciously planned to be two-way. The reporting system itself thus acknowledges the virtues of negative feedback. Constructive criticism is more valuable than praise.

Information

The store of information on hazardous substances held by a company can, if properly used, be an effective means of communication to works personnel. The information should be held in summary form as far as possible and in a standardised format. This makes feasible automated information flow and its analysis with computers (Wright, 1981). The occupational hygienist is often faced with a paucity of information (Oldershaw, 1989). Specialists within the company can hold the details and references to published literature (Hall, Jamieson and Taylor, 1971). These should be complemented with abstract journals and computerised databases such as OSH-ROM for its information on occupational hygiene and TOXLINE PLUS for its information on toxicology. The databases may be the only means of obtaining information about scores of thousands of minor hazardous substances which do not receive a mention in current reference books. Retrieval is effected by instructing the computer to search for substance names and key words occurring in titles of papers, abstracts, monographs and reports. Organisations with their own teletype-compatible computer terminal, and password, on-line, interactive search are available. These databases are limited historically, in the sense that their coverage has a definite start date, and their information is necessarily stereotyped, so that they should be supplemented with the services of a good library.

Organisations which can provide information on the toxicology of hazardous substances include the UK Chemical Industries Association, Health and Safety Executive and the Royal Society of Chemistry; USA Chemical Manufacturers Association, Occupational Safety and Health Administration, National Institute of Occupational Safety and Health, and Environmental Protection Agency. Similar organisations are found in many other countries.

Computers have brought a radical change in the analysis of monitoring data. Statistical calculations which previously were passed to a skilled statistician are now completed in a few minutes with the aid of a suitably programmed computer. Large sets of data involving thousands of measurements can be stored and analysed on a single disc. Packaged programmes have been developed to process, edit and analyse the data.

The risk assessments, monitoring and exposure controls should be recorded in a standard fashion and the results reported to appropriate persons. Careful consideration should be given to the design of the reports. They should include an executive summary, the objectives, methods, results, conclusions and recommendations. The technical information should be presented so that it is easy to read and understand, without unnecessary jargon. It has to be held in mind that the information they contain is likely to be transmitted to others in addition to the immediate recipients. The results may be kept for many years so that it is important that there is enough detail to enable the future reader to understand to what they relate and show why decisions about risks and precautions have been arrived at. Clarity is certainly as important as correctness. The quality of internal written reports in industry is often given too little attention but can be improved with practical guidance from expert texts (Crews, 1974; Bates, 1978; Cooper, 1981).

Hazard data sheets are the most useful form of information for the individual manager or employee. They need to be brought up to date periodically as more information comes to light. Labels on vessels and containers also need to be standardised in size, colour and form.

Write it up

There are certain essential data and documents which should be compiled, kept up-to-date, and retained as a permanent record. These include:

- results of assessments and re-assessments
- job-exposure groups, listed by group name, definition and exposure range
- identity of individuals in job-exposure groups with significant exposure
- identity of hazardous materials handled in significant quantity, their site inventory and annual throughput
- skin irritants
- identity of especially hazardous materials
- hazardous uses and operations, listed by location
- hazard data sheets
- identity of readily airborne substances listed by use, operation and location
- published exposure limits and their documentation
- in-house exposure limits and their documentation
- ventilation provisions; hoods, ducts, fans, air cleaners; flow rates and location
- atmospheric exposure data listed by job exposure group, employee name and substance
- risk monitoring procedures; atmospheric and biological
- engineering controls over exposure to hazardous substances
- personnel controls; respirators, protective clothing and management procedures
- air sampling reports

Air and biological sampling reports

These should include the following items;

- a statement of purpose
- date, day of week, shift and time of sampling
- place
- job-exposure group
- employee or employees under investigation
- investigator
- analyst
- reason for choice of
  - —date, day, shift and time
  - —plant, job-exposure group and employee(s)
- operating conditions of
  - —plant and process
  - —ventilation provisions
  - —other engineering controls
  - —personnel controls
- air and biological sampling details
  - —substance(s) measured
  - —medium (air; blood, urine or exhaled air)
  - —method of sampling
  - —duration
  - —analytical method
- results
- exposure limits and biological exposure indices
- summary of results
- interpretation of results
- conclusions
- recommendations
  - —actions recommended and why
  - —who takes the action
  - —when

Labelling

Labels identify the materials held in containers and provide the individual employee with the essential information concerning the hazards associated with the materials. Labels may be useful as guidelines for creating occupational exposure limits (Gardner and Oldershaw, 1991). All containers, including waste bins, must be properly labelled. Unlabelled or incorrectly labelled materials may be mishandled or misused.

All vessels or containers which are used to store hazardous substances should be labelled. As a general rule unlabelled substances should not be used. If a new substance is transferred to a cleaned bottle, drum or other container which previously held a different substance the old label must be completely removed or covered over and a new label added to identify the new contents.

Hazard data sheets

Material hazard data sheets, sometimes called safety data sheets, should be maintained for all hazardous chemicals to which employees may be exposed. They should be kept up-to-date, in a standardised format and be accessible to all employees (Health and Safety Executive, 1988; Akerman, 1989). Hazard data sheets provided by suppliers are invaluable when well written but are often lacking in detail and may need to be supplemented with additional information available in-house or found by a search of published literature (Churchley, 1977; Cumberland and Hebden, 1975; Ross, 1978; Wright, 1981). They are, unfortunately sometimes inaccurate, inadequate and unintelligible (Wood, 1989). Material hazard data sheets should include at least the following items about hazardous substances:

- date sheet compiled
- chemical name
- trade name, ingredients, chemical constitution, formula, molecular structure, in-house code
- exposure limits, principle route(s) of exposure (inhalation, ingestion, skin, other)
- melting point, boiling point, specific gravity
- physical form, appearance, odour, vapour pressure, vapour density (air = 1), solubility in water
- general chemical properties
- reactivity, fire and explosion hazards associated with the material
- storage procedures, handling and disposal precautions
- personal protection (respiratory, eye, hand and arm, other)
- toxicological data; acute, sub-acute, chronic; oral, inhalation and dermal
- medical treatment; emergency actions in the event of extreme exposure, first-aid, further treatment.
- safety data; flash point, flammable limits, extinguishing media, special fire fighting procedures, intrinsic safety provisions, flame proof requirements, stability, hazardous decomposition products and so on
- spill or leak procedures, disposal of wastes
- references

Toxicity tests should relate to the actual formulation rather than its pure, active ingredients. A high toxicity impurity may be present in variable amounts in a technical grade material. The toxicity may vary fourfold according to the precise formulation (Ball, Kay and Sinclair, 1953; Friess, Jenden and Tureman, 1959).

A pocket guide to chemical hazards of the work place, including notes on treatment and toxicology for some 800 chemicals has been prepared by Proctor, Hughes and Fischman (1988).

Worked example of a health risk assessment and control system

Methods of making a health risk assessment vary according to the size and complexity of the company, the nature of its products, its organisation and so on. A comprehensive scheme to assess the magnitude of exposure to hazardous substances and ascertain its importance is outlined next. It may be expanded or re-shaped to suit local circumstances. A logic flow-diagram for gathering the necessary data is presented in Figure 14.1 and one for assessing exposure is in Figure 14.2. The following is a summary of the system. These notes are based on the principles introduced earlier and are elaborated as appropriate in later chapters.

Layout plans (Chapter 10)

Outline plans are obtained of each building, floor, room or work area. Outlines are included of fixed machines and equipment.

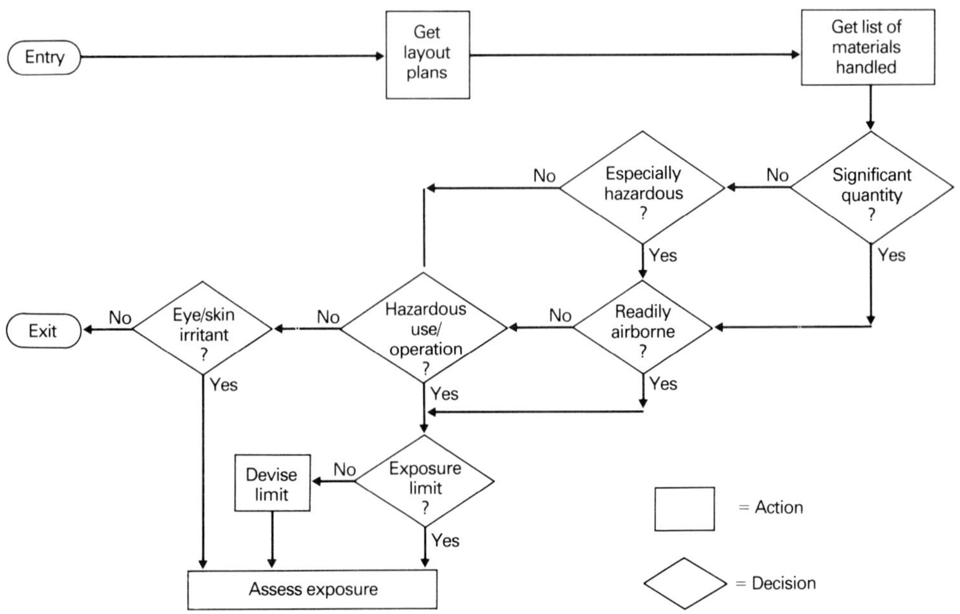

FIG. 14.1. Logic diagram for gathering data to assess health risk.

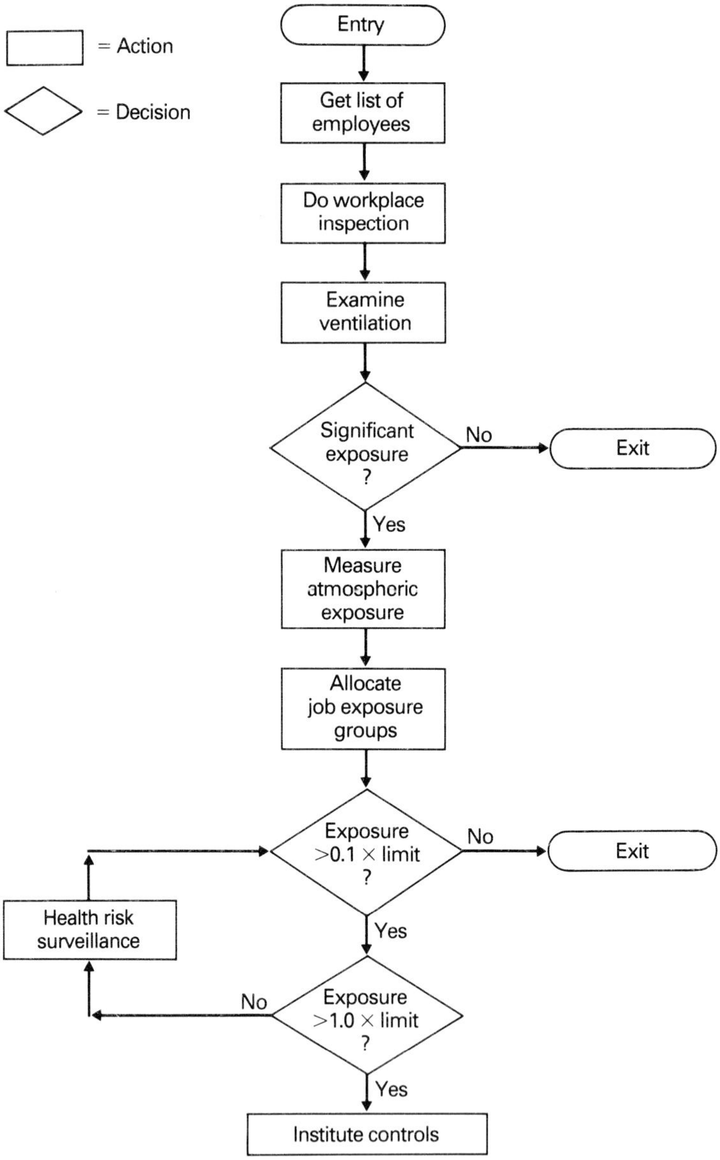

FIG. 14.2. Logic diagram for assessing exposure.

Transparent overlays are prepared, marked with factors relevant to employee exposure, such as:

- employee density
- walk-ways
- pipe-ways
- conveyors
- ventilation supply and exhaust

List of materials handled (Chapter 1)

All hazardous materials handled in significant quantities are listed, that is those in excess of 50 kg per annum throughput of a process or 1 kg inventory in a given work area, including:

- raw materials
- ancillary chemicals
- intermediates
- final products
- by-products
- wastes

If none, consideration is given to their hazardous nature, below.

Especially hazardous materials (Chapter 1)

Materials which are not handled in significant quantities but which are known to be especially hazardous are added here. This includes gases and vapours which have an acute inhalation LC50 in rats or mice which is less than 40 mg m^{-3}, liquids which have an acute toxicity to rats or mice equivalent to LD50 less than 10 mg per kg body weight, and solids which have an LD50 in rats or mice less than 2.5 mg per kg body weight or LC50 less than 10 mg m^{-3}.

Materials which are readily airborne

For materials so designated (details under Chapter 1) the assessor proceeds to consider appropriate exposure limits. Otherwise the possibility of hazardous use is considered next.

Hazardous uses and hazardous operations (Chapter 1)

Materials in small quantities which are handled in such a manner as to render them hazardous are added to the main list. Some processes and operations involve the emission of relatively large quantities of air contamination from a small quantity of material. Materials subject to such emissions should also be added to the list. Typical operations when this is liable to happen and should be considered in this regard include:

- boiling
- drilling
- filling containers
- filling vessels
- hand grinding
- melting
- sampling reactor vessels

- sanding
- sawing
- sieving
- spraying
- sweeping up
- welding

Eye and skin (Chapter 1)

Materials which are notable eye or skin irritants, skin sensitisers or especially prone to skin absorption are added here.

Exposure limits (Part II)

Current lists of atmospheric exposure limits for hazardous substances include:

- UK Health and Safety Commission MELs and OESs
- American Conference of Governmental Industrial Hygienists TLVs
- American Industrial Hygiene Association WEELs
- German MAK values
- Other published exposure limits
- In-house limits
- Suppliers data sheets

New exposure limits and exposure limits for mixtures are devised, as appropriate to fill any gaps (see Chapters 6–9).

Evaluation of employee exposure and its control

An up-to-date list of all employees on the books is obtained and the evaluation proceeds as follows.

Workplace inspection

A walk-through survey is conducted of hazardous substances, emissions to air, ventilation provisions, protective clothing and exposure conditions, including use of respiratory protection where provided. Where it is judged that atmospheric exposure is 0.1–1.0 × limit regular monitoring is instituted (Chapter 13). Where it is judged that exposure is greater than 1.0 × limit or otherwise excessive, appropriate controls are instituted immediately (Part IV). Direct reading instruments for measuring air contamination are valuable aids for checking the performance of control equipment (Chapter 11).

Ventilation examination (Chapter 13)

The proper functioning of the ventilation system is checked. The flow rates and integrity of systems are examined. Some of the ventilation may be through local exhaust hoods which are providing containment of sources within the hoods. Analogy may be sought between the hoods as found and standard hood types. Local exhaust ventilation requirements are specified for many standard hoods in the ACGIH manual on Industrial Ventilation (1986). Recent evidence should be sought or new evidence obtained that the volume flow rates meet the design specifications. Objective means to confirm the ventilation performance include the use of manometers, anemometers, smoke tubes and dust lamps (Chapter 18).

A dossier is compiled on the rates of emission of hazardous substances from the processes to workshop atmosphere. It might include, for example:

- The volume flow rate of extracted or discharged air from a process, work room or building, multiplied by the concentration of contamination in it.
- The quantity collected per hour, week, month, etc, by dust arrestment plant, scrubber or other air cleaning equipment.
- The loss, when making a materials balance of incoming and outgoing chemicals, after all other losses have been taken into account. Thus, the amount of organic solvent lost by evaporation per shift may be calculated from records of supplies and losses in some processes. The average volume flow rate of vapour released from a solvent is given by $0.9\ W/M$ ml s^{-1}, where W is the average weight in grams of solvent vaporised per shift and M is the gram molecular weight of the substance.
- The quantity deduced from the difference in atmospheric concentration downstream and upstream of a source of air contamination in a draught, multiplied by the air volume flow rate. This requires careful mapping of average concentration contours and velocity contours. The rate of emission of atmospheric contamination is CQ ml s^{-1} or mg s^{-1}, where C is the increment in concentration in ppm or mg m^{-3}, respectively and Q is the air volume flow rate downstream in $m^3\ s^{-1}$.

Insignificant exposure

Where it is assessed that the exposure to a substance by all routes is insignificant (see Chapter 10) the assessment is complete and the interval before re-assessment is agreed, with dates (see below).

Measure exposure (Chapter 12)

Where classification of the exposure of a job-exposure group is uncertain sufficient measurements are made of atmospheric exposure and/or biological indices to remove the uncertainty. Skin inspections should also be made of those employees exposed to skin irritants.

Job-exposure groups (Chapter 10)

Job titles are listed of employees expected to have a similar exposure to hazardous substances (Austin and Phillips, 1983). They are also given codes according to the areas within which they work, identified on the layout plans, above. Three standard exposure bands are recommended here; $<0.1 \times$ exposure limit, $0.1–1.0 \times$ exposure limit and $>1.0 \times$ exposure limit. Employees having the same job title, exposure band and area code constitute a job-exposure group. The number of employees in each job-exposure group is indicated.

Health risk surveillance (Chapter 13)

When exposure of a job-exposure group is between $1.0 \times$ exposure limit and $0.1 \times$ exposure limit atmospheric exposure is monitored as one element of the continuing health risk surveillance.

When it is found that exposure is repeatedly exceeding exposure limits further controls are instituted as necessary.

When exposure is repeatedly less than $0.1 \times$ exposure limit air monitoring ceases until a re-assessment is triggered.

Instituting controls

An assessment is made of the precise preventative steps that are necessary including:

- engineering controls (Chapter 15—18)
- personnel controls (Chapter 19)

Re-assessments and review of controls

An assessment should include a statement of the appropriate interval before it will need up-dating. It is suggested that where significant quantities of hazardous substances are handled the situation is reviewed annually, to decide the re-assessment programme. Otherwise a review every third year should be sufficient. The review should highlight shortcomings, seek agreement on their rectification and where agreed, bring the following up-to-date:

- Listing of hazardous substances, inventory and throughput.
- Job-exposure groups and employee numbers.
- Exposure limits, both external and in-house.
- Hazard data sheets.
- Health risk monitoring systems; exposure measurements, biological and medical examinations and record keeping.
- Engineering controls.
- Personnel controls.

- Labels and signs.
- Allocation of responsibility for implementing the above.
- Amendments to the above.
- Date for next review.

A re-assessment should be instituted in any of the following circumstances:

(a) After allowing a reasonable and agreed interval of time for commissioning remedial or corrective measures recommended in the previous assessment. The re-assessment is distinct and separate from examination and testing involved in the commissioning.

(b) When new plant or processes for hazardous substances are installed or there are changes in throughput, inventory, methods of manufacture or handling.

(c) When there is a change in the exposure limit of one or more of the hazardous substances in use.

(d) When a hazardous substance is replaced by one which is less so (or more so).

Re-assessment will be required upon:

- instituting controls
- change of methods of control
- change of plant or processes or materials handled
- change of volume or rate of production
- lowering or raising occupational exposure limits
- job-exposure group moving into different exposure zone

Re-assess the monitoring requirements, testing requirements and re-assessment schedule.

So you want to be an assessor

The duty to make assessments rests with the employer. A responsible employer will naturally want enquiries, investigations and assessments of health risk to be undertaken competently. Some aspects of an assessment are best known by local staff and shop floor employees with a detailed knowledge of day-to-day occurrences. They may also be able to identify in broad terms the types of changes needed. Other aspects may require skills in operating occupational hygiene instruments. Still others require knowledge of the principles and practice of exposure control. However, it would be wrong to suppose that a professional hygienist is always needed (Kirkwood, Trenchard, Uzel and Colby, 1991). Occupational hygiene encompasses knowledge and skills over a broad spectrum of subjects only a fraction of which is needed for assessing health risks from hazardous substances. Staff at a given works or process could have separately or jointly that minimum of knowledge and experience to competently perform a suitable assessment of exposure to hazardous substances in that process.

Assessments would normally be undertaken by, or under the supervision of local or company specialists, possibly assisted by an outside consultant attending personally at the company. Consultants act as agents for change, and when they have done the job they can be taken out and shot, which is their great advantage over getting new staff. Where there are no significant quantities of hazardous substances on site a safety officer should certainly be able to verify this and thereby complete the assessment. Someone with a good safety qualification having a strong occupational hygiene component such as that provided by the Examining Board in Occupational Safety and Health might well be sufficiently knowledgeable for deciding whether atmospheric exposure measurements are needed. Where such exposure measurements are necessary someone with skills in this area is needed. Similarly ventilation examinations require someone with different skills.

People carrying out assessments and consequential tasks must be given the necessary facilities and authority to do the work competently. They need to have enough time and status to gather the information, discuss it with the appropriate people, study existing records, examine the work place and, above all, realise the implications of what they find. A number of different people may be assigned different tasks within the assessment, but there must be reporting arrangements so that their findings are brought quickly to the attention of the employer or to some other specially delegated representative of the employer who can coordinate the conclusions and can directly authorise any necessary action (Health and Safety Executive 1988).

A person competent to make assessments of risk to health based on atmospheric exposure measurements of hazardous substances at work would be one who is a Registered Professional Hygienist or has a Certificate of Operational Competence of the British Examining Board in Occupational Hygiene or is an Associate Member of the Institute of Occupational Hygienists or is Certified in Comprehensive Practice by the American Board in Industrial Hygiene. Others having equivalent minimum qualifications would also be satisfactory for this purpose. For example, a person with a Bachelor's or Master's degree in Occupational/Industrial Hygiene with industrial experience should be competent for these purposes.

Companies of any size would benefit from having occupational hygiene staff although those with fewer than 500 employees are unlikely to find it imperative to have someone full time with these qualifications unless they have pressing occupational hygiene problems and companies with more than 5,000 employees are the ones most likely to employ a full time Registered Professional Hygienist. Most work places and most employees are in small 'plants'. At the present time, outside agencies and services will generally be employed for atmospheric exposure measurements. This may gradually change as development of inexpensive, reliable air sampling devices becomes more widespread.

Many health practitioners are able to diagnose the adverse effects from excessive exposure at work and, with experience and training, thereby make valid conclusions about the risk to health. Doctors competent to advise generally on health surveillance at work and to develop arrangements for its implementation would include Fellows, Members or Associates of the Faculty of Occupational Medicine in the Royal College of Physicians of London. Others having equivalent minimum specialist qualifications would also be satisfactory for this purpose.

Except for the very simplest of routine examinations it has to be recognised that a nurse with specialist training and experience is the very least that is needed for the conduct of an effective medical surveillance program for employees exposed to hazardous substances. A qualified general nurse may, of course, with specialist training, acquire the skills necessary to undertake health surveillance in the work place. For example, a qualified general nurse who has successfully completed a specialist course in occupational health nursing recognised by statutory nursing bodies would have the necessary qualifications.

References

Akerman T. Occup. Health Rev. **17** 14 (1989).
Austin W B and C F Phillips. Am. Ind. Hyg. Ass. J. **44** 638 (1983).
Ball W L, K Kay and J W Sinclair. AMA Arch. Ind. Hyg. and Occup. Med. **7** 292 (1953).
Baram M S. Am. J. Public Health **74** 1163 (1984).
von Bertalanffy, L. General system theory—A critical review. General Systems VII: 1 (1962).
von Bertalanffy, L. Science, Jan 13 (1950).
Bridge D P. Am. Ind. Hyg. Assoc. J. **40** 255 (1979).
Cavello V T and J Mumpower. Risk Analysis **5** 103 (1985).
Christensen J M. The human factors profession. In: Handbook of human factors (Edited by G Salvendy). John Wiley & Sons, New York (1987).
Churchley A R. Chemy. Ind. **13** 425 (1977).
Cumberland R F and M B Hebden. J. Hazard. Mater. **1** 35 (1975).
Friess S L, D J Jenden and J R Tureman. AMA Arch. Ind. Health **20** 253 (1959).
Gardner R J and P J Oldershaw. Ann. Occup. Hyg. **35** 51 (1991).
Hall S A, V Jamieson and P J Taylor. Ann. Occup. Hyg. **14** 295 (1971).
Health and Safety Executive. Substances for use at work: The provision of information. HS(G)27(2nd Edn.). H M Stationery Office, London (1988).
International Labour Office. World labour report. ILO, Geneva, Switzerland (1985).
Kirkwood P, P J Trenchard, A R Uzel and P J Colby. Ann. Occup. Hyg. **35** 233 (1991).
Molyneux M K and H G E Wilson. Ann. Occup. Hyg. **34** 177 (1990).
Oldershaw P J. Occup. Health Rev. **21** 14 (1989).
Rest K M and W B Patterson. Sem. Occup. Med. **1** 49 (1986).
Ross S S. Toxic substances sourcebook. pp. 143–147. Environment Information Center, New York (1978).
Schroy J M. Ann. Occup. Hyg. **30** 231 (1986).
Western N J. Ann. Occup. Hyg. **30** 237 (1986).
Wood S. Ann. Occup. Hyg. **33** 607 (1989).
Wright R B. Ann. Occup. Hyg. **24** 313 (1981).

Bibliography

Bates J D. Writing with precision—How to write so that you cannot possibly be misunderstood. Acropolis Books Ltd., Washington DC, USA (1978).
Cooper B M. Writing technical reports. Penguin Books, Harmondsworth, Middlesex, UK (1981).

Crews F. The Random House handbook. Random House, New York (1974).
Garrett J T, L J Cralley and L V Cralley (Editors). Industrial hygiene management. John Wiley & Sons, New York (1988).
Hammer W. Handbook of system and product safety. Prentice-Hall. Englewood Cliffs, New Jersey, USA (1972).
Hosmer L T. The ethics of management. Dow Jones Irwin, Momewood, Il, USA (1987).
Karnopp D and R Rosenberg. Systems dynamics : A unified approach. John Wiley & Sons Inc., New York (1975).
Kast F E and J E Rosenzweig. Organisation and management—A systems approach. McGraw-Hill (1970).
Katz D and R L Kahn. The social psychology of organisations. John Wiley & Sons Inc., New York (1966).
Proctor N H, J P Hughes and M L Fischman. Chemical hazards of the work place. J B Lippincott Company, Philadelphia, USA (1988).

PART IV

GETTING CONTROL OVER HEALTH RISKS

Introduction

The first line of defence against health risks from exposure to hazardous substances is maintenance of the engineering and personnel controls. The cause of an adverse trend and its rectification lies in the working of the controls. Surveillance of the engineering and personnel controls is in principle superior to regular measurement of the exposure of employees and to medical examinations. This does not mean that exposure measurement should not be done nor that employees should not have medical examinations, but it does mean that the greatest effort should be given to surveillance of the controls (Gerhardsson, 1988).

When an adverse situation occurs the first priority goes to consideration of substitution of a hazardous material by a less hazardous one. Failing this, modification to the premises, processes and plant so as to prevent or reduce emissions is considered. Otherwise, total enclosure may be feasible. The enclosure may be closely fitted to a machine and exhausted or the employees may be enclosed in a ventilated room or cabin. Where complete enclosure is not practicable partial enclosure of sources of air contamination is arranged together with exhaust of the enclosure to prevent escape of contamination. These possibilities are considered in Chapter 15.

Wherever the exposure of employees is found to be excessive engineering and personnel controls are instituted as necessary. The cycle is completed by making a re-assessment of health risk and, hopefully, confirming that there is no significant risk remaining.

1. Identification of engineering and other controls in plant, processes and buildings where exposure is excessive.

2. Installation of controls and re-assessment of remaining risks.

The principle control over air contamination is ventilation; dilution ventilation and local exhaust ventilation. The underlying principles of contamination transfer by convective flow and by turbulent diffusion are introduced in Chapter 16. These factors have an important bearing on the efficiency of exhaust hoods.

The prime means of preventing hazardous substances being dispersed into work-room air is local exhaust ventilation. The most important element of any local exhaust system is the inlet since if this fails to capture the air contamination the efficiency of the rest of the system is of no avail. Unfortunately, it is probably this part of the system that has all too frequently in the past received the least attention. The configuration of exhaust hoods is matched to the source of contamination and the volume flow rate of extracted air is adjusted to secure containment of the hazardous substances within the confines of the hood. Local exhaust ventilation provisions for gases, vapours and dusts are described in Chapter 17. The analysis of ventilation systems so as to locate and rectify the faults is in Chapter 18. So also is the detailed mapping of the ventilation achieved.

Engineering control may be supplemented with personnel control. Education and training in the prevention of exposure to hazardous substances has the highest priority. Adequate protective clothing must be provided to prevent contamination of the skin as far as possible. This is supplemented with suitable washing facilities, commensurate with the needs. The limitation of duration of exposure should also be encouraged. Technical improvements are realised only gradually because of their cost and limited feasibility. Respiratory protection is needed in the transitional period. Respiratory protection should be provided as long as there is a risk that employees may be exposed above the atmospheric exposure limit. These matters are considered in Chapter 19.

Reference

Gerhardsson G. Ann. Occup. Hyg. 32 1 (1988).

CHAPTER 15

Physical Environment Control

The control of exposure to hazardous substances at work is conveniently subdivided into control relating to the physical environment, which includes engineering controls, and control relating to personnel. The two kinds are generally managed by different persons and require different skills. The two approaches need to be integrated. Control of the physical environment takes precedence where feasible. It is the preferred means by which the highest practical standard of occupational hygiene is maintained in the work-place.

No exposure to hazardous substances should be permitted without considering the limited benefits associated with that exposure and without weighing the relative risks of alternative approaches. The possibility of substitution by a relatively innocuous substance should always be considered first. Often enough there is not a feasible alternative but where there is one the result can be a dramatic reduction in risk. Changing a process so as to eliminate the use of a particular substance or changing the layout so as to restrict its use to certain buildings or rooms can be equally effective (Mutchler, 1970). If this cannot be done, then limited but total enclosure of individual processes, plant or handling systems should be considered. Failing this, partial enclosure and exhaust of the enclosure is necessary.

A logic flow diagram is given in Figure 15.1, which summarises the preferred sequence in the decision-making procedure. In keeping exposure to hazardous substances to a minimum precedence goes successively to:

(i) Substitution of a less hazardous substance.

(ii) Modification of premises, processes and plant. This includes
  (a) Limiting exposure to specific localities by segregation of hazardous processes and plant.
  (b) Eliminating or reducing emissions by modification of processes and plant.
  (c) Controlling exposure by means of changes to the ventilation system, employing dilution ventilation principles.

(iii) Total enclosure.

(iv) Partial enclosure of sources of air contamination and exhaust of the enclosure.

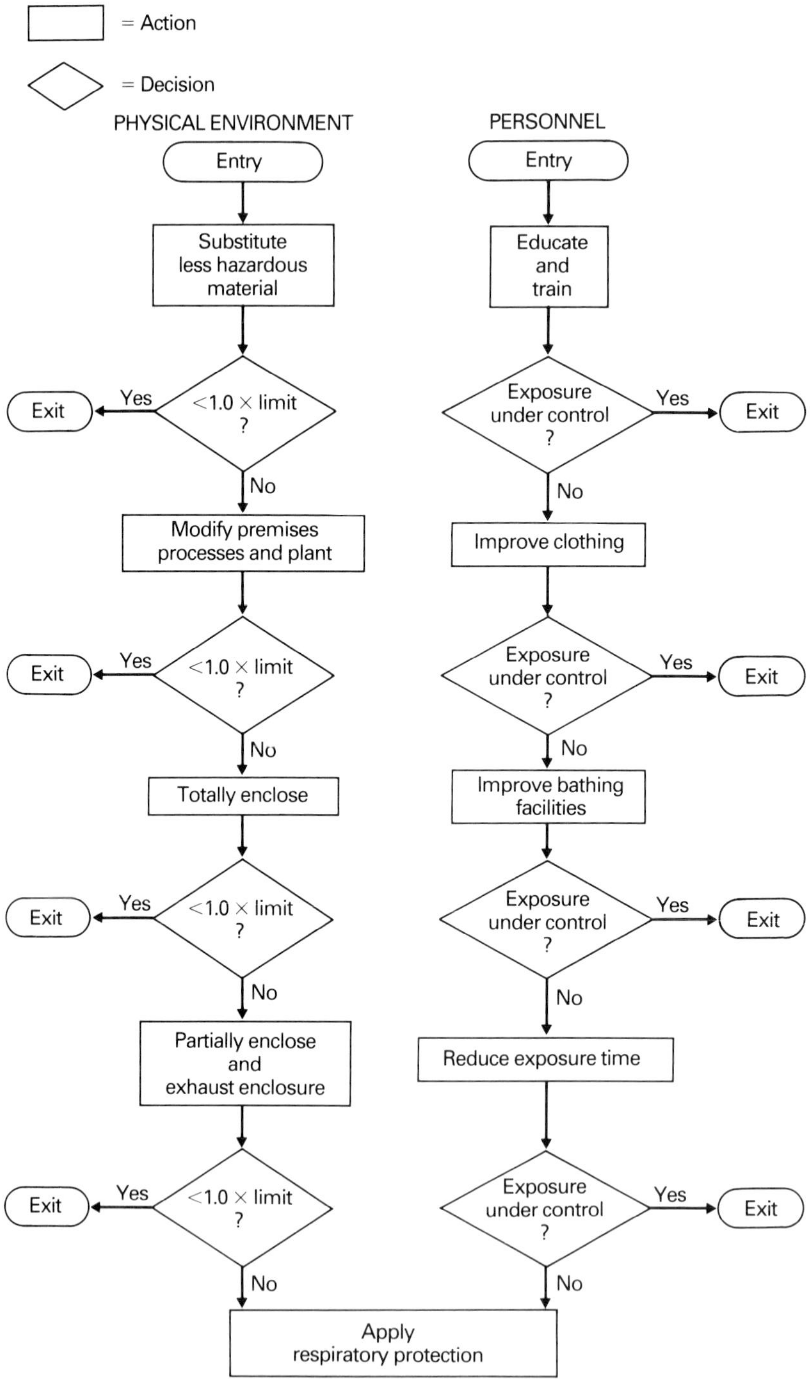

FIG. 15.1. Physical environment control. Logic diagram for assessing the required controls. The instructions are: Fulfil each step to the extent reasonably practicable. If the residual exposure is likely to be excessive proceed to the next step.

The criterion at each stage and in each work place is not simply that some action be taken, however well meant. The action must be heavily qualified by strictures about its effect on exposure to hazardous substances of specific employees at work. The action taken should be shown to reduce employee exposure to a level below the exposure limits. Otherwise the sequence continues until this requirement is shown to have been fulfilled in each work place.

In practice several or all of these control measures will be adopted at the same time since different aspects of a process are easier to control than others. This applies to different types of ventilation in particular.

Contamination from sources far away from the exhaust openings mixes into the general body of air and its control relies principally on dilution by thorough mixing with uncontaminated air. On the other hand, air contamination from sources inside or very close to exhaust openings is mostly extracted from the work-room before getting the opportunity to become mixed into the general body of air. Ventilation of air contamination relying essentially on the dilution effect is termed a dilution ventilation system, whereas one relying principally on locating exhaust openings over or very close to contamination sources is known as a local exhaust ventilation system. Both principles may, of course, be operating in one work room. Dilution systems are employed to good effect for controlling temperature and humidity. Standard air conditioning plant can, *inter alia*, be very effective for air contamination liberated in work rooms from sources having a low emission rate or where the materials being handled are of low toxicity. Otherwise local exhaust provisions are necessary to keep control over employee atmospheric exposure levels.

The air extracted from work rooms may be heavily contaminated with hazardous substances. Indeed the more efficient the system the more heavily contaminated will be the exhausted air. Air cleaning of its dust contamination and increasingly of its gaseous contamination is commonplace. The air cleaner and fan are integral parts of the ventilation system. There is not the space in this book to cover industrial air cleaning practices in great detail but the principle types will be outlined. Similarly air conditioning of the supply air to work rooms is another specialised field. The act of extracting contaminated air from a building necessarily means that it is replaced with fresh air by natural ventilation or, preferably, by a controlled supply of make-up air.

Substitution of a hazardous substance by a less hazardous one

The most effective and economical solutions to health hazards are ones which precede final process and design decisions. The first and very effective method to consider for controlling exposure to hazardous substances is replacement or substitution of a hazardous substance by a less hazardous one. The health risk is thereby removed or at least considerably reduced. There are often considerable long term financial benefits flowing from substitution.

Historical examples of substitution include a notorious one in the felt hat industry, where mercurial poisoning which was once very prevalent has been eliminated by employing mercury-free materials for treating the fur felt (Beal, McGregor and Harvey, 1941). White phosphorus is no longer used in the manufacture of matches. Foundries nowadays prevent much silicosis by using silica-free parting powders (Davis, 1979). The substitution of alumina for silica in the pottery industry to reduce the risk of silicosis is a classic case of prevention by substitution (Meiklejohn, 1963). Similarly, sandstone grinding wheels high in quartz content have been largely replaced by alumina synthetic abrasive wheels and ones made of carborundum, that is, silicon carbide. Paints containing lead compounds have been replaced by lead free paints; common white lead in paint pigments has been replaced by zinc, barium or titanium oxides. Lead silicate used as 'frit' in the ceramic industries offers very little hazard as compared with the litharge formerly employed. Sand blasting has been replaced by grit blasting and iron or steel shot. Ammonia has been replaced by freons as a cooling liquid in refrigerators. This particular substitution may prove to be a mixed blessing in view of the possible phasing out of chlorofluorocarbons (CFCs) to protect the Earth's ozone layer. In recent times asbestos as an insulator and binder has been replaced by a variety of asbestos-free materials such as glass wool, mineral wool and carbon fibre (Hodgeson, 1989). Mercury gauges have been replaced by mechanical pressure gauges. Despite these brilliantly successful examples it has to be cautioned that exposure to a high enough concentration of a less hazardous substance could be worse than exposure to a low concentration of the original substance.

Less hazardous solvents can in many cases be substituted for a more hazardous one. The potential hazard is reflected by the exposure limit and rate of evaporation. A good rule-of-thumb is that, other things being equal, the least hazardous solvent is the one with the lowest ratio;

$$\frac{\text{vapour pressure}}{\text{OEL}},$$

where vapour pressure is the partial vapour pressure of the solvent at working temperature and OEL is the occupational exposure limit (Popendorf, 1984). Consistent units should, of course, be employed when making comparisons between solvents. Thus:

(a) benzene is a good solvent but should be replaced where feasible by the less hazardous toluene or xylene,

(b) perchloroethylene should be replaced by methylene chloride,

(c) detergent and water based cleaning solutions can sometimes also be considered for use in place of organic solvents,

(d) water based latex paints are increasingly used in place of ones based on organic solvents,

(e) coatings employing organic solvents should be replaced by water based coatings on medicine tablets.

Modifications to premises, processes and plant

A change of premises, processes or plant often offers the best opportunity to improve working conditions by elimination or reduction of exposure. Most such changes are initiated for reasons of production and any reduction in employee exposure to hazardous substances incidental, so that a conscious effort must be made to bring reduction of health hazards to the fore. There are countless examples of process changes which are accompanied by reduced exposure to hazardous substances; brush painting instead of spray painting, riveting in place of arc welding, automatic degreasing in place of hand washing, low-speed sanders in place of high-speed types, keeping powder in suspension in a liquid and so on. Some techniques have only limited application outside of the industry in which they have been developed, as the infusion of coal with water before commencing coal-getting operations, but many operations and techniques for reducing employee exposure to hazardous substances are common to a wide spectrum of industries (Burgess, 1981).

In the chemical industry the manufacturing processes may usually be resolved into a coordinated series of unit physical operations and unit chemical processes held in common by many different products (Cralley and Cralley, 1982). A typical chemical process has a relatively short life (5–10 years). There are few processes which remain unchanged for many decades. Indeed, in batch processes the process duration may be for a few weeks or even less. This is a feature of chemical engineering generally (Perry and Chilton, 1973). The services, including ventilation supplied to buildings are also changed periodically. There is consequently constant opportunity at the design stage to re-arrange processes and plant in such a way as to reduce employee exposure. The opportunity should be taken to reduce subsequent employee exposure since process and product changes introduced later just to reduce exposure to hazardous substances can occupy considerable manpower and capital.

Many of the associated fittings of chemical plant such as valves, flanges, seals, drains and vents are liable to give rise to fugitive emissions (Brief, Lipton, Amarnani and Powell, 1983; Lipton and Lynch, 1987). Careful selection of new fittings with regard to sealing provisions, preventive maintenance of existing fittings and their replacement as necessary can make significant contributions to reducing employee exposure (Hoyle, 1978; Ramsden, 1978; Wu and Schroy, 1979; Schroy, 1981; Powell, 1984; Flitney and Nau, 1987).

In designing new plants and modifying existing ones the following principles are among the most helpful for reducing the amount of gas, vapour or dust generated by processes and operations in the chemical industry:

- Screw or pneumatic conveying of granular non-combustible materials is favoured over belt, scraper or shaker conveyors. Conveyor beds are chosen which are designed to contain spills.
- Solid substances are kept in solution or as a slurry, a paste, flakes or in pelletised form as far as the process will allow.
- Transfer points feature specific provisions to reduce the free fall of materials and to shield the falling material from draughts.

- Fixed bed mixers are utilised with bottom, 'gravity' discharge to reduce the manual handling of powders and liquids.
- Tight covers are fitted over screens and over all belt, scraper or shaker conveyors of powdery materials, having minimal crack length around cover sections and kept under negative pressure. In doing this due care has to be taken with combustible materials to avoid risk of fire and explosion within the enclosure.
- Manual emptying of bags of solids is replaced by automatic emptying under negative pressure.
- Weighing ingredients out with a hand scoop is replaced by automatic feed and hopper.
- Operations are conducted with liquid temperatures held as low as the process will allow. Drip trays are employed to contain spills.
- Paint and glaze dipping are substituted for spraying as far as reasonably practicable.
- Plastic balls, chips or foam floating on the surface of liquid treatment tanks are introduced to reduce spray and liquid evaporation. The same effect on evaporation is sometimes feasible by floating a layer of low vapour pressure liquid over the surface.
- Manual liquid transfer is replaced by automatic metered transfer.
- When a sample is taken from a process stream, a common operation liable to spills, the sample outlet is drawn from a closed loop which is itself enclosed and exhausted (Lovelace, 1979).
- As a matter of policy hazardous waste is kept to the minimum consistent with efficient production. Given the cost and trouble of treating and storing hazardous waste it is obviously better not to generate it in the first place.
- Cleaning with compressed air is barred. Brush sweeping is replaced by methods which do not themselves raise dust. Cleaning dust with a wet method is preferred, where feasible. Otherwise, portable vacuum cleaners are employed provided with efficient filters or a central vacuum cleaning unit with fixed piping around the building.

Many other examples are given by Cralley and Cralley (1985a).

The wetting of dust with water or other liquids is one of the oldest methods of control and may be very effective if properly used. Dust dispersion is prevented by wet drilling in mines and quarries; by infusion of coal with water before coal-getting; by water sprays at blasting, crushing, conveying operations and foundry shakeout; by wet grinding and machining; and by oiling of cotton. The application of water should be designed to blanket the dust source completely. The dust is thoroughly wetted by means of high pressure sprays, wetting agents, deluge sprays and other means.

Segregation of processes and plant

Ventilation problems often result from a combination of technological and economic changes that have resulted in production facilities becoming larger and more complex and caused them to concentrate production in a small area. When planning the layout of a new building or process the general principle applies that the overall ventilation requirements can be reduced by segregating those plants and processes which are liable to liberate hazardous substances freely into the air (Cralley and Cralley, 1985b). Segregation may be by space or time. To achieve segregation

(1) Highly hazardous materials are stored in a separate building.

(2) Employees are placed in an enclosed control room, wherever possible. This is aided by automation and robotics.

(3) Special rooms, areas within rooms and exhaust hoods are dedicated to the handling of highly toxic substances and hazardous operations. Good examples are plating tanks, lead melting pots, paint dipping operations. Enclosed structures require, in addition to adequate ventilation for normal operations, means for generous ventilation in emergencies, such as quick opening doors and windows.

(4) In plants where most of the processes produce air contamination it is preferable to isolate the clean operations in a separate building or room. Such a room is maintained at slightly higher pressures than adjoining rooms.

(5) To encourage 'plug' flow by the ventilation, the supply and discharge points are arranged at opposite sides of a building, with the hazardous processes located in the vicinity of the discharge openings.

(6) Sometimes the hazardous operations can be conducted during the evening or night shift when the least number of other employees are in the vicinity. Examples are foundry shakeout, shot-firing in mines and cleaning generally.

Enclosure

Enclosure is an effective means of putting a physical barrier between a source of air contamination and the employee. Plant, processes and equipment which would otherwise emit air contamination can sometimes be totally enclosed (Jones, 1984). Where this is feasible the control is very efficient. A total enclosure completely envelops the operation. Nevertheless, few such enclosures are completely air-tight and they are often exhausted with a modest inward flow of air to keep the enclosure under negative pressure.

The opposite form of enclosure is enclosure of employees in a ventilated room or cabin for most of the work shift (Lunchick, Nielsen and Reinert, 1987). Again, where this is feasible the control is very efficient. The room is in this instance kept under slight positive pressure to prevent ingress of air contamination.

The following features exemplify the trend to total enclosure:

- In-line cleaning facilities which allow decontamination without opening equipment.
- Double mechanical seals on pumps and valves replace single mechanical seals and packing glands.
- Totally encapsulated or 'canned' pumps replace pumps with seals.
- Welded construction replaces flanging.
- Totally enclosed drainage systems are fitted incorporating minimal make-and-break couplings.
- Materials are transported in tight, closed containers. This applies to powders, liquids, paints and preparations.
- Near-total enclosure of items of plant such as conveyors, feeders, hoppers, and other elements of the transport system. Items under this heading also include storage tanks, with a vent to a discharge stack or fume extraction system to absorb pressure fluctuations.
- Clean-up of liquids and solids vented to a fume extraction system.
- Processes located in an enclosed room under negative pressure.
- Employees located in a control room ventilated with fresh air. This is applicable in automatic or semi-automatic processes, such as stone crushing, grinding, conveying and in many chemical processes.
- Employees operating cranes, trucks, tractors and so on located in closed cab systems ventilated under positive pressure with filtered, tempered air.

The very highest rates of release of hazardous substances and the most toxic substances are controlled by complete or nearly complete isolation and enclosure of a process and by exhausting the enclosure. Generally, enclosing hoods should not enclose the worker, but should be interposed between the worker and the operation. In industrial practice an enclosure has small openings to allow the entry and removal of materials and so that the process or machine can be manipulated through them. To prevent air contamination from leaking from the enclosure it is only necessary to apply to it an exhaust volume flow rate which exceeds the rate at which the contaminated air would otherwise escape from the openings in the absence of the ventilation. Provided there is little air disturbance within the enclosure and little air movement through it a low exhaust volume flow rate of the order of 0.5 m^3 sec^{-1} per square metre of opening may suffice to prevent any escape. Manipulations inside such an enclosure may be performed by remote control or through gloves in the side. Otherwise, the operatives don fresh air fed helmets or suits to enter the enclosure. Grit blasting is a common process demanding complete enclosure

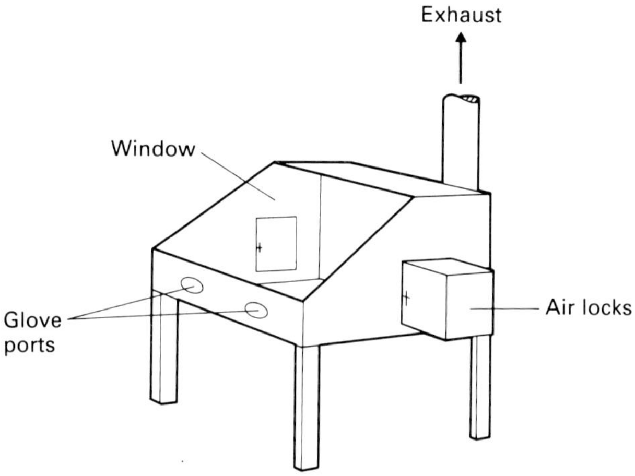

FIG. 15.2. Glove box.

due to the extremely high dust emissions. Radioactive materials and those of the very highest toxicity are often handled in glove boxes.

A glove box is a complete enclosure carefully sealed around all openings and provided with gloves having the gauntlet sealed to the enclosure for manipulation inside the box (Figure 15.2). To ensure containment of air contamination released inside, glove boxes are kept under negative pressure by connecting them to exhaust fans. They are usually provided with air locks for transferring materials and equipment in and out of the box.

Ventilation systems

The most important exposure to hazardous substances at work is usually through work room air contamination and ventilation is the most important means of controlling health risk from exposure to these substances. Ventilation reduces the concentration of air contamination in work spaces from residual gas, vapour or airborne dust given off the processes. Atmospheric exposure limits for different substances range over more than a millionfold in concentration in air. Moreover the rate of emission of contamination can also vary over a similar range. Consequently the ventilation system needed to control the emission of a hazardous substance will correspondingly differ very substantially from one substance and process to another.

Once hazardous substances are released into the air the atmospheric concentration of contamination to which the employees are exposed is governed largely by the ventilation system. There may be enough ventilation without calling on the assistance of fans. In a small work room without significant atmospheric contamination the ventilation may be sufficient by infiltration through cracks, assisted by opening windows and doors. There is more ventilation when it is assisted by heat convection from process machinery and hot processes. Thus ventilation may be with or without the assistance of fans.

TABLE 15.1. *Typical air velocities in workrooms*

Metres per second	Where found
0.003	Terminal falling velocity of particles 10 microns aerodynamic diameter
0.2	Terminal falling velocity of particles 100 microns aerodynamic diameter
0.2	Random air velocities
0.25	Minimum capture velocity for cold open surface tank
0.5	Minimum capture velocity for laboratory hood
0.5	Minimum air velocity felt on dry skin
1.0	Maximum acceptable air velocity for 8 hours at a fixed work station
1.5	Eddy velocity in wake of someone walking briskly
1.5	Eddy velocity in wake of arm movements
3.5	Average wind velocities in open air
9.0	Maximum acceptable air velocity for short periods
10.0	Duct velocities for gases and vapours
15.0	Stack exit velocities
20.0	Minimum transport velocity for average dusts
25.0	Minimum transport velocity for dense dusts

A single fan may be supplying air to a room for general dilution ventilation or exhausting it. In systems which include some recirculation of extracted air both supply and extract openings in the workshop may both be connected by ducts to either side of a single fan. A supply fan could be fitted with air cleaners, a humidifier and heating and cooling sections upstream for complete air conditioning. The air passes from the supply fan to the building through duct work and emerges from grilles, diffusers and other air outlets in the workrooms. Typical air velocities in various settings are illustrated in Table 15.1.

The air passes out of the building through any openings, some of which may be made for this purpose or it is exhausted by a fan connected by duct work to return grilles and the exhaust hoods in the rooms. In the event that extract air is moderately contaminated the extract fan may be preceded by an air cleaner; cyclone, bag house, electrostatic precipitator, or a wet cleaner. The fan is preceded by the air cleaner in order to maintain the air cleaner housing under negative pressure (Harwood, Oestreich, Siebert and Stockham, 1975). It is not feasible to make air cleaners 100% leak tight. The outlet from a fan extracting contaminated air should in any event be connected to a stack with an outlet well above roof height to prevent adventitious recirculation back into the building.

The fresh air supply openings into a room are located near the employees if possible and on opposite sides to the air return openings so as to encourage 'plug' flow and keep air contamination in the occupied zone to a minimum (Breum, 1988; Breum, Helbo and Lausten, 1989). Care is taken to locate and direct the supply outlets in such a way that the region of high air velocity does not penetrate the occupied zone and cause discomfort from draughts (Croome-Gale and Roberts, 1975). These outlets are nevertheless directed towards or over the occupants so that air moves from the fresh air supply openings in the direction of the occupants and then out of the returns and exhaust hoods (Baturin, 1972; Health and Safety Executive, 1988). An exhaust system operates as a functional entity, and the performance of all parts is affected by the design and performance of all other parts (Piney, Alesbury, Fletcher, Folwell, Gill, Lee, Sherwood and Tickner, 1987).

Wind

The ventilation of work stations outdoors is very much dependent upon the wind velocity, its direction, gustiness and turbulence. Indoors the ventilation system at its simplest may also be one relying on wind forces, supplemented by convection currents or other natural sources of draught. Natural ventilation is caused by the wind forcing ventilation through open windows and doors or the internal heat inducing ventilation through roof ventilators. It suffers from the disadvantage of being variable with wind velocity and direction and upon the temperature difference between inside and outside the building. Also, such ventilation is usually under the control of many different individuals and cannot therefore be relied upon to be effective at all times.

Natural ventilation

There is a certain amount of dispersion and dilution of contaminated air caused by natural ventilation. At the surface boundaries of a room the air inside will eventually pass through any openings or cracks and be replaced by fresh air infiltrating through the same or other openings, thus diluting any contaminated air inside. The amount of natural ventilation produced by such infiltration is difficult to predict with accuracy since it varies according to external wind pressure, wind direction, solar radiation, the number and size of openings or cracks in the walls and roof and the internally induced convection currents. It is not feasible to rely upon infiltration to control high emissions of dust and fumes.

Natural ventilation induced by the heat given off by processes within the building is under better control. When the heat liberation is constant this means of inducing ventilation is sufficient for control of low levels of air contamination by hazardous substances. Low emissions of gases and vapours may be effectively controlled by such ventilation. The maximum degree of ventilation in such cases is usually controlled by the open area of permanent roof ventilators and adjustment is afforded by variable opening louvres in the side walls at or near floor level (American Society of Heating Refrigeration and Air-Conditioning Engineers, 1978).

Fan assisted ventilation

More commonly than reliance on natural ventilation, where hazardous substances are liable to become airborne the ventilation within buildings is assisted by or may be totally dependent upon mechanical means employing fans of various kinds. Full regard has to be paid to the energy costs incurred through the removal of air at a high rate and, more importantly, the resulting necessity to temper the complementary incoming air (Department of Energy, 1980). The

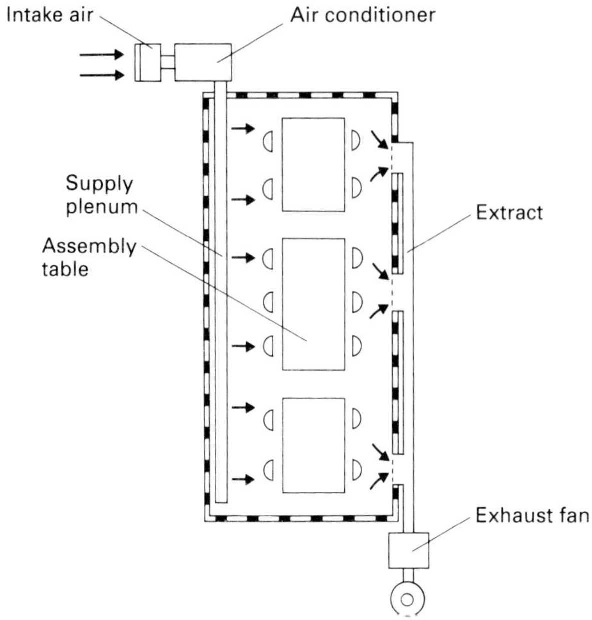

FIG. 15.3. Dilution ventilation system of a hand assembly gluing operation—plan view.

direct and indirect power costs of ventilation may be high (Gill, 1980). Scattered processes or equipment operated infrequently should be provided with separate exhaust systems.

Moderate emissions of gases and vapours are kept under control by supplementing the natural ventilation with forced or induced dilution ventilation. Forced ventilation is performed by blowing fresh air into a workshop, usually through diffusers. Induced ventilation is performed by extracting room air with fan units in the walls or roof (Figure 15.3). Besides its common use to control heat loads and humidity, dilution ventilation is often sufficient, for example, to control the vapours from organic solvents at room temperature (ACGIH, 1986). It is very effective for gases and vapours from sources having a low emission rate, of less than about 100 ml min^{-1}, also airborne dust from sources having a low emission rate of less than about 20 mg min^{-1} or where the materials being handled are of low toxicity. Otherwise local exhaust provisions are necessary to keep control over atmospheric exposure levels. As just one example of the benefits of dilution ventilation, painters in maintenance painting operations in work-rooms with forced ventilation enjoy reduced exposure to paint solvents; and a reduction by a factor of 2- to 10-fold is quite common (Riala, Kalliokoski, Pyy and Wickstrom, 1984).

Aside from dilution ventilation of low or moderate emissions of solvents, the engineering control of hazardous substances in the work environment is very dependent on good local exhaust ventilation. Where emissions of air contamination into the occupied zone in buildings are at a high rate it becomes too expensive in heating and ventilation costs to rely on dilution ventilation close to the source of the contamination (McDermott, 1976). The high degree of control obtainable with local exhaust ventilation is needed for all but the lowest emission rates of dusts and fumes found in practice. Furthermore, the quantity of air required by local exhaust appliances is made minimal by placing them directly over or very close to the sources of emissions. More importantly, if an exhaust hood entrance is close enough to the source of the contamination, the velocities induced by the exhaust may exceed the dispersion velocity at the source and most of the contamination may thereby be extracted before becoming dispersed into the work place.

The required quantity of the ventilation and the manner of its distribution so as to keep exposure to air contamination below exposure limits depends on the one hand, on the rates of emission of hazardous substances and their subsequent dispersion and, on the other hand, on the dilution and removal that may be achieved by ventilation.

Besides the common principles governing ventilation systems, whether designed to control gases, mists or dusts, there are others which assume special importance depending on the distinct physical properties of dust laden or hot air. The deposition of dust on the floor, benches, ledges and other horizontal surfaces is obvious in any dusty process. By weight this is mostly dust larger than 20 microns diameter as the falling velocity of smaller particles is too low to be a significant factor in comparison with their removal by ventilation. Aside from the effects of gravity, particles which deposit on a surface adhere by the attraction caused by inter-molecular van der Waals' forces. Electrostatic forces may also be contributory.

There is a marked decrease in density caused by even small quantities of heat in air, which results in the substantial rising convectional air currents over hot bodies. Canopy hoods are therefore erected over warm or hot processes so as to 'receive' the rising air. For the same reason the associated fresh air supply openings are kept low. By contrast, it is easy to demonstrate that hazardous gases and vapours which, while in the pure state, are mostly markedly heavier than air, nevertheless, when mixed with air in concentrations which may be inhaled by the employees, cause a density difference which is so small as to be negligible (Hemeon, 1963).

Air cleaning

Air contaminated at a level in excess of 10 times the exposure limits should preferably be cleaned before discharge to atmosphere. Thus, dilution ventilation systems do not normally include air cleaning arrangements whereas it is accepted that local exhaust ventilation systems normally do. Moreover, air cleaning provisions are required in law for specific hazardous substances discharged at a rate exceeding certain limits.

Many believe that the problems of community air pollution have only recently taken prominence, but historical records show as long ago as the 13th century an Act was passed prohibiting the burning of bituminous coals in London. However, it was not until 1863 when the first Alkali Works Regulation Bill was passed that any really serious attempt was made to control industrial air pollution by regulation in UK. That Act related to the provision of packed towers to collect hydrochloric (muriatic) acid emitted in the manufacture of soda by the Leblanc process. In the succeeding years other types of works were added and eventually consolidated in the 1906 Alkali etc Works Regulation Act. Since the beginning of the century there has been progressively broader and tighter regulation over the mass and concentration of emissions from stacks and other discharges, culminating in the Environmental Protection Act 1990. This mirrors the developments in other industrialised countries and applies to smoke, dust, grit, combustion gases, process tail gases and discharges from ventilation systems loaded with hazardous substances in suspension.

There is a wide variety of means available for cleaning exhaust air and only the principle types can be mentioned here, divided broadly into ones for gases or vapours and ones designed primarily for aerosols, sub-divided according to the loading of the contamination in exhaust air and the cleaning efficiency. The types employed are generally simpler and smaller versions of those employed for cleaning process gases.

Choice of a particular type of cleaner is based in the first instance upon the efficiency required.

$$\text{Cleaning efficiency} = \left[1 - \frac{\text{effluent concentration}}{\text{inlet concentration}}\right] \times 100$$

In some cases there may be a technical or statutory limit on the effluent concentration and on the mass emission rate. Exhaust air may be combined with process gases for cleaning. In other cases, particularly for smaller plant where the provision for air cleaning is little more than precautionary, it may be enough merely to specify a high, medium or low efficiency without going to the greater expense of designing for a precise performance.

In the case of aerosols the overall cleaning efficiency in terms of mass concentration is critically dependent upon the particle size distribution since the cleaning efficiency is higher for coarse particles and the mass of particles is proportional to the cube of their diameter. The effect of these considerations is that for fine particles the choice of air cleaner is restricted. Should the contamination be fume or dust mostly below 10 microns in diameter, for example, the simple settling chamber or cyclone would be unsuitable and a fabric filter, electrostatic precipitator or wet cleaner would be employed. Typical efficiencies according to particle size are given in Table 15.2. In some cases a primary and secondary air cleaner may be in series. Other factors such as chemical composition and flammability of the contamination, its stickiness, hygroscopicity, electrical properties and flammability, coupled with humidity and temperature of the exhaust gases, limit the choice still further. The final choice will be made on a more careful consideration of other factors, such as economics, space requirements, reliability needed, and the expected working life of the process which the equipment serves.

TABLE 15.2. *Collection efficiency of air cleaners for aerosols*

Air cleaner	Efficiency for particles of diameter 1 micron	5 microns	50 microns
Settling chamber	—	2%	50%
Cyclone	5%	15%	80%
High efficiency cyclone	25%	75%	95%
Electrostatic precipitator	90%	98%	99%
Orifice wet cleaner	50%	95%	100%
Venturi scrubber	98%	99%	100%
Fabric filter	99%	99%	100%

Note. The values given in the table are typical for the various forms of equipment and should not be taken as strictly applicable to particular makes.

Absorber for gaseous contamination alone or mixture of gaseous and solid contamination

Rather than separate gases and vapours from the exhaust air it is easier to disperse them in the general atmosphere through a stack but in principle it is preferable to clean air containing high concentrations of hazardous materials whether they be dusts, gases or vapours. Soluble and chemically reactive gases and vapours are generally removed in packed towers employing water alone or with chemical additives (Figure 15.4). Ordinarily the air and liquid streams are made to flow countercurrently past each other through the equipment in order that the greatest rate of absorption may be obtained. Such a packed tower is

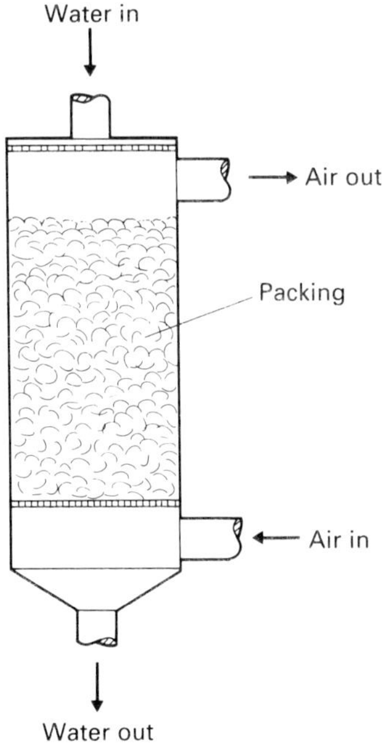

FIG. 15.4. Absorber tower. Used to separate gases and vapours.

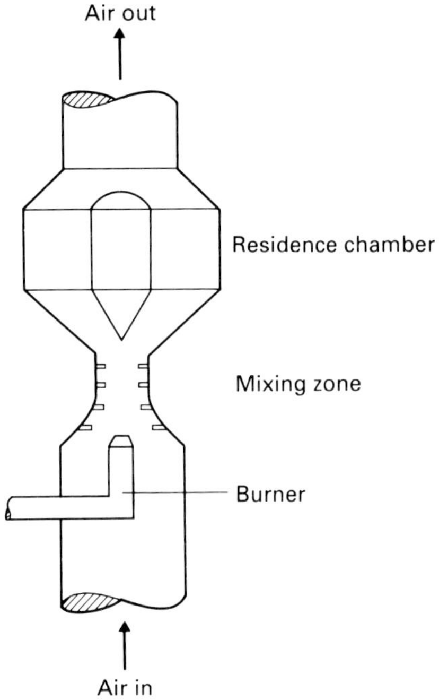

FIG. 15.5. Afterburner for combustible vapours in process air.

also a good dust collector so may be used for mixtures of gaseous and solid contamination in modest concentration. Water flows downwards as the contaminated air is made to rise through it. Water is generally used for gases fairly soluble in water, straw oil for hydrocarbons and special chemical solvents for acid gases such as sulphur dioxide, and hydrogen sulphide. Solid adsorbent such as activated charcoal is often used for organic solvent vapour.

Afterburners for combustible gases and vapours

Afterburners may be employed for high concentrations of combustible gases and vapours downstream of the fan. The air is first taken up to 500–750C to speed the oxidation in the turbulent residence chamber before discharge to atmosphere (Figure 15.5). Alternatively the contaminated gases may be introduced into a gas or oil burner itself so as to take part in direct combustion.

Settling chamber for coarse dust

The simplest form of air cleaner is the settling chamber, used for coarse dust larger than about 100 microns mass median diameter. The principle of this method of collection is to pass the air laden with dust through a large chamber at very low velocity so that the larger dust particles fall under gravity to the bottom of the chamber and can be removed periodically. Its great virtue is in extreme simplicity and low operating costs. However the chamber size required increases rapidly with decreasing size of particles and the method finds little application for particles smaller than 100 microns diameter.

Cyclone for dust or mist

Process air with a high dust loading, up to 50 g m^{-3}, and consisting of coarse solids, consisting mostly of particles in excess of 20 microns median mass diameter, is typically first cleaned with a cyclone. This is an inertial collector and is much more compact than a settling chamber for the same performance. The cleaning efficiency is in excess of 95% for particles larger than 100 microns diameter although efficiency falls to below 50% for particles smaller than 20 microns diameter. The chief advantage of common cyclones is the low first cost, moderate pressure loss and constant performance characteristics. Large diameter cyclones separating easily wetted dusts may frequently be improved by the use of water sprays or a water curtain on interior surfaces.

In use, air enters the cyclone tangentially, swirls round in a spiral path at high speed and exits from the top centre. The dust is thrown on the inside wall of the cyclone by centrifugal action into a zone of relatively still air and slides down the sides under the influence of gravity into a collecting hopper at the base (Figure 15.6).

There are no moving parts in a cyclone, but as it operates in turbulent conditions eddy currents distort the flow pattern thereby limiting the efficiency. The centrifuged dust is liable to be re-entrained by the inner vortex at the bottom of the cyclone, when it may be carried out to atmosphere with the outgoing air, and particles may also be lost by bouncing from the inside surface of the wall.

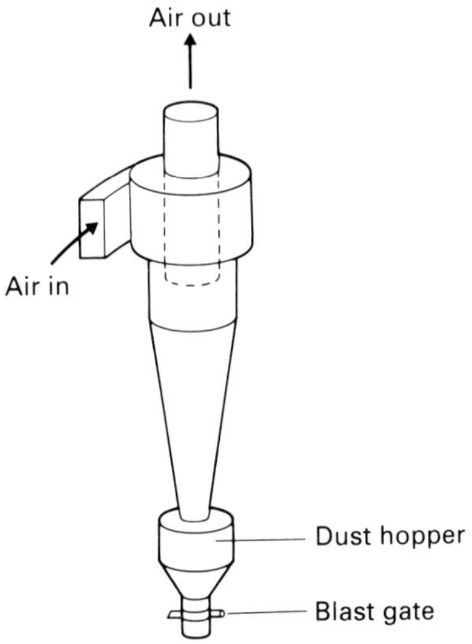

FIG. 15.6. Cyclone collector. The most widely used separator for large quantities of coarse solids, either alone or as a primary cleaner ahead of high efficiency collectors.

The efficiency of cyclones improves as the diameter of the body decreases. Small, high efficiency cyclones used in parallel extend the useful particle size range of contamination collection by cyclone down to particles about 10 microns diameter. This is only achieved, however, at the expense of high pressure losses requiring large power fans with high operating costs. In addition the plant becomes more complicated since the gas volume handled by each individual cyclone is small and hence the overall plant requires a large number of small cyclones, which complicates not only the ducting, but also the deposited dust removal system, resulting in a relatively high capital cost.

The cyclone and centrifugal dust collectors in general are suitable for dry dusts or free flowing liquids or slurries, but are not suitable for hygroscopic or adhesive materials, Because of the high velocities at the periphery of the cyclone, they are also unsuitable for large particles of abrasive materials. This effect is more pronounced for the higher efficiency designs and dust such as that from sinter machines which rapidly wears holes in the cyclone walls. A cyclone is often followed by a fabric air cleaner or one of the other high efficiency types to remove the fine dust.

Since cyclones normally operate under negative pressure to protect the fan, it is imperative that air tight dust valves and inspection covers are used to prevent air ingress.

Electrostatic precipitator for all aerosols

The forces of separation employed in an electrostatic precipitator are essentially electrical in nature as distinct from inertial or mechanical forces on particles employed with other air cleaners.

Historically, the electrostatic attraction of rubbed amber for small particles and fibres was known to the Greeks, probably as early as 600 BC, and the first recorded reference to the electrical attraction of smoke particles first appears in the famous work, 'De Magnete', of the English court physician, William Gilbert in the year 1600 (English edition, 1900). The experiment of clearing fog in a jar containing an electrified point appears to have been first reported in Germany in 1824 (Hohlfeld, 1824). The electrostatic precipitator which established this type of air cleaner as a practical industrial process in USA was the Cottrell precipitator (Cottrell, 1911). The first applications were for strictly commercial reasons; the collection of sulphuric acid or lead and zinc oxide fume in zinc-lead smelting. Nowadays electrostatic precipitation is increasingly used for the prevention of community air pollution by particulate; dust from flue gases in power stations, cement works, briquette manufacture and so on. Small precipitators are used for dust and oil mist in local exhaust ventilation systems.

The dust loading of particles smaller than 10 microns mass median diameter and in concentration between 5 mg m^{-3} and 500 mg m^{-3} would typically be cleaned with these devices. The electrostatic precipitator makes use of the fact that electrically charged particles in an electric field are attracted towards and deposited on the electrodes creating the field.

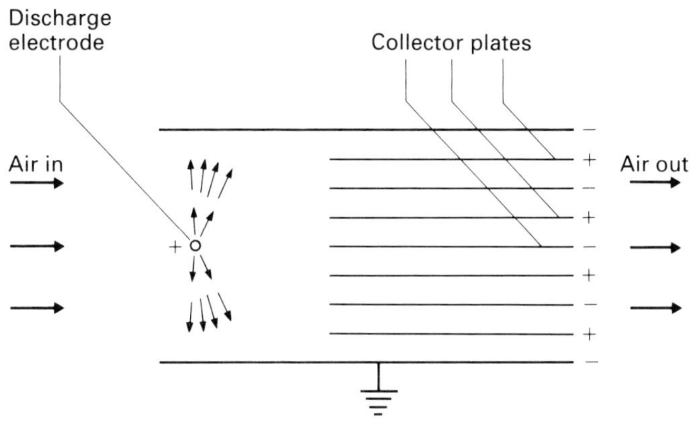

FIG. 15.7. Principle of the electrostatic precipitator. This type of air cleaning has very high efficiency for very fine dusts, mists and fumes from either hot or cold gases.

The most commonly used form for exhaust ventilation air is the double stage type. This consists of vertical plates and has discharge electrodes in the form of wires or serrated strips hanging midway between the plates. The plates are normally connected to earth, while the discharge electrodes are connected to a DC supply of positive polarity. The highest voltage precipitators applied to process air with the highest dust loading ($\gg 250$ mg m^{-3}) generally employ negative polarity at the discharge electrode. In operation, wires at high DC voltage of 10,000 to 75,000 volts ionize the surrounding air entering the precipitator and thereby charge the particles. The particles then migrate under the very high field of force to earthed plates or tubes under the action of the electrostatic field (Figure 15.7). The plates may be regularly shaken, flooded, or continually washed to remove the collected dust.

The efficiency of any form of electrostatic precipitator is governed by its operating voltage and for a given electrode configuration the maximum efficiency is limited by the highest voltage that can be maintained across the electrodes. The efficiency of electrostatic precipitators is normally in excess of 95% for particles of 1 micron mass median diameter and tends to reduce only slightly with reducing particle size. This accounts for its popularity for the very finest aerosols.

The electrostatic precipitator is probably the most versatile of all filtering plant. The pressure drop across a precipitator is low, being of the order of 5 mm water gauge. Its working range of temperature is very great and precipitators have been designed to operate successfully on process air up to 800C. Owing to its greater complexity than other types of air cleaner it has a high capital cost, offset to some extent by low running costs.

Wet cleaner for gas, vapour, mist or dust

This name is given to cleaning devices in which air contamination, whether it be gas, vapour, mist, dust or a mixture is removed by contact with a liquid.

Some of the more common wet types of collecting devices include gravity spray towers, packed and fluidised bed scrubbers, self induced spray type collectors (Figure 15.8), venturi scrubbers and disintegrators. Dust particles coated with liquid are more readily removed from suspension in air by centrifugal force or gravity. The contact between gaseous contamination and aerosols can be achieved in a variety of ways, ranging from simply bubbling the air through the liquid, allowing it to flow through an irrigated packed tower or even just spray water droplets falling down the tower. More sophisticated, high efficiency forms of accelerated impact type of device such as an orifice scrubber use finely atomised water mixed violently with the air stream to achieve absorption or wetting.

Wet collectors are able to handle high temperature and high humidity gases, mists and sticky dusts. One class of wet collectors is wet operation of similar types operated dry, such as irrigated cyclones, coarse filters operated wet, irrigated electrostatic precipitators employing water to remove collected dust, and so on. The sludge can be dealt with in settling tanks or mechanically.

Scrubbers designed specifically for wet cleaning with water employ air bubbling through water, impaction under water, sprays in various configurations; water sprays alone, sprays employing centrifugal deposition or sprays in venturi-shaped passages operating at high velocity. Collection efficiency of wet cleaners for aerosols is on a par with cloth filters. Typical ranges of overall efficiencies are some 80–95% for the low energy spray tower, packed bed and self-induced spray type collectors up to 99.9% in the case of high energy scrubbers.

The advantage of the scrubber are its simplicity of design together with low capital costs of installation. Against this must be offset the high cost of fan power to overcome the pressure loss, up to 100 cm water gauge for venturi scrubbers, pumping power for the liquid and the cost of the water treatment or liquid recovery plant which is also required. The disposal of large quantities of effluent water presents special problems, as does freezing in winter. The humidity of the discharged air is also sometimes a disadvantage.

Fabric filter for dry dust

The highest efficiency of collection for fine particles suspended in air is obtained with fabric filters. Separation of dust under this heading is largely self-explanatory. Basically, the dust laden air is passed through a porous medium which retains the dust and allows the cleaned air to pass through. Contaminated air having a dust loading in the range 10 mg m^{-3} to 1 g m^{-3} may be cleaned with any of an immense variety of fabric cleaners. The dust laden air is drawn through long, closely woven, fabric tubes, bags, envelopes, or pleated cartridges mounted in parallel (Figure 15.9). The material used for the filter may be either felt or woven cloth of cotton, wool or synthetic fibre. Felt has a better permeability and a higher collection efficiency; the woven material generally has greater strength and flexibility. Synthetic cloths are used for

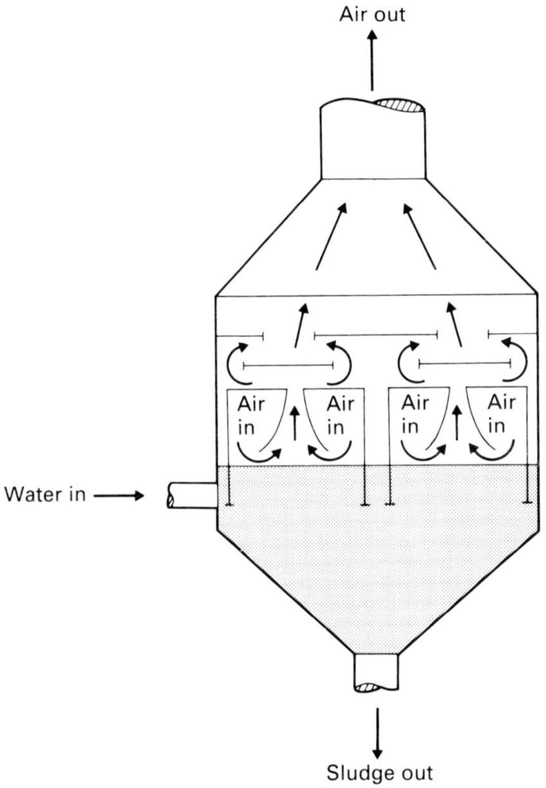

FIG. 15.8. Wet cleaner—simple orifice type—low energy scrubber used in foundries. Such scrubbers collect dust by centrifugal and inertial means, the air flow making several sharp changes in direction under a spray of water and foam.

abnormal temperature or corrosion problems. The so called 'absolute' filters use paper as the filter medium and attain very high efficiencies.

The precise mechanism of filtration is extremely complex, involving a combination of impaction, diffusion, molecular attraction and electrostatic forces. The dust rapidly builds up a dust cake which subsequently acts as the primary filter. As the collected dust layer increases in thickness, the collection efficiency also increases. This build up of the dust layer increases the resistance to flow and the pressure drop across the filter increases. Consequently, in order to maintain air throughput it is necessary to remove this layer of accumulated dust periodically.

The filters are cleaned periodically by shaking or by rapping the frame supporting the filter or by employing reverse air jets. The tops of the bags are closed and the lower ends remain open. These lower open ends fit into a series of holes in a fixed metal plate over a hopper. The bulk collects in the hopper beneath the bags. There are two main types of unit; intermittent and continuous. Intermittent operation is used when the pattern of work allows the whole unit to be closed down for cleaning the filter, as at shift changes or week-ends,

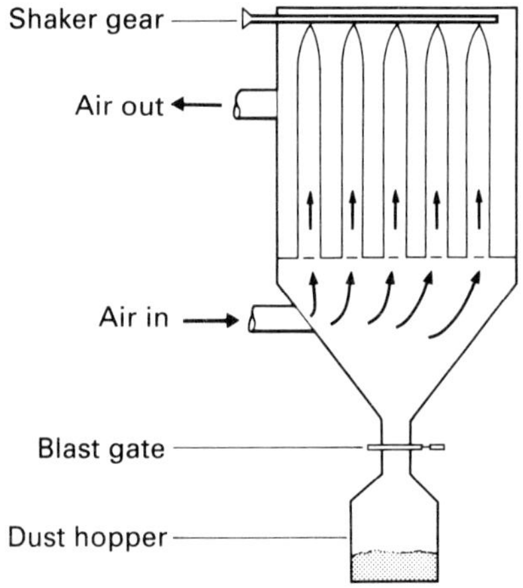

FIG. 15.9. Simple cloth type fabric filter. A high efficiency cleaner for moderate loads of dry aerosols.

and the dust load is within the capacity of the filters. Continuous operation is employed by splitting the unit into sections, allowing one section at a time to be automatically isolated from the fan for a few minutes for cleaning, while the remainder of the filter sections continue with their normal function. Another method which allows continuous operation is to employ the reverse air jet system in which high velocity jets of air on the clean side slowly traverse the area of the filter during operation.

The efficiency of fabric filters is in excess of 99% and sometimes in excess of 99.9% for particles of 10 microns mass median diameter. This is despite the fact that the openings in the fabric itself may be many times the size of the smallest particles being collected. Particles deposit initially on the weave material, bridge the gaps and form a layer through which the air has to pass.

Fabric filters generally have very high collection efficiencies which are less dependent on air flow rate than are other forms of collector. On the other hand, the pressure drop across the filter is directly related to the gas volume flow rate and the greater the volume flow rate the greater the area of filter employed and the greater the corresponding size of the complete cleaner. The pressure drop of bag filters is normally in the range 50–150 mm water gauge dependent on their cleanliness. A disadvantage of fabric filters is that they are unable to handle hygroscopic dust, liquid droplets or sub-dewpoint conditions.

Although the foregoing factors relate primarily to woven cloth or bag filters, collectors have also been manufactured from porous ceramic and plastic materials as well as sintered and woven metal gauzes. This type of filter has also employed pulped materials such as paper and slag wool. Some of these unusual materials for filtering have found specific uses in certain industries.

Fans

A fan is chosen for its reliability, efficiency, quietness and dimensions in relation to the specific application plus a balance of first cost and operating

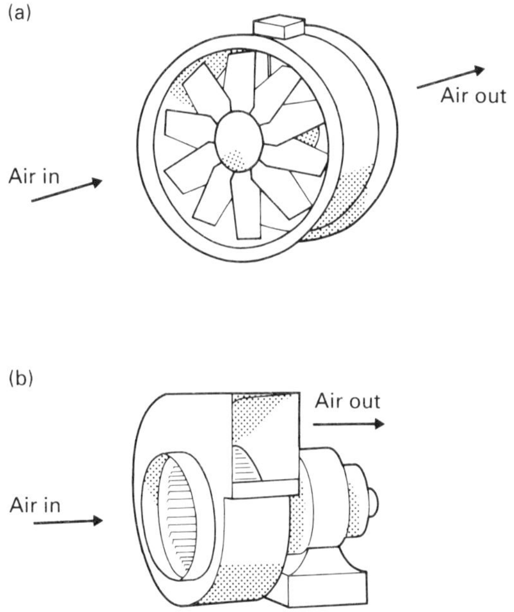

FIG. 15.10. Fans. (a) Axial flow fan. (b) Centrifugal fan.

costs. Numerous fan types are available. There are two main classes used in industrial ventilation systems; axial flow fans and centrifugal fans (Osborne, 1978). Axial flow types, although noisier, are more efficient for moving high volume flow rates against low resistances. Axial flow fans are more usual when the pressure drop divided by the volume flow rate is less than 20 mm water gauge for each $m^3 s^{-1}$ of volume flow rate.

The axial flow, propeller fan is used principally for spray booths and similar applications which employ large volume flow rates of air at relatively low pressure difference across the fan. The costlier but more efficient vane or tube axial fan, shown in Figure 15.10a, is similar, but it can generate higher pressures and is used where higher pressures must be maintained through the ventilation system and where an in-line fan would be advantageous over a centrifugal fan. Vane or tube axial fans are sometimes arranged in series to create the highest pressure. Least noise is produced by disc type propeller fans and the highest pressures by aerofoil blade, vane axial fans. Axial flow fans made with thermoplastic or similar blades have high inherent damping and are thereby much quieter.

A centrifugal fan consists of a fan rotor or wheel set within a scroll shaped housing. Air enters a centrifugal fan at the centre of the impeller from one side (Figure 15.10b). When used for handling large volume flow rates they may be supplied in double inlet form, in which the air enters the impeller housing from both sides. The impeller spins the air and centrifugal force causes a radial flow of air which is funnelled by the volute form of housing into a discharge set at

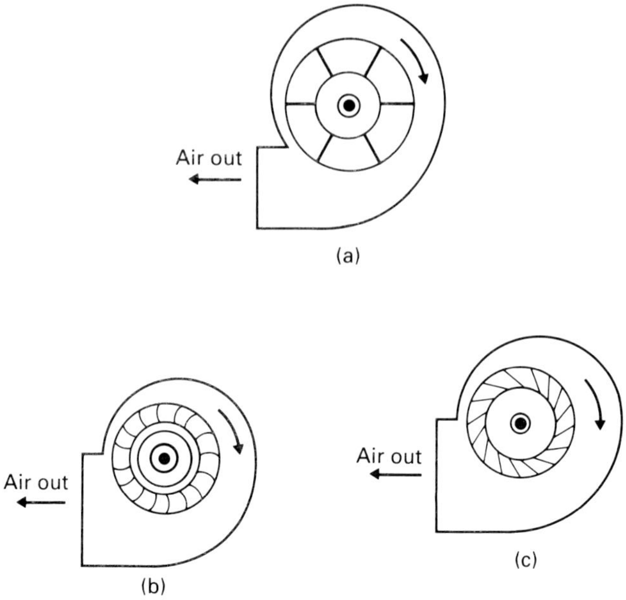

FIG. 15.11. Types of centrifugal fan. (a) Straight, radial blades. (b) Forward curved blades. (c) Backward curved blades.

right angles to the shaft. There are three main types of centrifugal fan; the radial blade or paddle wheel type, the forward curved blade and the backward curved blade.

The straight, radial bladed centrifugal fan is commonly used for dust handling systems, sawdust, metal turnings and other applications where a simple, high pressure fan is needed (Figure 15.11a). Fans of this type are rubust and easy to clean, repair or reblade. The forward curved blade type or squirrel cage wheel has the outer edges of its fan blades curved towards the direction of rotation (Figure 15.11b). It produces higher pressure than the equivalent radial bladed fan turning at the same tip speed, is quieter and takes up less space. The backward curved blade type has the outer edges of its blades angled or curved in a direction opposite the direction of rotation (Figure 15.11c). This type is more efficient than other types of centrifugal fan. It is used for handling gases, vapour and light fumes but is not very suitable for dust systems as small quantities of material tend to build up on the blades.

Turbo-compressors are multistage centrifugal fans and are employed to handle air at especially high static pressures. A typical use is for an industrial vacuum cleaning system or similar systems involving high velocities through small diameter ducts or pipes.

Fans may have direct drive or belt drive. Direct drive is more efficient but belt drive is more flexible; a distinct advantage if there should be a requirement for a higher fan speed.

The power consumed by a fan is given by

$$\text{Input power at shaft (kW)} = \frac{0.0098 \times Q \times \text{TP}}{\text{ME of fan}},$$

where kW = kilowatts power
Q = volume flow rate in $m^3\ s^{-1}$
TP = total pressure in mm water gauge
ME = mechanical efficiency (0.3—0.7)

It is prudent, when calculating the plant required, to allow 10% additional air entrainment to cater for leakages which may develop during the life of the installation or for small increases in extraction found necessary at commissioning.

Wherever possible the exhaust system and process machinery are interlocked so that the process machinery can operate only when the exhaust system is in operation.

Make-up air

It is important that the amount of air that is extracted from a room be replaced through properly located inlets of adequate size. Air simply extracted from a building is replaced by outside air entering windows, grilles, doors and by natural infiltration through the numerous cracks in any building. This is acceptable when the volume flow rates involved are small but as the volume flow rates get larger this brings with it increasing difficulties from cold draughts and uneven distribution of the ventilation. Make-up air is clean outdoor air supplied by fans and distributed to a work space in a controlled manner to replace air removed by exhaust ventilation or some industrial process. The supplied make-up air is usually heated in winter in countries with a temperate climate and may be fully tempered all year round in large modern buildings, that is, with heating, cooling and humidifying as necessary.

The power required to heat make-up air is given approximately by:

$$\text{Heat required (kW)} = 1.22\ Q\ (t_i - t_o),$$

where, kW = kilowatts power
Q = volume flow rate in $m^3\ s^{-1}$
t_i = inside temperature in Celsius
t_o = outside temperature in Celsius.

The make-up air supply volume flow rate normally equals the total extracted by exhaust systems, by plant and processes and by convection. Multiple air registers and inlets in the building are distributed so that as far as possible air contamination liberated in the work space is swept away from employees and towards the exhaust hoods and air outlets. A perforated ceiling as a supplied air source to a room is almost ideal in its non-interference with exhaust hood operations.

References

Beal G D, R R McGregor and A W Harvey. Chem. Eng. News. **19** 1239 (1941).
Breum N O. Ann. Occup. Hyg. **32** 481 (1988).
Breum N O, F Helbo and O Laustsen. Ann Occup. Hyg. **33** 321 (1989).
Brief R S, S Lipton, S Amarnani and R W Powell. Ann. Am. Conf. Gov. Ind. Hyg. **5** 121 (1983).
Cottrell F G. J. Ind. Eng. Chem. **3** 542 (1911).
Davis J W. Am. Ind. Hyg. Assoc. J. **40** 609 (1979).
Flitney R K and B S Nau. Ann. Occup. Hyg. **30** 241 (1987).
Gill F S. Ann. Occup. Hyg. **23** (1980).
Harwood C F, D K Oestreich, P Siebert and J D Stockham. Am. Ind. Hyg. Assoc. J. **36** 395 (1975).
Health & Safety Executive. Ventilation of the workplace. Guidance Note EH22/88. H M Stationery Office, London (1988).
Hohlfeld M. Archiv. fur die gesammte Naturlehre **2** 205 (1824).
Hoyle R. Chem. Eng. **85** 103 (Oct. 1978).
Jones A L. Ann. Occup. Hyg. **28** 211 (1984).
Lovelace B G. Chem. Eng. Progress **75** 51 (Nov. 1979).
Lunchick C, A Nielsen and J C Reinert. Engineering controls and protective clothing in the reduction of pesticide exposure to tractor drivers. Presented at 2nd International Symposium on the performance of protective clothing, American Society for the Testing of Materials, Tampa, Florida, USA (1987).
Meiklejohn A. Brit. J. Ind. Med. **20** 255 (1963).
Mutchler J E. Am. Ind. Hyg. Assoc. J. **31** 233 (1970).
Osborne W C. Fans. Pergamon Press, Oxford, UK (1978).
Popendorf W. Am. Ind. Hyg. Assoc. J. **45** 719 (1984).
Powell R W. Am. Ind. Hyg. Assoc. J. **45** (1984).
Ramsden J H. Chem Eng. **85** 97 (Oct. 1978).
Riala R, P Kalliokoski, L Pyy and G Wickstrom. Scand. J. Work Environ. Health **10** 263 (1984).
Schroy J M. Prediction of workplace contaminant levels. In: NIOSH Symp. Proc: Control technology in the plastics and resins industry. p. 190. NIOSH Publication No. 81–107, National Institute for Occupational Safety and Health, Cincinnati, Ohio, USA (1981).
Wu J M and J M Schroy. Emissions from spills. Proc. on Control of Specific (Toxic) Pollutants. Florida Section, Air Pollution Control Association, p. 377 (1979).

Bibliography

American Society of Heating, Refrigeration and Air Conditioning Engineers. Heating, ventilating, air conditioning guide. 62 Worth Street, New York (1978).
Baturin V V. Fundamentals of industrial ventilation. Pergamon Press, Oxford, UK (1972).
Burgess W A. Recognition of health hazards in industry. John Wiley and Sons, New York (1981).
Cralley L V and L J Cralley. Industrial hygiene aspects of plant operations. Vol 1: Process flows. Macmillan (1982).
Cralley L V and L J Cralley. Industrial hygiene aspects of plant operations. Vol. 2: Unit operations and product fabrication. Macmillan (1985a).
Cralley L V and L J Cralley. Industrial hygiene aspects of plant operations. Vol. 3: Equipment selection, layout and building design. Macmillan (1985b).
Croome-Gale D J and B M Roberts. Air conditioning and ventilation of buildings. Pergamon Press, Oxford, UK (1975).
Department of Energy. Energy management. H M Stationery Office, London (1980).
Gilbert W. De Magnete, London (1600). English edition prepared by S P Thompson, pp. 24–25 (1900).
Hemeon W C L. Plant and process ventilation. Industrial Press, New York (1963).
Hodgeson A A. (Editor). Alternatives to asbestos—The pros and cons. John Wiley & Sons (1989).

Lipton S and J Lynch. Health hazard control in the chemical process industry. John Wiley & Sons, New York (1987).

McDermott H J. Handbook of ventilation for contaminant control. Ann Arbor Science Publishers, Ann Arbor, Michigan (1976).

Perry R H and C H Chilton. Chemical engineers handbook. McGraw-Hill Book Company, Inc. New York (1973).

Piney M, R J Alesbury, B Fletcher, J Folwell, F S Gill, G L Lee, R J Sherwood and J A Tickner. BOHS Technical Guide No. 7: Controlling airborne contaminants in the workplace. British Occupational Hygiene Society, Georgian House, Great Northern Road, Derby, UK (1987).

CHAPTER 16

Ventilation Basics

The dispersion of air contamination into a work-room and its removal by being carried in the air leaving the room is a variable blend of bulk air movement or convective flow and turbulent diffusion. The transport of contamination by convective flow is essentially ordered; it follows the direction and net velocity of the air and its velocity may be measured with standard anemometers. The transport by air turbulence is essentially disordered and follows mixing laws. In real work-rooms the two phenomena are both present in varying proportion.

Fresh air enters a work-room, mixes with contaminated air and leaves from discharge openings. Gas, vapour and aerosols emitted from a process become intimately mixed with the air. Dispersion takes place by mixing and the turbulence induced by many factors including, amongst others:

- Thermal convection currents from space heaters, hot processes, ovens, driers, motors, people, weather conditions
- Wind forces causing gusting air currents through doors, windows and other openings
- Fans and air stirrers
- The motion of rotating machinery, inherent in production operations and of people
- Ventilation purposely introduced

The laws governing convective flow along pipes have been researched in depth and are well understood. The fundamental laws governing the dispersion of air contamination in a work-room are complex and poorly understood. Except for confined spaces, where the air is stagnant, the air of work places is generally in a state of turbulence. A partial analysis of the role of turbulence can be made using the principles of fluid mechanics (Roach, 1981). For the most part the solution of practical problems relies heavily on common sense and experience. In advanced studies tracer gases may be employed (Niemela, Toppila and Tassavainen, 1984; Lefevre and Muller, 1987; Breum, 1988).

A substance liberated at low velocity continuously into a region of uniform mixing, causes the concentration of contamination of the air around the source to rise sharply. The ventilation disperses and dilutes the contaminated air. This process tends rapidly to a steady state in which the concentration of contamination falls steeply with distance from the source, as will be shown

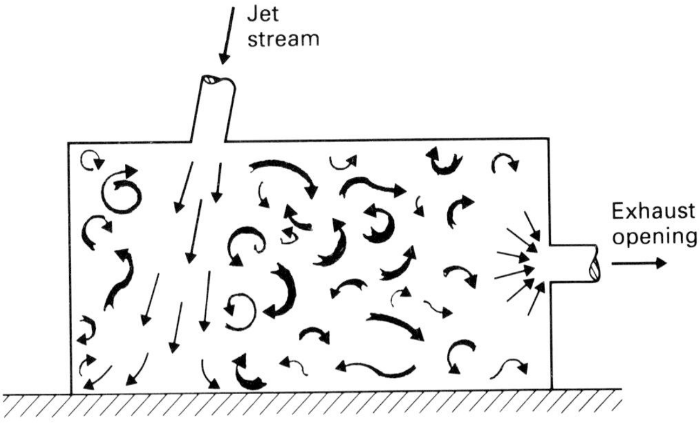

FIG. 16.1. Three flow regions in a room. 1. Jet streams, characterized by linear flow. 2. Exhaust openings, characterized by radial flow. 3. Random mixing, in the main body of a room.

shortly. By the same token the atmospheric concentration is lower the further an employee is removed from the source. Exposure can sometimes be satisfactorily controlled solely by designing the tasks to be performed so as to keep a sufficient distance between the employee and the source. Dilution ventilation used in this way finds considerable use throughout industry for ventilation of control rooms, photographic laboratories, offices, restaurants and printing rooms where the release rates of gases and vapours are not high and their toxicity is low. Nevertheless this method of ventilation is not normally very satisfactory for the control of dust, mist or fume, or for gases and vapours of moderate to high toxicity, or for situations where the rate of generation of air contamination is varied or high. For all these, local exhaust ventilation is essential.

Local exhaust ventilation and dilution ventilation operate together. To understand where in a room local exhaust ventilation takes over from dilution ventilation it is necessary to take account of the manner in which contaminated air becomes mixed with the general body of air and how fresh air is similarly mixed to dilute and remove it. There are three different regions; first is the region where there are coherent streams of air into or within a room, that is, the region of convective flow, second is the region of more or less random, turbulent mixing and third is the region of local exhaust where the dominant air movement is directed towards an opening through which air leaves a room (Figure 16.1).

Air streams

The definition of the boundaries of the convective flow and local exhaust regions in a room is somewhat arbitrary as they merge imperceptibly into one another. For the purposes of the present discussion the boundary of a stream is where the convective air velocity caused by the stream has fallen to 0.25–0.5

m s^{-1}, and the boundary of the local exhaust region is where the velocity induced by the exhaust has reached 0.25–0.5 m s^{-1}.

The air in a stream is carried forward by its momentum for considerable distances and the stream becomes larger and slower because of the increasing amounts of surrounding air which become entrained by it. Its identity becomes lost in the region where its forward speed is similar to random air currents. The distance an air stream travels from an outlet to a position at which air motion along the axis is similar to random air currents is the 'throw' of the stream. There are two main types of stream and they are treated differently. One type contains hot or contaminated air from machines, processes or operations. The other type contains fresh air from openings and ventilation grilles. It is usually desirable to discourage dispersion of a stream of contaminated air since such a stream may then be readily contained by erecting a hood to transect the stream and receive it by exhausting the hood with an air volume flow rate sufficient to extract all the contaminated air. On the other hand, a stream from a fresh air supply opening near operatives may cause complaints of draughts unless it is dispersed. For this reason fresh air supply fittings are commonly shaped so as to induce rapid mixing or they incorporate enlargements to reduce the reach of the high velocity region.

Convective flow

The convective flow of air between two points in a ventilation system, work-room or open air is caused by the pressure difference between the two points. The volume flow rate of air, Q, through an imaginary surface of area A is given by

$$Q = AV, \tag{1}$$

where V = air velocity (m s^{-1})
Q = volume flow rate (m^3 s^{-1})
A = area of the surface (m^2)

The volume flow rate of air through a duct or tunnel is given by the product of its cross sectional area and the time-mean air velocity over the cross section. The rotating fan in the duct shown schematically in Figure 16.2 creates a pressure difference between the two ends of the duct. The left hand end is at a lower pressure than the work-room and air enters from all directions. The air

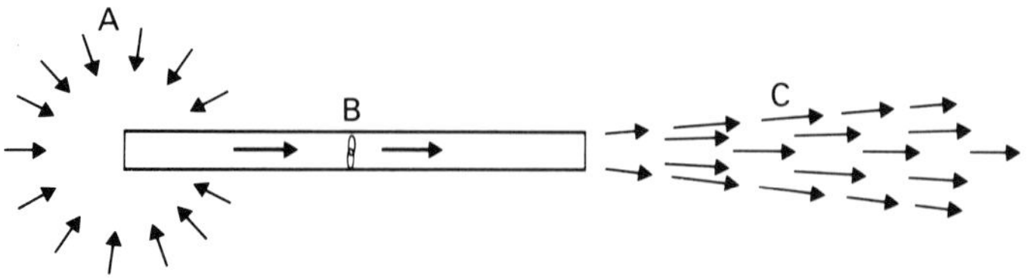

FIG. 16.2. Air flow entering and leaving a ventilation system. A: Entry. Air enters from all directions. B: Fan. This provides the motive force. C: Exit. Air leaves as a jet.

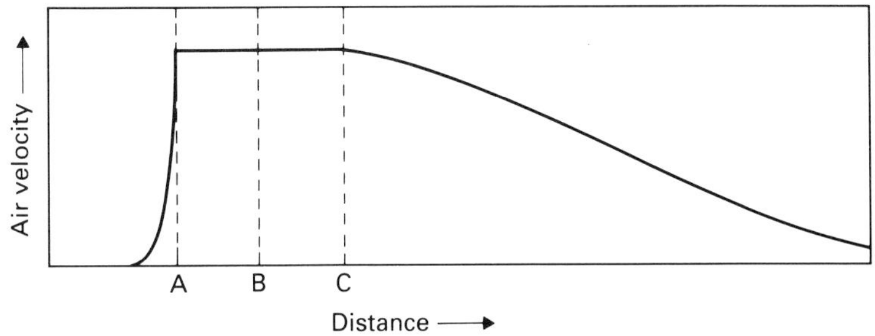

FIG. 16.3. Centre line air velocity through the system depicted in Figure 16.2. A: Entrance to duct. B: Fan. C: Exit from duct.

flow in the duct is turbulent. The air proceeds along the duct at a mean velocity, V. This is equal to the volume flow rate, Q, divided by the cross-sectional area of the duct, A. The convective flow leaving the duct is uni-directional and the subsequent stream of air entrains increasing amounts of surrounding air as it proceeds forwards until at a distance of about 30 duct diameters from the duct exit the maximum forward velocity has fallen to $0.1V$. In contrast, at the inlet end the air accelerates towards the entrance, coming from all directions and the maximum induced velocity towards the duct entrance is $0.1V$ at about 1 duct diameter from the inlet. The centre line, time-mean air velocity is indicated in Figure 16.3.

The pressure relationships which hold between different points in the ventilation system and work-room atmosphere require careful thought. Pressure in the duct could be measured by drilling a hole in the side and connecting it to a U-tube manometer. This is the static pressure, measured relative to room pressure. The static pressure in the air stream can be found by a suitably designed tube with a closed end facing the air stream and with one or more holes drilled into it at an appropriate distance from the closed end. The right hand half of the duct would be under positive pressure relative to the work-room air and the left hand half would be under negative pressure relative to work-room air. Now, were an open ended tube to be bent at right angles, inserted in the hole and pointed with the open end facing squarely upstream the pressure in the tube would be raised due to the impact of the air on the open end. This increment in pressure over and above the static pressure is called the 'velocity' pressure. By connecting the other side of the manometer to the first point, which registers 'static' pressure the manometer registers the velocity pressure directly. The relationships are indicated in Figure 16.4. At any single point in the air stream,

$$\mathrm{TP} = \mathrm{SP} + \mathrm{VP} \tag{2}$$

This is an algebraic relationship.

The velocity pressure (VP), total pressure (TP) and static pressure (SP) at different points along the duct in Figure 16.2 are indicated in Figure 16.5. Within a duct system such as this the total pressure and static pressure are both negative at points upstream of the fan and positive at points downstream of the

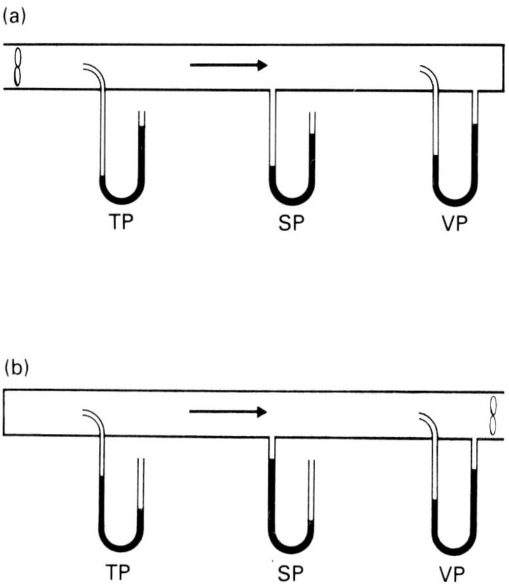

FIG. 16.4. Total pressure, static pressure and velocity pressure measurements (idealised). (a) Downstream of fan. (b) Upstream of fan.

fan. Velocity pressure is positive, both upstream and downstream of the fan. It is proportional to the square of the air velocity.

Velocity of air at standard temperature and pressure is given by,

$$V = 4.04 \times (\text{velocity pressure})^{1/2},$$

where V is in m s^{-1} and velocity pressure is in mm of water gauge (mm WG).

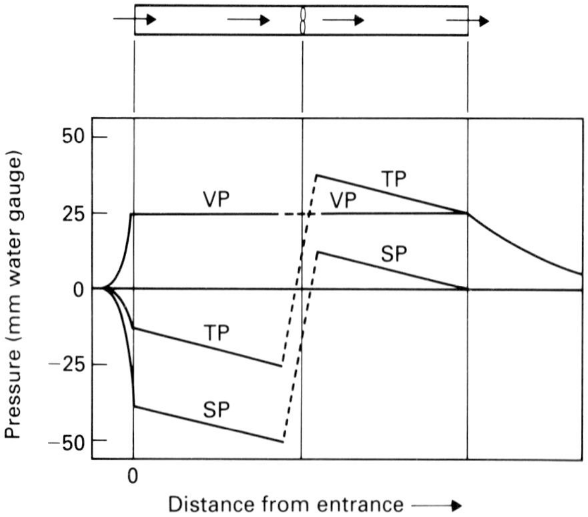

FIG. 16.5. Velocity pressure, static pressure and total pressure through the system depicted in Figure 16.2.

For very accurate work, or for work above sea level on a mountain or below it in a deep mine corrections are made according to the precise temperature and pressure.

Problem 16.1

At a point in the centre of a duct downstream of a grinder exhaust hood the static pressure is −90 mm water gauge and the total pressure is −60 mm water gauge. Calculate the air velocity.

Solution

From equation (2), and inserting the given values:

$$\begin{aligned} \text{VP} &= \text{TP} - \text{SP} \\ &= -60 + 90 \\ &= 30 \text{ mm water gauge.} \end{aligned}$$

From which, the velocity is given by:

$$\begin{aligned} V &= 4.04 \times (\text{VP})^{1/2} \\ &= 4.04 \times 30^{1/2} \\ &= 22.1 \text{ m s}^{-1}. \end{aligned}$$

When air is extracted from a point in space remote from any bounding surface, stream or exhaust hood the flow field has spherical symmetry. The point exhaust acts as an air sink towards which air flows equally from all directions, and the locus of points of equal velocity is the surface of a sphere centred on the point exhaust. At a given distance r from the point exhaust, the time-mean air velocity V, induced through extracting volume flow rate Q, is given in spherical polar coordinates, by

$$V = -\frac{Q}{4\pi r^2}. \qquad (3)$$

Total pressure of the air would not be expected to change as the point exhaust was approached along a streamline and changes in static pressure at any point, caused by the air flow, would be expected to equal and be opposite to the changes in velocity pressure caused by the air flow.

Extraction from a practical exhaust hood suspended in the middle of a large room induces air velocities at points remote from the hood close to those predicted by such simple geometrical considerations and there is a loss of static pressure head approximately equal and opposite to the changes in velocity pressure head there. Close to the entrance, as air enters an exhaust hood it changes direction, drag is caused by the hood surfaces, velocity contours are distorted and considerable turbulence is developed. Because of these factors and also because of its finite size, when air is extracted through an exhaust hood the convective air velocity induced in the vicinity of the entrance departs from equation (3). Further downstream, in the ducting, the streamlines become parallel. The velocity field induced in the transitional region close to hood entrances having a simple configuration has been studied experimentally, notably by Silverman (1942) and Dallavalle (1962).

By the principles of fluid dynamics, whatever the size and shape of a hood, when air is extracted through it, the convective velocity field thereby induced is geometrically similar to the velocity field of all other hoods of similar shape.

Moreover, where Q is the volume flow rate extracted through a hood and V is the mean air velocity thereby induced at some point in space, for similar hoods Q/V is the same at geometrically similar points for all practical values of Q. These principles strictly apply only to similar hoods in regions whose outer boundaries also hold to the same geometrically similarity. In small rooms or corridors this limitation becomes an important consideration at points whose distance from the hood is comparable with the distance from the bounding surfaces.

Turbulent flow

From the time of Osborne Reynolds in the late 1800s two different types of motion have been recognised in the flow of fluids and discussed in elementary texts on hydraulics. They are streamline and turbulent flow. The type of flow is largely determined by the magnitude of the Reynolds number of the flow system. The higher the velocity, the higher the Reynolds number. The frictional resistance between fluid and solid surfaces follows different laws in each case. In a given case, if the mean linear velocity of a fluid exceeds a value known as the critical velocity, the motion is no longer found to be streamline throughout the stream. On the contrary, either locally or everywhere in the moving fluid, the flow becomes turbulent. Then, elementary parcels of air follow erratic and continually varying courses. The flow becomes turbulent.

The alternative types of flow, streamline and turbulent, are found in fluid streams through tubes or ducts and around particles moving through fluids. A Reynolds number is any of several dimensionless quantities of the form $du\rho/\eta$, that occur in the theory of fluid motion. Here d is a characteristic diameter of the tube through which the flow is taking place or of the particle moving through the fluid; u is the linear velocity; ρ is the fluid density; and η is the absolute viscosity.

When air flows through a tube or streams past a particle the magnitude of the drag or resistance is also determined by the character of relative air flow. The types which may be identified include streamline flow, fully developed turbulent flow and a transitional region intermediate between these two. The magnitude of the Reynolds number, Re, defines which of these types of flow is to be expected within a given tube or around a given particle.

Under streamline flow conditions the pressure drop along a tube and drag on a particle are proportional to velocity, whereas under fully developed turbulent flow conditions the pressure drop and drag are proportional to velocity squared. In the transitional region the power law increases gradually from 1 to 2. Air flow through straight tubes and ducts ceases to be streamline at $Re > 2{,}000$ and for all practical purposes is effectively fully turbulent at $Re > 5{,}000$. This is the case for all practical ventilation systems and their components, save for bag filters, whose interstices are so small that the flow is streamline. Similarly, in the design of horizontal elutriators for sampling respirable dust the plate spacing is kept low enough in relation to the air velocity to maintain the Reynolds number well below 2,000 to preserve streamline flow.

Air flow around particles ceases to be streamline at Re > 5 and is effectively fully turbulent at Re > 500. Flow is streamline around particles falling in air under gravity whose aerodynamic diameter is less than 75–100 microns and turbulent around particles whose aerodynamic diameter is in excess of 1.5 mm.

Turbulent diffusion

It is important to remember that ambient air found in industrial settings is almost always in a constant state of movement. Random mixing velocities are almost always present. Turbulent and unpredictable velocities of 0.1–0.25 $m\ s^{-1}$ are common. This turbulent mixing disperses released air contamination throughout the space and keeps small particles from settling out. Visualisation of air flow patterns by releasing white smoke in a room or around an exhaust hood reveals an ever-changing and complex pattern of large and small transient eddies which, inside the hood, are quite commonly seen to be rotating briefly about horizontal or vertical axes and give the impression of having originated at the front edges of the hood openings. The practice of visualisation of hood containment by generating smoke from a small source such as a smoke tube, pellet or continuous generator and then by watching its dispersal is a quick way of showing up poor ventilation of an exhaust hood and this technique has often been employed in the past (Schulte, Hyatt, Jordan and Mitchell, 1954).

In any practical ventilation system hood turbulence causes diffusion as does the motion of molecules cause molecular diffusion on a smaller scale and ensures, for example, that some air contamination always escapes an exhaust hood, thus reducing its effectiveness. Further, the resulting concentration of air contamination in the body of a room is dependent on the diffusion from turbulence in the room itself, number of room air changes per hour and other parameters of general ventilation. In order then to estimate, for example, what would be the influence on the exposure of employees in the vicinity of an exhaust hood, it is necessary to have an appreciation of how the amount of diffusion in the room influences the outcome as well as other parameters of ventilation.

Diffusion in work-room air arises principally from the turbulent motion of the air and only secondarily from molecular diffusion. The energy in the turbulence originates partly at obstacles in the path of moving air and at surface roughness over which the air passes. It comes also from local convection currents induced by small temperature differences in the surrounding surfaces and is caused mechanically by the moving parts of machines and through the activities of employees in the vicinity. Eddies in work-room air vary enormously in size, in persistence and in translational and rotational velocity.

Sutton called attention, long ago, to the very wide spread in frequency and amplitude of eddies in natural wind (Sutton, 1947). These will have repercussions on the air flow in and out of buildings and, indeed, the forces causing air turbulence inside buildings are not too dissimilar from those at work outside. The movement of turbulent eddies is largely random and is characterized by its disorder. Chaotic movement of eddies mixes the air, thus tending to reduce any differences in concentration. In air conditioning practice, turbulence is often

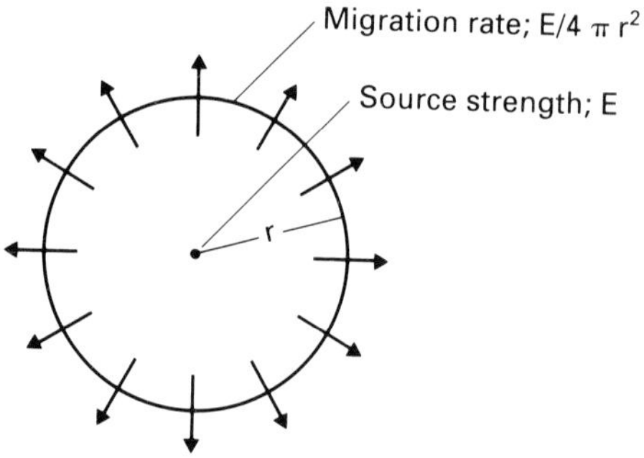

FIG. 16.6. Liberation of hazardous substance from a point source at a constant rate, E, into turbulent air. At equilibrium the rate of migration outwards through the surface of a sphere, radius r, about the source is $E/4\pi r^2$.

deliberately introduced in the air being supplied to a room through diffusers to encourage rapid mixing of supply air with room air. Similarly, in dilution ventilation practice a source of air contamination located in a relatively stagnant part of a room will give rise to a high atmospheric concentration in that region.

As to what is the effect of turbulent diffusion in the ideal case of liberation of contamination at a constant rate from a point source, firstly, the time-mean concentration of the contaminating substance in air would be uniform at a given distance from the source, that is at the imaginary surface of a sphere centred on the source (Figure 16.6). Because of diffusion the contaminating substance would pass through unit area of this surface at a rate dependent on the concentration of contamination and the radial concentration gradient. The latter is directly proportional to the rate of transfer by turbulent mixing. There is eventually a state of equilibrium when the rate at which contaminating substances are removed from a room is equal to the rate at which they are being liberated inside.

Suppose then, air contamination issues from an isolated point source remote from any boundary, is released with negligible momentum and at a constant rate into air which is turbulent but has no overall directional velocity. The turbulence disperses the contamination and spreads it rapidly in all directions. In the equilibrium state what then would be the concentration in air at a given distance from the source?

A locus of points of equal concentration will form a roughly spherical surface or envelope centred on the source. At the start, if contamination begins to issue at a steady rate, a given concentration locus will form an envelope around the source which expands outwards with time, rapidly at first and more slowly later, ceasing to expand when the total rate of loss of contamination through its surface becomes equal to the rate of releases of contamination at the source. The average position of such a locus over an extended period of time tends to become stationary. Like velocity contours in a smaller time-frame, the position of concentration contours averaged over a short period of time wander about their position averaged over a longer period owing to there being sensible fluctuations in air movement of short periodicity. These fluctuations cannot be ignored and dismissed as 'background grass', since they can give rise to variations in concentration with a periodicity considerably in excess of 1 second.

Migration of contamination is due to a combination of diffusion caused by air turbulence and molecular diffusion. Across unit area of the 'stationary' envelope the amount of migration will relate to the gradient of atmospheric contamination concentration along the normal to the surface. The units of mass migration, mass per second per square metre, are $MT^{-1}L^{-2}$ and of mass concentration gradient, change of concentration per metre, are ML^{-4}. Consequently, the units of diffusivity, which is the ratio of the two, reduce to L^2T^{-1}, or square metres per second, and the same is true if migration of contamination is expressed in terms of volume units. This expression of the magnitude of migration due to diffusion may be termed the 'effective' diffusivity when embracing both air turbulence and molecular diffusion.

Suppose contamination were issuing from a point, the origin, at constant flow rate E units of mass or volume in unit time. The mathematics is simplified by employing spherical polar coordinates. Diffusivity is assumed to be the same throughout the region of interest. The surface area of a sphere of radius r is $4\pi r^2$. Ignoring the volume of air displaced by the contamination, at equilibrium the contamination must be migrating across unit area of the surface of an imaginary sphere centred on the origin at a rate $E/4\pi r^2$ units of mass per second. This mass flux vector points in the radial direction. Where c is the stationary concentration at radius r and D is the diffusivity, then,

$$-D\frac{dc}{dr} = \frac{E}{4\pi r^2}, \tag{4}$$

by Fick's law of diffusion (Fick, 1855).

From which, if $c \to 0$ as $r \to \infty$,

$$c = \frac{E}{4\pi D r} \tag{5}$$

(see Appendix 1, pp. 489–90).

The concentration would be double this value around a point source situated on an infinite plane (hemisphere), quadruple around a point source at an internal edge between a wall and the floor (quarter sphere) and eight times around a point source in a corner (eighth sphere).

Equation (4) has a central place in understanding the spatial distribution of time-mean concentration of air contamination in the vicinity of its source. Particularly important is:

—That random diffusion by itself would give rise to the long term atmospheric concentration of contamination varying inversely as distance from the source. The phrase 'long term' is to cover two aspects; one is the time it takes to reach an equilibrium balance between input and output, whereas the other is the averaging time necessary to even out the fluctuations due to uneven turbulence. This very long term average is known as the 'stationary' concentration.

—The stationary concentration at any point would be directly proportional to the time rate of release of contamination from the source. This factor is central in calculations concerning dilution ventilation but is sometimes forgotten in local exhaust ventilation systems. The greater the rate of release of contamination, the higher must be the exhaust flow rate, even though not necessarily in direct proportion.

—The stationary concentration at any point would be inversely proportional to diffusivity. The lower the diffusivity, that is, the more stagnant the air, the higher the concentration.

The exact equivalent to equation (4) occurs in steady state thermal conductivity problems to derive the temperature distribution around a point source of heat in terms of the thermal conductivity of the medium (American Society of Heating, Refrigeration and Air-Conditioning Engineers, 1978; Carslaw and Jaeger, 1978). By Fourier's law of heat conduction the heat conduction vector is proportional to the negative temperature gradient, the constant of proportionality being in that instance the thermal conductivity.

Dilution ventilation

The fixing of a standard of dilution ventilation is a matter of much complexity. De Chaumont (1875) advocated 50 cfm (23.6 l s^{-1}) per person should be provided to remove the 'organic putrescent matter' in the case of soldiers' sleeping rooms. More recently, and in relation to office areas a standard minimum nominal ventilation rate of 10 l s^{-1} of outside air per occupant has been recommended (American Society of Heating, Refrigerating and Air Conditioning Engineers, 1986). This is met by natural ventilation in small buildings, supplemented by mechanical ventilation in larger office blocks (Persily and Grot, 1985). As regards the release of hazardous substances in an industrial setting, where local exhaust ventilation is absent or negligible possible air contamination levels in closed rooms or halls may be estimated with the aid of material emission/evaporation rates, room size and the general ventilation rates. Data may be sought on materials losses by evaporation available from in-house information or published evaporation rates in certain circumstances (National Printing Ink Research Institute, 1974).

Dilution ventilation consists of general ventilation of a work room so designed that the contaminating substance released into the atmosphere is continuously diluted by the introduction of fresh air to levels well below the exposure limit. It is usually applied to the control of contamination released over a large area such as the vapour evaporated from solvents or thinners. Examples of operations at which dilution ventilation may be used include dip or roller coating and air drying, use of volatile liquids in open surface tanks and cementing.

A suitable analogy to the situation of interest may, for example, be the release of air contamination from a single, point source in the centre of the work-room. At the other extreme the situation may be analogous to many small sources of contamination spread evenly over the room. Where air is supplied at one end or the centre of a room and the hazardous substance is being liberated from numerous places within the room the concentration of contamination builds up to its highest value at the discharge end or room periphery, respectively. From the total rate of release of contaminating substance into room air the volume flow rate of dilution ventilation may be calculated which would keep the concentration differential between the inlet and return air below fixed limits. The inlet 'fresh' air is normally, but not always uncontaminated. In any event the aim is to maintain working conditions in the occupied zone well below the relevant atmospheric exposure limit of the appropriate substance or mixture.

The advantages of dilution ventilation are its simplicity, relatively low capital cost and low fan power requirements. The main disadvantage lies in the high volume flow rates required which result in the need for heating and tempering large volumes of make-up air for efficient operation.

Single point source

Suppose contamination were issuing from a point at constant flow rate E units of mass or volume in unit time and the point source is in the centre of the floor of a hemispherical room. The mathematics is simplified by employing spherical polar coordinates. Suppose next, the convective flow is negligible and ignoring the volume of air displaced by the contamination, where c is the stationary concentration at a distance r from the centre of the floor and D is the diffusivity, assumed for the moment to be the same everywhere in the room, then,

$$-D\frac{dc}{dr} = \frac{E}{2\pi r^2} \tag{6}$$

From which,

$$c = \frac{E}{2\pi Dr} + A_1, \tag{7}$$

where A_1 is an arbitrary constant, to be determined from the boundary conditions.

Work-rooms are of finite size with definite boundaries and the modifications they might make to equation (7) need to be considered carefully. Plausible boundary conditions may be introduced into solutions (7) by adding ventilation to the model boundaries. The requirement for some ventilation is fairly obvious since otherwise if the source were in a hermetically sealed room the concentration would increase without limit.

A system of general ventilation in a room may rely, for example, simply upon supply of fresh air and discharge of contaminated air through openings distributed around the periphery. An analogous model of ventilation possessing symmetry about a source would be a hollow hemisphere, radius R, having

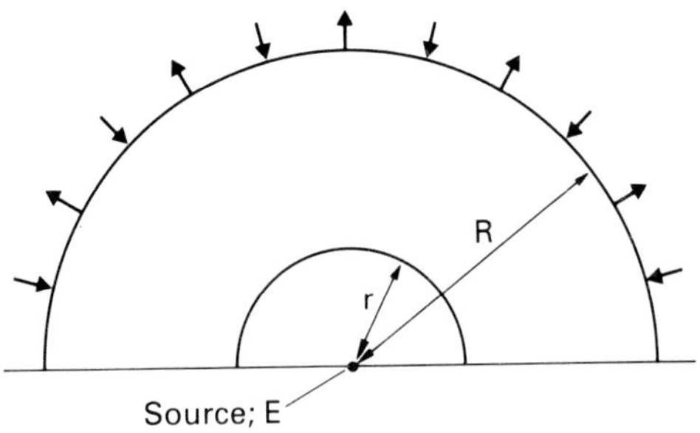

FIG. 16.7. Contamination liberated from a point source on a plane. The hemisphere, radius R, is a region of uniform mixing. The time-rate of liberation of contamination is denoted E. Air leaves the periphery of the hemisphere at time-rate Q, and is replaced by fresh air.

air seeping through numerous very small air supply and discharge points spread evenly over the curved surface (Figure 16.7).

The volume flow rate of fresh air being supplied and contaminated air being discharged is denoted Q so that the stationary concentration of contamination in the discharge air from the periphery is E/Q. From equation (7), if $c = E/Q$ when $r = R$,

$$c = \frac{E}{Q} + \frac{E}{2\pi D}\left[\frac{1}{r} - \frac{1}{R}\right], \tag{8}$$

where c = concentration of contamination in air (ppm vapour or mg m^{-3} aerosol),
E = source strength (ml s^{-1} vapour or mg s^{-1} aerosol),
R = radius of hemisphere (m),
Q = volume flow rate out of room ($m^3\ s^{-1}$),
D = diffusivity ($m^2\ s^{-1}$),
r = distance from source (m).

This expression forms a useful starting point in practice by equating R with the mean distance from a source to the centres of the walls and ceiling of a room and equating Q with the volume flow rate of air out of the room. Where E is gas or vapour release in ml s^{-1} or dust release in mg s^{-1} the corresponding concentration of contamination in air would be in ppm or mg m^{-3} respectively.

To crystallise ideas it is helpful to plot a trial graph. Entering equation (8) with $R = 5$, say, $E = 1$ and $Q = 1$, the calculated stationary concentration is plotted against distance from the centre of the floor in Figure 16.8 for example values of diffusivity, D. Consistent units are given in the legend. At the periphery, the stationary concentration is the same, whatever the diffusivity, but in the interior the lower the diffusivity the higher is the stationary concentration. The volume of the contamination could, of course, be included in such calculations, but its effect on the conclusions drawn is negligible for all realistic values of E and will be ignored here.

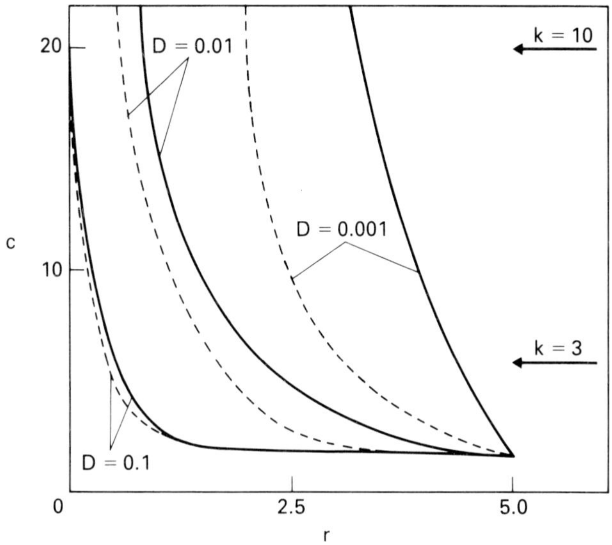

FIG. 16.8. Concentration of air contamination in the model hemispherical room (see text). A source of air contamination is located at the centre of the floor. $R = 5$ m. $E = 1$ mg s^{-1} or 1 ml s^{-1}. $Q = 1$ m^3. r = distance from centre of floor. c = concentration of air contamination (mg m^{-3} or ppm). D = diffusivity (m^2 s^{-1}). —— = air supply at room perimeter. ---- = air supply in body of room.

In a real room, in the absence of forced ventilation air contamination often becomes removed principally by the action of the turbulence and other air movement through grilles, cracks, windows, doors and similar openings to the outside. Such 'infiltration', strongly influenced as it is by the weather, can be very unpredictable and in industrial ventilation practice it is customary to put most reliance on forced ventilation by fans.

Problem 16.2

In a proposed gluing operation 10 litres of toluene is expected to be used each 8-hour work shift in a room 25 m long, 10 m wide and average height 3 m. The process will be 10 m from one end wall along the centre line of the room. Machine attendants work stations will be 1.5–3.0 m from the source of vapour. Packers will be working at the bench along one wall. The present volume flow rate of general ventilation supplied by mechanical ventilation through overhead diffusers is 20 m^3 s^{-1}. The air leaves the room through grilles in the room. Estimate the exposure of the employees.

Solution

The vapour equivalent of the emission is found first (see Appendix 2). Toluene has specific gravity 0.866, molecular weight 92.13 and exposure limit 100 ppm 8-hour time weighted average concentration.

The volume of toluene vapour liberated per shift is given by:

$$E = \frac{10{,}000 \times 0.866 \times 24.45}{92.13} = 2{,}300 \text{ litres per shift}$$

$$= \frac{2{,}300 \times 1{,}000}{(8 \times 216{,}000)} \text{ ml s}^{-1}$$

$$= 1.33 \text{ ml s}^{-1}.$$

The mean distance from the source to the centres of the walls and ceiling is given by:

$$R = \frac{(10 + 15 + 5 + 5 + 3)}{5} = 7.6 \text{ m}, \quad \text{approx.}$$

The volume flow rate out of the room is equal to the volume flow rate supplied:

$$Q = 20 \text{ m}^3 \text{ s}^{-1}.$$

Therefore, the concentration of contamination, C ppm, at X m from the source is given very approximately by equation (8). Take $D = 0.05$ in the vicinity of low velocity operations.

$$C = 0.067 + 4.2\left[\frac{1}{X} - \frac{1}{7.6}\right] \text{ppm}, \quad \text{approx.}$$

For the machine attendants take $X = 2$ m. The conclusion is that their exposure will be of the order of 2–3 ppm toluene vapour, leaving a good margin below the exposure limit of 100 ppm. Nevertheless, their exposure should be checked at the earliest opportunity. The expected concentration near the walls, where the packers worked, would be 0.07 ppm.

Where an operation is performed for only part of a work shift, the time during which the vapour is being released is used to determine the rate of emission in the above example. Where there is no mechanical fan and ventilation relies on natural infiltration a value for Q may be found from $Q = VA$ (approx.) where A is the external wall area and the infiltration is U m^3 s^{-1} per square metre of wall area. The value of U from infiltration by natural ventilation generally lies between 5×10^{-4} m^3 s^{-1} per square metre and 5×10^{-5} m^3 s^{-1} per square metre.

Problem 16.3

Evaluate the operation described in Example 16.2 with the difference that there is no mechanical fan in operation.

Solution

External wall area,

$$A = (2 \times 25 \times 3) + (2 \times 10 \times 3) = 210 \text{ m}^2$$

Natural infiltration, take $U = 5 \times 10^{-5}$ m s^{-1}, then

$$Q = (210 \times 5)(10^{-5}) = 0.0105 \text{ m}^3 \text{ s}^{-1}.$$

Inserting this value in equation (7),

$$C = 130 + 4.2\left[\frac{1}{X} - \frac{1}{7.6}\right] \text{ppm}, \quad \text{approx.}$$

For the machine attendants take $X = 2$ m. From which, their exposure would be of the order of 130–135 ppm toluene vapour in air. The conclusion is that without mechanical ventilation the exposure of both the machine attendants and the packers would be expected to exceed the exposure limit. Mechanical ventilation is clearly essential.

When a general ventilation system features air discharge at the room periphery but air supply through diffusers located in the body of a room rather than at the periphery, concentration of contamination tends to be lower because of this. In the model hemisphere this feature is created by having numerous small air supply points to the sphere as before, but this time they are spread evenly throughout the volume, whilst air discharge points are still spread over the periphery. The stationary concentration in the air discharged from the hemisphere is again E/Q. However, since the location of the air supply gives rise in this instance to a bulk movement of air towards the periphery, movement of contamination by this means must be added to the diffusive flow.

Employing spherical polar coordinates, again, when the contamination is issuing from the origin and where u is the radial velocity at radius r, the combined radial flow rate of contamination across unit area is given by

$$\frac{E}{2\pi r^2} = -D\frac{dc}{dr} + uc. \tag{9}$$

Diffusivity and convective flow rate are, in practice, not completely independent variables, as high diffusivity is associated with high velocity. However, for the purpose of establishing the main principles diffusivity is presently assumed to be the same everywhere, irrespective of the local convective velocity. In the present instance, the air velocity at a point, u in equation (9), is $Qr/2\pi R^3$.

Substituting this value in equation (9) and solving the first order differential equation by standard methods, at equilibrium, the solution is

$$c = \frac{E}{2\pi D}\left[\frac{1}{r}\left\{1 - A - \frac{A^2}{3.2!} - \frac{A^3}{5.3!}\cdots\right\} - \frac{1}{R}\left\{1 - B - \frac{B^2}{3.2!} - \frac{B^3}{5.3!}\cdots\right\}\right]\exp -A, \tag{10}$$

where $A = \dfrac{Qr^2}{4\pi R^3 D}$, $B = \dfrac{Q}{4\pi RD}$ and $c = E/Q$ when $r = R$

(see Appendix 1, Problem A.2).

As an illustration, the concentration in the body of the hemisphere, calculated from equation (10), is plotted against radius in Figure 16.8 (broken lines) for the example values given there. Concentrations in the body of the hemisphere are lower than when the fresh air was supplied at the periphery and more markedly so when diffusivity is low.

Within general ventilation design practice, in order to make some allowances for concentrations being higher near the source, it is quite usual to aim for a value of E/Q which when multiplied by a 'K' factor between 3 and 10 gives the maximum concentration desired in inhaled air (American Conference of Governmental Industrial Hygienists, 1988). In the parallel models just presented, such a K factor is aligned with expectations from diffusivity. It is the ratio of the stationary concentration at a point in the interior of the hemisphere to the stationary concentration at the periphery. The concentration values for $K = 3$ and $K = 10$ are arrowed in Figure 16.8. The hemisphere radii at which they cut the curves mark the inner limit of the regions where such values would be suitable. The specific K design value adopted is based on the designer's judgement. The following instances would indicate that a high value of K should be used:

- poor mixing in the ventilated space,
- the possibility of seasonal changes in the natural ventilation component,
- location of employees especially close to the sources of air contamination,
- location of employees work downstream of the sources.

The following generalisations emerge from this analysis:

—Wherever the sampling inlet is located the stationary concentration of air contamination is proportional to its mass or volume flow rate from the source.

—When the results of different air sampling regimens known as personal and fixed position sampling are compared, the ratio of the average concentrations may be anything between zero and infinity according to the proximity of the sources of contamination.

—The stationary concentration tends to E/Q near the periphery and to $E/2\pi Dr$ near the centre. In any event the concentration is infinite at the source.

—The effectiveness of the ventilation is influenced by volume flow rate rather than by air changes per hour, although, of course, these two are directly related through the volume of the hemisphere.

—The higher the diffusivity, the further from the perimeter does the stationary concentration nearly equal E/Q. Thus, there is a premium on high volume flow rate and high diffusivity in the models as in general ventilation practice.

Multiple sources evenly spread about a work-room

The consumption of solvents in some processes comes about solely by evaporation directly into work-room air. In these circumstances, given the rate of consumption of solvent per week or hour, it is possible to make a simple estimate of the volume flow rate of air that is necessary to keep the concentration of air contamination in the work-room below the exposure limit.

In the first example the rate of evaporation is assumed to be constant over time. At the start of the process, when evaporation begins, the concentration of vapour in work-room air builds up until the quantity of vapour in air leaving the work-room equals the rate of evaporation inside. The concentration then levels off and stays constant until the process ceases. When this happens, the concentration dies away, tending ever closer to zero over a period of time.

Build-up. In a room where there are very many small sources of air contamination spread throughout the room, each emitting contamination and where the supply of air, Q, enters the room at many points the assumption may be made that the solvent vapour emitted is thoroughly mixed into all the air in the work-room. The average build-up of concentration in air leaving the room, C, can be calculated from the combined rate of emission of air contamination, E. If no air was entering or leaving the room the concentration of contamination would increase at rate E/p, where p is the volume of the room.

The ventilation will remove the contamination from the room at an instantaneous time-rate directly proportional to the concentration. The contamination is removed by the ventilation at a time rate QC. This would act to reduce the concentration at rate QC/p.

$$a = \frac{Q}{p},$$

where a is the number of room air changes in unit time. Consequently,

$$\frac{dC}{dt} = a\left[\frac{E}{Q} - C\right]. \qquad (11)$$

It is assumed that the air supply and discharge points are favourably located, in the sense that there is no recirculation of discharge air back into the intake.

The solution of such a differential equation as (11) is discussed in the Appendix (Linear differential equations, Case 4). Where the rate of emission of contamination, E, is constant it follows that if the initial concentration in air is zero,

$$C = \frac{E}{Q}\,[1 - \exp(-at)], \qquad (12)$$

and if the initial concentration is C_o,

$$C = \frac{E}{Q} + \left[C_o - \frac{E}{Q}\right]\exp(-at). \qquad (13)$$

Or,

$$\frac{E - QC}{E - QC_o} = \exp(-at), \qquad (14)$$

where E = rate of emission of air contamination (ml s^{-1})
Q = supply air volume flow rate (m^3 s^{-1})
t = time in hours since start of emission (h)
C_o = initial concentration of contamination in air (ppm)
C = concentration of contamination in air at time t (ppm)
a = number of air changes per hour (h^{-1})

This standard formula shows how the number of air changes per hour, a, is critical in determining the rate at which the concentration increases with time, where at the start, concentration is zero or at a low level. Other things being equal (Q and E) the larger the room the fewer the air changes per hour and the slower the build-up.

Levelling-off. As t increases $\exp(-at)$ tends to zero, and the rate of removal of contamination tends to the rate of emission,

$$C = \frac{E}{Q} \quad \text{(approx.)} \tag{15}$$

Now, if room air was always perfectly mixed employees working at the periphery of a work-room would not be exposed to a concentration of air contamination greater than C. In practice, because of uneven mixing, some parts of a room are well mixed whereas others are relatively stagnant. Employees at the periphery and others in the main body of the room may be exposed to a somewhat higher concentration and a safety factor, K, is introduced to be sure, if anything, to err on the side of safety. It is customary to use a safety factor, K, of 3 to 10. The fresh air inlets and return points should be so distributed in relation to sources of contamination that there are no highly contaminated spots where employees may found within the room and no 'short circuiting' between the supply and return openings. Further safety factors are sometimes introduced if the conditions of the calculations are not met precisely.

Problem 16.4

A materials balance between input and output of materials in an assembly room shows that the use of acetone amounts to 25 litres per 40-hour week. The volume of the room is 1,000 m^3, and the volume flow rate of air into the room is 1.2 $m^3\ s^{-1}$. The occupational exposure limit of acetone is 750 ppm. Calculate firstly, the expected atmospheric concentration of acetone 15 minutes after switch-on and secondly the expected average concentration over the week.

Solution

The vapour equivalent of the acetone usage each week is found from its specific gravity, 0.79, and molecular weight, 58 (see Appendix 2).

The volume of acetone vapour liberated each week,

$$E = \frac{(25{,}000)(0.79)(24.45)}{58} = 8326 \text{ litres per week}$$
$$= 8.33 \times 10^6 \text{ ml per week}$$
$$= 58 \text{ ml s}^{-1}.$$

Number of room air changes per hour,

$$a = \frac{(1.2)(3600)}{1000} = 4.32.$$

From equation (12) the average concentration 15 minutes after switch-on,

$$C_{15} = \frac{58}{1.2}[1 - \exp(-4.32/4)] = 32 \text{ ppm}.$$

The volume of air passing through the room during working hours each week,

$$Q = (1.2)(40)(3{,}600) = 173{,}000 \text{ m}^3 \text{ per week}.$$

The average concentration of acetone in the air leaving the room,

$$C = \frac{E}{Q} = \frac{(8.33)(10^6)}{173{,}000} = 48 \text{ ppm}.$$

Introducing a generous safety factor, K, of 10 the average exposure of employees over 40 hours each week is estimated to be at a level of

$$< 480 \text{ ppm acetone vapour.}$$

It is concluded that there is an ample margin between the expected exposure and the exposure limit of acetone.

Die-away. A similar treatment may be applied to the case where the air of the room is initially charged with contamination and it is desired to estimate the rate of decrease of concentration over a period of time, by some fixed rate of ventilation.

Substituting $E = 0$ in equation (14):

$$C = C_o \exp(-at). \tag{16}$$

Again, the number of air changes per hour, a, is critical in determining how rapidly concentration changes with time. The concentration of contamination in the work-room falls exponentially from the value C_o at the start of the period of zero emission, tending to a concentration of zero as t increases.

The time it takes for the concentration to die away to half its initial value is the half time, T, which bears a simple relationship to the number of air changes per hour,

$$T = 0.693/a,$$

as may be verified by substitution back into equation (16).

Problem 16.5

In the room pictured in problem 16.4, assuming the air contaminating process is interrupted but ventilation continues at the same rate, calculate the time required for the concentration to decrease from 50 ppm to 25 ppm.

Solution

The number of air changes per hour, a, is 4.32. The half time,

$$T = 0.693/a = 0.146 \text{ hours} = 8.75 \text{ minutes.}$$

This is the time it would take for the concentration to reduce from 50 ppm to 25 ppm, half its initial value.

An important practical example of vaporising solvents, also amenable to analysis, is painting and such like operations in which the painted area from which vapour is being emitted increases with time. This situation is examined next.

During painting, spraying, gluing, cleaning and similar operations the rate of emission of vapour increases from zero as the area of the treated surface increases; and keeps on increasing until the area first painted has dried. The rate of emission of vapour then stays constant as long as painting continues. During the period of increasing rate of emission suppose $E = bt$, where b is a constant, approximately correct for drying paint, at least in the early stages (Sletmoe, 1970). Then,

$$\frac{dC}{dt} = a\left[\frac{bt}{Q} - C\right]. \tag{17}$$

Solving for C,

$$C = \frac{b}{Q}\left[t - \frac{[1 - \exp(-at)]}{a}\right] \tag{18}$$

The rate of emission increases and the concentration follows equation (18) until t is equal to the drying time. From then on, as long as painting continues, the rate of emission remains constant until painting ceases. After painting has stopped the rate of emission then decreases steadily until all the paint has dried. These exponential equations predict within a factor of 3 the concentration in rooms the size of laboratory chambers (Hallin, 1975; Hansen, 1988; Bjerre, 1989).

Minimum capture velocity

The basic idea behind replacing dilution ventilation with local exhaust ventilation is not hard to understand. Instead of relying upon air leaving the perimeter of a room to remove contaminated air a portion may be removed directly from a point of high concentration inside a room close to the source of contamination and thereby reduce the build up of high concentrations through the remainder of the workshop. The air in the vicinity of a source of contamination is subject to a combination of random turbulence and ordered, convective air currents. A small exhaust hood through which air is extracted acts as an air sink towards which air will tend to flow equally from all directions. In a certain region close to the hood the induced air velocity caused by extraction through the hood may be higher than most of the random air currents and the convective flow. The air contamination in this region may thereby be effectively contained (McDermott, 1976). This delineates a region where local exhaust ventilation is the dominant control mechanism (Dallavalle, 1933).

The parts of a workshop where air movement is dominated by currents directed towards openings is limited to those regions close to exhaust hoods or partial enclosures and to roof ventilators and return grilles. Individual windows, doors and other openings are also in this category when wind pressure difference favours discharge. Each such opening acts like a sink for air being extracted, which enters from all directions. While the air volume flow rate through an opening may be high, the induced velocity towards the opening decreases as the inverse square of the distance from it. Consequently, the size of such a region within which a directional air current can be identified is quite small. For this reason exhaust openings seldom, by themselves, create draughts over the occupants. For the same reason, exhaust hoods must surround or be adjacent to a source of air contamination to reduce escape into the general body of air to a minimum. The minimum 'capture' velocity is a term commonly employed in ventilation parlance when referring to the induced velocity at

TABLE 16.1. *Recommended minimum capture velocities*

Process/Operation	Capture velocity $m\ s^{-1}$
Open surface tank—cold	0.25
Open surface tank—hot	0.5
Plating tank	0.5
Degreasing tank	0.5
Laboratory hood	0.5
Soldering	0.5
Barrel filling	0.5
Bagging	1.0
Conveyor transfer point	1.0
Welding	1.0
Conveyor loading point	2.0
Shakeout—cool	2.0
Shakeout—hot	4.0
Abrasive blasting	6.0

some point in the vicinity of an exhaust hood produced by the exhaust volume flow rate believed to be necessary to do four things:

(a) overcome opposing directional stream velocities generated at the source of air contamination,

(b) overcome opposing convective air flow streams and the influence of stray or secondary air currents in the vicinity,

(c) capture the cloud of air contamination issuing from the source and being dispersed by turbulent mixing,

(d) carry away all air, steam, gas or vapour introduced into or created within the enclosure.

Guidance as to the induced capture velocity that is found by experience to give satisfactory results by these standards is in Table 16.1. By following these recommendations capture is highly probable but not guaranteed (Heinsohn, Johnson and Davis, 1982). The state of turbulence in a work-room and adventitious draughts in the region of a hood are impossible to predict very accurately. Confirmation that capture is sufficient to prevent excessive exposure by employees should be sought by direct measurement.

Efficiency of capture

The ultimate test of the effectiveness of ventilation equipment is the atmospheric exposure of individuals in the vicinity. However there are often many very variable sources of air contamination, other hoods in operation and the movements of employees are somewhat unpredictable so that a simple test of hood efficiency has an important role in control engineering. An anemometer, by itself, is not sufficient.

Anemometers register mean air velocity not capture efficiency. The key point to observe is that concentration of air contamination may be measured at a point in the duct downstream of an exhaust hood and when this is multiplied by the volume flow rate through the duct the result is the rate of extraction of contamination. At such a position, provided the contamination is well mixed in the duct and all the contamination is being extracted by the exhaust, so the amount of contamination exhausted will equal the amount emitted at the source. On the other hand if the capture is less than 100% the amount exhausted through the exhaust hood will be less than the amount emitted. The ratio of the two is a quantitative measure of the efficiency of use of the exhaust volume flow rate employed. When tracer gas is experimentally emitted at a constant rate from some point in the vicinity of the hood and a steady state is attained,

$$\text{Capture efficiency } (\%) = \frac{\text{tracer gas captured} \times 100}{\text{tracer gas released}},$$

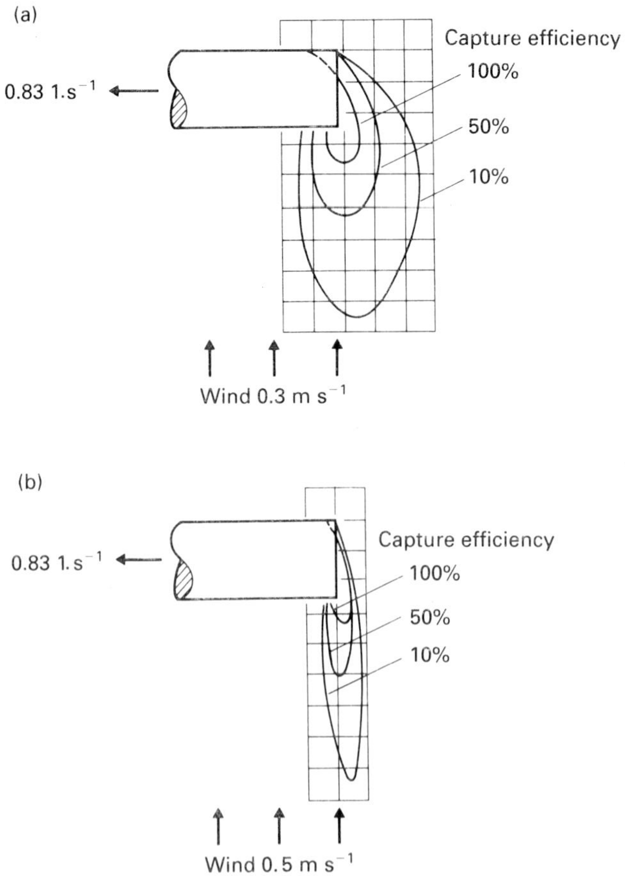

FIG. 16.9. Mapping contours of capture efficiency of an exhaust hood with respect to exterior locations. Plain pipe, 2 in. dia. (a) Lateral wind 0.3 m s^{-1}. (b) Lateral wind 0.5 m s^{-1}.

where

tracer gas captured by exhaust =

volume flow rate ($m^3\ s^{-1}$) × tracer gas concentration (ppm)

tracer gas released = tracer release rate (ml s^{-1}).

Measurements of capture efficiency of industrial hoods have been reported, as, for example, in grinding, drilling and welding operations and when using low-volume-high-velocity exhaust hoods (Burgess and Murrow, 1976; Jansson, 1980).

The capture efficiency is a function not only of the configuration of the hood and the volume flow rate of air exhausted through it. The surrounding conditions of ventilation and the objects in the surroundings also influence the efficiency obtained. An illustrative example of mapping capture efficiency by this means is a bench study conducted over 20 years ago by the author around a plain pipe, 2 inch diameter through which air was drawn variously at 25 lpm, 50 lpm or 100 lpm. Capture efficiency was measured by employing ammonia gas as a tracer in this instance. The tracer gas was emitted at a volume flow rate of 100 ml min^{-1} from a small spherical spray head placed in the vicinity of the tube entrance. The concentration in extracted air was measured by having a probe in the exhausted air leading to a midget impinger provided with 10 ml of N/100 sulphuric acid with methyl orange indicator. A measurement was the time taken to neutralise the acid when the spray head was placed in position. By

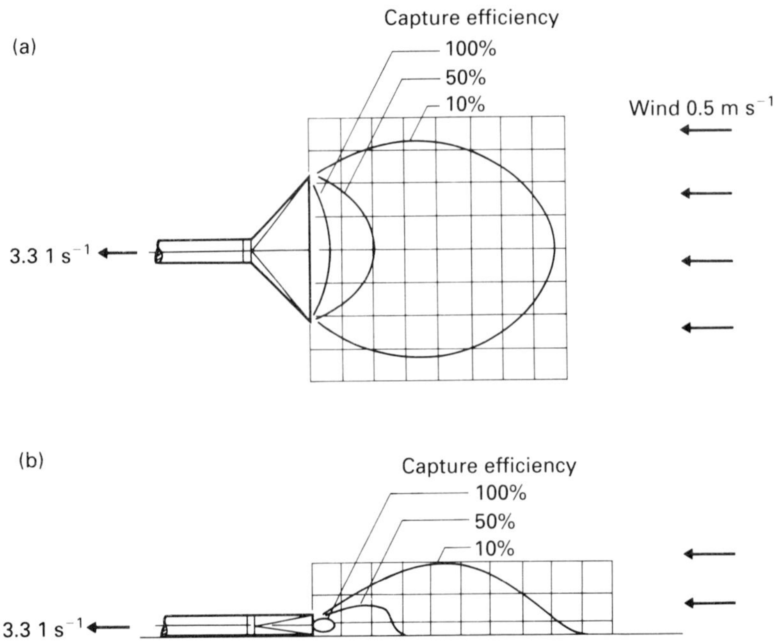

FIG. 16.10. Contours of capture efficiency of an exhaust hood with respect to exterior locations. Fish-tail slot, 4 in × ½ in. Head wind 0.5 m s^{-1}. (a) Plan view of horizontal cross section through the centre line. (b) Elevation of vertical cross section through the centre line.

varying this position and repeating the measurements contours of equal capture efficiency could be located. This was repeated at different wind speeds and direction. The 'wind' was controlled by conducting the experiments in a walk-in booth with variable ventilation. Example results are plotted in Figure 16.9.

A similar series was undertaken but this time using a 'fish-tail' hood having an entrance 4 inch × 1/4 inch flat on a bench. Example results are plotted in Figure 16.10. Both these experiments typify the procedure that is used to map capture efficiency in practical situations. Fletcher and Johnson have employed a visual judgement of the capture of released smoke to investigate the capture 'envelope' of a hood in a wall subject to cross-flow velocities 0.81–11.26 m s^{-1} (1986).

References

American Society of Heating, Refrigeration and Air Conditioning Engineers. Ventilation for acceptable indoor air quality—Standard 62–1981R. The American Society of Heating, Refrigeration and Air Conditioning Engineers, Atlanta, GA, USA (1986).

Bjerre A. Ann. Occup. Hyg. **33** 507 (1989).

Breum N O. Ann. Occup. Hyg. **32** 481 (1988).

Burgess W A and J Murrow. Am. Ind. Hyg. Assoc. J. **37** 546 (1976).

Dallavalle J M. J. Ind. Hyg. **15** 18 (1933).

Fick A. Ann. Physik. **94** 59 (1855).

Fletcher B and A E Johnson. The capture efficiency of local exhaust ventilation hoods and the role of capture velocity. In: Ventilation '85, p. 369 (Edited by H D Goodfellow). Elsevier Science Publishers B.V., Amsterdam (1986).

Heinsohn R J, D Johnson and J W Davis. Am. Ind. Hyg. Assoc. J. **43** 587 (1982).

Jansson A. Staub-Reinhalt. Luft **40** 111 (1980).

Lefevre A and J P Muller. Application of the tracer gas method to the evaluation of local and general ventilation a workshop—a case study. Proc. Roomvent-87, pp. 1–21. Stockholm (1987).

Niemela R, E Toppila and A Tossovainen. Ann. Occup. Hyg. **28** 203 (1984).

Persily A K and R A Grot. ASHRAE Trans. **91** 488 (1985).

Roach S A. Ann. Occup. Hyg. **24** 105 (1981).

Schulte H F, E C Hyatt, H S Jordan and R N Mitchell. Amer. Ind. Hyg. Quart. (Sept. 1954) 195 (1954).

Silverman L. J. Ind. Hyg. Toxicol. **24** 259 (1942).

Sletmoe G M. J. Paint Technol. **42** 246 (1970).

Sutton O G. Quart. J. R. Met. Soc. **73** 257 (1947).

Bibliography

Alden J L. Design of industrial ventilation systems. Industrial Press, New York (1982).

American Conference of Governmental Industrial Hygienists. Industrial ventilation: A manual of recommended practice. ACGIH Committee on Industrial Ventilation, PO Box 16153, Lansing, Michigan 48901, USA (1988).

American Society of Heating, Refrigeration and Air Conditioning Engineers. Heating ventilating air conditioning guide. 62 Worth Street, New York (1978).

Carslaw H S and J C Jaeger. Conduction of heat in solids. Oxford University Press, Oxford UK (1978).

De Chaumont F S B F. Lectures on State medicine. Smith, Elder and Co., London (1875).

Hallin N. Arbetshygieniska problem vid maleriarbete—loosningsmedesangor fran olika fargprodukter. Byggforlaget, Stockholm (1975).

Hansen T B. Organiske oplosningsmidler i husholdningsprodukter. miljoprojekt nr. 101. Miljostyrelsen, Kobenhavn, Denmark (1988).
McDermott H J. Handbook of ventilation for contaminant control. Ann Arbor Science Publishers Inc., Ann Arbor, Michigan, USA (1976).
National Printing Ink Research Institute. Raw materials data handbook. Organic solvents, Vol. 1. Lehigh University, Bethlehem, Pennsylvania, USA (1974).

CHAPTER 17

Mostly about Local Exhaust Ventilation

Sources of air contamination are usually physically small in relation to the size of processes and plant in a workshop. Small sources, emitting air contamination at moderate or high rates are more efficiently controlled by local exhaust ventilation than by dilution ventilation. The required volume flow rate of local exhaust ventilation is usually much less and there is consequential saving in heating and air conditioning costs.

The design of local exhaust hoods is best left to the specialist. Local exhaust ventilation is usually targeted on the sources of emissions into the air from processes or machines which are fixed in place. Examples of fixed process equipment employing liquids which are liable to release air contamination are common mixing vessels, degreasing tanks and plating baths. Equipment handling granular solids which are apt to give off airborne dust includes conveyors, screens and bagging-off points, in particular. Sometimes the exhaust ventilation has to be mobile; located in relation to hand held tools wielded by the employees, such as a welding torch, pneumatic pick, spray gun, saw, drill, shovel, and so on.

The heart of the system is the hood. The exact design of local exhaust hoods and their ventilation in a given application is dependent upon the configuration of the source of air contamination; also whether the contamination is a dust, fume, smoke, mist, vapour or gas. The manner in which the substance is liberated into the atmosphere is influential, as is the air temperature, velocity and direction with which it is released. Above all else, the time rate of release of contamination in relation to the atmospheric exposure limit of the hazardous substances is crucial for setting the required ventilation flow rates.

In the absence of turbulence a gas emitted from a point would follow a single streamline into the nearest exhaust opening, irrespective of the induced velocity, but wherever turbulence is significant some contamination is dispersed in every direction. Nevertheless, even with dispersion operating, the higher the induced velocity, the higher is the probability of contamination entering the nearest exhaust opening (Figure 17.1).

The types of hood employed for gases and vapours are generally more flexible than those for dusts and are found in greater variety. Exhaust

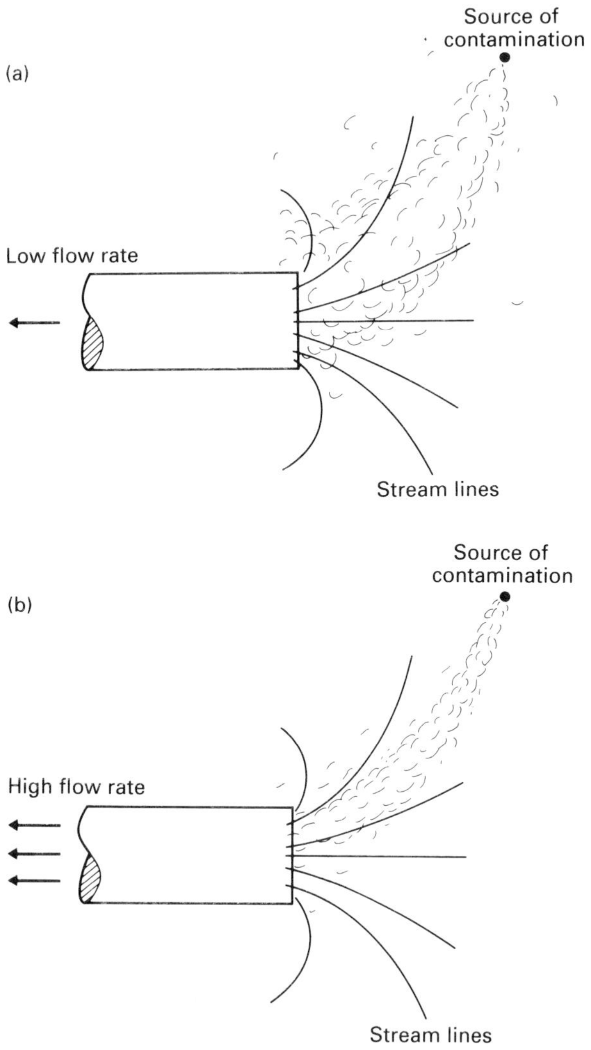

FIG. 17.1. Dispersion of air contamination entering an exhaust opening. (a) Low air volume flow rate. (b) High air volume flow rate.

ventilation systems for dusts have to be more specialised. They must be designed with regard to the intrinsic momentum of individual large particles and bulk powders in motion as well as their natural falling velocity under gravity.

When large amounts of air are extracted from a workshop to control air contamination this will be replaced with noticeably cooler outside air unless steps are taken to avoid this happening by recirculating warm, cleaned air. This strategy is not always feasible as the efficiency of air cleaning devices is limited. The use of at least some outside air is almost certainly accompanied by increased costs to heat the incoming air in winter. To reduce heating costs a part of the air discharged from the air cleaners, particularly fabric filters, may be recirculated in the make-up air. There must nevertheless be adequate safeguards against the build up of excessive concentrations of atmospheric contamination in workshop air.

Extraction through a small opening in a wall

Escape from an extraction point in a room by turbulent mixing and dilution may be examined, according to the system of ventilation, by theoretical models. These show the principle features of general systems and indicate which variables should be studied in practical investigations. These are especially relevant in large rooms or halls, where there may be up to 50 employees or more.

In the high velocity region between a source and an extraction point contamination will not have sufficient time to become evenly dispersed despite the turbulence and the relative location of the individual sources, the extraction points and the employee in the vicinity become crucial. These circumstances are found especially in small rooms where there may be just one or two employees in close proximity to the source of contamination.

Large rooms

Suppose contamination is issuing from some point in a room, distance X from an opening in a wall through which air is being extracted. The portion, p, of the contamination which fails to be extracted through it will be said to have 'escaped' that exhaust and the remaining portion, $1 - p$, which leaves with the air extracted through the opening will be said to be the portion 'captured'. The air in the room is turbulent, but in the first instance it is supposed there is no bulk movement of air in the vicinity of the hood entrance other than that induced by the exhaust.

A source of contamination may indeed be sufficiently far away for the opening to be acting as a point exhaust. Now, the locus of point sources of air contamination from which an identical fraction of the contamination would be captured by an isolated point exhaust on an infinite plane is a hemispherical surface or envelope about the entrance. Indeed, the contamination could have been issuing from many points evenly spread over this envelope, radius X, say, about the exhaust point and the apportionment of contamination between that which is captured by the exhaust and that which escapes capture would be the same. This concept allows considerable simplification in the development of the background theory.

In the steady state, where E is the mass or volume flow rate at which the contamination is issuing from the source and employing polar coordinates, at radius r about the exhaust point, in the region $X < r < \infty$, the combined radial flux of contamination from convection and diffusion would be given by

$$\frac{pE}{2\pi r^2} = -D\frac{dc}{dr} + Vc. \qquad (1)$$

In this equation,

p = portion of contamination escaping extraction (dimensionless),
E = rate of liberation of air contamination (mg s^{-1} or ml s^{-1}),
c = atmospheric concentration of air contamination (mg m^{-3} or ppm, respectively),
r = distance from source of contamination to exhaust point (m),
D = diffusivity (m^2 s^{-1}),
V = velocity (m s^{-1}).

In the region $0 < r < X$, the net mass flow rate of the contamination would be inwards and equal to the amount being captured by the exhaust, $E(1-p)$.

The two regions meet at a common border, $r = X$, at which the concentration of contamination in air on either side would be identical. When the volume flow rate of air being extracted is denoted q, the velocity, V in equation (1) is $-q/2\pi r^2$, neglecting drag by the wall.

Now, if $c \to 0$ as $r \to \infty$, as might be envisaged in a large hall ventilated at the periphery, solving equation (1) yields

$$p = \exp\left[-\frac{q}{2\pi DX}\right]. \qquad (2)$$

Thus, in such a case, if q were increased, p would decrease and tend to zero, but would always be finite. Otherwise expressed, no matter how high the extracted volume flow rate of air, diffusion should ensure that capture by the exhaust would never quite be total. More importantly, the amount of the contamination escaping the exhaust is not determined by knowing only the induced velocity at the source of contamination, $q/2\pi X^2$. To solve for p, the distance from the source of contamination to the exhaust entrance and diffusivity would also need to be known. Further, if diffusivity were held constant, escape would be a function of the ratio q/X, not of induced velocity. For this reason large hoods are more effective than small hoods, even if the air velocity at the point of release of air contamination is the same in both instances. Small hoods require higher capture velocities to be equally effective. This concept is sometimes referred to as the 'defence in depth' concept.

Large hoods have another advantage in the control of air contamination, in addition to the 'defence in depth' advantage. The larger air volume flow rate usually induced by hoods of large size provides a higher degree of dilution ventilation in the vicinity of the hood than is accomplished by a small hood, and thereby dilutes to a lower concentration any contamination that may escape. This factor is frequently quite important in the design of paint spray booths and the like where contaminated air may 'spill' beyond the confines of the hood. The 'spills' are diluted rapidly by the large volume flow rate of dilution ventilation created by the exhaust hood.

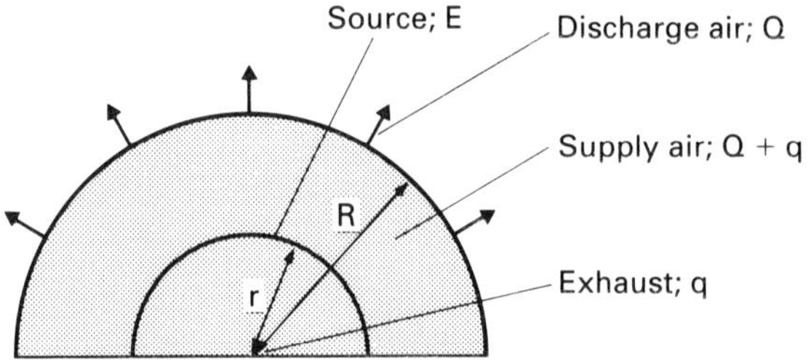

FIG. 17.2. Model hemisphere having air partially extracted through an exhaust point at the centre and the remainder discharged at the periphery. Situation similar to extraction through a hole in a wall of a room in the vicinity of a source of air contamination. Air supplied evenly throughout the body of the hemisphere. R = radius of hemisphere. X = distance from exhaust point to source of contamination. E = rate of emission of contamination. Q = air volume flow rate discharged at the periphery. q = air volume flow rate extracted at the centre. $Q+q$ = air volume flow rate supplied to the body of the hemisphere.

Escape from an extraction point in a well ventilated room may be examined, according to the system of ventilation, by theoretical models as before. For example, the system of ventilation may include a substantial quantity of air being extracted through a single opening in one wall as well as air being supplied and discharged in small quantities through many openings in the other walls. An analogue model for this case is the hollow hemisphere, radius R, having many very small air supply and discharge points evenly spread over the curved surface, the air being partly discharged there, volume flow rate Q, as before and partly extracted through a point at the centre of the plane face, volume flow rate q, made up by a total fresh air supply volume flow rate of $Q+q$ (Figure 17.2). The convective air velocity, V, would then be $-q/2\pi r^2$. Employing the procedure providing hemispherical symmetry as before, at radius X, the stationary concentration of air contamination is $E(1-p)/q$, and at the periphery, radius R, the stationary concentration is Ep/Q. From which,

$$p = \frac{Q}{Q+q} \exp -\frac{q}{2\pi D}\left[\frac{1}{X} - \frac{1}{R}\right]. \qquad (3)$$

Thus, in this case, the portion of the contamination escaping the point exhaust is between zero and $Q/(Q+q)$, at a value determined by five variables. For example, in Figure 17.3, equation (3) is entered with $R = 5$, supply air volume flow rate, $Q+q = 1$, and while holding X constant, at unity, the proportion of supply air leaving the hemisphere at the periphery is varied. The amount of the contamination escaping the exhaust increases as the proportion of supply air leaving at the periphery increases, although, should diffusivity be small, the amount escaping is also small. It is known by practical experiment that under certain conditions, hood extract volume flow rate is more significant than the velocity induced at the point of release of air contamination (Fletcher and Johnson, 1986).

In the model, should no air be leaving at the periphery, that is, when all the supply air is leaving at the centre of the hemisphere, none of the contamination

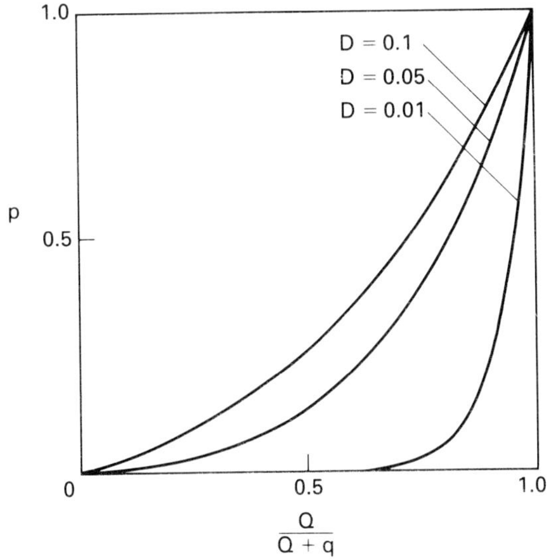

FIG. 17.3. Model hemisphere in Figure 17.2. Graph showing the fraction of the contamination escaping exhaust, that is, the fraction leaving the hemisphere at the periphery. $R = 5$ m. $E = 1$ mg s^{-1} or 1 ml s^{-1}. $Q + q = 1$ m^3. $X = 1$ m. D = diffusivity (m^2 s^{-1}).

would escape being extracted at the centre. Capture of contamination would be 100%. The condition for this is $Q = 0$. But this does not mean the stationary concentration of contamination in the body of the hemisphere would be zero. At $r \leqslant X$ the stationary concentration would be E/q and at $r > X$ the stationary concentration would be less than this but positive and would drop to zero only at the perimeter, radius R. There would be a state of dynamic equilibrium between contamination moving away from and towards the exhaust entrance.

The term capture 'velocity' could, perhaps, be employed to mean that velocity induced by extracting sufficient air through an exhaust to capture a certain percentage, say, 90, 99 or 99.9% of the contamination being released, but it would have limited usefulness, as it would be expected to have a fixed value only for a fixed diffusivity, fixed distance from the exhaust entrance and fixed room dimensions (R). The exhaust entrance dimensions would also have to be fixed. Capture velocity is a term best reserved for reference to the region so close to an opening that the induced flow equals or exceeds the average dispersion velocities in turbulent eddies.

Other variations of the supply and exhaust arrangements may be explored as in the above examples. However, the conclusions are much the same and these examples suffice to show that, when armed with an appreciation of the role of diffusivity, one can gain a better understanding of the relationship between induced velocity, supply air distribution and distance to extract openings and how they influence effectiveness of ventilation. Established design principles

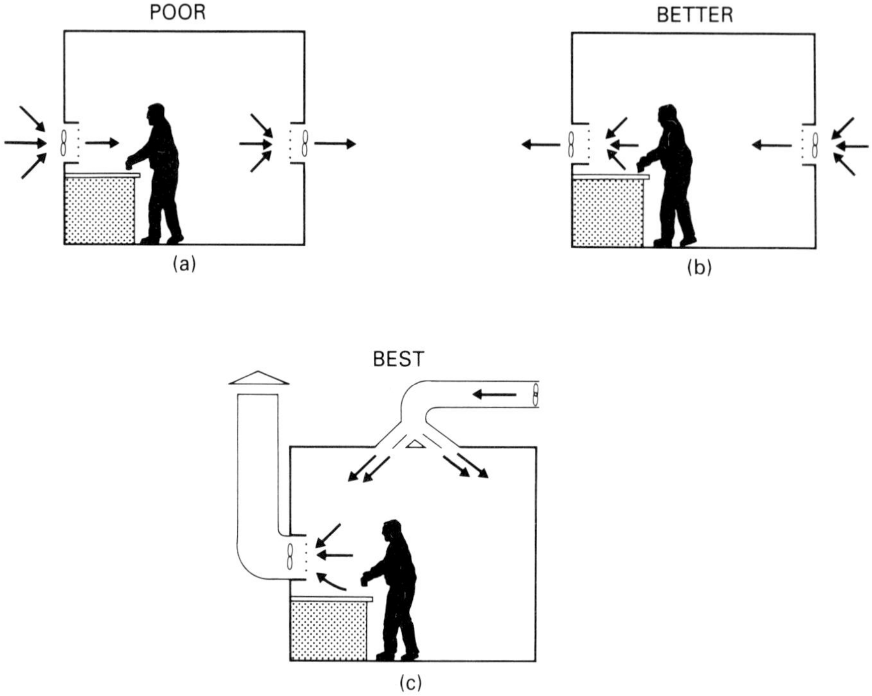

FIG. 17.4. Distribution of ventilation in a small work room. (a) Poor distribution illustrated by locating source of air contamination between the air supply and the employee. (b) Better distribution obtained by reversing the fans. (c) Best distribution obtained by locating air supply behind a ceiling diffuser.

for ventilation systems are seen to have a rational explanation in terms of a combination of diffusion by turbulence and convective air flow.

Small rooms

A single exhaust point in a small room under negative pressure extracts all the contamination emitted in the room. Capture of contamination by the exhaust is 100%. However this does not mean that employees will not be exposed to air contamination. The exact placement of the air supply points in relation to the extraction provisions may be the dominant factor in determining the exposure of the employee. This is illustrated in Figure 17.4a, which shows bench work ventilated with an exhaust fan behind the employee and a supply fan directly in front of the employee. The following features are especially noteworthy:

(i) Bench work in which the contamination source may be anywhere over the bench top area.

(ii) Employee working position which is located directly between the source and the exhaust point.

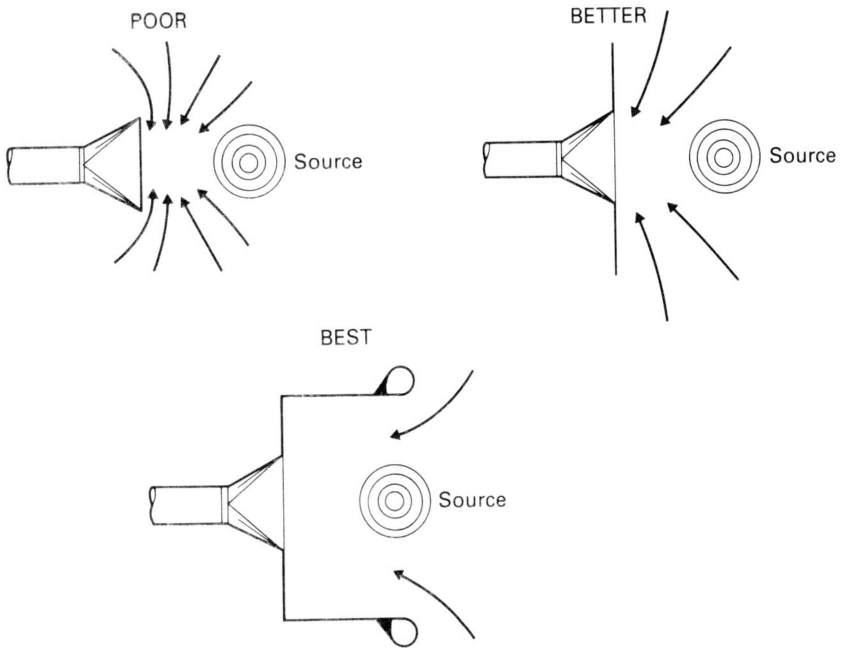

FIG. 17.5. Pictorial representation of how the efficiency of use of exhaust volume flow rate is governed by the degree of enclosure of the source of contamination. The greater the degree of enclosure, the less is the volume flow rate of extracted air required to prevent escape of contamination.

(iii) Supply air stream which is badly positioned, blowing contamination directly at the employee.
(iv) Corners of the room which are poorly ventilated.

Reversal of the fans is shown in Figure 17.4b. This arrangement is clearly a considerable improvement, although the stream of supply air is liable to cause complaints of draughts, and the corners of the room are still poorly ventilated.

The arrangement illustrated in Figure 17.4c is the best of the three. The following features are especially advantageous:

(i) Source located between the employee and the exhaust point.
(ii) Supply air distributed with a ceiling diffuser.
(iii) Supply air heated in winter.
(iv) Exhaust air discharged above roof height.

Custom built exhaust hoods

It is possible to obtain some types of exhaust hood to standard designs, as, for example:

Grinder wheel hood
Portable welding hood

Ventilated chipping and grinding table
Lip slot ventilation for tanks
Booth
Laboratory hood
Bin filling hood
Canopy hood
Glove box

Nevertheless, in many applications a standard hood is not suitable. In such applications the exhaust hood is custom-built for the operations. Enclosure of the operations, however desirable, may have to be limited in order to gain ready access. Work pieces may have to be lowered into place or slid sideways. Complete or large processes may employ several hoods and different hood types in combination. The units should be designed to have a degree of flexibility to meet the future needs of the operation and extract the air at minimum cost. The design and selection of exhaust ventilation equipment should, where possible, be performed by a specialist.

Exhaust hoods should as far as is reasonably practicable enclose the zone where air contamination is released, otherwise the exhaust volume flow rate necessary to control the contamination would be excessive. This is illustrated in Figure 17.5. The sides of the hood surround the source as far as reasonably practicable so that all dispersive actions take place within the confines of the hood. The aim of the exhaust flow rate is to create enough air flow through the release zone to convey the contamination into the exhaust system. When the configuration of a hood is settled the volume flow rate of air exhausted from a partial enclosure determines its containment. Because of the variability of industrial processes and operator movements it is virtually impossible to ensure 100% containment all the time. The precise volume flow rate needed in a given hood application depends on the rate of emission of air contamination in the hood and its exposure limit. Generally, the design volume flow rate is chosen so as to confer adequate containment for the maximum credible release rate of hazardous substances. An example of finding this experimentally is given by Roach (1981). The rate of air exhausted, in the case of flammable gases and vapours, should be sufficient to result at all times in a concentration, in the hood, duct work and subsequent equipment, below 25% of the lower explosive limit.

The practical necessities of operating the process must, of course, be taken into account. Too complete enclosure of a process may result in economic penalties in terms of operating efficiency, or capital expenditure greater than that for the larger exhaust system required by a less complete enclosure; an economic balance should be maintained between the two extremes.

The types of hood employed for gases and vapours are generally more flexible than those for dusts. Hoods for dusts are so designed as to collect particles thrown out by the process and deposit them in a hopper provided for the purpose or convey them into the ventilation system by providing sufficient transport velocity.

A sphere radius X m centred on the hood entrance has a surface area of $12.6X^2$ m^2. Thus, the air velocity induced at a distance X from the hood entrance extracting a volume flow rate Q m^3 s^{-1} is approximately $Q/12.6X^2$ m s^{-1}.

By exhausting air at a sufficient air volume flow rate an air velocity may be induced at a distance X which exceeds values of the same order as found in air currents induced by low velocity processes (<0.5 m s^{-1}). Examples are plating tanks, degreasing tanks, soldering and laboratory hoods. A more extensive listing of the minimum capture velocity for specific operations is in Table 16.1 (p. 409). Thus most air contamination released into air at low velocity within a distance of 0.4 $Q^{1/2}$ m in front of the entrance of a small, freely hanging hood would be captured in this way. This is known as the maximum capture 'distance'. Hoods of different configuration may be compared in terms of this distance. For low velocity processes (<0.5 m s^{-1}) the maximum practical capture distance from a small, compact opening in a wall is approximately 0.56 $Q^{1/2}$ m. The maximum capture distance for a narrow slot length L, on a bench, is approximately $Q/1.7L$ m for such processes (Figure 17.6). Nomograms for calculating more exact values have been given by Fletcher and Johnson (1982). The volume flow rates required would be proportionately greater to overcome steady side draught in addition to random air currents.

The velocity induced by an exhaust opening falls off rapidly with distance from the opening. Consequently considerable savings are made by arranging that an exhaust hood is as close as reasonably practicable to the source of air

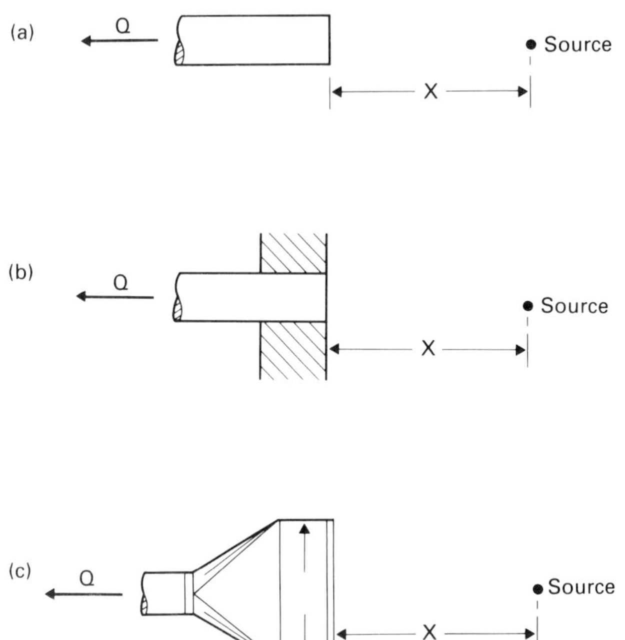

FIG. 17.6. Capture distances. X = maximum distance for capture. Q = exhaust volume flow rate. (a) Free opening; $X = 0.4Q^{\frac{1}{2}}$. (b) Opening in a wall; $X = 0.6Q^{\frac{1}{2}}$. (c) Thin slot, length L; $X = 0.6Q/L$.

contamination and preferably within a distance of 1.5 hood diameters of the hood entrance. In this region, along the centre line of a hood with a square or circular entrance, the induced velocity, U, at distance X from the hood is given approximately by,

$$V = \frac{Q}{10X^2 + A} \quad \text{(approx.)} \tag{4}$$

(Dallavalle, 1930; Pruzner, 1939; Brandt, 1947). This is sufficiently accurate for most assessment purposes. Formulae for calculating more exact values of centre line velocity have been reported by Fletcher (1977). Amendments for flanged hoods, which have greater effectiveness than those without flanges have been given in a later paper (Fletcher, 1978). A wide flange of width one half the entrance diameter will increase the centre line induced velocity by about one third, up to a distance of one diameter from the entrance.

Problem 17.1

A small, counterweighted hood with flexible trunking is available for use to control fumes from a soldering operation. The volume flow rate of air extracted by the hood is 0.1 $m^3\,s^{-1}$. The hood entrance is circular. Where should the hood be placed in relation to the source?

Solution

The maximum capture distance for a small, freely hanging hood is 0.4 $Q^{1/2}$ for low velocity processes. Substituting the extracted volume flow rate, the maximum capture distance is,

$$0.4\,Q^{1/2} = 0.12 \text{ m.}$$

Thus, such a hood should be held within 120 mm of the work piece. Since there will be a convection current above the soldering iron the hood would as nearly as practicable be located above the work and angled towards it so as to 'receive' the fume.

Problem 17.2

Assembly of small parts on a bench-mounted turntable may be conveniently ventilated by arranging an exhaust hood stood on the bench. The exhaust volume flow rate of an available hood is 0.1 $m^3\,s^{-1}$. Where should the hood be placed. Compare the solution with problem 17.1.

Solution

The maximum capture distance for a hood on a bench or from an opening in a wall is 0.56 $Q^{1/2}$, provided the contamination is released at low velocity. Substituting the extracted volume flow rate, the maximum capture distance is,

$$0.56\,Q^{1/2} = 0.18 \text{ m.}$$

Thus, such a hood should be held within 180 mm of the work piece. A hood on a bench has a higher 'reach' than does a freely hanging hood because of the partial enclosure afforded by the bench.

The minimum volume flow rate required into an exhaust hood may be reduced considerably if the hood can be placed close to the source. This principle is employed in the next example problem.

Problem 17.3

A hood entrance is 150 mm diameter. What is the exhaust volume flow rate required through the hood for effective control of a low velocity process emitting air contamination at a distance of 50 mm from the hood entrance, along the centre line?

Solution

Rearranging equation (4)

$$Q = U(10X^2 + A).$$

Substituting the required capture velocity, $U = 0.5\ \mathrm{m\ s^{-1}}$, $X = 0.05$ m, and the area of the hood entrance, $A = 0.018\ \mathrm{m^2}$:

$$Q = 0.5[(10)(0.05)^2 + 0.018] = 0.021\ \mathrm{m^3\ s^{-1}}.$$

Thus, the same degree of ventilation control as in problem 17.1 would be obtained with a fraction of the original exhaust volume flow rate by taking advantage of close positioning of an exhaust hood.

Where it is imperative to keep the space above the work free the most unobtrusive hood shape is a thin slot. A slot also gives width to the capture region obtained from a given exhaust volume flow rate, although with some sacrifice in capture distance.

Problem 17.4

In this application a thin slot 1 metre in length, on the work bench, would be ideal. Assume the available exhaust volume flow rate is $0.1\ \mathrm{m^3\ s^{-1}}$. How close should the slot be located relative to the work piece?

Solution

The maximum capture distance for a thin slot, length L, on a bench is $Q/1.7L$. Substituting the extracted volume flow rate, the maximum capture distance is,

$$\frac{Q}{1.7L} = 0.06\ \mathrm{m}.$$

The slot should be within 60 mm of the work piece. Otherwise a higher exhaust volume flow rate must be provided, commensurate with the maximum capture distance required.

If the hood entrance is made larger than the source of air contamination and the velocity through the entrance exceeds the required capture velocity the hood may be placed completely over the source to gain control. Higher entrance velocities would be employed to counter air velocities in excess of $0.5\ \mathrm{m\ s^{-1}}$ induced at the source by the process, nearby machines, the occupants or general draughts from doors and windows. The average velocity over the plane of the entrance of such a hood and measured in the direction perpendicular to that plane is known as the 'face' velocity. Local exhaust hoods are such partial enclosures of a process with provisions to exhaust air at a volume flow rate which is sufficient to capture contaminated air under the most adverse conditions likely to be encountered.

Because of the variety of industrial processes such local exhaust hoods vary correspondingly in design and ventilation requirements (Dallavalle, 1947). The hood may have extensions to shield the process itself from side draughts and thereby reduce the exhaust ventilation requirements. The following is a summary of the general principles:

—Operations should be fully enclosed as far as possible. Flanges should be fitted to the remaining openings where feasible. Encouragement should be given to the adoption of modifications to the operations which allow the most complete enclosure.

—Cross draughts in front of hood entrances should be avoided. Hoods should be moved, if necessary, to situations without cross draughts.

—Operations involving high volume gaseous emissions should be segregated from dusty operations.

—Hoods with high heat loads should have most air exhausted through a top slot. The volume flow rate of air extracted should exceed both the volume flow rate of air contamination liberated and the additional volume flow rate of air induced by that liberation.

—Hoods and ducts should be made accessible for decontamination.

—The exhaust fan should be located outside the confines of the building so that the ducts inside the building are under negative pressure.

—Tempered make-up air should be supplied to a room to replace exhausted air and located at a position remote from the hoods.

Exhaust ventilation for gases and vapours

There are three main classes of exhaust hood used for gases and vapours:

Exterior hood
Slot
Booth

These are indicated in Figure 17.7.

A remote exterior hood, such as a canopy hood, would be located at or beyond the confines of a machine or process and would normally be made about 10% longer and wider than the source of air contamination. A canopy hood is used, for example, where it is necessary to keep a tank free of obstructions immediately above it. However it would not be used where it is necessary for the employee to lean over the tank to use it since the employee would thereby be exposed to heavily contaminated air. Where possible, the hood is provided with one or more sides between the tank and hood entrance to provide better containment.

For cold materials a canopy hood is suitable at heights up to 0.75 m above the emission. An exhaust volume flow-rate of about 0.5 $m^3 s^{-1}$ per square metre of open side area of the column between the top of the tank and the hood entrance is generally sufficient to control emissions. When substances which have a low exposure limit are being liberated (<100 ppm gas or vapour) a volume flow rate of 0.75 $m^3 s^{-1}$ per square metre of open side area may be advisable. A tapered transformation from the face of a hood to the duct inlet, having an included

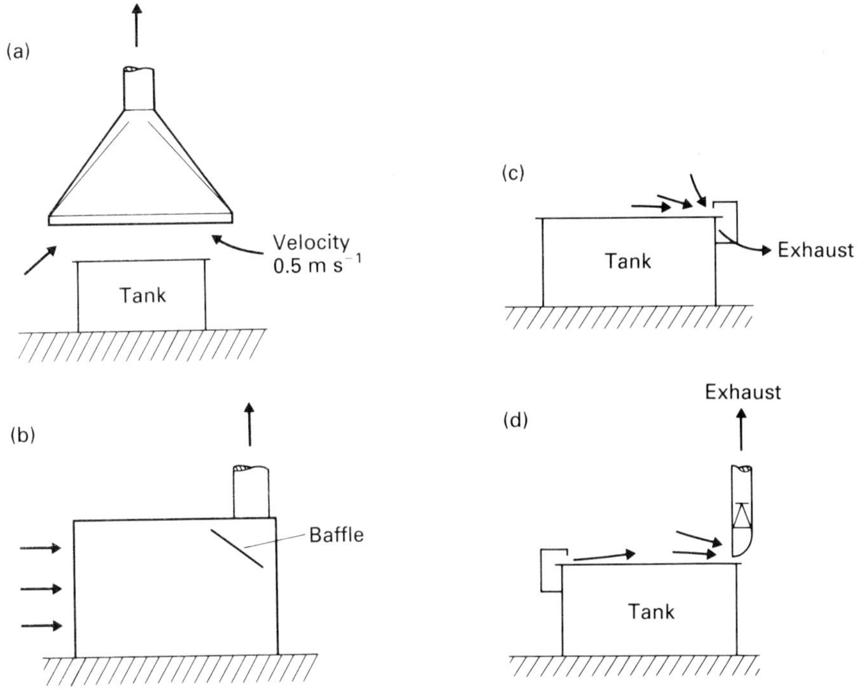

FIG. 17.7. Common types of exhaust hood. (a) Exterior canopy hood. (b) Booth. (c) Slot fitted to an open surface tank. (d) Push-pull slots over an open surface tank.

angle 90 degrees or less, helps to promote a uniform face velocity. Canopy hoods are often used over hot processes to take advantage of the thermal up-draught. The heated air rising into the hood creates a secondary draught so that the exhaust volume flow rate has to be increased to prevent contaminated air spilling out of the hood (Hemeon, 1963). It is important that the hood over a hot process be as low as possible as the rising column of hot air entrains and mixes with the surrounding air thus increasing the required exhaust volume flow rate. A deep skirt around the edges of the canopy will improve its effectiveness.

An in-line exterior hood is also a type used to exhaust a jet or stream of contaminated air from a process. The air flow of the stream of air produced by a jet is proportional to the jet flow rate, Q, at the source, the distance X from the jet source and the inverse of the jet diameter, L, at its source. In these terms the volume flow rate of the jet and entrained air at distances remote from the source, is approximately $0.05XQ/L$ m^3 s^{-1} and an exhaust volume flow rate exceeding this is required. If this is impractical, a hood held at right angles to the stream captures it, provided the air velocities it would induce in the absence of the stream exceed the air velocities in the stream of the contaminated air.

An exhaust slot would normally be located around the opening or area of a machine or process from which the gas or vapour is emitted. A slot is useful where it is necessary to allow room for cranes or other objects vertically above a tank and, indeed, is more efficient than a canopy hood for small tanks up to 1.5 m square. Slots are commonly used on degreasing tanks, cleaning baths and electroplating tanks. The slot height is made sufficiently narrow to keep the slot velocity over 10 m s^{-1}. This should ensure even ventilation along its length. The exhaust volume flow-rate usually required is about 0.5 m^3 s^{-1} per square metre of tank area. Larger tanks may have exhaust slots on opposite sides. Some plating tanks employ a push-pull system in which fresh air is supplied through a slot on one side of the tank and extracted from another slot along the opposite side (Ege and Silverman, 1950). The jet from the slot acts rather like an air curtain and improves containment of atmospheric contamination from the tank (Nitsu and Katoh, 1963). The extracted air volume must exceed the supplied air plus its entrained air by up to 100% (Huebner and Hughes, 1985; Hampl and Hughes, 1986). This implies that the supply jet air volume flow rate should be no more than about 0.05–0.1 × (exhaust volume flow rate) of most tanks or the jet ceases to be helpful. Each such installation should be adjustable and the jet volume adjusted empirically to give the best performance as the balance is very critical. In operation objects placed in the path of the jet can badly upset the performance of the system.

The third common type of exhaust hood is one in which the machine or process emitting the air contamination is placed in a booth or box-shaped enclosure, open on one side and exhausted from the back or roof. The entrance provides access for the operator but also provides an escape route for the air contamination. A sliding door or sash may be provided to improve the containment. The standard spray booth is typical, where the worker sprays objects inside the enclosure through an open front or from within the enclosure. Here, the flow of air through the front effectively acts as a curtain between the spray point and the work room. Such a booth is often provided with internal baffles near the back of the booth, designed to produce even flow through the booth entrance. A volume flow rate of 0.5 m^3 s^{-1} per square metre of booth opening is usually sufficient to contain the air contamination in a large, walk-in booth. This presupposes no fortuitous air currents outside the booth nor high velocity release of contamination within it. When substances which have a low exposure limit are being liberated (<100 ppm gas or vapour or <0.1 mg m^{-3} aerosol) a volume flow rate of 1.0 m^3 s^{-1} per square metre of booth opening may be advisable. In operations where the contamination is released from the source at high velocity, such as spray painting, a volume flow rate in excess of 1.5 m^3 s^{-1} per square metre of booth opening may be required. Smooth, even flow in a small booth is promoted by aerofoil shaped entrance edges.

Another example of a booth, albeit a small one, is the common laboratory hood or fume cupboard. Laboratory hoods are very common in the chemical industry and are found universally in chemical laboratories in industry, teaching institutions and research laboratories. Many are old and poorly designed (British Occupational Hygiene Society, 1975). Hoods with sliding sashes, as in laboratory hoods, should have controlled face velocity, by an arrangement for varying the hood volume flow rate on lowering and raising the sash or by fitting a proportional bypass (Garrison, Dong, Lengerich and

Rabiah, 1989). The volume flow rate may be varied by a mechanical connection between the sash and a damper or by switching a two-speed motor. Some employ an air flow sensor registering face velocity, the output from which controls a damper or a variable frequency fan motor (Bentsen, 1985; Wiggin and Morris, 1985). These controls involve substantial installation costs but there are considerable reductions in running costs by way of energy conservation and better control over air contamination (Neuman and Rousseau, 1986; Davis and Benjamin, 1987; Maust and Rundquist, 1987; Moyer and Dungan, 1987). A volume flow rate of 0.5 $m^3 s^{-1}$ per square metre of entrance area is recognised good practice although in favourable circumstances adequate control can be maintained with less. Where the operations being undertaken in the hood may cause moderate disturbance, as with hand seiving, shaking or spraying, a volume flow rate of 1.0 $m^3 s^{-1}$ per square metre of entrance area may be advisable.

Exhaust ventilation for dust and fumes

When designing the control of dust emissions special attention is given to the momentum possessed by the particles and their falling velocity under gravity. Individual particles thrown out by a process have a trajectory through the air determined principally by the balance between their residual momentum, gravity and air resistance or drag. This combination of horizontal and vertical motion through the air is the dominant factor governing the motion of particles larger than 1 mm diameter. The trajectory of such particles itself induces an air stream and smaller particles are carried in the wake of larger ones. The air velocity required to capture these may be many times the initial velocity of the particle. To prevent the escape of large sized solid and liquid aerosols, the hood or part of it should be located in the path of the escaping material if practicable. In lieu of this, baffles or shields may be located across the line of throw of the particles to absorb their kinetic energy, after which lower air velocities will suffice to capture them and carry them into the hood.

The horizontal component of motion through the air is less important for smaller particles. They are slowed by the air resistance, but their falling velocity persists and they deposit quite rapidly because of this. Particles smaller than 5 microns in diameter rarely have a horizontal movement of more than 5 mm through the air in which they are suspended and they have a downward velocity which is small in comparison with the air velocities of natural turbulence. Consequently, the movement of such fine dust and its extraction closely matches that of gaseous contamination.

Down-draught hoods which take advantage of the falling velocity of fine particles are particularly useful for grinding and chipping. They are also advantageous where the operation must be unencumbered on all sides. The work piece is placed on a grille or grid and air is extracted downwards through the grid. Chippings collect in the base and are removed periodically from clean-out drawers. Down-draft is also used in the foundry industry where it is necessary to work over and around castings. Other examples of this kind of operation are found in spray painting, solvent drying, welding and soldering.

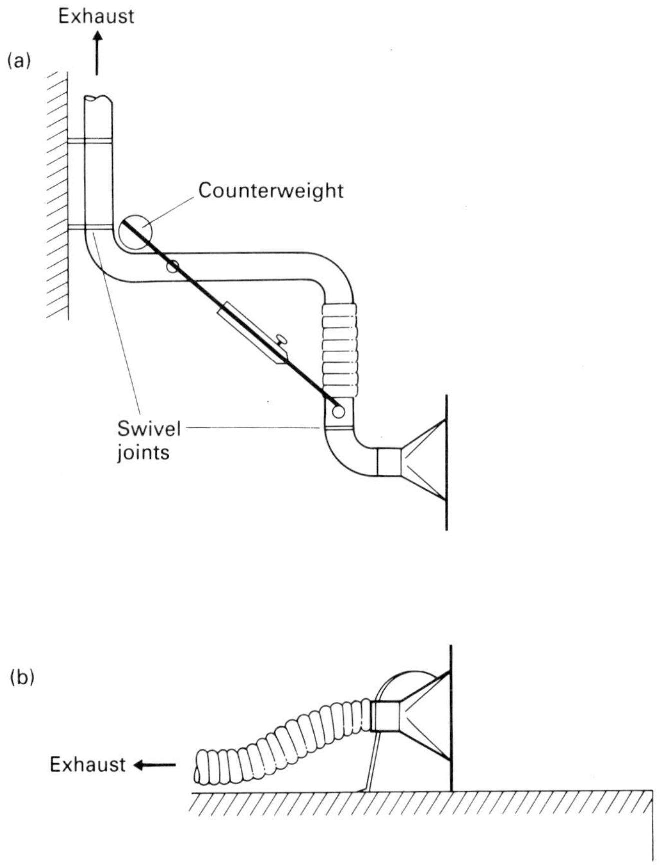

FIG. 17.8. Exhaust hoods employed when welding. (a) Counterweighted, wall mounted welding hood. (b) Flanged, bench mounted welding hood.

The down-draught is often supplemented with side-draught provisions to improve the dust control (Ayalp and Myroniuk, 1982).

Remote hoods are employed in operations where it is not feasible to enclose the source within the boundary of the hood as when welding large pieces (Faggetter and Freeman, 1983). A remote hood is one in which the work-piece is partially or wholly outside the face of the hood. An example of this is the counterweighted, wall-mounted welding hood shown in Figure 17.8 which can be placed close to the welding point on a large item, or the common flanged welding hood which, by means of flexible trunking, can be set at a convenient position on a bench for smaller work.

Processes which allow a degree of enclosure of the source of contamination can be controlled with less exhaust air volume flow rate than would be required with a remote hood. The common booth is an example of partial enclosure. In controlling dust emissions booths are used where the dust is emitted at relatively low velocity. The enclosing walls serve to permit control through inducing an air velocity past the source and through acting as baffles against room draughts. Booths are often employed for spray painting in industrial applications and are represented in large numbers in the form of laboratory hoods or fume cupboards.

Special types of hoods developed for dust emissions include hoods over grinding and polishing wheels, circular saws and band saws. These machines act like a crude fan, and the guard like a fan casing. In consequence, the dust which is produced leaves the machine in an air stream which may be moving at considerable velocity. The ordinary grinding wheel hood assumes the contours of the wheel so as to limit the volume flow rate of air induced by the rotating parts and has an off-take shaped to receive the induced air stream. The hood is provided with a minimum working opening through which the velocity is high enough to capture the fine particles thrown off by the machine in use. In such hoods the pipe take-off is set tangential to the point of operation so that the large particles are projected at high velocity directly into the pipe. Inner baffles are used to slow these particles.

Routing machines are difficult to provide with a standard hood form due to the wide range of operations performed on them. For small, repetitive work a close proximity hood fitted with a 'bristle curtain' may be applied. The open area face velocity should not be less than 3 $m\ s^{-1}$. For large and varied operations, a booth type of enclosure on the machine table and around the cutter is employed with a minimum face velocity of 1.25 $m\ s^{-1}$.

Milling and turning of items which yield airborne dust is controlled with close proximity hoods exhausted with face velocities of 10 $m\ s^{-1}$. It may be necessary to supplement the local exhaust points with booth form around the machining area, themselves exhausted so as to induce a minimum face velocity of 2 $m\ s^{-1}$.

The process of filling bins with powders is another common one in a wide variety of industries. Appropriate exhaust hoods are designed to prevent dust being liberated into the work-room air while avoiding losses of powder to the exhaust system. This may be done by having a peripheral exhausted semicircular slot matching the tops of the bins (Chang, 1988), by feeding the powders through a pipe or sock in a conical lid provided with extract ventilation or by filling the bins in an exhausted cabinet. The required extract air volume flow-rate from such hoods is usually about 0.5 $m^3\ s^{-1}$ per square metre of open area of the bin top or of the enclosure (Figure 17.9).

Low volume–high velocity (LV–HV) systems are an extreme type of exhaust arrangement, especially adapted for portable hand tools, machining and other operations where the dust is emitted from a small source at high velocity. The tool is fitted with a hood making a small clearance to the moving parts so as to induce high exhaust velocities very close to the point of dust generation (Figure 17.10). Control is obtained with relatively low volume flow rate, using close-fitting, custom-made hoods and nozzles. Capture velocities are high (50 to 60 $m\ s^{-1}$), but the volume flow rate of air used is low (0.005 to 0.125 $m^3\ s^{-1}$) because of close proximity of the nozzle to the dust source (5 mm). Portability is maintained by employing small diameter flexible hose as a connection to the tool with the remainder of the system utilising fixed piping and fittings. The

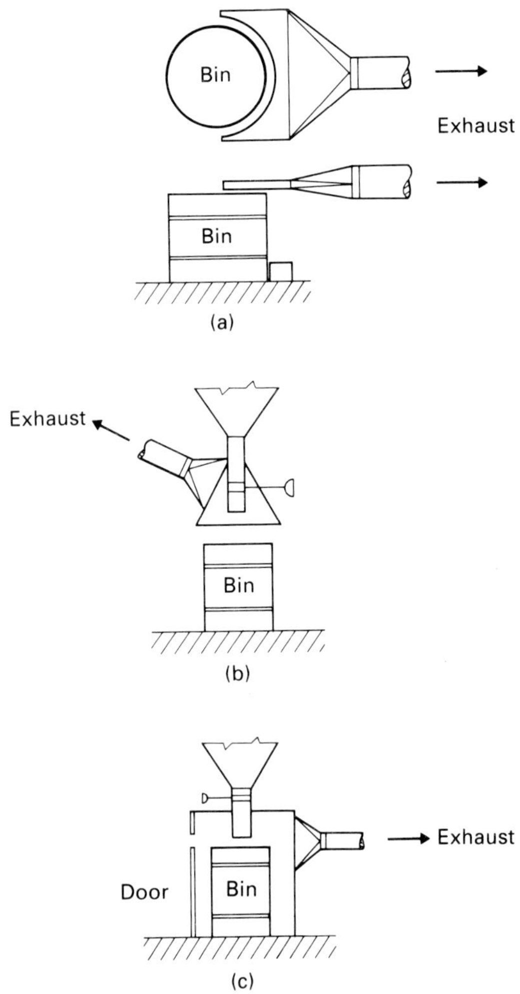

FIG. 17.9. Bin filling exhaust hoods. (a) Semi-circular slot around bin top. (b) Conical exhausted lid. (c) Exhausted bin cabinet.

dust is conveyed at high speed through the flexible hose and exhaust power needs to be provided to produce about 125 mm mercury at the nozzle opening. A multi-stage centrifugal turbine pump is used to provide the necessary high negative pressure through the system (Chamberlin, 1959; American Conference of Governmental Industrial Hygienists, 1988; The Black and Decker Tool Company, 1987).

Designs for many common operations in industry are given by the American Conference of Governmental Industrial Hygienists (1988). A companion step-by-step guide to the study of industrial ventilation is given by Burton (1984).

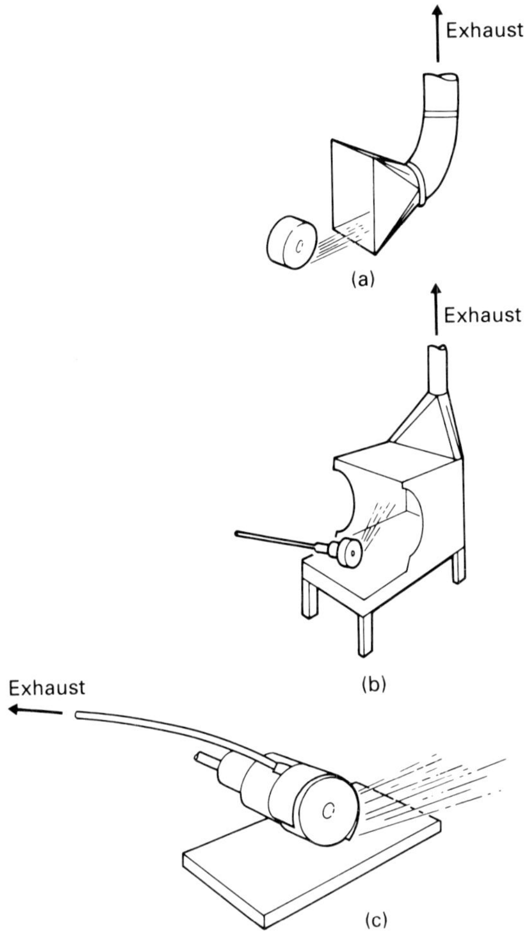

FIG. 17.10. Partial enclosure of a surface grinder. (a) Receiving hood, requiring the highest volume flow rate to control the fine dust. (b) Booth type. (c) High velocity, low volume (HV–LV) type, requiring the least volume flow rate to control the fine dust.

Duct systems

There are two basic methods employed in designing a duct system; the velocity method and the 'equal pressure' method . The velocity method is based upon choosing duct diameters in each branch so as to maintain a desired velocity in relation to the volume flow rate. The volume flow rate is adjusted by the use of blast gates. The equal pressure method is used to provide a 'balanced' system in which the duct diameters in each branch are so chosen that the system operates with the correct volume flow rate in each branch without the use of blast gates or dampers. A balanced system is used with dusts to avoid problems with build up around blast gates, although the velocity system is more flexible.

Problem 17.5

The duct from a hood over a powder filling point is 7.5 m length and 200 mm diameter. The air volume flow rate is 0.6 $m^3 s^{-1}$, controlled by a blast gate. Because of problems from dust blockages around the blast gate it is desired to replace the duct with one of smaller diameter but without a blast gate. Calculate the required diameter of the new duct. The hood static pressure is 20 mm WG and the static pressure at the junction with the main is 80 mm WG.

Solution

The pressure drop along the duct is

$$80 - 20 = 60 \text{ mm WG.}$$

The pressure drop per metre length of appropriate duct is

$$60/7.5 = 8 \text{ mm WG per metre length.}$$

Reference to a friction loss chart shows that the diameter of duct with this pressure drop for a volume flow rate of 0.6 $m^3 s^{-1}$ is 155 mm diameter. The duct velocity would be 34 $m s^{-1}$, ample transport velocity.

Recirculation

When large amounts of air are extracted from a workshop to control air contamination, it follows that this will almost certainly be accompanied by increased costs to heat the incoming air in winter. Where these heating costs are severe a part of the air discharged from high efficiency air cleaners, particularly fabric filters, may be recirculated in the make-up air to reduce heating costs. This might be acceptable provided there are adequate safeguards against the build up of dangerous concentrations. The local regulatory authority should be consulted. Where some recirculation is permissible an on-off or modulating damper is set to recirculate a portion of the air from the air cleaner when the outside temperature falls below a certain level and set to discharge completely to the outdoors at other times. The following rules would be advisable:

(1) No more than 50% exhaust air should be recirculated after cleaning.

(2) The level of contamination in recirculated air should be at low concentration; that is, less than 1/5th its exposure limit. A secondary air cleaner for the recirculated air may be necessary to guarantee this condition.

(3) The control system should be fail-safe with respect to concentration in air returned to the building and with respect to failure of the power supply.

(4) Recirculation should be avoided for the following hazardous substances:
   (a) Carcinogens
   (b) Reproductive toxins
   (c) Dusts which cause pneumoconiosis
   (d) Acute systemic poisons
   (e) Chemical asphyxiants
   (f) Cumulative systemic poisons
   (g) Respiratory sensitisers

In the absence of air cleaning facilities, or where the cleaning performance cannot be guaranteed, heat recovery from discharged air may be undertaken and employed to heat incoming air. This is accomplished with rotating heat exchangers, fixed plate systems or heat exchange coils by which 40% to 85% of the heat is exchanged (Weldler and Youle, 1981). Recirculation of contaminated air back into the work place should be discouraged by having discharge stacks located well above roof peak height.

A form of recirculation regularly occurs with unit air cleaners, which are small, free standing air cleaners with a capacity less than 1 $m^3\ s^{-1}$. They are normally used intermittently and may be moved from one place to another. The air cleaning element is usually a fabric filter and the cleaned, discharged air normally passes back to the work-room. These are usually acceptable provided the concentration of air contamination in the discharge air is less than 1/5th the exposure limit. Otherwise they should be fitted with ducts leading outside the building.

Portable vacuum cleaners used for floor cleaning also discharge directly into the work-space. Their capacity is usually less than 0.1 $m^3\ s^{-1}$. The dust loading of the air inlet may be high in use for floor cleaning so that when hazardous dusts are being cleaned up the vacuum cleaner should be one with a secondary high efficiency back-up filter. The period of use is usually limited and the discharge air considerably diluted before reaching breathing height. Nevertheless the maximum concentration in the discharge air should be less than the relevant exposure limit. It may be difficult to find a suitable portable vacuum cleaner meeting this performance criterion (Myers and Hunckler, 1987; Wake, Gray and Brown, 1991).

References

Ayalp A and D Myroniuk. Am. Ind. Hyg. Assoc. J. **43** 825 (1982).
Bentsen L. Heat./Piping/Air Cond. 57 67 (1985).
British Occupational Hygiene Society. A guide to the design and installation of laboratory fume cupboards. Ann. Occup. Hyg. **18** 273 (1975).
Chamberlin R I. Am. Med. Assoc. Arch. Ind. Health **19** Feb. (1959).
Chang S N. Am. Ind. Hyg. Assoc. J. **49** 367 (1988).
Dallavalle J M. Studies in the design of local exhaust hoods. Sc. D. Thesis, Harvard Engineering School (1930).
Davis S J and R Benjamin. Heat./Piping/Air Cond. **59** 75 (1987).
Ege J F and L L Silverman. Heating and Ventilation **47(10)** 73 (1950).
Faggetter A K and V E Freeman. Am. Ind. Hyg. Assoc. **44** 316 (1983).
Fletcher B. Ann. Occup. Hyg. **20** 14 (1977).
Fletcher B. Ann. Occup. Hyg. **21** 265 (1978).
Fletcher B and A E Johnson. Ann. Occup. Hyg. **25** 365 (1982).
Fletcher B and A E Johnson. The capture efficiency of local exhaust ventilation hoods and the role of capture velocity. In: Ventilation '85, p. 369 (Edited by H D Goodfellow). Elsevier Science Publishers B.V., Amsterdam (1986).
Garrison R P, Y Dong, S A Lengerich and T L Rabiah. Am. Ind. Hyg. Assoc. J. **50** 501 (1989).
Hampl V and R T Hughes. Am. Ind. Hyg. J.**47** 59 (1986).
Huebner D J and R T Hughes. Am. Ind. Hyg. J. **46** 262 (1985).
Maust J D and R P Rundquist. Am. Soc. Heat. Refrig. Air Conditioning Engrs. Trans. **93** 1813 (1987).

Moyer R C and J O Dungan. Am. Soc. Heat. Refrig. Air Conditioning Engrs. Trans. **93** 1822 (1987).
Myers G E and C A Hunckler. Appl. Ind. Hyg. **2** 71 (1987).
Neuman V A and W H Rousseau. Am. Soc. Heat. Refrig. Air Conditioning Engrs. Trans. **92** 330 (1986).
Nitsu Y and T Katoh. Trans. Society of Heating, Air-Conditioning and Sanitary Engineers—Japan **1** 1 (1963).
Pruzner A S. Otoplenie i ventilyatsiya **3** 13 (1939).
Roach S A. Ann. Occup. Hyg. **24** 105 (1981).
The Black and Decker Tool Company, Towson Md. USA (1987).
Wake D, R Gray and R C Brown. Ann. Occup. Hyg. **35** (1991).
Wiggin M E and R H Morris. Heat./Piping/Air Cond. **57** 59 (1985).

Bibliography

American Conference of Governmental Industrial Hygienists. Industrial Ventilation—A manual of recommended practice. Committee on Industrial Ventilation, ACGIH, P O Box 16153, Lansing, Michigan 48901, USA (1988).
Brandt A D. Industrial health engineering. John Wiley and Sons, New York (1947).
Burton D J. A self-study companion to the ACGIH Industrial ventilation manual. ACGIH, 6500 Glenway Ave., Bldg. D-5, Cincinnati, OH 45211, USA (1984).
Dallavalle J M. Exhaust hoods. Industrial Press, New York (1947).
Health and Safety Executive. Principles of local exhaust ventilation. H M Stationery Office, London (1975).
Hemeon W C L. Plant and process ventilation. Industrial Press, New York (1963).
Weller J W and A Youle. Thermal energy conservation: Building and services design. Applied Science Publishers Limited, London (1981).

CHAPTER 18

Ventilation Investigations

When undertaking an assessment of possible health risks to employees from atmospheric exposure to hazardous substances a systematic check on the ventilation provisions is essential. To do this it is not necessary to be able to design a new system but it is important to know enough about the principles to be able to examine an existing system intelligently.

A ventilation and air cleaning system has a number of interrelated components in series. The inadequacy of any one of these components because of faulty design, the failure of one of them, or poor maintenance will result in a loss of efficiency and the possible introduction of a health risk.

The performance of a ventilation system may be judged in health hazard control terms by its ability to keep atmospheric concentration of hazardous substances below that specified in exposure limits. Where it is evident that the hazard control performance is poor or deteriorating the first investigation is to measure the ability of the ventilation system to provide the volume flow rates for which it was designed.

Where the volume flow rates do meet design performance but it is not certain which hood or which aspect of the ventilation system is at fault investigation of containment by means of the controlled release of tracer gases may help (Sandberg and Sjoberg, 1983; Cameron, Johnston and Konzen, 1987; Davidson and Olsson, 1987; Niemela, Toppila and Tossavainen, 1987). Indeed, the direct assessment of the concentration field from a known rate of release of contamination is advocated wherever possible, as this is greatly influenced by local dispersion, diffusivity and a host of other factors whose combined effect cannot, in truth, be accurately predicted at this time (Ivany, First and Diberardinis, 1989). Measurement of the same concentration field under a range of ventilation volume flow rates is the means of direct quantitative assessment of the ventilation required. These techniques are invaluable on introduction of a new substance, new plant or new process.

The aim is to predict the maximum credible concentration to which employees will be exposed for a defined period of time. Expecting a supplier of ventilation equipment to do the predicting, to know all the vagaries of emissions peculiar to a particular process and to know the consequences on the health of operatives of inhaling air contamination is expecting too much. The greatest responsibility for selecting effective ventilation provisions really should be borne by the local occupational hygienist or other officer who serves the same purpose.

Early signs of poor performance

A loss of containment may result from a change in materials handled, a change in the processing of the materials or a deterioration of the ventilation system. An analysis of the ventilation system would be needed to establish which if any part is faulty. Common indications that the performance of a system as a whole or of an individual exhaust hood may be unsatisfactory include:

(1) Dust and fumes seen escaping the hood.

It is often claimed, quite rightly, that individual particles smaller than 20 microns diameter are too small to be seen, but such fine particles are almost invariably accompanied by larger particles or aggregates which are visible. In any event a cloud of fine particles is visible en masse, especially when viewed towards a light source, along a beam of light a few degrees out of line.

A portable light source may be used for this. A lamp having a parabolic reflector is preferred, such as a long range driving lamp, having a beam divergence of less than 3 degrees. Background light should be turned off or shaded to show the aerosol to best effect. Direct light from the light source should be shielded and fine, respirable dust is best seen when viewed towards the light source, between 5 and 15 degrees from the direct line, against a dark background (Figure 18.1). In complete darkness a miner's lamp or a hand torch will suffice. A powerful tungsten halogen lamp, 100 watts or more, is necessary for daylight use (Lawrie, 1951; Health and Safety Executive, 1990).

(2) Odours becoming noticeable or objectionable.

Nevertheless, the absence of any noticeable odour does not necessarily mean containment is adequate.

(3) Loss of containment revealed with the aid of commercial smoke tubes, pellets or candles or by the use of swabs of titanium tetrachloride.

Smoke generated close to the hood entry, just inside the plane of the entrance

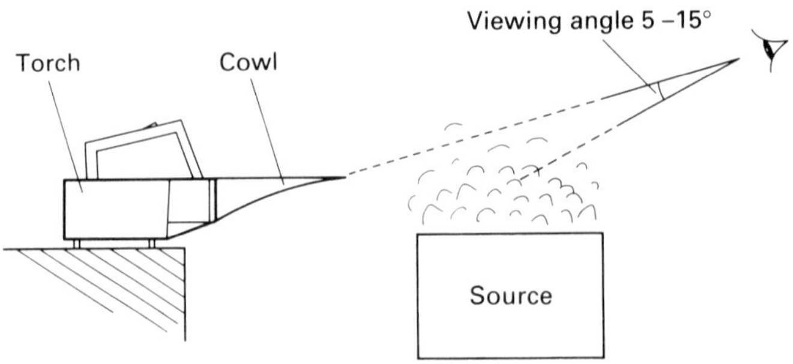

FIG. 18.1. Viewing a cloud of dust with a portable lamp.

and close to the sides enables the critical air flow patterns to be studied (Redfearn, 1979).

Several commercial smoke tube kits are available which produce a cloud of white smoke. The proprietary ones consist of a glass tube, sealed at either end and filled with a chemical which produce the smoke when the tube ends are broken and air is blown through. Leaks in joints and cracks in walls can also be tested using smoke clouds.

(4) Reduced air velocity into an exhaust hood indicated by fixed hood static pressure gauges reading below normal.

Readings of velocity by an anemometer may otherwise be available. An anemometer capable of reading to 0.25 m s^{-1} should suffice for hoods with a clearly defined opening.

(5) Abnormal pressure drop across the air cleaning plant.

There should always be a gauge across a filter to indicate the limits of the working pressure drop, and it is convenient to mark these limits with red lines on the dial, so that a glance at any time would indicate whether or not the filter is working within its prescribed limits.

(6) An adverse trend indicated by the results of regular measurements of air contamination.

Static sampling tests made at one position or by automatic, multi-point sampling may reveal an unfavourable trend associated with a particular hood or ventilation system. The ultimate test is exposure of employees.

Ventilation volume flow rate reconnaissance

A sketch plan of each room should be drawn up. It could be based on the ventilation dossier prepared for a risk assessment (Chapter 10). A separate sketch is made of each independent system. The sketch of an exhaust system includes outline diagrams of hoods, branches, elbows, expansion pieces, valves, blast gates, air cleaner, fan and discharge stack. The sketch of a supply system includes the principle features of diffusers, registers, branches, elbows, expansion pieces, valves, blast gates, heater, humidifier, air cleaner and fan. The ventilation system should have been surveyed in detail when first installed and commissioned. Any subsequent ventilation reconnaissance is essentially to check that the ventilation provisions have not deteriorated and are adequate for any changes of plant, processes and equipment that have been introduced since the previous ventilation reconnaissance.

Examination of the plant ventilation would be greatly aided by a plan or drawing showing the position of the principal features to be scrutinised. To this end a simple layout plan is prepared for each work area to describe the arrangement of processes, location of employees and hazardous substances. Pertinent elevations should be included. Transparent overlays could be marked up with the ventilation arrangements and other factors relevant to employee exposure. An overlay should identify where the employees make key operations on the plant like sampling, adding chemicals and drawing/blowing down.

The *dossier on the mechanical and natural ventilation arrangements should include*:

- Location, size, configuration and last known volume flow-rate from air supply openings, including re-circulated air where present.
- Location, size and configuration of high louvres and gravity air stacks, where reliance is placed on natural ventilation.
- Location, size, configuration and last known volume flow rate into extract openings. Hood static pressures.
- Static pressure upstream and downstream of fans. Normal values of fan power and/or fan amps and location of meters.
- Static pressure upstream and downstream of air cleaners.

A ventilation reconnaissance is primarily concerned with an assessment of the volume flow rate entering and leaving the various work places. Care should be taken to assess the flow rate through each local exhaust hood since this is critical in assuring containment of air contamination. A semi-quantitative estimate of the flow through visible cracks around doors and windows should be made. In small rooms the flow rate should be measured through all openings down to 2 cm diameter. When the reconnaissance has been completed it should be possible to make a rough air balance between air entering and leaving each room.

The air volume flow rate through air registers, doors, windows and other openings should be measured with an anemometer. Three types of anemometer are in common use; heated head, swinging vane and rotating vane. The heated head probe may have a resistance coil, a thermocouple junction or other temperature sensitive element which responds to heating from the battery pack in the meter or mains supply and cooling from the passage of air over the probe. The head is relatively insensitive to direction but by the same token is partly responsive to random turbulence as well as convective flow. These instruments are flexible, easy to use and favoured in non-flammable atmospheres. By switching ranges air velocities from 0.1 to 30 m s^{-1} may be measured with good accuracy.

The swinging vane 'Velometer' is a portable instrument which is fitted with a series of probes and calibrated to read velocity directly on some scales and pressure directly on other scales. It has an internal vane which swings through an angle in a shaped cavity, according to the air flow through the instrument. The instrument is itself held in the air stream and, where this is not feasible, comes with a variety of attachments for grilles, diffusers or duct measurements. The instrument is surprisingly versatile and robust but is not ideal for small hood entrances.

The rotating vane anemometer was originally developed for use in mines and is still the most common device used in tunnels. The rotating vane is not unlike a child's windmill, shrouded and geared to a meter. The instrument is strongly directional and has to be handled with care. It registers the number of meters of air and the velocity is calculated by switching the instrument on for a timed period. It can be used to average velocities over areas traversed at a uniform rate, and has an approximately straight-line calibration. This instrument is particularly adapted to measuring average velocities through large openings, such as spray booths, doorways and discharge grilles, but is not adapted for small areas where the size of the instrument offers appreciable obstruction in relation to the size of the opening. Small, low velocity, rotating vane anemometers are commercially available.

Hood static pressure should be measured to check the volume flow rate through hoods. Where this is not feasible there are three possible alternatives; anemometer traverse across the hood entrance, dilution technique downstream of the hood or, thirdly, a pitot-static traverse downstream of the hood.

Hood static pressure. The air volume flow rate at the entrance to all hoods should be checked. All branches should be exhausting within 10% of the design volume flow rate. By far the most convenient method for regular checks is to measure hood static pressure or 'throat suction'. This is the static pressure in the duct connected to the hood, at a position chosen for convenience, between 1 and 5 duct diameters downstream of the hood. A permanent pressure tap should be fitted. In critical installations it is useful to have instruments permanently fixed to the ventilation system at suitable points so that the static pressure can always be seen whenever the plant is operating and give an immediate indication if the plant fails to function for any reason.

In the absence of friction losses and turbulence created by the imperfect entry the velocity pressure achieved in the duct would, ideally, be equal and opposite to the static pressure. In practice there are friction losses and the volume flow rate is significantly less than the ideal. The percentage of ideal flow that enters a given hood in practice is its coefficient of entry, C_e.

Coefficient of entry has been tabulated for many standard hood shapes (American Conference of Governmental Industrial Hygienists, 1988). It is a function of the shape of the hood but is independent of the exhaust volume flow rate. Exhaust hoods are often custom built or heavily modified to suit local circumstances. If the coefficient of entry is measured at the time a hood is built and put into service, it can be used later with hood static pressure measurements to estimate flow rates directly. Where Q is the volume flow-rate ($m^3\ s^{-1}$), A is the cross sectional area of the duct (m^2) and SP_h is the (negative) hood static pressure (mm water gauge);

$$Q = 4.043 A C_e (SP_h)^{1/2}. \quad (1)$$

Problem 18.1

A paint spray booth is extracted by a propeller fan connected by ducting 450 mm diameter. The booth was tested on installation and has a coefficient of entry 0.76. The design volume flow rate is 1.9 $m^3\ s^{-1}$. At a subsequent ventilation reconnaissance the hood static pressure is 17 mm water gauge. Calculate the volume flow rate extracted.

Solution

The cross sectional area of the ducting,

$$A = 0.785\ (D)^2 = 0.16\ m^2.$$

Substituting the example values in equation (1),

$$Q = (4.043 \times 0.16 \times 0.76)\ (17)^{1/2} = 2.0\ m^3\ s^{-1}.$$

The conclusion is that the ventilation volume flow rate performance of the spray hood meets the design specification.

Anemometer traverse. Hood static pressure readings may not be feasible for some hoods due, for example, to the duct being inaccessible. In this event, ventilation performance may be assessed by checking the entrance velocity with an anemometer at a characteristic point, such as the centre of the entrance or in the centre of one quadrant. A Pitot-static tube is not suitable for entrance velocities below about 3 m s^{-1}. The normal air velocity at the characteristic point of the hood entrance must be known. Otherwise it will need to be calibrated against a known air volume flow rate through the hood established either by a complete traverse across the hood entrance or, if this is too complex, or not accessible, by a Pitot-static tube traverse in the branch connected to the hood (Air Moving and Conditioning Association, 1980; British Standards Institution, 1983).

A velocity traverse across the entrance to a hood is made by dividing the face area into 16 or more equal areas by eye and by measuring the air velocity in the centre of each small area. The volume flow rate through an exhaust hood fluctuates due to pressure fluctuations in the work-room and at the discharge stack opening. The readings should be repeated at least twice, with an interval of at least 1 hour between each set. The volume flow rate is the grand average of all velocity measurements multiplied by the hood entrance area (BOHS, 1975).

Dilution technique. Where the duct cross section is complex or partially obstructed so that a Pitot-static tube traverse is impossible or very inaccurate the 'dilution' technique may be used, in which a known volume flow rate of tracer gas is released at the hood entrance and the concentration is measured 8–10 duct diameters downstream, where mixing is complete (Hatch and

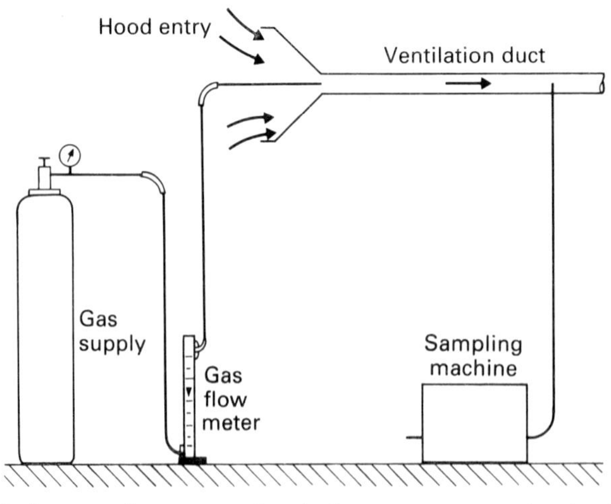

FIG. 18.2. Layout of apparatus for air flow measurement by the dilution method.

Walpole, 1942; Hemeon, 1963). The principle features of the method are illustrated in Figure 18.2.

$$\text{Volume flow rate,} \quad m^3\,s^{-1} = \frac{\text{rate of gas release, ml s}^{-1}}{\text{concentration in duct air, ppm}}$$

Thus, the flow rate can be determined absolutely by this method. Provided the flow is turbulent, as is invariably the case in practice, the method is independent of air velocity, duct dimensions, or any variation in these dimensions.

Pitot-static tube traverse. In order to find the volume flow rate through ducts with a Pitot-static tube traverse it is important to select a location where the velocity profile across the diameter is symmetrical. The location should be at least eight duct diameters downstream of the hood, elbow, obstruction or other disturbance of the air flow and at least two duct diameters upstream of any obstruction, branch or elbow. A hole is drilled for the Pitot-static tube and velocity pressure measurements are made at standard positions across a diameter of ducting circular in cross section, each position representing the velocity through an annulus of equal area. A 10-point traverse is usual. These positions are in Table 18.1. In a rectangular duct the divisions should be equal areas geometrically similar to the section of the airway. Readings are taken in the centre of each area. The velocity pressure readings are converted to velocity before averaging.

Velocity pressure is exerted in the downstream direction. It is the difference between total pressure and static pressure at the same point.

$$\text{total pressure} = \text{static pressure} + \text{velocity pressure}$$

The Pitot-static tube is designed for measuring the total and static pressure in ducts. Velocity pressure is readily found by sensing these pressures either side of a manometer. Velocity of air at standard temperature and pressure is given by,

$$V = 4.04\,(\text{velocity pressure})^{1/2},$$

where V is in m s^{-1} and velocity pressure is in mm of water gauge (mm WG).

TABLE 18.1. *Pitot-static tube 10-point traverse across round pipe. D = internal pipe diameter*

Point	Distance from wall
R1	0.026 *D*
R2	0.082 *D*
R3	0.146 *D*
R4	0.226 *D*
R5	0.342 *D*
R6	0.658 *D*
R7	0.774 *D*
R8	0.854 *D*
R9	0.918 *D*
R10	0.974 *D*

The velocity contour at a point eight duct diameters downstream of any disturbance to the air flow is usually more or less symmetrical in a circular duct, with average velocity about 85–90% of the centre line velocity. For approximate measurements, the centre line velocity pressure is observed and the corresponding velocity multiplied by 0.90 to obtain average velocity. The centre line velocity pressure is multiplied by 0.81 to obtain the average velocity pressure.

Fault finding in ventilation systems

Deteriorating or poor ventilation performance may be due to any of a host of different factors, not all of which are obvious. Some of these factors are external to local exhaust hoods; alterations to the buildings, plant and process changes, new machinery, changes in supply air arrangements, personnel traffic past hoods and individual operator work activities. Deterioration in ventilation performance is indicated in the following ways;

- Hood static pressure readings below normal indicate a blockage or leak downstream in the exhaust ventilation system. Hood static pressure readings above normal indicate a blockage in another branch of the system or upstream of the pressure point. Permanent hood static pressure gauges are recommended wherever hazardous substances are being handled.
- Air contamination results from area sampling in the vicinity of hoods repeatedly above normal may be due to any of several factors, including poor ventilation. Check the ventilation. The ventilation may meet the ventilation design criteria but those criteria may be wanting. For example, a design based simply on achieving a specified capture velocity whilst usually satisfactory does not guarantee 100% capture. The amount of contamination which fails to be contained by a hood is a function of the rate of production of contamination and the ventilation conditions external to the hood as well as the velocity field induced by the exhaust volume flow rate on its own. The only fully satisfactory test of adequate ventilation is by measurement of work-room air contamination in the vicinity of employees and in the conditions of use.
- Exposure of individual employees being repeatedly above normal is a sign of poor engineering or personnel control. Check the ventilation.
- Visible airborne dust escaping a hood is a sure sign of insufficient ventilation. However, it may have been brought about by a change in process or throughput and not necessarily due to deteriorating ventilation provisions. Check the ventilation. More may be needed.
- Air velocity in the ducts must be high enough to carry dust to the filter or other air cleaning plant. Dust settlement in horizontal ducts indicates that the duct velocity has fallen below the minimum transport (conveying) velocity for the material. When this happens, dust collects in the duct, the effective cross-section of the duct is reduced, the resistance to air flow

TABLE 18.2. *Transport velocity of certain materials*

Material	Transport velocity ($m\ s^{-1}$)
Sawdust	6
Paper dust	8
Rubber dust	10
Zinc oxide fume	10
Lint	10
Grain	15
Starch dust	15
Wood chips	15
Coal dust	20
Metal turnings	20
Clay dust	20
Rock dust	25
Lead dust	25

Note. Design velocity is generally higher than transport velocity for the following reasons: (1) To provide a factor of safety to ensure the maintenance of transport velocity in the main in the event of closure of one or more branches, leakage of duct work, corrosion of the fan blades, increase in system resistance due to denting of duct work and other causes. (2) To balance systems having several branches. (3) To allow the use of smaller duct sizes.

increased and in consequence the air volume flow rate is generally reduced. The failure may have originated from adventitious entry of an off-cut, carton, paper or other rubbish causing a blockage. The system duct design velocity may have been insufficient at the design stage. The material being handled by the ventilation system may have changed. The minimum transport velocity of typical materials is in Table 18.2.

Deteriorating performance may be traced to any of six main elements of a local exhaust system:

(1) *Hood*

The entrance may be blocked with equipment, containers or rubbish. The hood may be physically damaged, rusted or corroded. It may be poorly placed in relation to the sources of contamination.

The performance of the hood may be reduced by draughts from the passage of people or vehicular traffic in front of the hood or by the proximity of moving machinery. The disturbance might be reduced by fitting shields to screen the hood or by routing the traffic away from the hood entrance. It may suffice to locate the hood closer to the source of contamination. Otherwise an increase in volume flow rate may be necessary.

(2) Duct

This may be blocked with settled dust, off-cuts or rubbish, especially around blast gates and dampers. The use of balancing cones rather than slide dampers has the advantage of them being less likely to accumulate dust and they are not subject to unauthorised alteration.

Ducts may be leaking due to rusting, corrosion or physical damage. Faulty sections should be replaced. Joints may be leaking.

Adjustable blast gates and dampers may have been set wrongly. Extra branches may have been added. The need for continuing extra branches may be reviewed. Otherwise it may be feasible to increase the volume flow rate by increasing the speed of the fan, depending upon the design capacity. Increase in fan speed may also be required due to process changes or reduction in exposure limits necessitating increased volume flow rate.

There are certain common principles governing the operation of those ventilation systems which are dependent upon the speed of a mechanical fan, known as the fan 'laws':

(i) Volume flow rate is proportional to fan speed.
(ii) Total pressure and static pressure are proportional to the square of fan speed. This follows from the first law.
(iii) Fan power expended is proportional to the cube of fan speed—a 10% increase in volume flow rate results in a 33% increase in power costs.

The fan laws are not exact, but are held closely enough for all practical purposes when adjusting fan speed.

When it becomes necessary to change the fan for a similar one of a larger diameter there are three more fan laws to remember:

(iv) The volume delivered is proportional to the cube of the diameter.
(v) The pressure developed is proportional to the square of the diameter.
(vi) Fan power expended is proportional to the fifth power of the diameter.

Problem 18.2

An exhaust ventilation system is powered by a belt driven centrifugal fan. An extra branch has been added and it is proposed to increase the fan speed by 10%. Before increasing the fan speed measurements show:

Volume flow rate	$Q_1 = 3\ \text{m}^3\ \text{s}^{-1}$
Hood static pressure	$SP_1 = 70$ mm w.g.
Fan power	$W_1 = 2.5$ kW

Calculate the expected new volume flow rate, hood static pressure and power expended. Compare this with the effect of replacing the fan with one of 10% larger diameter.

Solution

Given,

$$\frac{RPM_2}{RPM_1} = 1.1$$

By the fan laws (i), (ii) and (iii):

(i)
$$\frac{Q_2}{Q_1} = 1.1$$

(ii) $$\frac{SP_2}{SP_1} = 1.1^2 = 1.21$$

(iii) $$\frac{W_2}{W_1} = 1.1^3 = 1.33$$

Volume flow rate $Q_2 = 3.3\ \text{m}^3\ \text{s}^{-1}$
Hood static pressure $SP_2 = 85$ mm w.g.
Fan power $W_2 = 3.3$ kW

Given,

$$\frac{D_2}{D_1} = 1.1$$

By the fan laws (iv), (v) and (vi):

(iv) $$\frac{Q_2}{Q_1} = 1.1^3 = 1.33$$

(v) $$\frac{SP_2}{SP_1} = 1.1^2 = 1.21$$

(vi) $$\frac{W_2}{W_1} = 1.1^5 = 1.61$$

Volume flow rate $Q_2 = 3.3\ \text{m}^3\ \text{s}^{-1}$
Hood static pressure $SP_2 = 85$ mm w.g.
Fan power $W_2 = 4.0$ kW

(3) *Air cleaner*

Special tests of the working efficiency of an air cleaner are conducted by measurements of contamination at the inlet and outlet. Isokinetic sampling is necessary for aerosols.

In wet cleaners settled sludge is an obvious fault, when excessive. Regular hosing down inside is important. Orifices should be cleaned regularly so that they will not choke. The drag link conveyor or other means of removing sludge should be properly maintained.

Blocked bags are a common fault in a bag-house. Fabric filters generally indicate any falling off in performance by the change in pressure drop across the filter. If the static pressure drop increases, it might be due to increased humidity of the air, or to excessive dust loading on the filter, and this latter might have been caused by a failure of the automatic shaking system, where provided. If the shaking mechanism is one which is manually operated the heavy dust loading will probably have resulted from insufficient shaking. Should the pressure drop fall significantly, one or more of the seals in the filter may have failed, and indeed it is a matter of considerable importance to ensure that all the seals are tight whenever the filter is replaced after having been stripped down for examination.

(4) *Fan*

The fan's performance may be below that specified in the suppliers capacity rating tables. These tables give the performance in terms of 'fan' total pressure or 'fan' static pressure produced at a designated volume flow rate. Fan total pressure represents all the energy requirements from the fan for moving air through the ventilation system. Fan static pressure represents the system losses, which have to be made up by the fan. These terms are defined by the following equations:

$$\text{Fan } TP = TP_{\text{outlet}} - TP_{\text{inlet}}$$

$$\text{Fan } SP = \text{Fan } TP - VP_{\text{outlet}}$$

$$= SP_{\text{outlet}} - SP_{\text{inlet}} - VP_{\text{inlet}}$$

A slipping fan belt is a common problem in belt driven fan systems. Check the fan speed, rpm. Check the belt tension.

Worn and corroded blades or a build-up on the blades are a sign of poor performance.

It is not unknown to find a centrifugal fan turning the wrong way due to wiring of 3-phase motors having been inadvertently switched during routine maintenance. A centrifugal fan turning the wrong way will still extract air but at a markedly reduced rate. Direction of fan rotation should be checked after any repair work on motor or wiring.

The fan may be badly placed. If located inside an occupied building any leakage from the downstream, positive pressure side of the fan will contaminate the work place air.

If there is an excessive consumption of power, which may in extreme cases result in a trip-out, the cause is probably a lowering of the resistance of the system. This may be due to erosion of ducting or hoods to such an extent that holes have appeared in them, and because the system resistance has fallen the air volume flow rate has increased. The power taken by forward bladed and radial fans increases with volume. This effect does not occur with backward curved blades in the impeller because these fans have a non-overloading power characteristic so that there is a limit to the power they absorb however high the air volume might be.

(5) *Discharge stack*

This may be blocked at the base or bent due to wind damage. The discharge stack may also be poorly placed or too short, resulting in discharged air contaminating intake air to the work place in unfavourable wind directions. Check for contamination in the intake air.

(6) *Make-up air*

The need for an adequate supply of make-up air may have been forgotten. The need is most often felt in small rooms. The air extracted by an exhaust system must be replaced by make-up air through gaps around doors and windows, through wall grilles, louvres and other openings or supplied by fans installed for the purpose. When large air volume flow rates are extracted a good supply of tempered make-up air is needed for air conditioning purposes. Excessive negative pressure in a work-room is made evident by outward opening doors slamming shut and inward opening doors blowing open. Check the air supply arrangements for blocks.

Static pressure survey

When the precise location of the fault in an exhaust ventilation system is not obvious by visual inspection it may be isolated by a stepwise study of the static pressure losses through the system. The essential apparatus required is a portable manometer of the aneroid type or a liquid filled inclined gauge. The aneroid type is preferred for field use as it is not sensitive to orientation. It is also mechanical in operation, which is more convenient than a liquid filled gauge, but is, by the same token, liable to mechanical damage and aneroid types should be calibrated regularly against a liquid filled manometer. Two ranges of instrument, 0–25 mm and 0–250 mm water gauge, are sufficient for most ventilation investigations. A liquid gauge may combine a 1 in 10 inclined section with a vertical section in one instrument.

In the first step of the fault finding process the static pressure is measured at pressure points fitted by the supplier such as at the hoods, across the air cleaner and across the fan. This is best done when the production plant is working normally, because large open doors on windy days or strong convection currents from adjacent furnaces may well alter the distribution and volume flow rate of the air flowing into hoods. It may be evident at this stage precisely where the fault lies or in which section. Next, proceeding through the section in order, from the air entry downstream, the static pressure values should be measured systematically immediately upstream and downstream of elbows, junctions, fittings and changes in cross section insofar as these positions are accessible. A hole 1 to 2 mm diameter is drilled, if necessary, at the measuring points and the static pressure is read with the manometer. The hole should be perpendicular to the wall and free of burr for the greatest accuracy. The pressure losses found are checked against losses calculated for each straight length of duct, junction, elbow, contraction or expansion piece. Any imbalance found during these estimations should be adjusted, provided this is possible, by means of the blast gates or dampers, and if these have to be reset they should be locked into position, and this position should be marked on the dampers themselves and on the drawings. Agreement within 10% of the design volume flow rate and agreement within 5% of the design static pressure is considered satisfactory.

The friction loss per metre length in a straight run of round, galvanized metal duct is given by

$$\text{Friction loss in mm WG m}^{-1} = 5.1 \times V^{1.9} \times D^{-1.22}, \qquad (3)$$

where friction loss is in mm water gauge per metre run,
V = velocity in m s^{-1},
D = duct diameter in mm

(Wright, 1945).

The friction loss may be larger than this for ducts made of other materials with a rough finish or fractionally smaller for especially smooth materials. The friction loss across branch entries, elbows, contraction and expansion pieces is also approximately proportional to the square of the air volume flow rate. The approximation is sufficiently close that the square law may be assumed to apply for most practical purposes. The same is true of duct-work generally so that the static pressure at the fan is also proportional to the square of the volume flow rate. Air flow resistance across any element has been tabulated in terms of equivalent straight length for many standard duct pieces (American Conference of Governmental Industrial Hygienists, 1986). The square law holds for all parts of the ventilation system where the flow is fully turbulent. The principle exception is a fabric filter through which, because of the small dimensions of the interstices, the flow is laminar and the pressure drop is proportional to the first power of volume flow rate.

Problem 18.3

Two exhaust hoods over building board saws, H1 and H2, are connected into a balanced ventilation system as shown schematically in Figure 18.3. The normal hood static pressures are;

Hood H1; SP = 32 mm WG,
Hood H2; SP = 24 mm WG.

A ventilation reconnaissance reveals that the hood static pressures have changed;

Hood H1; SP = 40 mm WG,
Hood H2; SP = 12 mm WG.

Analyse the findings.

Solution

The investigation proceeds as follows.

STEP 1.

The hood static pressure in hood H1 is above normal, indicating that another branch is blocked or, possibly, that there is a block upstream. The hood static pressure in hood H2 is below normal, indicating that there is a fault downstream. Taken in conjunction with the above normal value in H1, this indicates that the fault is a block in the section a-b-c-d-e-f-g. In order to locate the source of the block static pressures were measured in that section.

Point a, beginning of straight run	12 mm WG
Point b, end of straight run, beginning of elbow	14 mm WG
Point c, end of elbow, beginning of straight run	17 mm WG
Point d, end of straight run, beginning of elbow	19 mm WG
Point e, end of elbow, beginning of straight run	39 mm WG
Point f, end of straight run, beginning of elbow	42 mm WG
Point g, end of elbow	45 mm WG

STEP 2.

Pressure drop calculations proceed as follows

Straight run a-b (length 3 m)	2 mm WG
Elbow b-c (equivalent length 5 m)	3 mm WG
Straight run c-d (length 4 m)	2 mm WG
Elbow d-e (equivalent length 5 m)	20 mm WG
Straight run e-f (length 5 m)	3 mm WG
Elbow f-g (equivalent length 5 m)	3 mm WG

CONCLUSION

Evidently elbow d-e is partly blocked.

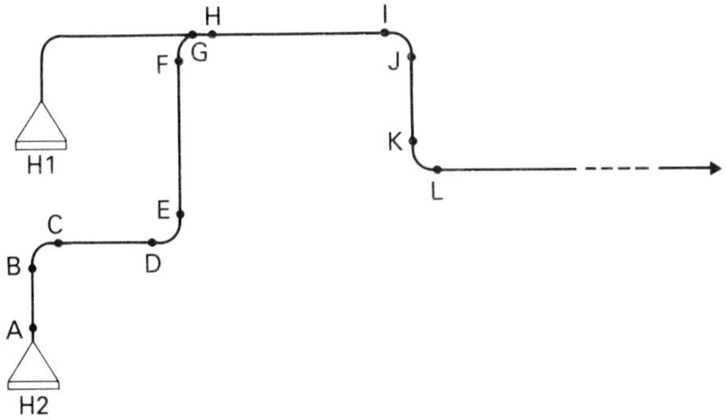

FIG. 18.3. Duct system in Example 18.2.

Problem 18.4

The same hoods as in problem 18.3, shown in Figure 18.3, but on this occasion a ventilation reconnaissance reveals that both of the hood static pressures are below normal.

The normal hood static pressures are;

Hood H1; SP = 32 mm WG,
Hood H2; SP = 24 mm WG.

The ventilation reconnaissance reveals that the hood static pressures have changed;

Hood H1; SP = 17.8 mm WG,
Hood H2; SP = 13.3 mm WG.

Analyse the findings.

Solution

The investigation proceeds as follows.

STEP 1.

The hood static pressure readings have both decreased by the same percentage. This indicates that the fault is downstream of the junction. If one hood static pressure had fallen by a small percentage and the other by a large percentage, such a finding would be indicative of a hole in the latter branch. Static pressures were measured in the section h-i-j-k-l, by which stage the following set had been accumulated:

Point h, beginning of straight run .... 38 mm WG
Point i, end of straight run, beginning of elbow .... 42 mm WG
Point j, end of elbow, beginning of straight run .... 60 mm WG
Point k, end of straight run, beginning of elbow .... 62 mm WG
Point l, end of elbow .... 65 mm WG

STEP 2.

Pressure drop calculations proceed as follows

Straight run h-i (length 6 m).. 4 mm WG
Elbow i-j (equivalent length 5 m)... 18 mm WG
Straight run j-k (length 2.5 m) .... 2 mm WG
Elbow k-l (equivalent length 5 m) .... 3 mm WG

CONCLUSION

Evidently elbow i-j is partly blocked.

At the completion of the investigation and after the correction of faults the information collected should be tabulated and preserved to form the basis of subsequent tests on the system.

Mapping ventilation system performance

The amount of contamination which is captured by an exhaust hood is dependent in part upon the ventilation system in the surroundings. The room plan in Figure 18.4 shows an exhaust hood at one end of the room and fresh air supply grilles at the other end, with no other means of ventilation. Gas or vapour liberated at the source point indicated would obviously all eventually leave the room from the exhaust trunking. The percent of the contamination which is 'captured' would be 100% whatever the induced air velocity and no matter how far the source of contamination was from the hood entrance. This illustrates how misleading can sometimes be a reported percent capture. It says nothing about the dispersion of contamination before it is captured (Hampl, 1984). Without dispersion all the contamination would follow a single streamline into the hood. A better test of hood ventilation performance is given by the atmospheric concentration field resulting from a known rate of liberation of the contamination from a fixed point in a standard environment or, preferably, in the actual operational conditions of use (Ellenbecker, Gempel and Burgess, 1983).

Tracer gases have been employed in recent years to investigate ventilation system performance of laboratory hood installations, in particular, by releasing the gas inside the hood and measuring the concentration at various locations in

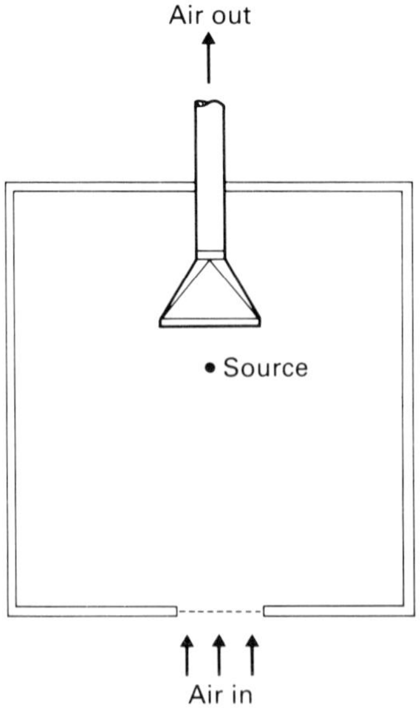

FIG. 18.4. A ventilation arrangement which ensures 100% capture of contamination by an exhaust hood.

front of the entrance (Fuller and Etchells, 1979; Caplan and Knutson, 1982; Macher and First, 1984). An oil mist generator using compressed air may also be used for this purpose (Hinds, 1982).

Physical intuition tells one that whatever the combination of ventilation conditions in the vicinity of a hood and its surroundings, wherever a source of contamination is positioned, stationary concentration of atmospheric contamination at a particular place in a work-room is directly proportional to mass or volume flow rate at which the contaminating substance is issuing from the source. This property is not registered by anemometers, which only register air velocity, not rate of emission of contamination, nor its dispersion.

Air velocity and air volume flow rates have distinct limitations as indices of hazard control by a ventilation system. They need to be coupled in some way with the corresponding rate of liberation of air contamination in the locality. Furthermore the coupling needs to be one which can be used to take into account the position of employees in the work place and their consequent exposure.

The ratio of the rate at which the contamination issues from a source at some point in space to the stationary concentration at some other point is a useful measure or 'yardstick' of the overall effectiveness of the ventilation system at the second point with respect to the source in question. A phrase coined by Lidwell (1960) for this concept is 'equivalent ventilation'.

$$\text{Equivalent ventilation in } m^3\,s^{-1} = \frac{\text{rate of release in mg s}^{-1}}{\text{concentration in mg m}^{-3}} = \frac{\text{rate of release in ml s}^{-1}}{\text{concentration in ppm}}$$

The units are cubic metres per second. It is also the volume flow rate of air in which all the contamination released at the first point would have to be evenly mixed to produce the stationary concentration at the measuring point. The higher this ratio or 'equivalent ventilation', the more effective is the ventilation of the measuring point with respect to the source.

The measuring point, A, may be a fixed or moving position and, in particular, could be the breathing zone of an employee. Thus the equivalent ventilation then measures the performance of the ventilation system in terms of the exposure of employees, the 'ideal' occupational hygiene test. The concept marries the ventilation system with the employee at the time of assessment.

A broad picture of the velocity distribution compared with equivalent ventilation contours in an experimental chamber is given in Figure 18.5. the chamber was (length) × (width) × (height) = 4.27 m × 2.44 m × 1.83 m, ventilated by 0.3 $m^3\,s^{-1}$ air. The contours of the type in Figure 18.5a are loci of equal velocity. This type of mapping may be used to decide where a source of given strength may best be placed to obtain the greatest containment by a local exhaust hood, for example. However, it does not, by itself, reveal what the concentration would be in the vicinity nor how much of the contamination would be captured.

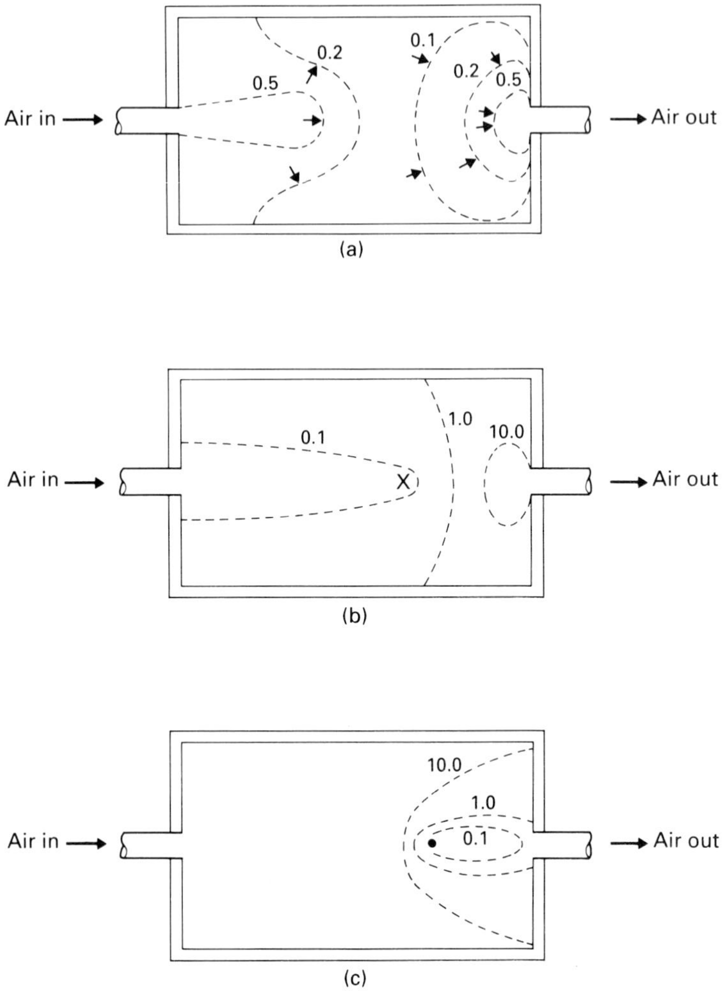

FIG. 18.5. Distribution of air velocity and of equivalent ventilation in an experimental chamber. (a) Air velocity field. Velocity given in m s^{-1}. (b) Equivalent ventilation contours. Stationary employee and mobile source of contamination. Equivalent ventilation given in m^3 s^{-1}. x = Position of employee. (c) Equivalent ventilation contours. Stationary source of contamination and mobile employee position. Equivalent ventilation given in m^3 s^{-1}. ● = Position of source of contamination.

The contours of equivalent ventilation in Figure 18.5b and 18.5c give a comprehensive description of the effectiveness of the velocity field. They were determined with 17% sulphur hexafluoride in helium, by allowing for a 15 minute build-up period and by employing a direct reading instrument, infrared spectrophotometer, adapted with a variable time constant read-out, normally set for 100 s for this work.

There are two main types of equivalent ventilation contours; fixed measuring point and fixed source.

—Fixed measuring point, roving source.

The loci of equivalent ventilation are located by measuring the concentration at the fixed point, marked by a cross in Figure 18.5b, when tracer gas is emitted by a probe which is moved about the room. They are illustrated by the example measuring point shown in Figure 18.5b. The contours would, of course, be different were the measuring point fixed elsewhere.

The contours show where a source of given strength may be located. Thus if the strength of the source in question is 10 ml s^{-1} and the exposure limit of that substance is 500 ppm then the minimum equivalent ventilation required is $10/500 = 0.05\ m^3\ s^{-1}$. The area where the equivalent ventilation exceeds $0.05\ m^3\ s^{-1}$ is the area where the source may be placed.

—Fixed source, roving measuring point.

The loci of equivalent ventilation in Figure 18.5c are located by fixing the tracer gas source in the position of the source under consideration and by measuring the atmospheric concentration at different points around the room. They show the area where an employee may work with respect to a fixed source of contamination. The area where the equivalent ventilation exceeds $0.05\ m^3\ s^{-1}$ is the area within which the employee may work without respiratory protection.

When specifying the equivalent ventilation both the measuring point and the source location must be specified. Thus the equivalent ventilation of A with respect to B refers to a measuring point at A and a source at B. In the same way, the equivalent ventilation of B with respect to A refers to a measuring point at B and a source at A. It may be remarked that in general these two will not be the same unless the ventilation field happens to have perfect symmetry about the plane perpendicular to the line AB through its mid point.

Tracer gases have long been employed for studies of ventilation in dwellings (Pettenkofer, 1858; Petri, 1889; Warner, 1940). Radioactive isotopes, Xenon-133, Argon-41 and especially Krypton-85 have been used more recently (Cadiergues and Leveque, 1952; Collins and Smith, 1956; Howland, Kimber and Littlejohn, 1960; Clayton, 1966). The present author has employed Arcton 12, ammonia, hydrogen, carbon dioxide, helium, nitrous oxide and sulphur hexafluoride on different occasions depending upon the commercial availability of cylinders of the gas and the sensitivity of available atmospheric concentration measuring instruments and procedures (Roach, 1981). Bromotrifluoromethane, chloropentafluoromethane, dichlorodifluoromethane and 1,2-dichlorotetrafluoroethane have been employed in office building ventilation studies (Fisk, Prill and Seppanen, 1987). There is not much technical advantage as between the different gases, but it is definitely advantageous if the use of the tracer gas does not present secondary safety and health problems. In this regard, sulphur hexafluoride and helium would be preferable as they are virtually inert. Rats kept for 24 hr in an atmosphere of 80% sulphur hexafluoride suffer no ill effects (Lester and Greenberg, 1950). Sulphur hexafluoride can be detected down to 1 ppm by a wide variety of methods, and even down to 1 part in 10^{11}, if necessary, by gas chromatography (Collins, Bartlett,

Turk, Edmonds and Mark, 1965; Clemons and Altshuller, 1966). It has been successfully used to trace the spread of community atmospheric pollution (Simmonds, Shoemaker and Lovelock, 1972; Dietz and Cote, 1973). Sulphur hexafluoride is heavier than air and helium is lighter. There are circumstances in industry where the density factor could be a drawback and the best solution is to employ a commercially available mixture of 17% sulphur hexafluoride in helium, which is the same density as air. This mixture is colourless, odourless, chemically inert, thermally stable up to 500C, non-explosive, non-flammable and seemingly non-toxic. Caution should be exercised if the tracer gas containing sulphur hexafluoride is liable to be exposed to temperatures above 500C or to sparks as toxic products may then result. Sulphur hexafluoride in air is normally less than 0.5 parts in 10^{12} parts of air so that interferences on that score are negligible (Ferber and Dietz, 1979).

Estimation of equivalent ventilation around industrial sources is best done by liberating the tracer gas in a steady stream over the source, employing a tracer gas different from process contamination or by liberating tracer gas over a model of the real source. The equivalent ventilation is dependent in part upon the geometry and temperature of the source. Use of a model could pose practical problems in extrapolation if the source of contamination to be imitated is a rotating or very hot or especially large source or releases its contamination with high momentum. In these circumstances it may be preferable to employ the source itself for the tests and employ other means to estimate the rate of release, taking steps to take enough samples to even out the variations in release rate.

The rate of release of contamination from a real source may be estimated by a variety of means, including;

- Where solvents are employed in large quantities, it may be feasible to estimate average rate of vapour release by making a materials balance of liquid used over a work-shift or work-week to estimate that lost by evaporation. The lengthy reference period for the air sampling that is required in this instance to calculate the equivalent ventilation is not without advantage, since occasional large fluctuations in air movement would, perforce, be allowed to exert due influence on the measurement of stationary concentration.

- The rate of release of contamination can, in principle, be determined by measuring the atmospheric concentration in the exhaust ducting, since even moderately successful exhaust hoods exceed 95% in capture. The concentration in exhaust ducting multiplied by the exhaust volume flow rate gives the rate of release. Contamination in exhausted air may be readily determined by sampling through a probe tube such as a standard Pitot tube. When assessing aerosols in ducting isokinetic sampling should be employed.

- Where the source is not near a local exhaust hood the rate of emission can be determined by summing the quantities in air being extracted and discharged from various openings, each individual quantity being calculated from the product of air volume flow rate and its concentration of contamination.

Correction for contamination from any other contributing source is necessary. The amount from such sources can sometimes be found by difference, that is, by also measuring in the absence of the source of interest.

—Adventitious recirculation of exhaust discharges.

Investigation of recirculation of discharge air through ventilation intakes, windows, doors and so on is most conveniently accomplished by employing tracer gases released at a controlled rate into the exhaust ventilation system and by measuring the concentration at the air supply points into workrooms. Sulphur hexafluoride has been employed in such investigations (Drivas, Simmonds and Shair, 1972). The underlying reason for such investigations is commonly the poor location of the discharge point on the roof or side of the building. Wind velocity and direction should be measured at the same time as they are primary variables affecting recirculation. Measurements should be repeated under a variety of wind conditions to establish the worst case. The equivalent ventilation at the work-room air supply points with respect to the exhaust discharge is liable to be variable because of wind direction variations and gustiness. There is also a variable delay between discharge and re-entry. Accordingly a 5-minute 'slug' of tracer gas may be liberated and be compared against the area-under-the-curve of the trace of concentration against time at the work-room supply openings.

The equivalent ventilation at the supply openings with respect to the discharge point should exceed the amount necessary to dilute the maximum discharge of normal exhaust air contamination to one tenth the exposure limit when the weather conditions are least favourable. Measurements may also be desired at the supply openings to rooms in adjacent buildings, to quantitatively investigate contamination reaching them from the discharge point.

References

Air Moving and Conditioning Association. AMCA Standard 210–67. 205 W.Toughy Ave., Park Ridge, Illinois 60068, USA (1980).

BOHS. A guide to the design and installation of laboratory fume cupboards. Ann. Occup. Hyg. **18** 273 (1975).

British Standards Institution. BS 1042: Part 2: Section 2.1: Methods of measurement of fluid flow in closed conduits: Method using Pitot-static tubes. British Standards Institution, London (1983).

Cadiergues R and P Leveque. Chaleur et Industrie, January (1952).

Cameron D B, W L Johnston and R B Konzen. Am. Ind. Hyg. Assoc. J. **48** 56 (1987).

Caplan K J and G W Knutson. Am. Ind. Hyg. Assoc. J. **43** 722 (1982).

Clayton C G. Isotop. Radiat. Technol. **4** 93 (1966).

Clemons C A and A P Altshuller. Anal. Chem. **38** 133 (1966).

Collins G F, F E Bartlett, A Turk, S M Edmonds and H L Mark. J. Air Pollut. Contr. Assoc. **15** 109 (1965).

Collins B G and Smith D A. J. Inst. Heat. Vent. Eng. **24** 45 (1956).

Davidson L and E Olsson. Build. Environ. **22** 111 (1987).

Dietz R N and E A Cote. Environ. Sci. Tech. **7** 338 (1973).

Drivas P J, D G Simmonds and F H Shair. Environ. Sci. Technol. **6** 609 (1972).

Ellenbecker M J, R F Gempel and W A Burgess. Am. Ind. Hyg. Assoc. J. **44** 752 (1983).

Ferber G J and R N Dietz. Perfluorocarbon tracer system for long-range transport and dispersion studies. Presented at Los Alamos Scientific Laboratory Tracer Workshop, Los Alamos, New Mexico, USA (1979).
Fisk W J, R J Prill and O Seppanen. Commercial building ventilation measurements using multiple tracer gases. Lawrence Berkeley Laboratory Report, LBL-25614, Berkeley, California, USA (1987).
Fuller F H and A W Etchells. Am. Soc. Heat. Refrig. Air Conditioning Eng. J. **43** 49 (1979).
Hampl V. Am. Ind. Hyg. Assoc. J. **45** 485 (1984).
Hatch T and R H Walpole. Ind. Hyg. Foundation Am. Preventive Engr. Ser., Bull. 3, Pt I (1942).
Howland A H, D E Kimber and R F Littlejohn. J. Ind. Htg. Engrs. May 1960, 57 (1960).
Ivany R E, M W First and L J Diberardinis. Am. Ind. Hyg. Assoc. J. **50** 275 (1989).
Lawrie W B. Observation of dust in foundry dressing operations (Part II). Proceedings of the Institute of British Foundrymen XLVI (1951).
Lester D and L A Greenberg. Arch. Ind. Hyg. Occup. Med. **2** 348 (1950).
Lidwell O M. J. Hyg. Camb. **58** 297 (1960).
Macher J M and M W First. Appl. Environ. Microbiol. **48** 481 (1984).
Niemala R, E Toppila and A Tossavainen. A multiple tracer gas technique for the measurement of air movements in industrial buildings. Roomvent-87, Stockholm, Sweden, pp. 1–19 (1987).
Petri R J. Z. Hyg. **6** 453 (1889).
Redfearn J. Nature **278** 384 (1979).
Roach S A. Ann. Occup. Hyg. **24** 105 (1981).
Sandberg M and M Sjoberg. Build. Environ. **18** 181 (1983).
Simmonds P G, G R Shoemaker and J E Lovelock. Anal. Chem. **44** 860 (1972).
Warner C G. J. Hyg. **40** 125 (1940).
Wright D K. A new friction chart for round ducts. ASHVE Research Report No. 1280. Trans. Amer. Soc. Heating and Ventilation Eng. **51** 303 (1945).

Bibliography

American Conference of Governmental Industrial Hygienists. Industrial Ventilation: A manual of recommended practice. Committee on Industrial Ventilation, PO Box 16153, Lansing, Michigan 48901, USA (1988).
Health and Safety Executive. The maintenance, examination and testing of local exhaust ventilation. H M Stationery Office, London (1990).
Hemeon W C L. Plant and process ventilation. Industrial Press, New York (1963).
Hinds W C. Aerosol technology. John Wiley and Sons, New York (1982).
Pettenkofer Max von. Uber den luftwechsel in wohnegebauden. Munich (1858).

CHAPTER 19

Personnel Control

Personnel control to limit employee exposure is a supplement to, not instead of the engineering control of the work place environment. Assurance of control of health risks cannot be based upon assuming every individual will use the personnel control provided on every occasion. Human lapses are inevitable. However, a planned approach to personnel control is an integral part of every hazardous substances control system. It includes such diverse topics as education about hazardous substances, training in safe work practices, protective clothing, the washing facilities, duration of exposure and respiratory protection.

Central to protection against excessive exposure to hazardous substances is the provision of items of special clothing to protect against skin contact, especially gloves, armguards, overalls, caps and footwear. Gloves or other forms of eye protection may also be required. In exceptional circumstances complete, impervious suits may be necessary.

The clothing must be chosen with regard to both the nature of the possible contamination and the kind of work to be undertaken (Mansdorf, Sager and Nielsen, 1986). The provision and use of appropriate washing and bathing facilities are also important, even though not necessarily the responsibility of health hazard control services. Redeployment of employees who are especially susceptible to particular hazardous materials may be necessary. Only where all these steps may be insufficient is respiratory protection contemplated (Figure 19.1).

There is a bewildering array of masks and other respiratory protection from which to choose and the type chosen must match the nature and intensity of atmospheric contamination in the conditions of use. The effective use of respirators requires an understanding of human behaviour. Respirators which are uncomfortable to wear are prone to be used casually or poorly fitted to the face or left hanging round the neck.

A systems approach is advocated when considering the introduction of items of protective clothing and equipment; that is the integration of all parts of the programme, including education and training, provision of equipment, its storage, cleaning, inspection and maintenance, provision of spare parts, washing facilities and their supervision.

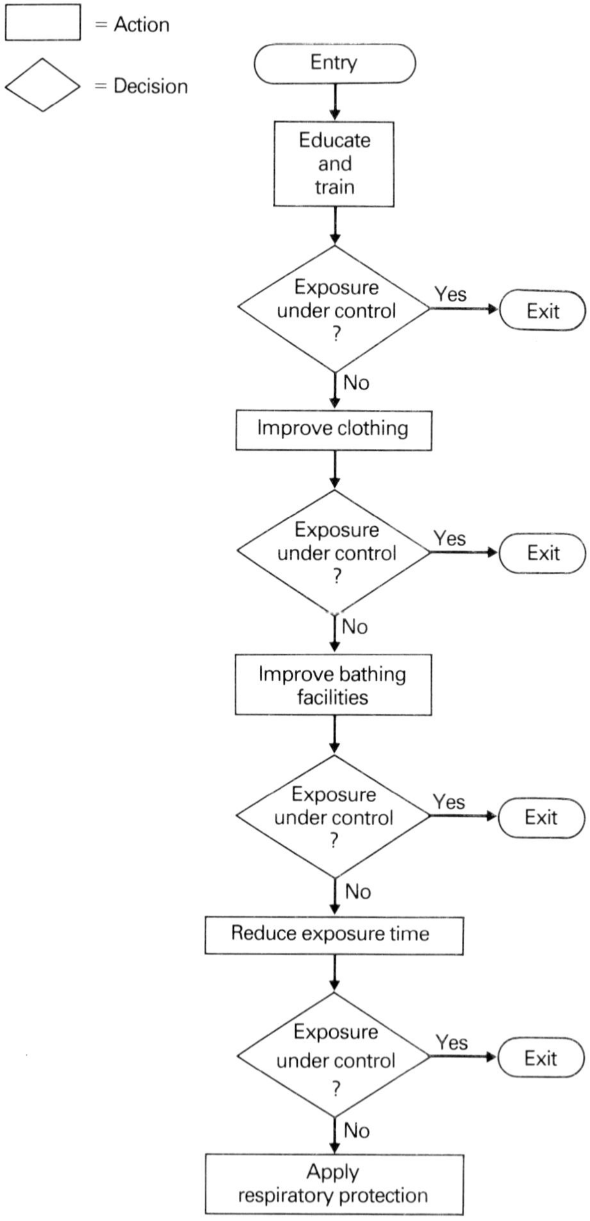

FIG. 19.1. Personnel control. Logic diagram for assessing the required controls. The instructions are: Fulfil each step to the extent reasonably practicable. If the residual exposure is likely to be excessive proceed to the next step.

Education and training

An assessment of risk to health from exposure to hazardous substances should include consideration of appropriate education and training of the employees. They should be kept informed about any hazardous materials which they may use, and be instructed about the importance of avoiding contact with them as far as possible. The employees should be informed about the nature and degree of the risks to health, and where and how they arise as a consequence of exposure. The health risk surveillance should be explained, including the arrangements for personal sampling and medical surveillance, as it affects the individual employee.

It is essential that a high standard of personal hygiene is inculcated so as to reduce absorption through the skin and prevent inadvertent ingestion of hazardous substances. Insistence on good housekeeping should be introduced to all new employees. They should be told when, where and how to wash; also when and where to change clothes. They should be told about factors such as smoking and dietary habits that may increase the risk to health from exposure to particular hazardous substances at work.

Safe practices by those who handle substances which are hazardous to health are particularly important. Before undertaking tasks that require the use of such substances employees are fully instructed in their characteristics, their hazards, physical properties, mode of entry into the body and methods of control adopted. Basic information is imparted on those aspects of chemical safety practice needed to perform the job in a safe manner. The particular safety rules which apply to their job should be explained. Employees should be told the reasons for personal protective equipment and clothing, and the jobs where these are necessary. They have detailed instruction in the wearing of the safety equipment provided; how it works, when to put it on, when to keep it on and how to recognise faulty functioning. They should know how to protect themselves, what protective garments should be worn and when; what to do in the event of an emergency or accident.

Practical training is needed to ensure the adoption of safe work practices and to develop skill in the use of personal protective equipment. Training in the proper use, maintenance and cleaning of protective clothing and equipment is an integral part of the programme. They should know what cleaning is required, why it is required and when it should be carried out. Health and safety officers bear due responsibility in this regard.

Employees who work at operations equipped with local exhaust systems should be instructed as to the proper operating method and as to the reason for the installation. Maintenance personnel receive similar instruction and the necessary additional training in fault finding in the event of malfunction.

Employees can be alerted to safe methods of work through printed matter, safety and health meetings, video tapes, films and other teaching aids. Edited film made with a video camera of the plant and processes where the employees will be working is especially instructive for showing the way atmospheric contamination spreads and how its emission may be avoided (Rosen, Andersson and Juringe, 1990). When filming dusty processes emissions of fine dust may be made visible with a portable lamp with a parabolic reflector to show good or bad control (Health and Safety Executive, 1990). Smoke tubes and canisters are employed to make the dispersion of gases and vapours visible for filming. Direct-reading instruments with a data logger or strip chart recorder are employed to produce an exposure profile and help the employee to better understand the relationship between work practices and atmospheric exposure.

Detailed practical instructions on chemical safety are generally provided in descriptions of specific procedures, tasks and jobs in Safety Data Sheets or Hazard Data Sheets and in instructions for use of equipment and devices. Particular attention should be given to practices which reduce exposure to air contamination from substances hazardous to health. Common examples are:

- Stay upstream of air contamination in so far as it is practicable to do so. For instance, go upstream of

  —spray in a spray booth,
  —dust from a grinder,
  —welding fume,
  —clean-up operations.

- Keep portable hoods within the prescribed capture distance for air contamination.
- Keep inspection doors shut. Always replace removable covers over conveyors, screens, enclosures and containers generally.
- Use the exhaust hoods provided.
- Report damage to the ventilation system promptly. Do not repair with string, wire or glue.

Teaching and training about safety from excessive exposure to hazardous substances should continue on the shop floor. Attention should be drawn to engineering controls and the provisions which require the cooperation of the individual employee.

(1) Exhaust hoods and the general ventilation arrangements are explained and detailed instructions given on their use, wherever adjustments are provided. This includes on-off switches, inspection ports, covers, adjustable blast gates, dampers, sliding doors, sashes, portable hoods and adjustable hoods. Instruction is given on how to keep sources of contamination within the minimum capture distance of exterior hoods and within the confines of partial enclosures.

(2) Conspicuous instructions, signs and labels are employed on containers in the plant. Employees should be instructed where they may obtain further written material and are urged to make full use of it. They should know the location of the department's hazardous substance information and be shown the location by their supervisor.

(3) The rules regarding food, drink and smoking are explained. It is carefully explained that food, drink and smoking materials should not be brought into any area where hazardous substances are being used because they may become inadvertently contaminated and the contamination swallowed. It is good practice to take the time and trouble to persuade all employees to wash their hands with soap and water before eating.

Responsibility for safe practices does not rest solely on individual employees. Supervisors should make frequent checks to ensure that the required safe practices are being followed consistently. The training of the supervisors themselves should not be forgotten. They frequently change jobs from one plant to another and require refresher courses and re-training from time to time. They should be told how exposure is measured and why, and about exposure limits. Relevant legislation, regulations and codes of practice must be consulted which may not only cover the degree of protection to be provided but the back-up facilities required in terms of changing rooms and cleaning. Information is required on the range of equipment or clothing available for dealing with the particular situation, with performance specifications, costs and relevant national standards. Engineers; process engineers, design engineers and maintenance engineers are also key specialists in the control of employee exposure to hazardous substances. It is particularly important to enlighten them about toxicity concerns (Schroy, 1986).

Controlling the duration of exposure

Time of exposure as an element of worker protection is almost self-explanatory. Other things being equal the shorter the time of exposure the smaller the dose of substance received and the less the likelihood of experiencing a short-term excessive exposure.

Consideration should be given to

(a) reduction of number of employees exposed,
(b) exclusion of non-essential access,
(c) stopping overtime in occupations liable to be subject to heavy exposure,
(d) rotation of employees so as to reduce time spent in high exposure jobs during the shift,
(e) organising the work in such a way as to reduce the period of exposure at the highest concentrations,
(f) introduction of ventilated booths to which employees retreat when it is not crucial for them to be physically present on the operations or process,
(g) re-deployment of the susceptible employees, that is those whose health is liable to be adversely affected by the work environment.

Protective clothing

Personnel control is aided in particular circumstances by providing clean protective clothing at regular intervals. In the context of an assessment of health risks aspects of the protective clothing should include;

—The type and quality of gloves, goggles, overalls, shoes, hats and all other forms of protective clothing.

—Special practices to protect against undue skin and eye contact.

—A scrutiny of the arrangements for the maintenance, inspection and storage of protective equipment and clothing. Contaminated protective garments should not be allowed to come into contact with personal clothing.

—Inspection of the provisions for supply, changing and the laundering arrangements for the protective clothing. Removal of superficial grime usually presents no problems but chemical residues may be retained in the body of the material and subsequently released. Even protective garments which are used infrequently will not last indefinitely. There should be a planned programme of inspection and replacement even when defects are not visually detectable.

Work where there is a possibility of skin contact with those substances which have significant potential for skin irritation should only be done with the protection of impervious clothing, including, where necessary, gloves, arm guards, overalls, caps and footwear (Barker and Coletta, 1986). Exposure can occur over the entire body. Hazardous substances may pass through fabric or through openings in the clothing. Protective clothing should be provided to all employees whose jobs entail risk of skin contamination by hazardous substances liable to significant absorption. But exactly which substances are these? As a working rule, a liquid or solid whose dermal LD50 in rabbits is less than 2 g per kg body weight is one which is liable to be hazardous to employees through skin absorption. In the absence of skin toxicity data the assumption should be that significant absorption will take place and skin protection should be provided as a precautionary measure wherever there is a risk of skin contamination.

Fabric resistant to penetration by the physical form of the hazardous substance in question should be employed, whether it be powder, liquid, spray, mist or vapour (Freed, Davies, Peters and Parveen, 1980; Moraski and Nielsen, 1985; Berardinelli and Roder, 1986). When working with flammable substances the clothing should be made from flame-proof fabrics and should be anti-static. Many people cannot tolerate excessive heat, sweating or generalised discomfort for long periods and items of protective clothing may be rejected or discarded in favour of comfort, thus leaving the skin vulnerable to contamination. Chemical protective clothing is normally of a loose, open design to facilitate air movement under the garment. This is to enable the wearer to lose body heat. On the other hand when dermal exposure is primarily to airborne hazardous substances the openings of the protective clothing provide a ready pathway for the substance to be carried to the skin (Crockford, Crowder and Pretidge, 1972; Olesen, Sliwinska, Madsen and Fanger, 1982; Vogt, Meyer and Candas, 1983; Fenske, 1988). Clothing in dusty environments should be close fitting at the neck, ankles and wrist. Overalls should be used in place of work-shirts. The clothing should be made of material such as one consisting of 60% polyester fibre and 40% cotton, which does not readily hold dust.

Rubber gloves of various types are in common use for hand protection from hazardous substances. When vulcanised, natural rubber gives a highly flexible and elastic material which is resistant to many chemicals. Butyl rubber has a greater resistance to chemical attack and is less permeable to gases than natural rubber. Neoprene is a synthetic rubber resembling natural rubber but having superior resistance to heat, oils and oxidising substances. Even so, in no circumstances should it be permitted to put a gloved hand in acid or alkaline solutions (BOHS, 1986). The gloves should be long enough to tuck under the sleeve or gauntlets should be used. Leather gloves are recommended for protection against scratches and rough material. Disposable, polythene gloves are used for light duties requiring resistance against solvents, oils, water and alkalis. Barrier creams are sometimes considered in place of gloves as they do not interfere with dexterity and they are easy to apply and wash off. They are not as protective as gloves. Some provide good protection against oil and selected creams may provide limited protection against mild skin contamination from specific substances. To use a protective cream correctly, it must be applied on clean skin.

There are many different types of PVC, chlorinated polyethylene and other glove materials for special applications (Figard, 1980). PVC has good resistance to water, acids, and oils. Chlorinated polyethylenes are characterised by improved resistance to heat, ozone and oils and have a higher tear strength than polyethylene. Fluorinated ethylene copolymers have exceptional chemical resistance and good insulating properties over a wide range of temperatures.

The resistance that a glove or other article of clothing poses to hazardous materials is a function of two things. One factor is the continued physical integrity of the garment under the physical stress of the work involved. A product's resistance to cuts, snags, abrasion, punctures or tears is a primary requirement. The other factor is the physical and chemical resistance of the garment to penetration and degradation of the material(s) by the chemicals to be handled (American Society for Testing and Materials, 1986; British Standards Institution, 1986a; Stull, 1987). The penetration of liquids through common glove materials tends to be very variable (Sansone and Tewari, 1980; Henry and Schlatter, 1981; Nelson, Lum, Carlson, Wong and Johnson, 1981; Forsberg and Faniadis, 1986).

Seamless gloves and other garments are preferred as failure is often through the seams. A rapid, pass-or-fail test for seam integrity has been developed by the American Society for Testing and Materials (ASTM, 1984; Mansdorf and Berardinelli, 1988). Penetration, that is, leakage through seams, tears and pin-holes or imperfections in the material is the most common type of fault in industrial use. Gloves used when handling hazardous materials should be discarded and replaced whenever they show signs of wear, cracking, swelling or leaks. Rubber gloves can be tested for leaks by trapping air in the glove and applying pressure. If any leaks are present the fingers of the glove will deflate.

Slow permeation of liquids and gases through the actual glove material is impossible to prevent indefinitely (Schwope, Goydan, Reid and Krishna-

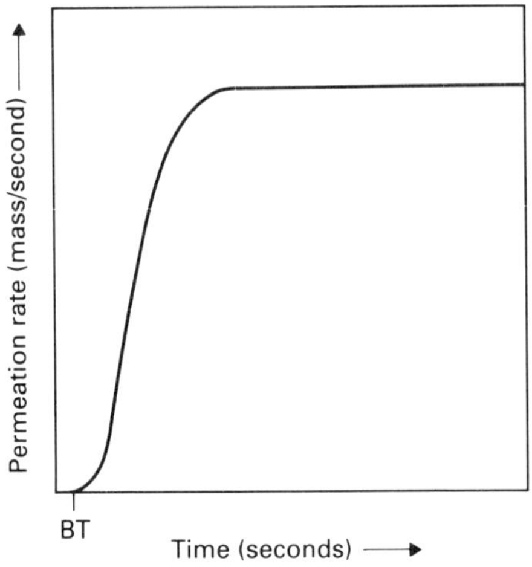

FIG. 19.2. Characteristic break-through curve of glove material. BT = Break-through time.

murthy, 1988). The typical curve of permeation rate of liquid through industrial glove materials subjected to standard tests is the S-shaped curve in Figure 19.2 (Henry and Schlatter, 1981; American Society for Testing and Materials, 1986; British Standards Institution, 1986b; ISO, 1988). The test is normally continued until a steady permeation rate has been established. The test may be discontinued if the steady state has not been reached in 24 hours or if the material fails completely before then. Such curves are obtained by clamping a sample of glove material across the middle of a standard cell and filling one side with the test liquid (Figure 19.3). Air, nitrogen or, sometimes, water is maintained on the other side and the concentration of challenge substance measured downstream from time to time to determine the permeation rate through the glove material. Breakthrough time is the time at

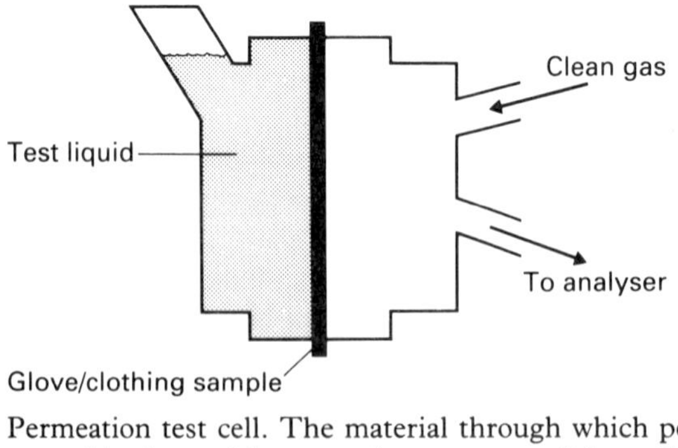

FIG. 19.3. Permeation test cell. The material through which permeation is to be measured is clamped between the two parts of the cell. One side is filled with test liquid and the permeation rate is measured by passing gas through the other side and determining the quantity of vapour in the gas.

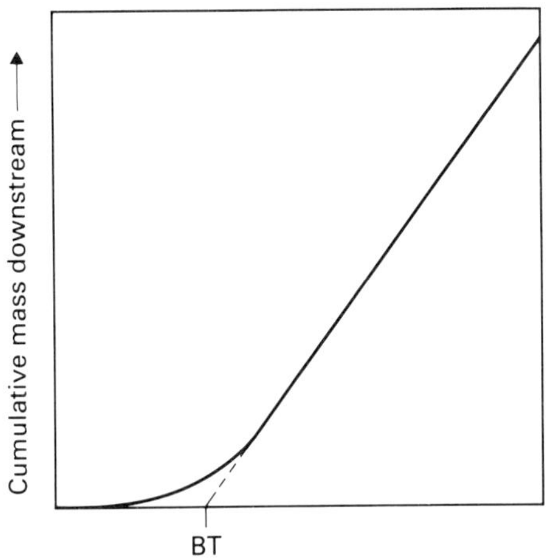

FIG. 19.4. Graphic method of determining the break-through time of a glove material as proposed by BOHS (see text). The curve shown is the time integral of the curve in Figure 19.2. BT = Break-through time.

which the challenge substance is first detected on the opposite side of the glove material. Since this depends in part upon the sensitivity of the measuring system this should be stated. Alternatively, the breakthrough time should be defined as the time the rate of permeation reaches a just measurable, but standard flow rate of 0.1 mg m^{-2} s^{-1}, say.

When comparing quoted results from different materials the procedure for determining breakthrough time should be taken into account as this often differs. The British Occupational Hygiene Society Technology Committee Working Party on Protective Clothing has proposed a graphic procedure based on a plot of the cumulative mass of challenge substance downstream, illustrated in Figure 19.4 (BOHS, 1986). This tends to give high results (Leinster, Bonsall, Evans and Lewis, 1990).

Breakthrough time is governed mainly by permeation through the material. Care is needed with liquid mixtures. Binary mixtures often exhibit a shortened breakthrough time and an accelerated permeation rate (Mickelson, Roder and Berardinelli, 1986). Partial degradation of the glove material by the challenge substance is also contributory and in some instances is the prime determinant (Coletta, Mansdorf and Berardinelli, 1988). Often both factors are significant. As a guide for interpreting the test results on specific hazardous materials, a suitable glove material for most tasks in the chemical industry would be one for which

(a) the breakthrough time is more than 1 hour, and
(b) the steady state rate of passage of substances by permeation or other means is less than 10 mg m^{-2} s^{-1} by the standard methods.

Such a material would be suitable for protection against adventitious skin contact in most applications. The less the frequency of splashing, the greater the protection afforded (Man, Bastecki, Vandal and Bentz, 1987). In jobs which involve prolonged contact, as for example immersion or wetting for more than 30 minutes the criteria would need to be strengthened proportionately. It has been found empirically that breakthrough time increases approximately in proportion to the square of the material thickness (Silkowski, Horstman and Morgan, 1984; Stampfer, McLeod, Betts, Martinez and Berardinelli, 1984; Mellstrom, 1991). A scheme for the systematic re-use or replacement of gloves according to the toxicity of the substances handled has been suggested by Leinster, Bonsall, Evans and Lewis (1990).

Dust- or splash-proof chemical eye cup goggles or face-shields should be worn if there is a possibility of eye contact with corrosive or irritant substances. Otherwise spectacle type eye protection or plastic eye shields should be worn when working with hazardous materials in accordance with national standards, as for example British Standards Institution BS 2092.

Protective clothing is a part of work-wear generally and as such should be designed for the purpose and made reasonably attractive to encourage its wear. A similar argument applies to gloves, which in some circumstances may be quite impracticable for tasks requiring great dexterity, flexibility or touch. Gloves should be available in five sizes. Guidelines for the selection of chemical protective materials have been given by Schwope, Costas, Jackson and Weitzman (1985). Completely suitable protective clothing may not be available. Substitution of less hazardous materials and changes in methods of working may be the most practicable solution.

Respiratory protective equipment

Primitive respirators have been used since ancient times. In the first century AD, the Roman writer Pliny the Elder told in his celebrated *Natural History* (AD 77) of 'minium' refiners wearing bladders over their faces to avoid inhaling lead-bearing dust (Legge, 1936). Minium was the erstwhile name for red-lead, Pb_3O_4. Nevertheless, it was after World War I before industrial respirators came into general use (Patty, 1948).

It is sometimes said that respiratory protective equipment is merely a lazy excuse for avoiding engineering control of air contamination. In truth the provision of such equipment is the final step in personnel control. There is no question that respirators and other respiratory protective equipment should serve as the last line of defence against excessive exposure to hazardous substances, but there are nevertheless plenty of jobs in which they can and should be used freely, not as a substitute for engineering control but as a necessary adjunct. Respirators enable jobs to be done which could not be done without them. Whenever concentrations of hazardous substances in work place air are liable to exceed exposure limits despite providing all other reasonably practicable control measures, it is sound policy to make full use of the appropriate respiratory protective equipment. The use of such equipment in jobs requiring the protection for short periods of time is current best practice.

There is an astonishing variety of equipment for respiratory protection to choose from, ranging between a simple disposable mask for reducing the nuisance of an innocuous dust to a sophisticated, self-contained, air-fed suit, necessary for survival in a high concentration of toxic gaseous contamination. A comprehensive guide to respirator usage is given by Rajhans and David (1985). A collection of papers on respiratory protection taken from the Journal of the American Industrial Hygiene Association with added introductory reviews and an extensive bibliography is in an Association Monograph (1985). A good discussion of the efficiency of chemical cartridges, which are used for gaseous contamination, is given by Nelson and Harder (1976). National standards for the selection, use and maintenance of respiratory protection are in British Standard 4275 (1974), American National Standards Institute Z88.2 (1980) and in the French AFNOR S 76–015 (1983), for example. Advice from an expert should be sought as the wrong choice could be very harmful (Crockford, 1976).

Appropriate respiratory protective equipment should be chosen with full regard to the following factors;

—nature of the contaminating material,
—respirator protection factor required,
—maximum length of time the respiratory protection will be worn,
—maximum permissible face leakage,
—discomfort involved,
—associated protective clothing, which may be necessary to protect against skin absorption when entering especially hazardous conditions,
—maintenance, inspection and cleaning arrangements,
—approval by acceptable national authority for the conditions of use: CEN, NIOSH/MSHA, UK(HSE), BSI, DIN, AFNOR, etc.

Respiratory protective devices fall into two classes: air 'purifying' devices and fresh air supplied breathing apparatus. Air purifying devices remove the contamination from inhaled air within specified limits. Fresh air supplied breathing apparatus includes a self contained supply of compressed air or oxygen to replace the contaminated air or an independent supply of fresh air through a lengthy hose connected to the wearer's respirator facepiece, hood or suit. Air purifying devices fall properly within the area of study introduced in this book whereas fresh air supplied breathing apparatus really falls within the orbit of safety professionals.

Air purifying devices

The physical and chemical nature of the air contamination obviously has a profound influence on the type of air cleaning unit that is chosen. The contents of chemical cartridges may be said to act like a filter for gases and vapours but they are quite different from particulate filters. They rely on different physical and chemical principles. Chemical cartridges and canisters contain solid adsorbent granules for gaseous contamination. Particulate filters consist of a thick mat of fibres or cloth upon which the particles deposit. As the deposit builds up the resistance tends to increase as also does the efficiency. A combination of chemical cartridge and particulate filter in series is commonly employed for a mixture of the two. Within the range of particulate respirators a type suitable for a coarse nuisance dust is simpler than one for a toxic dust of fine, respirable sizes. It is important to know the size distribution of an aerosol as well as the concentration when selecting a suitable respirator.

Cartridges, canisters and filters all have a limited service life depending, amongst other factors, upon the atmospheric concentration of hazardous substance in the vicinity when the respirator is donned. The maximum time specific chemical cartridges and particulate filters can be used has to be established before designing an appropriate respiratory protection programme. Solid adsorbent in chemical cartridges and canisters have a limited life because the quantity of adsorbent (activated charcoal) or chemical in a cartridge has only limited capacity. Except in certain specialised cases the employee is made aware that the cartridge is used up only by sensing the faint odour of the gas or vapour. Consequently air purifying respirators are not normally used to provide protection against gases and vapours with poor warning properties. Particulate filters also have a limited service life, but in this instance mainly because of the increased air flow resistance as dust builds up on the filter until a point is reached when breathing becomes difficult and the filter has to be discarded. A safety factor should be added to calculate replacement intervals allowing a margin of safety. In any event it is good practice to use new chemical cartridges and particulate filters each day irrespective of whether their full capacity has been reached.

The atmospheric exposure limit of the contamination must not be exceeded in the inhaled air and this determines the minimum required degree of protection of the respiratory protection. The degree of protection actually obtained in use is the ratio of the external concentration to the concentration inhaled. It includes the effect of total inward leakage from penetration through the body of the air cleaning unit, edge leakage, leakage around the face seal, valves and other components. The protection obtained is highly variable, and depends upon a multiplicity of factors related to the particular combination of atmospheric contamination, respirator, task undertaken and employee (Douglas, Lowery, Richards, Geoffrion, Yasuda, Wheat and Bustos, 1978; Hinds and Kraske, 1987; Myers, Allender, Iskander and Stanley, 1988).

The performance of respiratory protection may be quoted in terms of protection factor or maximum use concentration. Protection factor (PF) offered by a particular piece of equipment is a comparative term commonly measured at a 'test house' as the ratio of the concentration of contamination in the air outside the facepiece to the 'average' concentration of contamination of the air inside the facepiece, while the respiratory protection equipment is being worn by individuals undertaking certain standard tasks (Griffin and Longson, 1970; Hyatt, 1976; Willeke, Ayer and Blanchard, 1981). When protection factor is measured against aerosols, the size distribution, composition and manner of generation of the aerosols has to be strictly standardised, for obvious reasons, so that the protection factor obtained is somewhat artificial. Another problem with the use of protection factor derives from the fact that during the exhalation half of the respiratory cycle the air in a mask is mostly exhaled air. The alternative term, 'maximum use concentration', is less ambiguous and is expressed as a multiple of the exposure limit of specific substances. An element of judgement is implicit.

When choosing respiratory protection for a particular operation the maximum credible concentration in the work place should be below the maximum use concentration of the respiratory protection. The maximum credible concentration should be backed up by measurements, as a guess could be widely out.

The two main determinants of maximum use concentration of an air purifying respirator are first, the efficiency of the cartridge or filter employed for the particular atmospheric contamination in the vicinity, and second, the respirator face seal leakage. The latter is difficult to quantify, is highly variable between people not least because of their different shaped faces, and may become the dominant factor (Holton and Willeke, 1987; Xu, Han, Hangal and Willeke, 1991). For example, a respirator fitted with a filter unit which collects 99.95% of the contamination may seem at first sight to be highly efficient. However a face leakage of 5%, not at all extraordinary, would reduce the efficiency of the respirator to 95%.

Manufacturer's claims about protection factors of their products should be examined carefully as the protection achieved in the conditions of use may be very much less than obtained in favourable conditions in the laboratory (Harris, DeSeighard, Burgess and Reist, 1974; Toney and Barnhart, 1976; Smith, Ferrell, Varner and Putnam, 1980; Myers and Peach, 1983; Myers, Lenhart, Campbell and Provost, 1983; Lenhart and Campbell, 1984; Grauvogel, 1986; Tannahill, Willey and Jackson, 1990; Hery, Villa, Hubert and Martin, 1991). Test conditions for maximum use concentration should include realistic exercise levels and subjects with a representative variety of face shapes and sizes. Obviously, operations where there are cramped working conditions, or restricted vision, and places where maintenance of the respirators is rushed or their fitting is careless lead to less than perfect performance. A safety factor of 2–10 may be included to give some additional safeguard in rough working conditions.

Maximum use concentration for a particular hazardous substance varies according to the specific equipment, type and manufacturer. It does not take into account 'wearability', that is, the willingness of employees to wear the respiratory protection diligently. Some respirators, for example, with high maximum use concentration for the substance in question but which are relatively uncomfortable to wear and have a high breathing resistance may be ineffective because they are repeatedly removed to gain some temporary respite or worn strapped loosely to the face or even left off altogether.

It is not reasonable to expect production employees to wear respirators or masks continuously for a full work shift. Respirators should only be used when engineering controls are not possible or when controls are being established. They are onerous to use, communication becomes difficult, there is a loss of 'personal identity' and women employees sometimes object to wearing them simply for cosmetic reasons. Respirators which rely upon making a good seal to the face are not safe for use by bearded or unshaven employees, nor those with scars or facial deformities (Fergin, 1984). Spectacles may prevent a proper seal with some ori-nasal facepieces.

The most commonly used types of air purifying respiratory equipment fall into one of six classes, given below, set out broadly in order of increasing maximum use concentration, from approximately 5 × exposure limit to a maximum use concentration in excess of 500 × exposure limit:

(1) Lightweight nuisance dust mask (Figure 19.5).

Nuisance dusts are those which are essentially inert. Exposure limits for such dusts are set to keep the sheer nuisance within reasonable bounds. The traditional nuisance dust mask consists of a replaceable filter pad held to the face by a light alloy frame. The pads may be replaced several times during a shift. They are inexpensive, very comfortable and popular even though of limited efficiency for the finest dusts (Cherrie, Howie and Robertson, 1987).

(2) Disposable respirators for gas, vapour or aerosol (Figure 19.6).

The whole mask is disposable and is usually used on a one-shift basis. These masks are made of lightly compressed, preformed filter material and cover the nose and mouth. They are comfortable, have low resistance to breathing, are light, readily accepted, do not accumulate sweat and relatively inexpensive. Some have an exhalation valve which reduces the build-up of moisture on the filter. They have good protection factors when used properly (Clark, 1979; Wake and Brown, 1988).

(a) Disposable, 'single use' dust masks (Figure 19.6, *top*).

(b) Disposable, maintenance free gas and vapour respirators (Figure 19.6, *bottom*).

These incorporate carbon cloth, charcoal or other solid absorbent. Such products range from those suitable for odour nuisance relief through to those which offer good protection against specific gases and vapours equal to that achieved with a traditional reusable cartridge.

(3) Ori-nasal mask with replaceable filter (Figure 19.7).

This type of respirator, usually known as a rubber half-mask, is in very common use. The facepiece is made of natural black rubber or the more pliable silicone rubber. It covers the nose and mouth, is fairly rigid but has a soft face seal which in some versions is pneumatic or filled with glycerine for better comfort and sealing. Mounted on the facepiece is one or, in some versions, for longer life, two replaceable filter cartridges, one on either side. The cartridges are designed for dust, gas or vapour and sometimes all three. There is a one-way inhalation valve, and exhalation valve. The filters supplied are generally highly efficient against specified gases and vapours and particulate matter but such masks are still prone to face leakage. These masks rely for their performance upon obtaining good face fit. A variety of sizes should be made available to the individual employee, if possible, as the size and shape of faces differs considerably. They need to be strapped

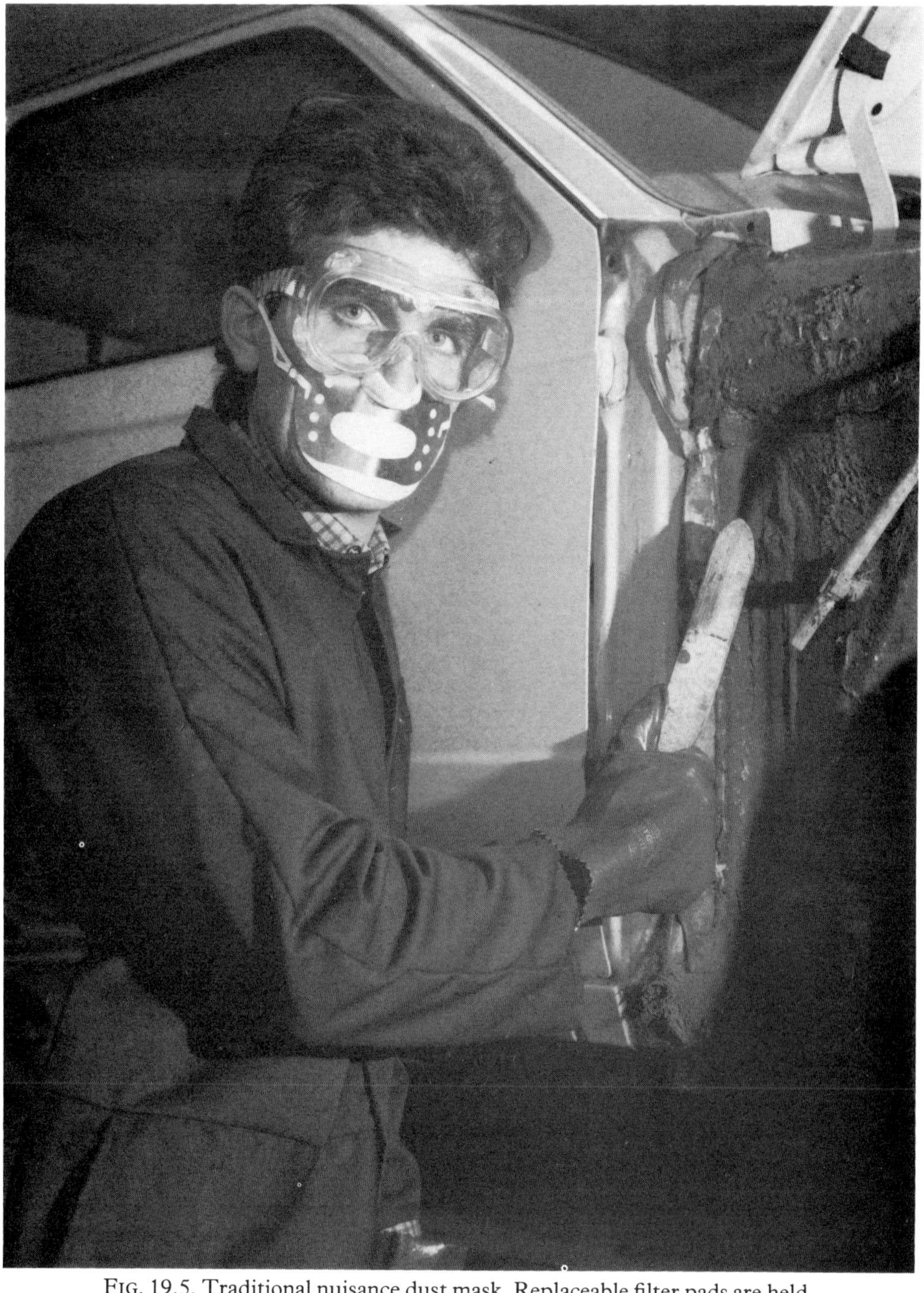

FIG. 19.5. Traditional nuisance dust mask. Replaceable filter pads are held to the face by a light alloy frame. These are inexpensive, very comfortable and popular even though of limited efficiency for the finest dusts. (By courtesy of James North & Sons Ltd.)

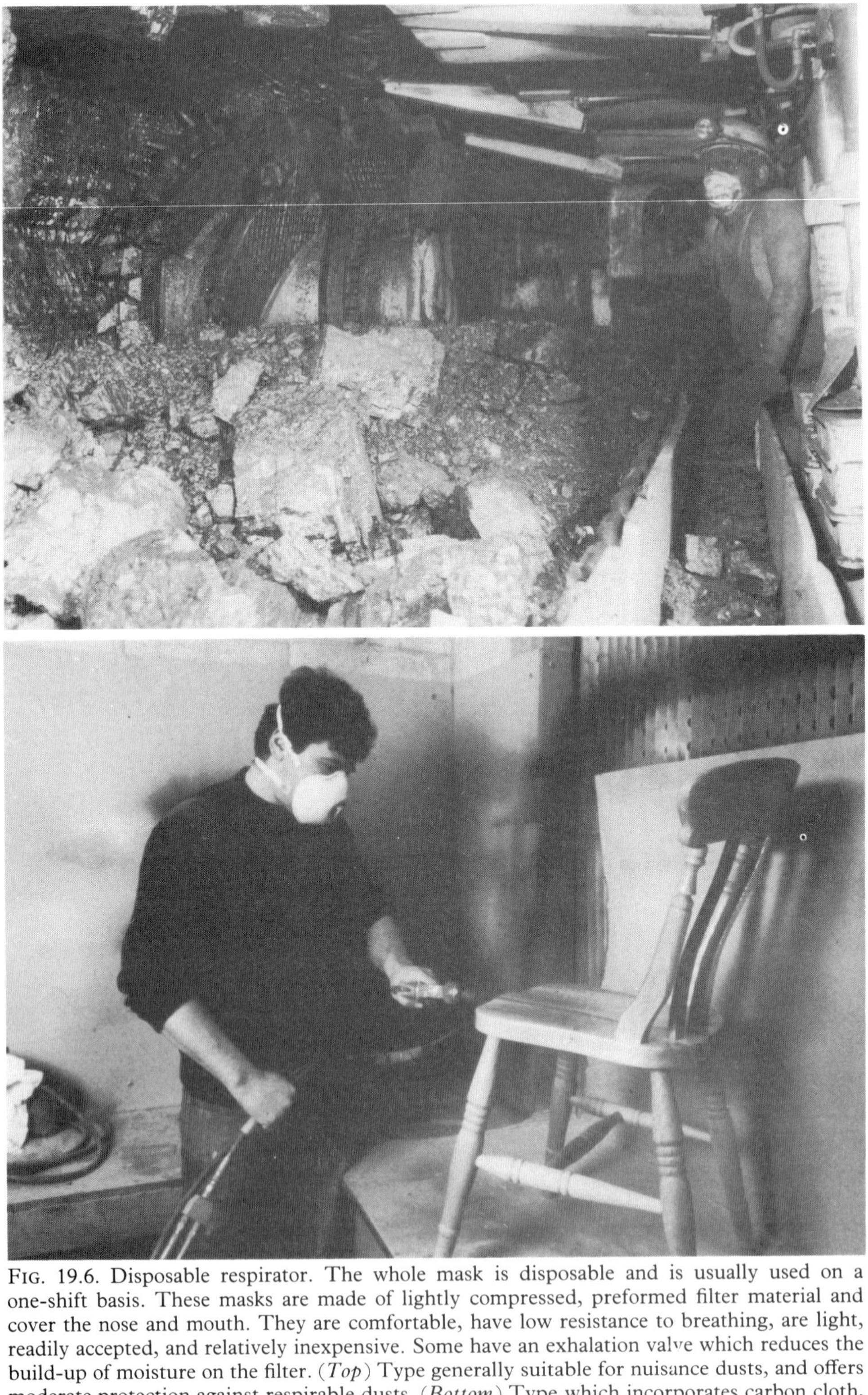

FIG. 19.6. Disposable respirator. The whole mask is disposable and is usually used on a one-shift basis. These masks are made of lightly compressed, preformed filter material and cover the nose and mouth. They are comfortable, have low resistance to breathing, are light, readily accepted, and relatively inexpensive. Some have an exhalation valve which reduces the build-up of moisture on the filter. (*Top*) Type generally suitable for nuisance dusts, and offers moderate protection against respirable dusts. (*Bottom*) Type which incorporates carbon cloth, charcoal or other solid absorbent for protection against specific gases and vapours as well as aerosols. (By courtesy of 3M United Kingdom PLC)

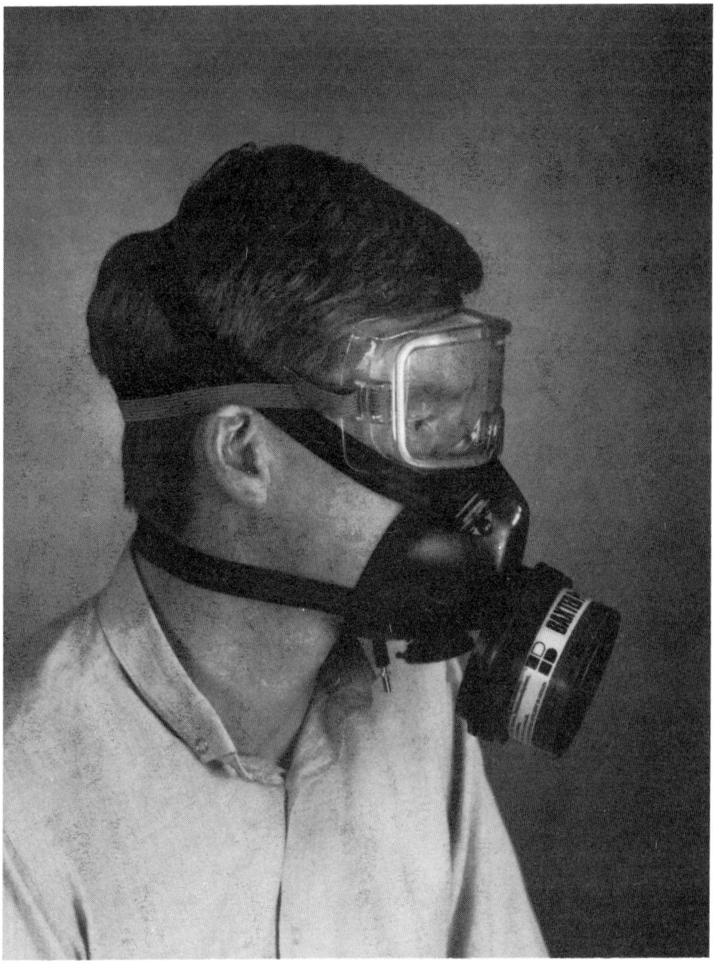

FIG. 19.7. Ori-nasal mask with replaceable filter. The facepiece covers the nose and mouth, is fairly rigid but has a soft face seal which in some versions is filled with glycerine or is pneumatic for better comfort and sealing, as in this version. Mounted on the facepiece is a replaceable filter cartridge designed for dust, gas or vapour and sometimes all three. (By courtesy of The Leyland & Birmingham Rubber Co Ltd)

quite tight to the face to be sure of a good face seal and they are consequently not very comfortable after long periods of use. The filters have a higher air flow resistance than most disposable respirators, making it a noticeable effort to breathe.

(4) Powered ventilated visor (Figure 19.8).

With one type of ventilated visor there is a small, nickel-cadmium battery powered fan in the back of a helmet which drives work-room air through a

FIG. 19.8. Ventilated visor. Good protection from aerosols is gained by providing excess air into a visor. The visor also offers head, eye and face protection. In this example the helmet and visor are one unit. The helmet is fitted with a battery powered fan which draws air from the rear of the helmet through a filter unit and blows it over the face. (By courtesy of Racal Safety Limited)

filter and over the face of the operator inside the hinged visor. Another version of the same principle has the filter and fan unit fastened to the belt. Supplying air to the face under slight positive pressure is much more comfortable for breathing than it is when inhaling air under negative pressure through a filter and much more comfortable to the face than a mask.

This type does rely on a filter for its performance but allows much more mobility than compressed air fed types. In practice it is leakage between the face and the visor rather than filter performance which is the most critical factor affecting overall protection against airborne dust (Lowry, Wheat and Bustos, 1979; Myers and Peach, 1983). Performance is liable to be upset in high ambient side draught (Cecala, Volkwein, Thomas and Urban, 1981; Myers, Peach and Allender, 1984; Myers, Peach, Cutright and Iskander, 1984). This may be mitigated by employing an elasticated face-seal that fits around the chin. Spectacles may be worn. Ventilated visors have a growing popularity with employees despite some complaints of noise and of draughts over the face. They have a good protection factor when used away from draughts.

(5) Full face mask with replaceable filter (Figure 19.9).

Types of respirator designed to fit the face and to cover the nose, mouth and eyes are known as full face respirators. A chemical cartridge containing absorbent granules for the vapour of interest, such as soda lime, silica gel or activated charcoal fits over the inlet or, in some types, is connected by a flexible coupling. Alternatively, or additionally a filter element removes airborne dust.

Full face respirators have a higher maximum use concentration than ori-nasal masks, principally because they make a better fit over the face. Nevertheless the degree of protection is not unlimited and they cannot safely be used in concentrations of air contamination in excess of those for which the filter cartridges are designed. They are not very suitable for employees who normally wear spectacles, although some are designed especially for spectacle wearers. As regards gases and vapours, they are most often used for protection against organic vapours although, with appropriate filters, they are sometimes used for acid gases, ammonia, carbon monoxide or mercury vapour. When used with an aerosol filter the protection factor is dependent upon the physical/chemical properties of the dust, its size distribution and so on.

These respirators also provide eye protection but against this the user feels more 'cut off'. Users should be repeatedly cautioned that although these devices remove air contamination they do not produce oxygen and must only be used in areas where there is known to be an adequate supply of oxygen. They are not suitable to wear for prolonged periods of work, because of discomfort and fatigue due to the resistance to breathing. With types supplied in multiple sizes it is noticeably easier to obtain one which is a comfortable fit.

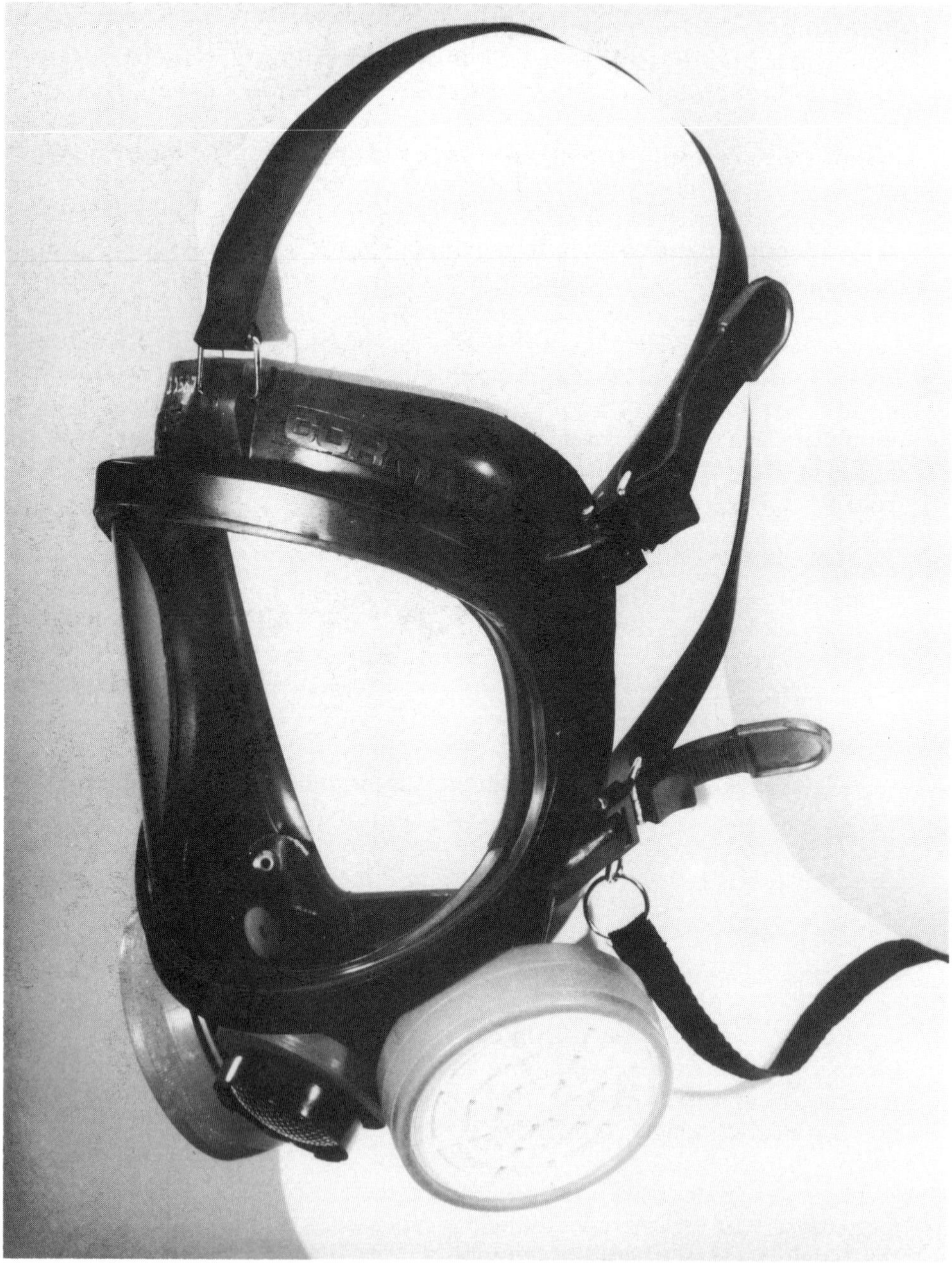

FIG. 19.9. Full face respirator. A lightweight, full face respirator with good visibility for use when full face and eye protection is required. Alternative filters are fitted, as appropriate for the gas, vapour or particulate air contamination. (By courtesy of Siebe Gorman & Co Ltd)

FIG. 19.10. Power assisted full face mask. Filtered air is delivered from waist level via a flexible breathing tube, keeping the respirator at positive pressure. This example of a powered respirator provides a constant air flow rate in excess of 120 litres per minute. A rechargeable battery pack, designed to last for 6.5 hours without recharging is supported on a waist-belt, with the fan and filter unit on the other side. The respirator mask has a pneumatic face seal. (By courtesy of Siebe Gorman & Co Ltd)

(6) Power assisted full face respirator (Figure 19.10).

A battery powered fan supplying filtered air to a face mask has the potential advantage over normal full face respirator of much reduced face leakage and greater comfort from breathing air under positive pressure. The fan drawing air through the filter may be belt mounted or, fitted directly to the mask. The protection factor is high but dependent upon whether the fan provides sufficient air to maintain positive pressure within the mask at high peak flow breathing volume flow rate. The power to the fan is normally from a nickel-cadmium or silver-zinc cell unit which clips to the waist and which may last 4–10 hours before recharging is required.

Fresh air supplied breathing apparatus

This class is associated with a higher maximum use concentration than is an air purifying respirator. Moreover a fresh air supply is essential where there is a likely to be a deficiency of oxygen. Fresh air supplied air devices offer essentially no resistance to breathing and the atmosphere supplied may be cool and more acceptable than that from other types of respirator. On the other hand when using self contained units the weight of the cylinder and other parts of the apparatus carried by the wearer is limiting. So much so that elaborate physical/chemical methods have been developed to conserve the air supply. Otherwise fresh air has to be supplied from a remote base by means of a long hose, which itself is an encumbrance.

Fresh air fed full face masks

A full face mask which is air-fed, having an independent air supply from a remote source of uncontaminated air and pressure-demand type flow regulation provides a high protection factor. The type may be used with confidence in adverse conditions up to 90 metres from an air supply at a pressure of 50–60 psi and include a regulated reducing valve at the user end. However, the air supply tubing limits operator mobility so that it is suitable only for jobs which are stationary or where the movements of the operatives are relatively limited. Maximum use concentration: $>2{,}500\times$ exposure limit.

Fresh air fed hoods

A hood over the head and shoulders is sometimes used when the concentration of hazardous substance in air is extremely high. It needs a reliable air supply, the trailing hose is limiting and the type is generally reserved for special operations. Can be light and in certain jobs where the hose is not an encumbrance is used routinely for up to 20% of the time (Figure 19.11). These hoods provide high protection against all air contamination. They can be excessively noisy from the air supply unless provided with a suitable silencer (Harrison and Stokes, 1971). The volume flow rate of air supplied to the hood should exceed the employee's respiratory requirements threefold or more to avoid 'inleaks' when a large breath is taken or draughts from the ventilation and wind are experienced. Maximum use concentration: $>5{,}000\times$ exposure limit.

Self-contained air-fed suits

The highest protection is obtained by supplying fresh air or oxygen for breathing and encasing the wearer in a suit of impervious material. Such

FIG. 19.11. Fresh air fed hood. This hood is made of a lightweight plastic resistant to common chemicals. It is comfortable to wear and gives all round vision. Spectacle wearers can use this option. It provides a constant flow of air, utilising an existing compressed air system to provide long duration breathing equipment capable of protecting against hazardous gases, vapours and aerosols. (By courtesy of Siebe Gorman & Co Ltd)

protection might be needed where the concentration of air contamination in the vicinity is exceedingly high or there is an oxygen deficiency such as to be immediately dangerous to life as there might be in a confined space or on vessel entry. Air supply from a small compressed air or oxygen cylinder carried in a harness on the wearer enables free movement in the toxic atmosphere and ready escape. Pressure-demand type of flow regulation is necessary to maintain a small positive pressure in the suit and thereby ensure a high protection factor. Movement is somewhat restricted by the suit. Normally they are reserved for emergency use in life-threatening atmospheres and thorough training to this end is essential. They provide the very highest protection to the whole body, where this is needed. They are, however, heavy and cumbersome. Regular inspection and check of the self-contained breathing apparatus is especially important. Maximum use concentration: $>5{,}000\times$ exposure limit.

Air-line breathing apparatus in general

An air compressor fully capable of supplying the fresh air requirements is generally required and there must be a suitable fresh air base within range of the work station. Where it is especially difficult to arrange a fresh air supply station air cylinders carried by the wearer and put on the ground nearby are sometimes used for brief periods, from 10 to 30 minutes, according to the size of cylinder.

The normal compressed air supply to the works is sometimes used with an appropriate air cleaner unit. Filters are available for dust, oil mist, water and hydrocarbons. Sampling and analysis of compressed air is in a BOHS Technical Guide (BOHS, 1988). If it is desired to use air from a unit employing internal lubrication automatic monitoring of carbon monoxide levels is advisable as high levels of carbon monoxide from such units are not unknown (Bernz and Drinker, 1936).

The supplied air must meet minimum air quality requirements;

Oxygen, 19.5–23.5%
Carbon monoxide, less than 15 ppm
Solids, less than 5 mg m^{-3}
Carbon dioxide, less than 1,000 ppm
No pronounced odour

Maintenance and testing of respirators

Respirators should be numbered and issued on an individual basis. Employees should be trained in their use. The employee should be shown how the respirator should be worn, how it is adjusted and how to determine if it fits properly.

Fault finding has top priority. The facepiece seal of ori-nasal and full face masks is the most common element to be deficient in those types operating under negative pressure. The exhalation valve is also occasionally at fault. The goodness of fit achieved by the wearer under training may readily be tested by waving a brush dipped in isoamyl acetate around the seal. A 'banana' oil smell detected by the wearer demonstrates a poor seal. Alternatively, a commercial smoke tube or saccharin aerosol connected to a rubber squeeze bulb may be employed (Marsh, 1984; Johnston, 1987). These tests are particularly useful for training the employee in respirator use and for inculcating confidence in the equipment.

A negative pressure test described in British Standard 5108 may be used daily.

> 'To ensure proper protection, the facepiece fit should be checked by the wearer each time he puts it on. This may be done in this way: Negative pressure test. Close the inlet of the equipment. Inhale gently so that the facepiece collapses slightly, and hold the breath for 10 seconds. If the facepiece remains in its slightly collapsed condition and no inward leakage of air is detected, the tightness of the facepiece is probably satisfactory. If the wearer detects leakage, he should re-adjust the facepiece, and repeat the test. If leakage is still noted, it can be concluded that this particular facepiece will not protect the wearer. The wearer should not continue to tighten the headband straps until they are uncomfortably tight, simply to achieve a gas-tight face fit.'

When using a respirator the user should be warned to leave the working area immediately breathing becomes difficult or dizziness is experienced or the contamination is smelt or tasted or the eyes become irritated.

All types of respiratory protection, save for disposable types, should be cleaned and inspected thoroughly after use, preferably daily, and the air cleaning elements replaced as necessary. Usage and sweat cause rubber to harden and crack. Plastic parts may become brittle or broken in use. Straps and harness suffer wear and tear. Parts which have become broken, defective or suffered deterioration should be replaced. Respiratory protective equipment used only occasionally should be examined at least quarterly. As a general rule, where many personal protective devices are in use, central maintenance facilities are far preferable to individual servicing, under a properly instructed supervisor, who can apply regular tests and maintain proper records of tests, inspections and maintenance.

Thorough examinations and, where appropriate, tests of items of respiratory protective equipment, other than disposable respirators, should be made at least once every month, and more frequently where the conditions are particularly severe. However, in the case of ori-nasal masks used only occasionally, for short spells, against dust or fume of relatively low toxicity, longer intervals between examinations may be suitable. In such cases, suitable intervals should be determined by the person responsible for the management of all aspects of the maintenance of respiratory protective equipment, but in any event, the intervals should not exceed 3 months. In the case of respiratory protective equipment incorporating compressed gas cylinders or electric motors, tests should be made of the condition and efficiency of those parts, including tests of the pressure in the cylinders. In the case of air-fed respiratory protective equipment, the volume, flow and quality of the supplied air should be tested. Any defects disclosed by the examination or test should, of course, be remedied before further use.

Washing facilities

When making an assessment of health risks from hazardous substances an assessment should be made of the essential features of the washing and changing facilities. Good bathing, washing and changing facilities should be provided, including soap and towels. Lockers for clean and dirty clothes should be provided in separate rooms laid out in such a way that contamination of the clean area is kept to a minimum. Laundry facilities should be provided for the clothing of employees who encounter heavy exposure.

Emergency showers and eye washes are generally regarded as a safety provision. They should be provided for use in the event of dangerous splashes or spills of hazardous materials or clothing fires. Employees should know the location of the nearest shower and eye wash and they should be instructed in how to use them in the event that materials are spilled on the body.

References

AFNOR S 76–015. French standard for respiratory protective equipment—tightness to the face and total inward leakage. Association Francaise de Normalisation, Paris (1983).

American Standard for Testing and Materials. Standard test method for resistance of protective clothing materials to penetration by hazardous liquid chemicals, F903–84. American Society for Testing and Materials, Philadelphia, USA (1984).

American Society for Testing and Materials. Performance of protective clothing STP-900. American Society for Testing and Materials, Philadelphia, USA (1986).

American Society for Testing and Materials. Standard Test Method F739–85: Resistance of protective clothing materials to permeation by liquids or gases. American Society for Testing and Materials, Philadelphia, USA (1986).

Berardinelli S P and M Roder. Chemical protective clothing field evaluation methods. In: Performance of protective clothing (Edited by R L Barker and G C Coletta), pp. 250–260. ASTM Special Technical Publication 900, American Society for Testing and Materials, Philadelphia, USA (1986).

Bernz N R and P Drinker. J. Ind. Hyg. & Toxicol. **18** 461 (1936).

BOHS Technology Committee Working Party on Protective Clothing. Ann. Occup. Hyg. **30** 381 (1986).

BOHS. The sampling and analysis of compressed air to be used for breathing purposes. BOHS Technical Guides, Science Reviews, PO Box MT27, Leeds, UK (1988).

British Standards Institution. Recommendations for the Selection, Use and Maintenance of Respiratory Protective Equipment. BS 4275. British Standards Institution, London (1974).

British Standards Institution. Specification for industrial gloves. BS 1651. British Standards Institution, London (1986a).

British Standards Institution. Resistance of clothing materials to permeation of liquids. Part 1. Method for the assessment of breakthrough time. Part 2. Method for the determination of liquid permeating after breakthrough. BS 4724. British Standards Institution, London (1986b).

Cecala A B, J C Volkwein, E D Thomas and C W Urban. US Bureau of Mines Report No 8591. Washington DC (1981).

Cherrie J W, R M Howie and A Robertson. Ann. Occup. Hyg. **31** 481 (1987).

Coletta G C, S Z Mansdorf and S P Berardinelli. Am. Ind. Hyg. Assoc. J. **49** 26 (1988).

Crockford G W. Ann. Occup. Hyg. **19** 345 (1976).

Crockford G W, M Crowder and S P Prestidge. Brit. J. Ind. Med. **29** 378 (1972).

Douglas D D, P L Lowery, C P Richards, L A Geoffren, S K Yasuda, L D Wheat and J M Bustos. Respirator studies for the National Institute for Occupational safety and Health LA-7317-PP). Lows Alamos, New Mexico, USA (1978).

Fenske R A. Appl. Ind. Hyg. **3** 207 (1988).
Fergin G S. Am. Ind. Hyg. Assoc. J. **45** 533 (1984).
Figard W H. Occup. Health and Safety **49** 30 (1980).
Forsberg K and S Faniadis. Am. Ind. Hyg. Assoc. J. **47** 189 (1986).
Freed V H, J E Davies, L J Peters and F Parveen. Residue Rev. **75** 159 (1980).
Grauvogel L W. Am. Ind. Hyg. Assoc. J. **47** 144 (1986).
Griffin O G and D J Longson. Ann. Occup. Hug. **13** 147 (1970).
Harris H E, W C DeSeighard, W A Burgess and P C Reist. Am. Ind. Hyg. Ass. J. **35** 159 (1974).
Harrison R and H J Stokes. Ann. Occup. Hyg. **14** 351 (1971).
Henry N W and C N Schlatter. Am. Ind. Hyg. Assoc. J. **42** 202 (1981).
Hery M, M Villa, G Hubert and P Martin. Ann. Occup. Hyg. **35** 181 (1991).
Hinds W C and G Kraske. Am. Ind. Hyg. Assoc. J. **48** 836 (1987).
Holton P M and K Willeke. Am. Ind. Hyg. Assoc. J. **48** 855 (1987).
Hyatt E C. Respiratory protection factors. National Technical Information Service, US Pepartment of Commerce, 5285 Port Royal Road, Springfield, VA 22151, USA (1976).
ISO. Protective clothing—protection against liquid chemicals—determination of resistance of air-permeable materials to permeation by liquids. Draft International Standard ISO/DIS 6529.4. International Organisation for Standardization, Zurich, Switzerland (1988).
Johnston A R. Appl. Ind. Hyg. **2** F18 (1987).
Legge R T. Ind. Med. Surg. **5** 30 (1936).
Leinster P, J L Bonsall, M J Evans and S J Lewis. Ann. Occup. Hyg. **34** 85 (1990).
Lenhart S W and D J Campbell. Ann. Occup. Hyg. **28** 173 (1984).
Lowry P L, L D Wheat and J M Bustos. Am. Ind. Hyg. Assoc. J. **40** 291 (1979).
Man V L, V Bastecki, G Vandal and A P Bentz. Am. Ind. Hyg. Assoc. J. **48** 551 (1987).
Mansdorf S Z and S P Berardinelli. Am. Ind. Hyg. Assoc. J. **49** 21 (1988).
Marsh J L. Am. Ind. Hyg. Assoc. J. **45** 371 (1984).
Mellstrom G A. Ann. Occup. Hyg. **35** 153 (1991).
Mickelson R L, M M Roder and S P Berardinelli. Am. Ind. Hyg. Assoc. J. **47** 236 (1986).
Moraski R V and A P Nielsen. Protective clothing and its significance to the pesticide user. In: Dermal exposure related to pesticide use (Edited by R C Honeycutt, G Zwieg and N N Ragsdale), pp. 395–402. ACS Symposium Series 273. American Chemical Society, Washington DC, USA (1985).
Myers W R, J R Allender, W Iskander and C Stanley. Ann. Occup. Hyg. **32** 345 (1988).
Myers W R, S W Lenhart, D Campbell and G Provost. Am. Ind. Hyg. Assoc. J. **44** B25–26 (1983).
Myers W R and M J Peach III. Ann. Occup. Hyg. **27** 251 (1983).
Myers W R, M J Peach III and J Allender. Am. Ind. Hyg. Assoc. J. **45** 236 (1984).
Myers W R, M J Peach III, K Cutright and W Iskander. Am. Ind. Hyg. Assoc. J. **45** 681 (1984).
Nelson G O and C A Harder. Am. Ind. Hyg. Assoc. J. **37** 514 (1976).
Nelson G O, B B Lum, G D Carlson, C M Wong and J S Johnson. Am. Ind. Hyg. Assoc. J. **42** 217 (1981).
Olesen B, E Sliwinska, T Masden and P O Fanger. Transactions of the American Society of Heating, Refrigerating and Air Conditioning Engineers **32** 791 (1982).
Patty F A. Respirators and protective devices. In: Industrial hygiene and toxicology (Edited by F A Patty). Vol. I. Interscience Publishers Inc., New York (1948).
Rosen G, I -M Andersson and L Juringe. Ann. Occup. Hyg. **34** 293 (1990).
Sansone E B and Y B Tewari. Am. Ind. Hyg. Assoc. J. **41** 527 (1980).
Sarner S F and N W Henry. Am. Ind. Hyg. Assoc. J. **50** 298 (1989).
Schroy J M. Ann. Occup. Hyg. **30** 231 (1986).
Schwope A D, R Goydan, R C Reid and S Krishnamurthy. Am. Ind. Hyg. Assoc. J. **49** 557 (1988).
Silkowski J B, S W Horstman and M S Morgan. Am. Ind. Hyg. Assoc. J. **45** 501 (1984).
Smith T J, W C Ferrell, M O Varner and R D Putnam. Am Ind. Hyg. Assoc. J. **41** 624 (1980).
Stampfer J F, M J McLeod, M R Betts, A M Martinez and S P Berardinelli. Am. Ind. Hyg. Assoc. J. **45** 642 (1984).
Stull J O. J. Hazard Mater. **14** 165 (1987).

Tannahill S N, R J Willey and M H Jackson. Ann. Occup. Hyg. **34** 547 (1990).
Toney C R and W L Barnhart. Performance evaluation of respiratory protection equipment used in paint stripping operations. NIOSH Technical Information, HEW Publication No. (NIOSH), 76–177. Cincinnati, Ohio, USA (1976).
Vogt J J, J P Meyer and V Candas. Ergonomics **26** 963 (1983).
Wake D and R C Brown. Ann Occup. Hyg. **32** 295 (1988).
Willeke K, H E Ayer and J D Blanchard. Am. Ind. Hyg. Assoc. J. **42** 121 (1981).
Xu M, D Han, S Hangal and K Willeke. Ann. Occup. Hyg. **35** 13 (1991).

Bibliography

American Industrial Hygiene Association. Respiratory protection monograph. American Industrial Hygiene Association, 475 Ledges Parkway, Akron, OH 44311–1087, USA (1985).
American National Standards Institute. Practices for respiratory protection. American National Standards Institute, 1430 Broadway, New York, USA (1980).
Barker R L and G C Coletta (Editors). Performance of protective clothing. ASTM Special Technical Publication 900, American Society for Testing and Materials, Philadelphia, USA (1986).
Clark J M. In: Advances in respiratory protection (Edited by B Ballantyre). Chapman and Hall, Bristol, UK (1979).
Health & Safety Executive. The maintenance, examination and testing of local exhaust ventilation. H M Stationery Office, London (1990).
Mansdorf S Z, R Sager and A P Nielsen (Editors). Performance of protective clothing: Second International Symposium on the performance of protective clothing, ASTM STP989. American Society for Testing and Materials, Philadelphia, USA (1986).
Rajhans G S and S L David. Practical guide to respirator usage in industry. Butterworth (1985).
Schwope A D, P P Costas, J O Jackson and D J Weitzman. Guidelines for the selection of chemical protective clothing. 2nd Ed., American Conference of Governmental Industrial Hygienists, Cincinnati, Ohio (1985).

APPENDICES

APPENDIX 1

Background Mathematics

The purpose of this appendix is to highlight those standard mathematical methods which have been found particularly useful in the development of the subjects covered in the book. Methods employed most frequently in the main text are introduced. A light treatment is given which will serve to jog the memory. The techniques are considered at length in books on engineering mathematics (see, for example, Spencer, Parker, Berry, England, Faulkner, Green, Holden, Middleton and Rogers, 1977). Statistical analysis is also outside the scope of this appendix and specialist books should be consulted for guidance in this area (see, for example, Caulcutt, 1983; Zar, 1984). In point of fact most of the methods used in this book are elementary. A few aspects are of intermediate difficulty.

Differentiation and Integration

Most of the physical laws that control the environment in which an occupational hygienist has to work are most conveniently expressed in terms of differential equations. The differential equations are not especially difficult although the concepts which they express sometimes require a little thought.

Example. A good example crops up in Chapter 15, where the role of turbulent diffusion in ventilation is considered. The diffusivity governs the rate of rise of the atmospheric concentration of contamination at a point in space due to contamination being liberated at some other point. Suppose contamination were issuing from a point, the origin, at constant flow rate E units of mass or volume in unit time. Eventually, when equilibrium has been attained, the rate of loss through a spherical envelope about the source will be equal to the rate of release of contamination. In polar coordinates and ignoring the volume of air displaced by the contamination, where c is the stationary concentration at radius r and D is the diffusivity, assumed for the moment to be the same everywhere, then,

$$-D\frac{dc}{dr} = \frac{E}{4\pi r^2} \tag{1}$$

From which, by use of a standard integral,

$$c = \frac{E}{4\pi Dr} + A, \tag{2}$$

where A is an arbitrary constant.

The value of the arbitrary constant is determined by the boundary conditions appropriate to the physical conditions. Thus, if the concentration is zero at a great distance from the source, evidently A in equation (2) would be zero and

$$c = \frac{E}{4\pi Dr}. \tag{3}$$

Equation (3) is the fundamental law of concentration distribution in a ventilated space with uniform diffusivity and no bulk air movement.

Natural logarithms

In purely arithmetical calculations it is most convenient to use common logarithms, that is, logarithms to base 10. In many other fields logarithms to base e have advantages. The exponent $e = 2.71828\ldots$. Logarithms to the base e are called natural or Naperian logarithms, denoted ln.

$$\log_{10} x = 0.4343 \ln x.$$

By differentiation,

$$\frac{d}{dx}(\log_{10} x) = \frac{0.4343}{x} \quad \text{approximately,}$$

but

$$\frac{d}{dx}(\ln x) = \frac{1}{x} \quad \text{exactly.}$$

It is evident that when using calculus e will be the most convenient base of logarithms. In calculus and in all more advanced mathematics it is customary to use natural logarithms and when no base is indicated it is understood that the base is e.

Exponential law of growth or decay

If

$$\frac{dx}{dt} = ax,$$

where a is a constant,

$$\frac{dt}{dx} = \frac{1}{ax}$$

and, integrating,

$$t = \frac{1}{ax}dx + c = \frac{1}{a}\ln x + c,$$

where c is an arbitrary constant.

Therefore $\ln x = a(t - c)$.
Therefore $x = \exp[a(t - c)]$
whence $x = A \exp(at)$,

where A is an arbitrary constant, equal to $\exp(-ac)$.

In the same way, if $dx/dt = a(x - b)$, where b is a constant,

$$x = b + A \exp(at).$$

Many physical quantities are related by an equation of this type, the quantity x tending to ∞ as t tends to ∞ if a is positive and tending to b if a is negative. If a is positive, this equation is the exponential law of growth of x with time t and if a is negative it is the exponential law of decay of x with time t.

In occupational hygiene the excretion of hazardous substances in the urine and the exhalation of others is often found to follow the exponential law of decay more or less closely. Another example is illustrated by many direct reading instruments for air contamination which rely upon the absorption of visible light, infra-red or ultra-violet. The absorption follows Lambert's law by which each layer of equal thickness absorbs an equal fraction of the light which traverses it. Consequently the transmitted light intensity through thick media falls exponentially with thickness.

First order linear differential equations

Problems in the real world can often be expressed with sufficient accuracy in terms of linear differential equations. This is as true of occupational hygiene problems as it is of those in other branches of science. Methods for finding the solutions have been developed for many different types.

First order linear differential equations appear frequently in the main text. Such equations can be written in the general form

$$\frac{dx}{dt} + Px = Q, \tag{4}$$

where P and Q are functions of t.

The general solution of this equation is

$$ux = \int uQ\, dt + A, \tag{5}$$

where A is an arbitrary constant and $u = \exp Pdt$.

Problem A.1

Find a solution to equation (4) when $P = 2at$, $Q = b/t^2$, where a and b are constants.

Solution

$$u = \int \exp 2at\, dt,$$

$$= \exp(at^2).$$

Substituting in equation (5), the general solution is

$$x \exp(at^2) = b \int \frac{1}{t^2} \exp(at^2) dt + A,$$

where A is an arbitrary constant.

Expanding the exponential term, integrating term by term and solving for x,

$$x = \frac{b \exp(-at^2)}{t} \left\{ At - 1 + \frac{at^2}{1.1!} + \frac{(at^2)^2}{3.2!} + \frac{(at^2)^3}{5.3!} + \cdots \right],$$

where A is an arbitrary constant.

There are four special cases in which the general solution is simplified.

Case 1. $P = f(t)$, $Q = bt$.

A general solution is sometimes needed for the case when $Q = bt$, where b is a constant. The general solution simplifies to

$$ux = b \int ut \, dt + A, \tag{6}$$

where b is the prescribed constant and $u = \exp\int P \, dt$.

Case 2. $P =$ constant, $Q = bt$.

In occupational hygiene problems P is usually constant. Then, when $Q = bt$, the general solution simplifies to

$$x = b \left\{ \frac{t}{p} - \frac{1}{p^2} \right\} + A \exp(-Pt). \tag{7}$$

Case 3. $P = f(t)$, $Q =$ constant.

In many occupational hygiene problems Q is constant, in which case the general solution simplifies to

$$ux = Q \int u \, dt + A, \tag{8}$$

where Q is the prescribed constant and $u = \exp\int P \, dt$.

Case 4. $P =$ constant, $Q =$ constant.

In simple toxico-kinetics and other areas P and Q may both be constants. Then the general solution reduces further to

$$x = \frac{Q}{P} + A \exp(-Pt), \tag{9}$$

where P and Q are the prescribed constants and A is an arbitrary constant. It will be recognised that this is the exponential law of decay.

Second order linear differential equations

Probably the most important single type of ordinary differential equation is of the form

$$\frac{d^2x}{dt^2} + \frac{adx}{dt} + bx = f(t), \tag{10}$$

where a, b are constants and $f(t)$ is a given function of t. Differential equations, like this, of second order, occur in multi-compartmental toxico-kinetics and elsewhere. Solutions where $f(t)$ is a constant will be considered here. Take first the simplest case, when $f(t) = 0$.

By analogy with the solution of first order differential equations a trial solution is

$$x = A\ \exp(mx),$$

where m and A are constants.

Whence

$$\frac{d^2x}{dt^2} + \frac{adx}{dt} + bx = (m^2 + am + b)x.$$

Hence

$$x = A\ \exp(mx)$$

is a solution, whatever the value of A, provided

$$m^2 + am + b = 0. \tag{11}$$

This is called the auxiliary quadratic equation, whose roots are, say, m_1 and m_2. Then

$$x = A_1 \exp(m_1 t) \quad \text{and} \quad x = A_2 \exp(m_2 t)$$

are both solutions.

Hence

$$x = A_1 \exp(m_1 t) + A_2 \exp(m_2 t), \tag{12}$$

where A_1 and A_2 are arbitrary constants, is a solution. It is the general solution.

The case where $f(t) = c$, a non-zero constant, is simplified by putting $y = bx - c$ in equation (10), which then reduces to

$$\frac{d^2y}{dt^2} + \frac{ady}{dt} + by = 0. \tag{13}$$

Hence, on employing the previous solution and expressing the result in terms of x,

$$x = \frac{c}{b} + A_1 \exp(m_1 t) + A_2 \exp(m_2 t) \tag{14}$$

is the general solution where $f(t)$ is a constant, where A_1, A_2 are arbitrary constants and m_1, m_2 are the roots of the auxiliary equation,

$$m^2 + am + b = 0.$$

Simultaneous differential equations

Pharmaco-kinetic models incorporating two or more compartments give rise to simultaneous differential equations. The easiest way to solve sets of two or three such equations is to use the D operator to denote differentiation and to proceed by eliminating the variables as with ordinary equations. Commercial software for solution of specific equations by numerical methods may be employed with microcomputers.

Problem A.2

Solve the simultaneous differential equations

$$\frac{dx}{dt} + 2x = 2y + 1$$

$$\frac{dy}{dt} + 3y = x$$

Solution

The simultaneous differential equations may be written

$$(D + 2)x = 2y + 1$$

$$(D + 3)y = x,$$

where D denotes the operator d/dt.

Eliminating x

$$(D + 2)(D + 3)y = 2y + 1,$$

from which

$$D^2y + 5Dy + 4y = 1,$$

which, from equation (14), has the general solution

$$y = 0.25 + A_1 \exp(-t) + A_2 \exp(-4t),$$

where A_1, A_2 are arbitrary constants.

Hence, by substituting back in the original equations

$$x = 0.75 + 2A_1 \exp(-t) - A_2 \exp(-4t).$$

The lognormal distribution

Skew distributions are found in three important areas of occupational hygiene: particle size of airborne dust, temporal variation of atmospheric concentration and biological tolerance to hazardous substances. The distribution of the logarithms of these data is often found to be approximately normal so that normal statistics can be employed on the logged data. The results of the analysis are converted back by taking antilogs.

The importance of the lognormal distribution in statistical analysis has been established for a century or more. In 1879 Galton showed that in certain cases the geometric mean is to be preferred to the arithmetic mean as a measure of the location of a distribution (Galton, 1879). Galton derived his ideas from a consideration of the Weber-Fechner law relating biological response to the magnitude of the stimulus. This law asserts that the response is proportional to the logarithm of the stimulus (Weber, 1834; Fechner, 1897).

The foundations of a theory accounting for the generation of lognormal frequency curves was laid by Kapteyn (1903). In more recent times Hatch and Choate (1929) and Hatch (1933) began investigating empirically the value of the lognormal distribution for displaying the statistics of the size distribution of dust particles. They studied the asymmetrical size-frequency curves obtained from measurements of particle size of crushed rock and airborne dust. They found that the asymmetrical size-frequency curves can generally be converted into symmetrical curves which follow approximately the normal probability curve when the logarithm of size is substituted in the graph for the size itself. At about the same time Gaddum (1933) and Bliss (1934, 1935) found the logarithmic transformation effective in normalizing the distribution of levels of tolerance to the action of drugs on living organisms. Oldham and Roach were amongst the first to examine the temporal and spacial lognormal character of atmospheric contamination concentration distribution in industry (Oldham and Roach, 1952; Oldham, 1953; Roach, 1959). Their studies were of airborne dust in coal mines. Lognormal distributions were soon also found in data from other dusty industries (Juda and Budzinski, 1964). Finally the same distribution was found in data of vapour concentration measured at intervals during a work shift, reported notably by Sherwood (1972). The lognormal distribution increasingly appeared to be a general property of air contamination in industry (Esmen and Hammad, 1977).

A variate subject to a process of change is said to obey the law of proportionate effect if the change in the variate at any step of the process is a random proportion of the previous value of the variate. Suppose the value of the variate is initially C_0 and that after the jth step in the process it is C_j. The change in the variate at each step is a random proportion of its value at the previous step. Algebraically,

$$C_j - C_{j-1} = E_j C_{j-1}, \tag{15}$$

where E_j is the proportionate error.

So that,

$$\frac{C_j - C_{j-1}}{C_{j-1}} = \sum_{j=1}^{n} E_j. \tag{16}$$

Now, supposing the effect of each step to be small,

$$\sum \frac{C_j - C_{j-1}}{C_{j-1}} \sim \int_{C_o}^{C_n} \frac{dC}{C} = \ln C_n - \ln C_o. \tag{17}$$

Thus,

$$\ln C_n = \ln C_0 + E_1 + E_2 + \cdots + E_n. \tag{18}$$

Given this relationship, by the additive form of the central limit theorem, $\ln C_n$ is asymptotically normally distributed and hence C_n is asymptotically lognormally distributed (Aitchison and Brown, 1976). Thus a physical or biological basis for the lognormal distribution exists whenever it can be shown that the law of proportionate effect applies (Koch, 1966, 1969).

The reason why the intensity of air contamination tends to follow the law of proportionate effect probably stems, in large measure, from the way turbulent air in a room mixes. The air contamination remaining at a particular location in a work place at a particular moment in time is the sum of many small residues of contamination originally emitted elsewhere and at earlier times, which became dispersed and the greatest percentage of it removed from the work place. Mixing in the air can be thought of as proceeding stepwise. By the mixing length theory of turbulence, in each element of the process a small parcel of air moves in a random direction along a path of some average 'mixing' length. It then disintegrates and mixes with its surroundings. Adapting this model to the stepwise dispersion of air contamination the small parcel of air could be supposed to have a certain mass of contamination in suspension. Thus, in the stepwise path from source to destination each time a parcel of air moves a mixing length, disintegrates and mixes with its surroundings it increases the mass of contamination at the new location by a small multiple of the previous concentration, that is, at the jth step, by a mass $C_j v$, where v is the volume of the parcel of air and C_j is the concentration of contamination in it. At the next step a parcel of the mixture, at concentration C_{j+1}, takes with it a mass of contamination, $C_{j+1}v$, and so on. Thus, by the mixing length concept of turbulence air from different sources would be expected to cause at each step a change in the mass of contamination at a given location proportional to the concentration at the previous location. This reasoning is short of being a rigid proof of the law of proportionate effect but it does show that it is justifiable on general grounds.

The probability density function of the lognormal distribution is

$$f(x) = \frac{f(C)}{x\sigma_{\ln x}}, \qquad (19)$$

where

$$C = \frac{\ln x - \mu_{\ln x}}{\sigma_{\ln x}},$$

and

$$f(C) = \frac{\exp(-C^2/2)}{(2\pi)^{1/2}},$$

the probability density function of the standardised normal distribution, with zero mean and unit standard deviation.

Statistical analysis of data from lognormal frequency distributions is performed on the logarithms of the data or, if graphic methods are preferred, by plotting the cumulative frequency on log-probability paper.

The layout and design features of lognormal probability paper compare directly with those of normal probability paper. A logarithmic scale is used instead of a linear scale to plot the values of the variable of interest. When a set of plotted data gives an acceptable straight line fit the distribution parameters may be easily estimated (King, 1971). The median, or 50% point is a good estimate of the geometric mean and can be read directly from the intersection of the fitted straight line and the 50% probability line.

The mean and standard deviation of the logged data are the logarithms of the geometric mean and geometric standard deviation, respectively. Where $\mu_{\ln x}$ is the geometric mean and $\sigma_{\ln x}$ is the geometric standard deviation the parameters of the unlogged data are given by:

$$\text{arithmetic mean} = \exp\left(\mu_{\ln x} + \frac{\sigma^2_{\ln x}}{2}\right) \tag{20}$$

$$\text{standard deviation} = (\text{mean}) \times [\exp(\sigma^2_{\ln x}) - 1]^{1/2} \tag{21}$$

$$\text{mode} = \exp(\mu_{\ln x} - \sigma^2_{\ln x}). \tag{22}$$

The geometric standard deviation (GSD) can be shown from the probability integral to have the following value:

$$\text{GSD} = \frac{84.13\%\ \text{value}}{50\%\ \text{value}} = \frac{50\%\ \text{value}}{15.87\%\ \text{value}}. \tag{23}$$

It is implicit in the aforementioned argument that the lognormal distribution starts at zero. Indeed there are good scientific reasons for employing lognormal distributions of this kind. But for many purely statistical purposes such a constraint may be a hindrance. Where there is no compelling reason to assume the distribution starts at zero a lognormal may be fitted assuming a non-zero (positive or negative) start point (Cohen, 1951; Hill, 1963).

References

Bliss C I. Science **79** 38 (1934).
Bliss C I. Ann. Appl. Biol. **22** 134 (1935).
Cohen A C. J. Amer. Stat. Assn. **46** 206 (1951).
Esmen N and Y Hammad. J. Environ. Sci. Hlth. **A12** 29 (1977).
Gaddum J H. Reports on Biological Standards III. Methods of biological assay depending on a quantal response. Special Report Series, Medical Research Council, London, No. 183 (1933).
Galton F. Proc. Roy. Soc. **29** 365 (1879).
Hatch T. J. Franklin Inst. **215** 27 (1933).
Hatch T and S P Choate. J. Franklin Inst. **207** 369 (1929).
Hill B M. J. Amer. Stat. Assn. **58** 72 (1963).
Juda J and K Budzinski. Staub-Reinhalt. Luft **24** 283 (1964).
Koch A L. J. Theor. Biol. **12** 276 (1966).
Koch A L. J. Theor. Biol. **23** 251 (1969).
Oldham P D and S A Roach. Brit. J. Ind. Med. **9** 112 (1952).
Oldham P D. Brit. J. Ind. Med. **10** 227 (1953).
Roach S A. Brit. J. Ind. Med. **16** 104 (1959).
Sherwood R J. Ann. Occup. Hyg. **15** 409 (1972).

Bibliography

Aitchison J and J A C Brown. The lognormal distribution. Cambridge University Press (1976).
Caulcutt R. Statistics in research and development. Chapman and Hall, London (1983).
Fechner G T. Kollektivmasslehre. Leipzig. Engelmann (1897).
Kapteyn J C. Skew frequency curves in biology and statistics. Astronomical Laboratory, Groningen, Noordhoff (1903).

King J R. Probability charts for decision making. Industrial Press Inc., New York (1971).

Spencer A J M, D F Parker, D S Berry, A H England, T R Faulkner, W A Green, J T Holden, D Middleton and T G Rogers. Engineering mathematics, Vols 1 and 2. Van Nostrand Reinhold, New York, Cincinnati, Toronto, London, Melbourne (1977).

Weber H. De pulsa resorptione auditu et tactu. Annotationes anatomicae et physiologicae. Leipzig, Koehler (1834).

Zar J. Biostatistical applications. Prentice-Hall, New Jersey, USA (1984).

APPENDIX 2

Conversion Factors

Mass and weight

To obtain → Multiply number of ↓ by ↘	milligrams (mg)	grams (g)	kilograms (kg)	pounds (lb)
milligrams	1	0.01	0.000001	0.000002205
grams	1,000	1	0.001	0.002205
kilograms	1,000,000	1,000	1	2.205
pounds	453,600	453.6	0.4536	1

Volume flow rate

To obtain → Multiply number of ↓ by ↘	ml s^{-1}	l s^{-1}	m^3 s^{-1}	ft^3 min^{-1}
ml s^{-1}	1	0.001	0.000001	0.002119
l s^{-1}	1,000	1	0.001	2.119
m^3 s^{-1}	1,000,000	1,000	1	2,119
ft^3 min^{-1}	471.8	0.4718	0.0004718	1

Volume and capacity

To obtain → Multiply number of ↓ by ↘	millilitres (ml)	litres (l)	cu. metres (m^3)	cu. feet (ft^3)
millilitres	1	0.001	0.000001	0.0000353
litres	1,000	1	0.001	0.03532
cu. metres	1,000,000	1,000	1	35.32
cu. feet	28,310	28.31	0.02831	1

Area

To obtain → Multiply number of ↓ by ↘	sq. cm (cm^2)	sq. metre (m^2)	sq. in (in^2)	sq. ft (ft^2)
sq. cm	1	0.001	0.1550	0.001076
sq. metre	10,000	1	1,550	10.76
sq. in	6.452	0.0006452	1	0.006944
sq. ft	929	0.0929	144	1

Velocity

To obtain → Multiply number of ↓ by ↘	$cm\ s^{-1}$	$m\ s^{-1}$	$ft\ s^{-1}$	$ft\ min^{-1}$	mph
$cm\ s^{-1}$	1	0.01	0.03281	1.968	0.02237
$m\ s^{-1}$	100	1	3.281	196.8	2.237
$ft\ s^{-1}$	30.48	0.3048	1	60	0.6818
$ft\ min^{-1}$	0.508	0.00508	0.01667	1	0.01136
mph	44.7	0.447	1.467	88	1

Length

To obtain → Multiply number of ↓ by ↘	centimetres (cm)	metres (m)	kilometres (km)	inches (in)	feet (ft)
centimetres	1	0.010	0.00001	0.394	0.0328
metres	100	1	0.001	39.37	3.281
kilometres	100,000	1,000	1	39,370	3,281
inches	2.54	0.0254	0.0000254	1	0.08333
feet	30.48	0.3048	0.0003048	12	1

1 angstrom = 0.001 microns = 10^{-7} millimetres
= 10^{-8} centimetres = 3.937×10^{-9} inches
1 micron = 10^{4} angstrom = 10^{-3} millimetres
= 10^{-4} centimetres = 3.937×10^{-5} inches

Air under standard conditions:

70F; 29.92 in mercury pressure
21C; 760 mm mercury pressure
density 0.075 lb ft^{-3}

Vapour equivalent of liquids

The gram-molecular weight of an ideal gas at 25C and 760 mm mercury pressure occupies 24.45 litres.

$$\text{Litres of gas or vapour} = \frac{24.45 \times \text{ml liquid} \times \text{S.G. liquid}}{\text{gram molecular wt}}$$

Example A2.1

A litre of toluene, specific gravity 0.866, molecular weight 92.13, is evaporated to dryness at 25C and 760 mm mercury pressure.

$$\text{Volume of toluene vapour} = \frac{24.45 \times 1{,}000 \times 0.866}{92.13} = 230 \text{ litres}$$

Conversion between mg m^{-3} of gas or vapour and ppm

Concentration of air contamination is usually standardised to a temperature of 25C and a pressure of 760 mm mercury.

To convert concentration of gas or vapour in ppm into concentration in mg m^{-3}, multiply by (molecular wt)/24.45.

To convert concentration of gas or vapour in mg m^{-3} into concentration in ppm, multiply by 24.45/(molecular wt).

Example A2.2

Given the concentration of butane is 600 ppm, convert this to mg m^{-3}.

The molecular weight of butane is 58.12, therefore,

$$\text{concentration in mg m}^{-3} = \frac{600 \times 58.12}{24.45}$$

$$= 1{,}425 \text{ mg m}^{-3}$$

Example A2.3

Given the concentration of carbon monoxide is 25 mg m^{-3}, convert this to ppm.

The molecular weight of carbon monoxide is 28.01, therefore,

$$\text{concentration in ppm} = \frac{25 \times 24.45}{28.01}$$

$$= 22 \text{ ppm}$$

Conversion between mg m^{-3} and grains per cubic foot

In air cleaning practice dust loading is often expressed in terms of grains per cubic foot.

1 pound = 7,000 grains

1 grain = 64.8 mg

1 grain ft^{-3} = 2,288 mg m^{-3}

Conversion of pressure units

In ventilation practice pressure is variously expressed in millimetres of water (mm WG), inches of water (in WG), millimetres of mercury (mm mercury), millibar (mb), or MKS pressure units, newtons per square metre (N m^{-2}).

1 in WG = 25.4 mm WG

1 mm mercury = 13.6 mm WG

1 mb = 10.2 mm WG

1 N m^{-2} = 0.102 mm WG

Conversion of power units

1 horse power = 746 watts

Index